Contents

5TH EDITION

GAS TURBINE
THEORY

PEARSON
Education

We work with leading authors to develop the strongest educational materials in engineering, bringing cutting-edge thinking and best learning practice to a global market.

Under a range of well-known imprints, including Prentice Hall, we craft high quality print and electronic publications which help readers to understand and apply their content, whether studying or at work.

To find out more about the complete range of our publishing please visit us on the World Wide Web at: www.pearsoneduc.com

By G. F. C. Rogers and Y. R. Mayhew
Engineering Thermodynamics Work and Heat Transfer (Longman)
Thermodynamic and Transport Properties of Fluids: SI (Blackwell)
By G. F. C. Rogers
The Nature of Engineering (Macmillan)

5TH EDITION

GAS TURBINE THEORY

HIH Saravanamuttoo
Professor Emeritus, Department of Mechanical and Aerospace
Engineering, Carleton University

GFC Rogers
Professor Emeritus, University of Bristol

H Cohen
Lately Fellow, Queen's College, Cambridge

 Prentice Hall
FINANCIAL TIMES

An imprint of **Pearson Education**
Harlow, England • London • New York • Boston • San Francisco • Toronto • Sydney • Singapore • Hong Kong
Tokyo • Seoul • Taipei • New Delhi • Cape Town • Madrid • Mexico City • Amsterdam • Munich • Paris • Milan

Pearson Education Limited
Edinburgh Gate
Harlow
Essex CM20 2JE
England

and Associated Companies throughout the world

Visit us on the World Wide Web at:
www.pearsoneduc.com

Published under the Longman imprint 1951, 1972, 1987, 1996

Fifth edition 2001
© Pearson Education Limited 1951, 2001

ISBN 0130-15847-X

British Library Cataloguing-in-Publication Data
A catalogue record for this book is available from the British Library

Library of Congress Cataloging-in-Publication Data
Saravanamuttoo, H.I.H.
   Gas turbine theory/H.I.H. Saravanamuttoo, G.F.C. Rogers,
     H. Cohen. – 5th ed. p. cm.
   Includes bibliographical references and index.
   ISBN 0-13-015847-X
   1. Gas-turbines. I. Rogers, G.F.C. (Gordon Frederick Crichton) II.
     Cohen, Henry, 1921 Sept. 29- III. Title.

TJ778.S24 2001
621.43'3–dc21

00-051594

10 9 8 7 6 5 4 3
07 06 05 04 03

Typeset in 10pt Times by 32
Printed and bound in Great Britain by T.J. International Ltd., Padstow, Cornwall.

Foreword

This book has a remarkable pedigree. The first edition, published in 1951, was one of the first books laying down the theory of this newly evolving machine, the gas turbine. The authors showed the basic aerodynamic and thermodynamic theories of this prime mover. They did this in a straightforward style which could be readily understood. Consequently the book was widely used, not only as a textbook for students, but also as source material for established engineers who were becoming involved in this rapidly expanding field.

As the gas turbine subsequently developed, so also has this book, with the later editions keeping abreast of practice. For example, while the first edition gave cycle performance plots assuming ideal (isentropic) compression and expansion processes, from the second edition, realistic process efficiencies have been used. New themes of relevance, such as transient performance, combustion emissions and coal gasification, have been introduced.

Whilst the book is entitled *Gas Turbine Theory* the authors have also given attention to the applications of the gas turbine. Simply but effectively the theory relating to the new situations has been introduced. This has been done for many developments – fan blades, engine growth by addition of compressor stages, turbine blade cooling etc.

As one who has been involved from the 1960's in teaching students the theory of gas turbines, this has been a marvellous textbook – and it has been continuously updated to keep up with the rapid developments in this propulsion device and power generator. I recommend it to students and also to those entering the gas turbine field. The book contains a wealth of knowledge about gas turbines and their development and is a goldmine of information. The book also retains the straightforward and lucid presentation of the earlier editions.

January 2001   Norman Maccallum

Preface to the fifth edition

This edition has been written to mark the 50th anniversary of the original publication of *Gas Turbine Theory* in 1951. The gas turbine was in its infancy when Cohen and Rogers laid the foundation of the basic theory of this new prime mover, including cycle design, aerodynamics and thermodynamics of the individual components and off-design performance. Fifty years later the layout of the book is essentially the same, but it has been greatly expanded to cover the continued development and widely increasing applications of the gas turbine.

In 1951 the gas turbine had a very limited role, primarily as a military jet engine with a 'big' engine having a thrust of about 15 kN. There were no civil aircraft applications and only a handful of experimental industrial engines had been built; these could not compete successfully against the established diesel engine and steam turbine. In the last fifty years the gas turbine has had an enormous impact, starting with the introduction of commercial jet transports in the early 1950s; this led directly to the demise of oceanic passenger liners and trans-continental trains in North America. The application of gas turbines has resulted in a revolutionary growth in the gas pipeline industry and, more recently, has had a similar impact on electric power generation. Industrial gas turbines are now approaching unit sizes of 300 MW and thermal efficiencies in excess of 40 per cent, while combined gas and steam cycles achieve efficiencies of close to 60 per cent. The beginning of the 21st century may see the widespread use of the microturbine in distributed power plants and also the combination of the gas turbine with fuel cells. The outstanding reliability of the gas turbine is clearly demonstrated by the large-scale use of twin-engined passenger aircraft on oceanic routes, with engines achieving as much as 40 000 hours without being removed from the wing.

In this edition the Introduction has been updated and expanded to deal with current new developments, the propulsion material has been extended considerably, a new section on performance deterioration has been added, and numerous minor improvements have been made. The aim of the book is unchanged, in that it gives a broad-based introduction to the key aspects of the gas turbine and its applications.

Today's vast body of knowledge on gas turbines makes it very difficult to keep abreast of developments on a wide front, and I am very grateful to the large number of gas turbine engineers who have provided me with information and help over the years. I have gained especially from my long association with the Advisory Group on Aeronautical Research and Development (AGARD) and the International Gas Turbine Institute of the American Society of Mechanical Engineers. The long-term financial support of the Natural Sciences and Engineering Research Council of Canada has been invaluable to my work.

As a boy growing up during World War II, I was fascinated by aeroplanes, and the appearance of the jet engine was a particularly exciting development. 1951 was also the year I started university and, in my final year, I was introduced to gas turbines using the book universally known as Cohen and Rogers. In 1964, after a decade in the Canadian gas turbine industry, I joined Professor Rogers on the staff of the University of Bristol and we have worked together since then. My first contribution to *Gas Turbine Theory* was with the second edition published in 1972.

It is my belief that several generations of gas turbine engineers, including myself, owe a deep debt of gratitude to Henry Cohen and Gordon Rogers for their pioneering efforts in gas turbine education, and I am very glad to have been able to continue this work to complete the half century in print.

Finally, I should like to thank my wife, Helen, for her continued support and encouragement for more than forty years.

September 2000   H.I.H.S.

A Solutions Manual for the problems in this book is available free of charge from the publishers to lecturers only.

The authors and publishers wish to thank the following companies for their permission to use illustrations and photographs within this text: Alstom Power UK, Bowman Power Systems, GE Aircraft Engines, Honeywell APU, Pratt and Whitney Canada Corporation, Rolls-Royce plc (not to be reproduced in any other publication unless with the permission of Rolls-Royce plc), Siemens Westinghouse Power Corporation, Solar Turbines Incorporated, TransAlta Energy Corporation.

Preface to the fourth edition

It is exactly 40 years since I was introduced to the study of gas turbines, using a book simply referred to as Cohen and Rogers. After a decade in the gas turbine industry I had the good fortune to join Professor Rogers at the University of Bristol and was greatly honoured to be invited to join the original authors in the preparation of the second edition in 1971. Sadly, Dr Cohen died in 1987, but I am sure he would be delighted to know that the book he initiated in 1951 is still going strong. Professor Rogers, although retired, has remained fully active as a critic and has continued to keep his junior faculty member in line; his comments have been very helpful to me.

Since the third edition in 1987, the gas turbine has been found to be suitable for an increasing number of applications, and this is reflected in the new Introduction. Gas turbines are becoming widely used for base-load electricity generation, as part of combined-cycle plant, and combined cycles receive more attention in this edition. With the increased use of gas turbines, control of harmful emissions has become ever more important. The chapter on combustion has been enlarged to include a substantial discussion of the factors affecting emissions and descriptions of current methods of attacking the problem. A section on coal gasification has also been added. Finally, the opportunity has been taken to make many small but significant improvements and additions to the text and examples.

As in the third edition, no attempt has been made to introduce computational methods for aerodynamic design of turbomachinery and the treatment is restricted to the fundamentals; those wishing to specialize must refer to the current, rapidly changing, literature. Readers wishing access to a suite of performance programs suitable for use on a Personal Computer will find the commercially available program GASTURB by Dr Joachim Kurzke very useful; this deals with both design and off-design calculations. The way to obtain this program is described in the rubric of Appendix B.

I would like to express my deepest gratitude to Gordon Rogers, who introduced me to a university career and for over 30 years has been boss, mentor, colleague and friend. Together with Henry Cohen, his early work led me to a career in gas turbines and it has been a great privilege to work with him over the years.

Thanks are once again due to the Natural Sciences and Engineering Research Council of Canada for financial support of my work for many years. Finally, sincere thanks to Lois Whillans for her patience and good humour in preparing the manuscript.

September 1995   H.I.H.S.

Preface to the third edition

The continued use of this book since the appearance of the first edition in 1951 suggests that the objectives of the original authors were sound. To the great regret of Professors Rogers and Saravanamuttoo, Dr Cohen was unable to join them in the preparation of this new edition. They would like to express their appreciation of his earlier contribution, however, and in particular for initiating the book in the first place.

Since the second edition was published in 1972 considerable advances have been made and the gas turbine has become well established in a variety of applications. This has required considerable updating, particularly of the Introduction. Sub-sections have been added to the chapter on shaft power cycles to include combined gas and steam cycles, cogeneration schemes and closed cycles, and many minor modifications have been made throughout the book to take account of recent developments. The chapter on axial flow compressors has been rewritten and enlarged, and describes in detail the design of a compressor to meet a particular aerodynamic specification which matches the worked example on turbine design. A section on radial flow turbines has been added to the turbine chapter.

In keeping with the introductory nature of the book, and to avoid losing sight of physical principles, no attempt has been made to introduce modern computational methods for predicting the flow in turbomachinery. Appropriate references to such matters have been added, however, to encourage the student to pursue these more advanced topics.

As a result of many requests from overseas readers, Professor Saravanamuttoo has agreed to produce a manual of solutions to the problems in Appendix B. The manual can be purchased by writing to: The Bookstore, Carleton University, Ottawa, Ontario K1S 5B6, Canada.

Finally, this edition would not have been possible without the generous support of Professor Saravanamuttoo's research by the Natural Sciences and Engineering Research Council of Canada. He would particularly like to express his thanks to Dr Bernard MacIsaac of GasTOPS and the many engineers from Pratt and Whitney Canada, Rolls-Royce, and the Royal Navy with whom he has collaborated for many years.

March 1986   G.F.C.R.
                                                                 H.I.H.S.

Preface to the second edition

At the suggestion of the publishers the authors agreed to produce a new edition of *Gas Turbine Theory* in SI units. The continued use of the first edition encouraged the authors to believe that the general plan and scope of the book was basically correct for an introduction to the subject, and the object of the book remains as stated in the original Preface. So much development has taken place during the intervening twenty-one years, however, that the book has had to be rewritten completely, and even some changes in the general plan have been required. For example, because gas dynamics now forms part of most undergraduate courses in fluid mechanics a chapter devoted to this is out of place. Instead, the authors have provided a summary of the relevant aspects in an Appendix. Other changes of plan include the extension of the section on aircraft propulsion to a complete chapter, and the addition of a chapter on the prediction of performance of more complex jet engines and transient behaviour. Needless to say the nomenclature has been changed throughout to conform with international standards and current usage.

The original authors are glad to be joined by Dr Saravanamuttoo in the authorship of the new edition. He is actively engaged on aspects of the gas turbine with which they were not familiar and his contribution has been substantial. They are glad too that the publisher decided to use a wider page so that the third author's name could be added without abbreviation. Dr Saravanamuttoo in his turn would like to acknowledge his debt to the many members of staff of Rolls-Royce, the National Gas Turbine Establishment and Orenda Engines with whom he has been associated in the course of his work. The authors' sincere appreciation goes to Miss G. M. Davis for her expert typing of the manuscript.

October 1971

H.C.
G.F.C.R.
H.I.H.S.

Preface to the first edition

This book is confined to a presentation of the thermodynamic and aerodynamic theory forming the basis of gas turbine design, avoiding as far as possible the many controversial topics associated with this new form of prime-mover. While it is perhaps too early in the development of the gas turbine to say with complete assurance what are the basic principles involved, it is thought that some attempt must now be made if only to fill the gap between the necessarily scanty outline of a lecture course and the many articles which appear in the technical journals. Such articles are usually written by specialists for specialists and presuppose a general knowledge of the subject. Since this book is primarily intended for students, for the sake of clarity some parts of the work have been given a simpler treatment than the latest refinements of method would permit. Care has been taken to ensure that no fundamental principle has been misrepresented by this approach.

Practising engineers who have been engaged on the design and development of other types of power plant, but who now find themselves called upon to undertake work on gas turbines, may also find this book helpful. Although they will probably concentrate their efforts on one particular component, it is always more satisfactory if one's specialized knowledge is buttressed by a general understanding of the theory underlying the design of the complete unit.

Practical aspects, such as control systems and mechanical design features, have not been described as they do not form part of the basic theory and are subject to continuous development. Methods of stressing the various components have also been omitted since the basic principles are considered to have been adequately treated elsewhere. For a similar reason, the theory of heat-exchangers has not been included. Although heat-exchangers will undoubtedly be widely used in gas turbine plant, it is felt that at their present stage of development the appropriate theory may well be drawn from standard works on heat transmission.

Owing to the rate at which articles and papers on this subject now appear, and since the authors have been concerned purely with an attempt to define the basic structure of gas turbine theory, a comprehensive bibliography has not been included. A few references have been selected, however, and are appended to appropriate chapters. These are given not only as suggestions for further reading, but also as acknowledgements of sources of information. Some problems, with answers, are given at the end of the book. In most cases these have been chosen to illustrate various points which have not appeared in the worked examples in the text. Acknowledgements are due to the Universities of Cambridge, Bristol and

Durham for permission to use problems which have appeared in their examination papers.

The authors have obtained such knowledge as they possess from contact with the work of a large number of people. It would therefore be invidious to acknowledge their debt by mention of one or two names which the authors associate in their minds with particular aspects of the work. They would, however, like to express their gratitude in a corporate way to their former colleagues in the research teams of what were once the Turbine Division of the Royal Aircraft Establishment and Power Jets (R. & D.) Ltd, now the National Gas Turbine Establishment.

In conclusion, the authors will welcome any criticisms both of detail and of the general scheme of the book. Only from such criticism can they hope to discover whether the right approach has been adopted in teaching the fundamentals of this new subject.

April 1950   H.C.
                                                         G.F.C.R.

1

Introduction

Of the various means of producing mechanical power the turbine is in many respects the most satisfactory. The absence of reciprocating and rubbing members means that balancing problems are few, that the lubricating oil consumption is exceptionally low, and that reliability can be high. The inherent advantages of the turbine were first realized using water as the working fluid, and hydro-electric power is still a significant contributor to the world's energy resources. Around the turn of the twentieth century the steam turbine began its career and it has become the most important prime mover for electricity generation. Steam turbine plants producing well over 1000 MW of shaft power with an efficiency of 40 per cent are now being used. Steam turbines were widely used in marine applications, but could not compete with the thermal efficiency of the diesel engine when fuel costs became important in the mid 1970s; they are still used, however, in nuclear-powered aircraft carriers and submarines. In spite of its successful development, the steam turbine does have an inherent disadvantage. It is that the production of high-pressure high-temperature steam involves the installation of bulky and expensive steam generating equipment, whether it be a conventional boiler or nuclear reactor. The significant feature is that the hot gases produced in the boiler furnace or reactor core never reach the turbine; they are merely used indirectly to produce an intermediate fluid, namely steam. A much more compact power plant results when the water to steam step is eliminated and the hot gases themselves are used to drive the turbine. Serious development of the gas turbine began not long before the Second World War with shaft power in mind, but attention was soon transferred to the turbojet engine for aircraft propulsion. The gas turbine began to compete successfully in other fields only in the mid 1950s, but since then it has made a progressively greater impact in an increasing variety of applications.

In order to produce an expansion through a turbine a pressure ratio must be provided, and the first necessary step in the cycle of a gas turbine plant must therefore be compression of the working fluid. If after compression the working fluid were to be expanded directly in the turbine, and there were no losses in either component, the power developed by the turbine would just equal that absorbed by the compressor. Thus if the two were coupled together the combination would do no more than turn itself round. But the power developed by the turbine can be increased by the addition of energy to raise the temperature of the working fluid prior to expansion. When the working fluid is air a very suitable means of doing this is by combustion of fuel in the air which has been compressed. Expansion of the hot working fluid then produces a greater power output from the turbine, so that it is able to provide a useful output in addition to driving the compressor. This represents the gas turbine or internal-combustion turbine in its simplest form. The three main components are a compressor, combustion chamber and turbine, connected together as shown diagrammatically in Fig. 1.1.

In practice, losses occur in both the compressor and turbine which increase the power absorbed by the compressor and decrease the power output of the turbine. A certain addition to the energy of the working fluid, and hence a certain fuel supply, will therefore be required before the one component can drive the other. This fuel produces no useful power, so that the component losses contribute to a lowering of the efficiency of the machine. Further addition of fuel will result in a useful power output, although for a given flow of air there is a limit to the rate at which fuel can be supplied and therefore to the net power output. The maximum fuel/air ratio that may be used is governed by the working temperature of the highly stressed turbine blades, which temperature must not be allowed to exceed a certain critical value. This value depends upon the creep strength of the materials used in the construction of the turbine and the working life required.

These then are the two main factors affecting the performance of gas turbines: component efficiencies and turbine working temperature. The higher they can be made, the better the all-round performance of the plant. It was, in fact, low efficiencies and poor turbine materials which brought about the failure of a number of early attempts to construct a gas turbine engine. For example, in 1904 two French engineers, Armengaud and Lemale, built a unit which did little more than turn itself over: the compressor efficiency was probably no more than 60 per cent and the maximum gas temperature that could be used was about 740 K.

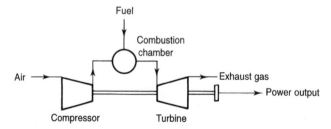

FIG. 1.1 Simple gas turbine system

It will be shown in Chapter 2 that the overall efficiency of the gas turbine cycle depends primarily upon the pressure ratio of the compressor. The difficulty of obtaining a sufficiently high pressure ratio with an adequate compressor efficiency was not resolved until the science of aerodynamics could be applied to the problem. The development of the gas turbine has gone hand in hand with the development of this science, and that of metallurgy, with the result that it is now possible to find advanced engines using pressure ratios of up to 35:1, component efficiencies of 85–90 per cent, and turbine inlet temperatures exceeding 1650 K.

In the earliest days of the gas turbine, two possible systems of combustion were proposed: one at constant pressure, the other at constant volume. Theoretically, the thermal efficiency of the constant volume cycle is higher than that of the constant pressure cycle, but the mechanical difficulties are very much greater. With heat addition at constant volume, valves are necessary to isolate the combustion chamber from the compressor and turbine. Combustion is therefore intermittent, which impairs the smooth running of the machine. It is difficult to design a turbine to operate efficiently under such conditions and, although several fairly successful attempts were made in Germany during the period 1908–1930 to construct gas turbines operating on this system, the development of the constant volume type has been discontinued. In the constant pressure gas turbine, combustion is a continuous process in which valves are unnecessary and it was soon accepted that the constant pressure cycle had the greater possibilities for future development. A key advantage of the constant pressure system was its ability to handle high mass flows, which in turn lead to high power.

It is important to realize that in the gas turbine the processes of compression, combustion and expansion do not occur in a single component as they do in a reciprocating engine. They occur in components which are separate in the sense that they can be designed, tested and developed individually, and these components can be linked together to form a gas turbine unit in a variety of ways. The possible number of components is not limited to the three already mentioned. Other compressors and turbines can be added, with intercoolers between the compressors, and reheat combustion chambers between the turbines. A heat-exchanger which uses some of the energy in the turbine exhaust gas to preheat the air entering the combustion chamber may also be introduced. These refinements may be used to increase the power output and efficiency of the plant at the expense of added complexity, weight and cost. The way in which these components are linked together not only affects the maximum overall thermal efficiency, but also the variation of efficiency with power output and of output torque with rotational speed. One arrangement may be suitable for driving an alternator under varying load at constant speed, while another may be more suitable for driving a ship's propeller where the power varies as the cube of the speed.

Apart from variations of the simple cycle obtained by the addition of these other components, consideration must be given to two systems distinguished by the use of *open* and *closed* cycles. In the almost universally used *open*-cycle gas turbine which has been considered up to this point, fresh atmospheric air is drawn into the circuit continuously and energy is added by the combustion of fuel in the working fluid itself. The products of combustion are expanded through the turbine and exhausted to the atmosphere. In the alternative, and rarely used, *closed* cycle shown in Fig. 1.2 the same working fluid, be it air or some other gas, is repeatedly circulated through the machine. Clearly in this type of plant the fuel cannot be burnt in the working fluid and the necessary energy must be added in a heater or 'gas-boiler' wherein the fuel is burnt in a separate air stream supplied by an auxiliary fan. The closed cycle is more akin to that of steam turbine plant in that the combustion gases do not themselves pass through the turbine. In the gas turbine the 'condenser' takes the form of a precooler for cooling of the gas before it re-enters the compressor. Although little used, numerous advantages are claimed for the closed cycle and these will be put forward in section 1.3.

Finally, various *combined gas and steam cycles* are now widely used for electric power generation, with thermal efficiencies approaching 60 per cent. The gas turbine exhaust, typically at a temperature of 500–600 °C, is used to raise steam in a *waste heat boiler* (WHB) or *heat recovery steam generator* (HRSG); this steam is then used in a steam turbine which drives a generator. Figure 1.3 shows such a system, using a dual-pressure steam cycle. With increasing cycle temperatures the exhaust gas entering the HRSG is hot enough to permit the use of a triple-pressure steam cycle incorporating a stage of reheat, and this is becoming common in large base-load power stations. It should be noted that the turbine exhaust has an unused oxygen content and it is possible to burn additional fuel in the boiler to raise steam output; this *supplementary firing* is most likely to be used with gas turbines operating at a relatively low exhaust gas temperature. The steam produced may be used in a

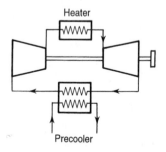

FIG. 1.2 Simple closed cycle

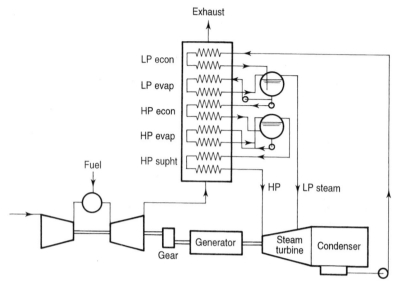

FIG. 1.3 Combined gas and steam cycle

process such as paper drying, brewing or building heating as well as producing electricity, and this is referred to as *cogeneration* or *combined heat and power* (CHP). Although the characteristic compactness of the gas turbine is sacrificed in *binary cycle* plant, the efficiency is so much higher than is obtainable with the simple cycle that such turbines are now widely used for large-scale electricity generating stations (see section 1.5).

The gas turbine has proved itself to be an extremely versatile prime mover and has been used for a wide variety of functions, ranging from electric power generation, mechanical drive systems and jet propulsion to the supply of process heat and compressed air, and the remainder of this Introduction is intended to emphasize the wide range of applications.† We commence, however, by discussing the various ways in which the components can be linked together to produce *shaft power*. Gas turbines for electric power generation, pump drives for gas or liquid pipelines and land and sea transport will be considered before turning our attention to aircraft propulsion. The vast majority of land-based gas turbines are in use for the first two of these, and applications to land and sea transport are still in their infancy, although the gas turbine has been widely used in naval applications since the 1970s.

† Some of the remarks about the 'stability of operation' and 'part-load performance' will be more fully understood when the rest of the book, and Chapter 8 in particular, have been studied. It is suggested therefore that the remainder of the Introduction be given a second reading at that stage.

1.1   Open-cycle single-shaft and twin-shaft arrangements

If the gas turbine is required to operate at a fixed speed and fixed load condition such as in base-load power generation schemes, the single-shaft arrangement shown in Fig. 1.1 is the most suitable. Flexibility of operation, i.e. the rapidity with which the machine can accommodate itself to changes of load and rotational speed, is unimportant in this application. Indeed the effectively high inertia due to the drag of the compressor is an advantage because it reduces the danger of overspeeding in the event of a loss of electrical load. A heat-exchanger might be added as in Fig. 1.4(a) to improve the thermal efficiency, although for a given size of plant the power output could be reduced by as much as 10 per cent owing to frictional pressure loss in the heat-exchanger. As we will see in Chapter 2, a heat-exchanger is essential for high efficiency when the cycle pressure ratio is low, but becomes less advantageous as the pressure ratio is increased. Aerodynamic developments in compressor design have permitted the use of such high pressure ratios that efficiencies of over 40 per cent can now be achieved with the simple cycle for power levels of 40 MW and higher. The Advanced Turbine Systems (ATS) programme conducted in the US in the 1990s has focused attention on the use of the heat-exchange cycle for lower power units, resulting in the emergence of the 4 MW Solar Mercury with an efficiency of 42 per cent. For many years, however, the heat-exchange cycle was not considered for new designs because high efficiency could be obtained either by the high pressure ratio simple cycle or by the combined cycle.

Figure 1.4(b) shows a modified form proposed for use when the fuel, e.g. pulverized coal, is such that the products of combustion contain constituents which corrode or erode the turbine blades. It is much less efficient than the normal cycle because the heat-exchanger, inevitably less than perfect, is transferring the whole of the energy input instead of merely a small part of it. Such a cycle would be considered only if a supply of 'dirty' fuel was available at very low cost. A serious effort was made to develop a coal burning gas turbine in the early 1950s but with little success. More success has been achieved with residual oil, and provided

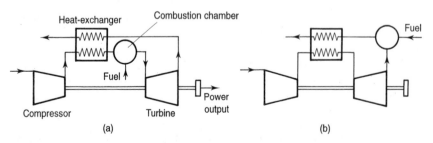

FIG. 1.4 Single-shaft open-cycle gas turbines with heat-exchanger

that the maximum temperature is kept at a sufficiently low level the straightforward cycle can be used.

When flexibility in operation is of paramount importance, e.g. when driving a variable speed load such as a pipeline compressor, marine propeller or road vehicle, the use of a mechanically independent (or *free*) *power turbine* is desirable. In this *twin-shaft* arrangement, Fig. 1.5, the high-pressure turbine drives the compressor and the combination acts as a *gas generator* for the low-pressure power turbine. For driving a compressor the turbine would be designed to run at the same speed as the compressor, making a gearbox unnecessary. The marine propeller, however, would run at a much slower speed and a reduction gearbox would be required. Twin-shaft engines may also be used for driving a generator; these would normally be derived from jet engines with the exhaust expanded through a power turbine rather than the original exhaust nozzle. The power turbine may be designed to run at the generator speed without a gearbox, as for the compressor drive. The twin-shaft engine has a significant advantage in ease of starting compared to a single-shaft unit, because the starter needs only to be sized to turn over the gas generator. The starter may be electric, a hydraulic motor, an expansion turbine operated from a supply of pipeline gas, or even a steam turbine or diesel; the last two are normally used only on large single-shaft machines. A disadvantage of a separate power turbine, however, is that a shedding of electrical load can lead to rapid overspeeding of the turbine, and the control system must be designed to prevent this.

Variation of power for both single- and twin-shaft units is obtained by controlling the fuel flow supplied to the combustion chamber. Although they behave in rather different ways as will be explained in Chapter 8, in both cases the cycle pressure ratio and maximum temperature decrease as the power is reduced from the design value with the result that the thermal efficiency deteriorates considerably at part load.

The performance of a gas turbine may be improved substantially by reducing the work of compression and/or increasing the work of expansion. For any given compressor pressure ratio, the power required per unit quantity of working fluid is directly proportional to the inlet temperature. If therefore the compression process is carried out in two or more stages with intercooling, the work of compression will be reduced.

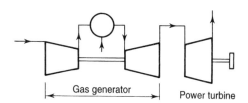

Gas generator   Power turbine

FIG. 1.5 Gas turbine with separate power turbine

Similarly, the turbine output can be increased by dividing the expansion into two or more stages, and reheating the gas to the maximum permissible temperature between the stages. One arrangement of a plant incorporating intercooling, heat-exchange and reheat is shown in Fig. 1.6. Complex cycles of this type offer the possibility of varying the power output by controlling the fuel supply to the reheat chamber, leaving the gas generator operating closer to its optimum conditions.

Complex cycles were proposed in the early days of gas turbines, when they were necessary to obtain a reasonable thermal efficiency at the low turbine temperatures and pressure ratios then possible. It can readily be seen, however, that the inherent simplicity and compactness of the gas turbine have been lost. In many applications low capital cost and small size are more important than thermal efficiency (e.g. electrical peaking, with low running hours), and it is significant that the gas turbine did not start to be widely used (apart from aircraft applications) until higher turbine inlet temperatures and pressure ratios made the simple cycle economically viable.

The quest for high efficiency as gas turbines became more widely used in base-load applications led to a revival of interest in more complex cycles in the mid 1990s. An important example is the reintroduction of the reheat cycle by ABB, referred to as 'sequential combustion'. This permitted the use of a very high cycle pressure ratio, resulting in a thermal efficiency of about 36 per cent, with no intercooling or heat-exchange. The use of reheat results in an exhaust gas temperature exceeding 600 °C, allowing the use of triple-pressure reheat steam cycles and a combined cycle efficiency approaching 60 per cent. Rolls-Royce developed the intercooled regenerative† (ICR) cycle for naval propulsion, giving both high thermal efficiency at the design point and excellent part-load efficiency, a very important feature for ships which generally cruise for extended periods at much lower power levels than the design value. A key design requirement was that the ICR engine, despite its additional components, should fit into the same footprint as current naval engines of similar power.

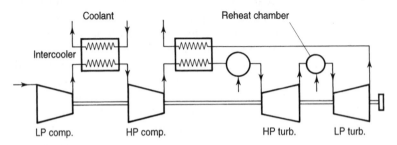

FIG. 1.6 Complex plant with intercooling, heat-exchange and reheat

† 'Regenerative' merely refers to the use of heat-exchange. It does not imply the use of a 'regenerator', which is a particular type of heat-exchanger wherein a rotating matrix is alternately heated and cooled by the gas streams exchanging heat.

1.2 Multi-spool arrangements

To obtain a high thermal efficiency without using a heat-exchanger, a high pressure ratio is essential. The nature of the compression process, however, limits the pressure ratio which can be achieved.

Non-positive displacement compressors are always used in gas turbines because of the high air mass flow rates involved. At full power the density at the rear end of the compressor is much higher than at inlet, and the dimensions of the compressor at exit are determined on the basis of the flow and density. When operated at rotational speeds well below the design value, however, the gas density at exit is much too low, resulting in excessive velocities and breakdown of the flow. This unstable region, which is manifested by violent aerodynamic vibration and reversal of the flow, is most likely to be encountered at low power or during start-up; this can result in either severe overheating in the turbine or a flame-out.

The problem is particularly severe if an attempt is made to obtain a pressure ratio of more than about 8:1 in one compressor. One way of overcoming this problem is to split the compressor into one or more sections. In this context we mean *mechanical* separation, permitting each section to run at different rotational speeds, unlike the intercooled compressor shown in Fig. 1.6. With mechanically independent compressors, each will require its own turbine and a suitable arrangement is shown in Fig. 1.7. The low-pressure compressor is driven by the low-pressure turbine and the high-pressure compressor by the high-pressure turbine (see also Fig. 1.12(a)). Power may be taken from either the low-pressure shaft, as shown, or from a separate power turbine. The configuration shown in Fig. 1.7 is usually referred to as a *twin-spool* engine. It should be noted that although the spools are *mechanically* independent, their speeds are related *aerodynamically* and this will be further discussed in Chapter 9.

The twin-spool layout was primarily developed for the aircraft engines discussed in section 1.4, but there are many examples of shaft power derivatives of these; most aero-derivative engines substituted a free power turbine in place of the propelling nozzle. Current examples include the Rolls-Royce RB-211 and the GE LM 1600. The GE LM 6000, on the other hand, uses the configuration of Fig. 1.7 and the generator and LP compressor run at constant speed. In smaller engines the HP compressor is

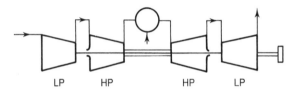

FIG. 1.7 **Twin-spool engine**

frequently of the centrifugal type, because at the high pressures and densities involved the volume flow rate is low and the blading required for an axial compressor would be too small for good efficiency. Twin-spool units were first introduced at a pressure ratio of about 10 and are suitable for cycle pressure ratios of at least 35:1. Triple-spool arrangements can also be used in large turbofan engines, where there is a requirement for both a very high pressure ratio and a low rotational speed for the large diameter fan.

As an alternative to multiple spools, a high pressure ratio can be safely employed with a single compressor if several stages of *variable stator blades* are used. General Electric pioneered this approach, which is now widely used by other companies; a pressure ratio of 30:1 on a single shaft is utilized in the ABB reheat gas turbine. It is frequently necessary to use *blow-off* valves at intermediate locations in the compressor to handle the serious flow mismatch occurring during start-up; as many as three blow-off valves are used. The single-spool variable geometry compressor is almost universally used in large electric power generation units.

Advanced technology aircraft engines usually employ combinations of multiple spools, variable stators and blow-off valves. This is particularly true for the high bypass turbofans discussed in section 1.4.

1.3   Closed cycles

Outstanding among the many advantages claimed for the closed cycle is the possibility of using a high pressure (and hence a high gas density) throughout the cycle, which would result in a reduced size of turbomachinery for a given output and enable the power output to be altered by a change of pressure level in the circuit. This form of control means that a wide range of load can be accommodated without alteration of the maximum cycle temperature and hence with little variation of overall efficiency. The chief disadvantage of the closed cycle is the need for an external heating system, which involves the use of an auxiliary cycle and introduces a temperature difference between the combustion gases and the working fluid. The allowable working temperature of the surfaces in the heater will therefore impose an upper limit on the maximum temperature of the main cycle. A typical arrangement of a closed-cycle gas turbine is shown in Fig. 1.8. The cycle includes a water-cooled precooler for the main cycle fluid between the heat-exchanger and compressor. In this particular arrangement the gas heater forms part of the cycle of an auxiliary gas turbine set, and power is controlled by means of a blow-off valve and an auxiliary supply of compressed gas as shown.

Besides the advantages of a smaller compressor and turbine, and efficient control, the closed cycle also avoids erosion of the turbine blades and other detrimental effects due to the products of combustion. Also, the need

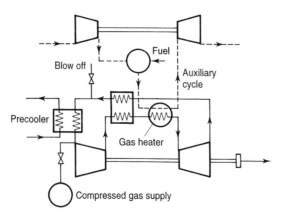

FIG. 1.8 Simple closed-cycle gas turbine

for filtration of the incoming air, which is a severe problem in the use of open-cycle units operating in contaminated atmospheres, is eliminated. The high density of the working fluid improves heat transfer, so that more effective heat-exchange is possible. Finally, the closed circuit opens up the field for the use of gases other than air having more desirable thermal properties. As will be seen in the next chapter, the marked difference in the values of the specific heats for air and a monatomic gas such as helium does not affect the efficiency as much as might be supposed. But higher fluid velocities can be used with helium and optimum cycle pressure ratios are lower, so that in spite of the lower density the turbomachinery may not be much larger. On the credit side, the better heat transfer characteristics of helium mean that the size of the heat-exchanger and precooler can be about half that of units designed for use with air. The capital cost of the plant should therefore be less when helium is the working fluid.

Only a small number of closed-cycle plants have been built, mostly by Escher–Wyss, and few are still in service. They were within the 2–20 MW power range. All used air as the working fluid, with a variety of fuels such as coal, natural gas, blast furnace gas, and oil. A pilot plant of 25 MW using helium was built in Germany, and it was thought that with this working fluid large sets of up to 250 MW would be feasible. They might have been required for use in nuclear power plant, if efforts to develop a reactor capable of operating at a sufficiently high core temperature had been successful. Considerable advantage accrues when the working fluid of the power cycle can be passed directly through the reactor core, because the reactor coolant circulating pumps are no longer required and the unwanted temperature drop associated with an intermediate fluid (e.g. CO_2 temperature – steam temperature) is eliminated. Helium is a particularly suitable working fluid in this application because it absorbs neutrons only weakly (i.e. it has a low neutron absorption cross-section). Attempts to develop the high temperature reactor (HTR) have been discontinued, however, and conventional nuclear reactors operate at much too low a temperature to be a possible source of heat for a gas turbine. It follows that gas turbines are unlikely to be used in any nuclear power plant in the foreseeable future.

A variety of small closed-cycle gas turbines (of 20–100 kW electrical output) have been considered for use in both aerospace and underwater applications. Possible heat sources include a radioactive isotope such as plutonium 238, combustion of hydrogen, and solar radiation. To date, none have been built.

1.4 Aircraft propulsion

Without any doubt the greatest impact of the gas turbine has been in the field of aircraft propulsion. The basic development of the jet engine took place during the period of World War II, with parallel development being carried out by Whittle in England and von Ohain in Germany; because of wartime secrecy neither was aware of the other's work. A Heinkel experimental aircraft powered by a von Ohain engine made the first flight of a jet-propelled aircraft on 27 August 1939. The Gloster-Whittle aircraft powered by a Whittle engine made its first flight on 15 May 1941. These developments were based solely on military requirements, and the initial applications of the jet engine were for high-speed military aircraft. The first engines had very short lives, poor reliability and very high fuel consumption. A great deal of development was necessary before the technology reached the level where it could be used in civil aviation, and the first civil aircraft applications appeared in the early 1950s; once the new power plant became established it rapidly made the aircraft piston engine obsolete, and today the piston engine is restricted to a very small market for light aircraft.

The mechanical layout of the simple *turbojet* engine is similar to that shown in Fig. 1.1 except that the turbine is designed to produce just enough power to drive the compressor. The gas leaving the turbine at high pressure and temperature is then expanded to atmospheric pressure in a propelling nozzle to produce a high velocity jet. Figure 1.9 shows a sectional view of the Rolls-Royce Olympus jet engine. This engine is of historical importance, being the first twin-spool engine in production; early versions were used in the Vulcan bomber and the advanced version shown powers the Concorde supersonic transport. (The Olympus has also been widely used as a gas generator to drive a power turbine, for both electricity generation and marine propulsion.)

For lower speed aircraft (up to about 400 knots)† a combination of propeller and exhaust jet provides the best propulsive efficiency. Figure 1.10

† 1 knot = 1 nautical mile/h = 0.5148 m/s.

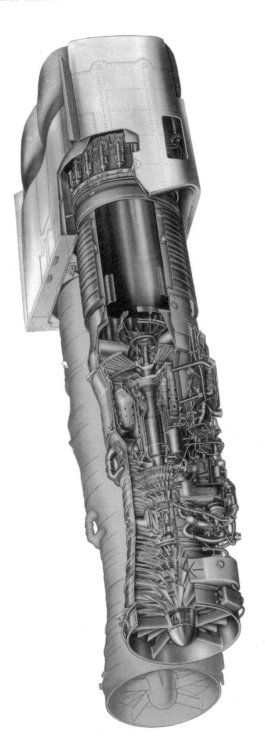

FIG. 1.9 Olympus turbojet engine [courtesy Rolls-Royce plc]

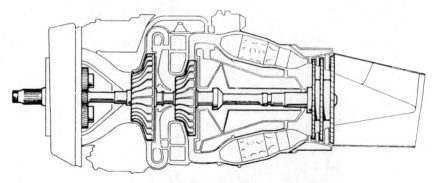

FIG. 1.10 Single-shaft turboprop engine [courtesy Rolls-Royce plc]

shows a single-shaft *turboprop* engine (Rolls-Royce Dart), illustrating the use of a two-stage centrifugal compressor and 'can'-type combustion chambers. It is notable that this engine entered airline service around 1953 at a power of about 800 kW and remained in production until 1986 with the final version producing about 2500 kW with an improvement in specific fuel consumption of about 20 per cent. Turboprops are also designed with a free turbine driving the propeller or propeller plus LP compressor. The Pratt and Whitney Canada PT-6, shown in Fig. 1.11, uses a free turbine; the use of a combined axial–centrifugal compressor and a reverse-flow combustor can also be seen. Development of the PT-6 started in 1956; it remains in production for a wide variety of applications ranging from single-engined trainers to commuter airliners, with versions ranging from 450 to 1500 kW. The *turboshaft* engine used in helicopters is similar in principle to the free-turbine turboprop, with the power turbine driving the helicopter main and tail rotors through a complex gearbox. In multi-engine helicopters two, and sometimes three, engines are connected to a single rotor system through a combining gearbox which shares the load equally between the engines.

At high subsonic speeds a propulsive jet of smaller mass flow but higher velocity is required. This was originally provided by the turbojet engine, but these have now largely been superseded by *turbofan* (or *bypass*) engines in which part of the air delivered by an LP compressor or fan bypasses the core of the engine (HP compressor, combustion chamber and turbines) to form an annular propulsive jet of cooler air surrounding the hot jet. This results in a jet of lower mean velocity which not only provides a better propulsive efficiency but also significantly reduces exhaust noise. Figure 1.12(a) is an example of a small turbofan engine, the Pratt and Whitney Canada PW 530. This is an extremely simple mechanical design giving good performance, intended for small business aircraft where capital cost is critical. A twin-spool arrangement is used, again with a centrifugal HP compressor and a reverse flow annular combustor. The reverse flow combustor is well suited for use with the centrifugal

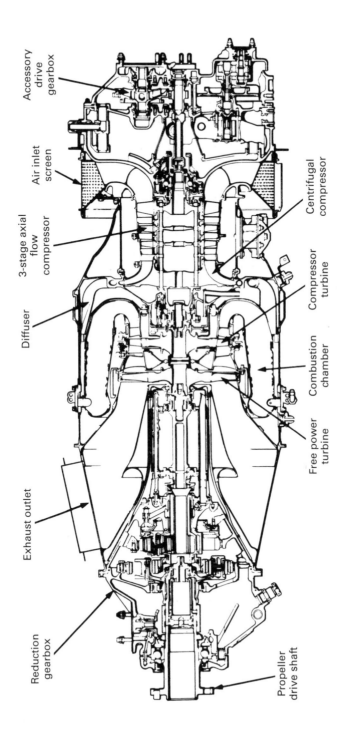

FIG. 1.11 Turboprop engine [courtesy Pratt and Whitney Canada]

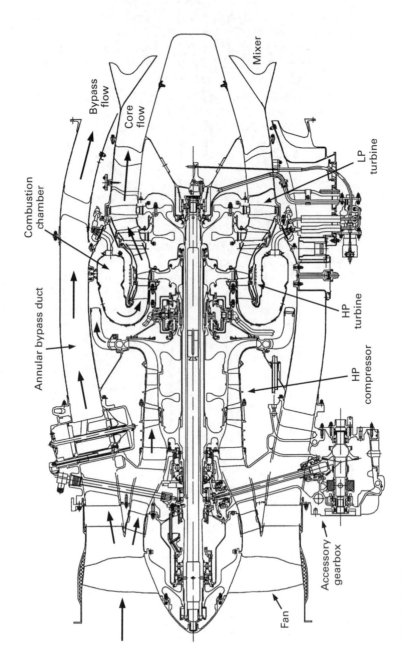

FIG. 1.12(a) Small turbofan engine [courtesy Pratt and Whitney Canada]

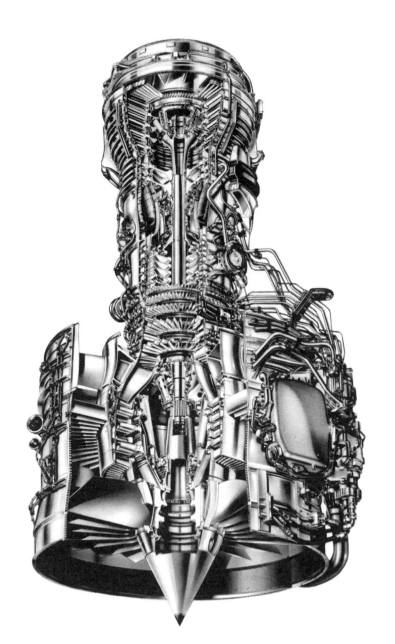

FIG. 1.12(b) Large twin-spool turbofan engine [courtesy International Aero-Engines]

compressor, where the flow must be diffused from a very high tangential velocity to a low axial velocity at entry to the combustor, and this configuration is widely used. Figure 1.12(b) shows a typical high bypass ratio turbofan designed for use with civil airliners; the engine shown, the V2500, was designed by a consortium of five nations. In the airline application fuel consumption is of paramount importance, requiring the use of high bypass ratio, pressure ratio and turbine inlet temperature. It can be seen that all of the turbomachinery is of the axial type and a straight-through annular combustion chamber is used. It is important to realize that good fuel economy requires advanced cycle conditions, but *reliability* cannot be compromised. With the rapidly increasing use of twin-engined aircraft for extended over-water operations (ETOPS), propulsion system reliability has a major effect on both aircraft safety and the marketability of an aircraft–engine combination; aircraft such as the Airbus 330 and Boeing 777 are available with a choice of GE, Pratt and Whitney or Rolls-Royce engines.

Heat-exchangers have as yet found no place in aircraft engines for reasons of bulk and weight, although they remain feasible for turboprop engines. This is because, with much of the net power output delivered to the propeller, the velocity of the gas leaving the turbine is relatively low and the frictional pressure loss need not be prohibitively high in a heat-exchanger of acceptable size. Around 1965 Allison† developed a regenerative turboprop for the US Navy, the object being to obtain an engine of exceptionally low specific fuel consumption for use on long-endurance anti-submarine patrols. In this kind of application it is the total engine plus fuel weight which is critical, and it was thought that the extra weight of the heat-exchanger would be more than compensated by the low fuel consumption. It was proposed that the heat-exchanger should be bypassed on take-off to give maximum power. The engine did not go into production, but it is not impossible that regenerative units will appear in the future, perhaps in the form of turboshaft engines for long-endurance helicopters.

1.5 Industrial applications

Sometimes in this book we shall find it necessary to use the distinguishing terms 'aircraft gas turbine' and 'industrial gas turbine'. The first term is

---

† Over the years there have been many mergers and changes of company names. Rolls-Royce, for example, took over Bristol-Siddeley, deHavilland and Napier in the mid 1960s; in the mid 1990s they also took over Allison. This has also happened in the industrial field; Brown Boveri and ASEA combined to form ABB, and then Alstom and ABB formed Alstom ABB, culminating in Alstom following the purchase of ABB. Siemens and Westinghouse combined in the late 1990s to form Siemens–Westinghouse. At various points in the text reference will be made to the company name under which developments took place.

self-explanatory, while the second is intended to include all gas turbines not included in the first category. This broad distinction has to be made for three main reasons. Firstly, the life required of an industrial plant is of the order of 100 000 hours without major overhaul, whereas this is not expected of an aircraft gas turbine. Secondly, limitation of the size and weight of an aircraft power plant is much more important than in the case of most other applications of the gas turbine. Thirdly, the aircraft power plant can make use of the kinetic energy of the gases leaving the turbine, whereas it is wasted in other types and consequently must be kept as low as possible. These three differences in the requirements can have a considerable effect on design and, in spite of the fact that the fundamental theory applies to both categories, it will be necessary to make the distinction from time to time.

Figure 1.13 shows the rugged construction employed in the Alstom (formerly Ruston) Typhoon, a 5 MW unit designed for long life with the capability of operating on either liquid or gaseous fuel. The Typhoon is used in both cogeneration and mechanical drive applications and can be built with either a fixed turbine or a free power turbine as shown. Several stages of variable stators are used and the actuators for these can be seen in Fig. 1.13. A large-scale single-shaft machine, the Siemens V94, is shown in Fig. 1.14. This machine is designed specifically for driving a constant speed generator, which is driven from the compressor end (cold end) of the unit; this arrangement also provides for a simple straight-through exhaust to the HRSG for combined cycle operations. The early version of the V94 shown uses two large off-board, or silo, combustors and was rated at about 150 MW. The latest versions, with increased airflow and cycle temperature, now incorporate an annular combustor, and are capable of more than 250 MW. Using aerodynamic scaling the V94, a 50 Hz machine running at 3000 rev/min, can be used as the basis for the V84 (60 Hz/3600 rev/min) or the smaller V63 (5600 rev/min, 50 or 60 Hz). Both GE and Westinghouse designs have been scaled in a similar way.

When gas turbines were originally proposed for industrial applications, unit sizes tended to be 10 MW or less and, even with heat exchangers, the cycle efficiency was only about 28–29 per cent. The availability of fully developed aircraft engines offered the attractive possibility of higher powers; the fact that a large part of the expensive research and development cost was borne by a military budget rather than an industrial user gave a significant advantage to the manufacturers of aircraft engines. The early *aero-derivative* engines, produced by substituting a power turbine for the exhaust nozzle, produced about 15 MW with a cycle efficiency of some 25 per cent. Modifications required included strengthening of the bearings, changes to the combustion system to enable it to burn natural gas or diesel fuel, the addition of a power turbine and a de-rating of the engine to give it a longer life; in some cases a reduction gearbox was required to match the power turbine speed to that of the

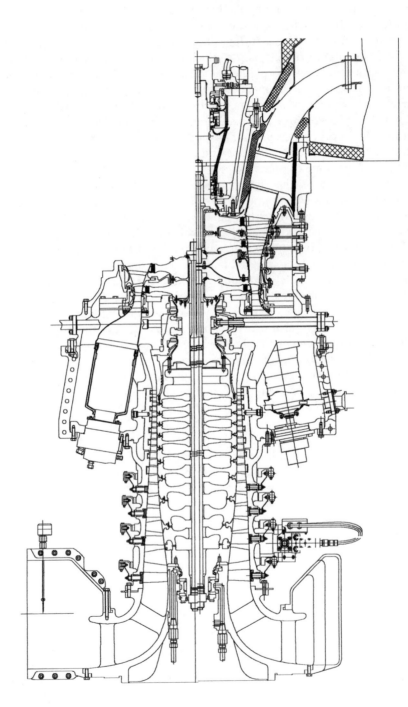

FIG. 1.13 Alstom Typhoon gas turbine; fixed turbine (top), free power turbine (bottom) [courtesy Alstom Power UK Ltd]

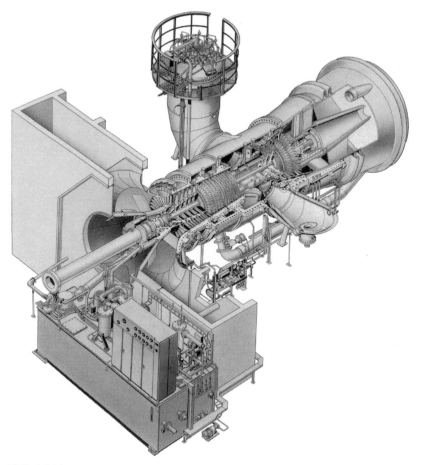

FIG. 1.14 Large single-shaft gas turbine [courtesy Siemens]

driven load, e.g. a marine propeller. For other types of load, such as alternators or pipeline compressors, the power turbine could be designed to drive the load directly. The Olympus, for example, had a single-stage power turbine for naval applications resulting in a very compact and lightweight design. For electric power generation a larger diameter two- or three-stage power turbine running at 3000 or 3600 rev/min was directly connected to the generator, requiring an increased length of ducting between the two turbines to allow for the change in diameter. Figure 1.15 shows a typical installation for a small power station using a single aero-derivative gas turbine; it should be noted that the air intake is well above ground level, to prevent the ingestion of debris into the engine. The aircraft and industrial versions of the Rolls-Royce Trent are shown in Fig. 1.16. The Trent is a large three-spool turbofan with the single-stage fan driven by a five-stage low-pressure turbine. The industrial version,

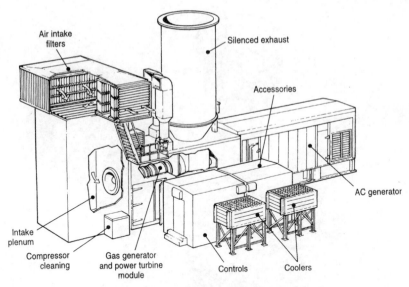

Air intake
filters

Silenced exhaust

Accessories

AC generator

Intake
plenum

Compressor
cleaning

Gas generator
and power turbine
module

Controls

Coolers

FIG. 1.15 Compact generating set [courtesy Rolls-Royce plc]

designed to drive a generator, replaces the fan with a two-stage
compressor of similar pressure ratio but much lower flow; as a result, the
low-pressure turbine can provide a large excess of power which can be
used to drive the generator. The low-pressure rotor speed of the aircraft
engine is restricted by the tip speed of the fan to about 3600 rpm; this
permits the shaft of the industrial version to be directly connected to a
60 Hz generator, avoiding the need for a gearbox. The Trent can also be
coupled to a 50 Hz generator running at 3000 rev/min by re-staggering the
blades of the two-stage LP compressor. For this reason, the compressor
blades are located in a turned annular fixing rather than the normal
broached fixings; this permits the same discs to be used for both
applications. The industrial version of the Trent is capable of 50 MW at a
thermal efficiency of 42 per cent, resulting from the high pressure ratio and
turbine inlet temperature. Figure 1.16 also shows the major changes in the
design of the combustion system; the aircraft version uses a conventional
fully annular combustor while the industrial engine uses separate radial
cans. This radical change is caused by the need for low emissions of oxides
of nitrogen, which will be dealt with further in Chapter 6.

Aero-derivative gas turbines have been widely used in pumping
applications for gas and oil transmission pipelines, peak-load and
emergency electricity generation, off-shore platforms and naval propulsion.
Turbines operating on natural gas pipelines use the fluid being pumped as
fuel and a typical pipeline might consume 7–10 per cent of the throughput
for compression purposes. When major pipeline systems emerged in the
mid 1950s gas was so cheap that little attention was given to thermal

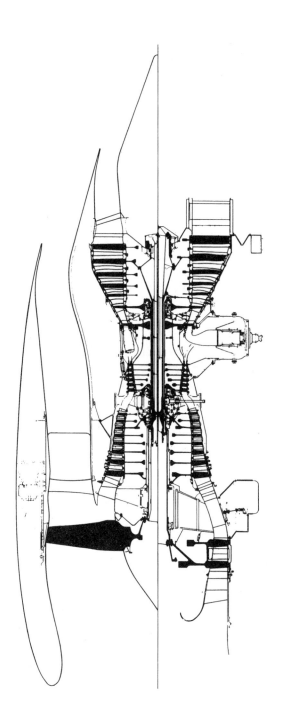

FIG. 1.16 Comparison between turbofan and industrial versions of R-R Trent [courtesy Rolls-Royce plc]

efficiency. In recent years, however, the value of natural gas has increased dramatically and this has led to a demand for high-efficiency pumping units. A major pipeline might have as much as 3000 MW of installed power and the fuel bills are comparable to those of a medium-sized airline. Pumping stations are typically about 100 km apart and the gas turbines used range in power from about 5 to 25 MW. Many compressor stations are in remote locations and aircraft derivatives of 15 to 25 MW are widely used; the light weight and compact size of these units greatly simplifies transportation to the site and permits easy replacement of a gas generator for overhaul. Other operators may prefer industrial engines and many 5–10 MW units built by Solar, Alstom and Nuovo Pignone are in service. Ease of automation and reduction of labour requirements are extremely important for remote stations; combined cycles, although offering higher efficiency, are seldom used in these applications because of the round-the-clock requirement for staffing steam plant. In the early days of pipelines a number of regenerative gas turbines were used, because of the low efficiency of the simple cycle. In the late 1970s some of the original heat-exchangers were replaced by modern units of higher efficiency, but the rapid improvement in the performance of simple-cycle machines has largely eliminated the use of the regenerative cycle. With oil pipelines the oil is often not suitable for burning in a gas turbine without expensive fuel treatment and it becomes necessary to bring a suitable liquid fuel in by road.

The use of gas turbines for electrical power generation has changed dramatically in recent years. In the 1970s, gas turbines (particularly in Great Britain and North America) were primarily used for peaking and emergency applications; aero-derivative units with a heavy-duty power turbine were widely used. One of the outstanding advantages of this type was its ability to produce full power from cold in under two minutes, although this capability should be used only for emergencies because thermal shock will greatly reduce the time between overhauls. In the mid 1960s, a major blackout of the eastern seaboard of the USA resulted in investment in gas turbines capable of a 'black' start, i.e. completely independent of the main electricity supply. In Great Britain, over 3000 MW of emergency and peak-load plant based on the Rolls-Royce Avon and Olympus engines was installed; these formed a key part of the overall electricity generating system, but only ran for a very small number of hours. Similar plants were built in large numbers in North America using the Pratt and Whitney FT-4. The aero-derivative units had a maximum rating of about 35 MW: their efficiency was about 28 per cent and they burned expensive fuel so they were not considered for applications involving long running hours.

In marked contrast, countries like Saudi Arabia, with a very rapidly expanding electrical system and abundant cheap fuel, used heavy-duty gas turbines for base-load duty; a particular advantage of the gas turbine in desert conditions is its freedom from any requirement for cooling water. Initially, the ratings of the aero-derivative and heavy-duty units were similar, but as cycle conditions improved over the years, the designers of industrial gas turbines were able to scale up their designs to give much higher power. The principal manufacturers of large gas turbines are Alstom, General Electric and Siemens–Westinghouse, all of whom design single-shaft machines which are capable of delivering over 250 MW per unit; the upper limit is fixed by considerations such as the size of disc forgings and the maximum width to permit transport by rail. Unlike steam turbines, gas turbines are not often erected on site but are delivered as complete packages ready to run; obviously this could not be done for turbines with silo combustors, which would have to be installed separately. Single-shaft units running at 3000 to 3600 rev/min respectively can drive 50 or 60 Hz generators directly without the need for an expensive gearbox. Compressor designs suitable for operation at these speeds result in 60 Hz machines of around 175 MW, and 50 Hz machines of around 250 MW, with the power largely determined by the airflow and turbine inlet temperature. North America is standardized on 60 Hz, while Europe and much of Asia operate at 50 Hz; Japan uses both 50 and 60 Hz. Smaller machines may be designed to run at about 5000–6000 rev/min with ratings from 50 to 100 MW, with gearboxes capable of either 3000 or 3600 rev/min output speeds to meet market requirements. Many heavy-duty units have run well in excess of 150 000 hours, a substantial number have exceeded 200 000 hours and a few have reached 300 000 hours.

Another major market for electricity generation is the provision of power for off-shore platforms, where gas turbines are used to provide base-load power. Many Solar and Ruston units of 1–5 MW have been used, but for larger powers aero-derivatives such as the Rolls-Royce RB-211 and General Electric LM 2500 have been installed at ratings of 20–25 MW; a big platform may require as much as 125 MW and both surface area and volume are at a premium. The installed weight is also of prime importance because of cranage requirements, and considerable savings accrue if the rig's own cranes can handle the complete machinery package. The aero-derivative dominates this market because of its compact nature.

The availability of gas turbines with an output of 100–250 MW has made large combined cycle plant a major factor in thermal power generation. Japan, because of its total dependence on imported fuel, was the first large-scale user of combined cycles, building several 2000 MW stations burning imported liquefied natural gas (LNG). A typical installation may be made up of 'blocks' consisting of two gas turbines with their own waste heat boilers and a single steam turbine; in general, using an unfired boiler the steam turbine power is about half that of the gas turbine. Thus, a single block of two 200 MW gas turbines and a 200 MW steam turbine provides 600 MW; a complete station may use three or four blocks. Large-scale combined cycle plants with ratings of up to 2000 MW have been installed in many countries, including Korea, Malaysia, Hong Kong, Singapore, USA and Argentina, with efficiencies in excess of 55 per cent.

Privatization of the electricity supply in Great Britain led to the installation of a large number of combined cycle plants of 225–1850 MW burning natural gas. Concern about the long-term availability of natural gas resulted in a temporary moratorium on the building of gas-fired plant, but a small number of plants continue to be built following a rigorous assessment of the particular project. In the longer term it is quite feasible that units burning natural gas could be converted to gas obtained from the gasification of coal.

Gas turbine power stations are remarkably compact. Figure 1.17(a) shows the comparison between the size of a 1950s era steam turbine station of 128 MW and a peak-load gas turbine plant of 160 MW shown ringed; the latter used eight Olympus 20 MW units. The steam plant required three cooling towers to cope with the heat rejected from the condensers. This plant has now been decommissioned and replaced by a 700 MW combined cycle base-load plant, shown in Fig. 1.17(b). The new plant consists of three blocks, each comprising a Siemens V94 gas turbine of 150 MW and a waste heat boiler, and a single steam turbine of 250 MW. An air-cooled condenser is used in place of the three cooling towers, because of restrictions on the use of river cooling water; the condenser is the large

FIG. 1.17(a) Comparative size of steam and gas turbine power stations [courtesy Rolls-Royce plc]

FIG. 1.17(b) Combined cycle plant with air-cooled condenser [courtesy Siemens]

rectangular structure to the left of the picture, and can be seen to be much less visually intrusive than the cooling towers. A small performance penalty is paid, because the condenser temperature (and hence back pressure on the steam turbine) is higher than could be obtained with a river-cooled condenser. The station, however, has a thermal efficiency of 51 per cent, which is much higher than that of a conventional steam turbine plant.

Another area of rapidly growing importance is that of the combined production of heat and power, in what are variously known as *cogeneration* or CHP plant. The gas turbine drives a generator, and the exhaust gases, typically at a temperature of 500–600 °C, are used as a source of low-grade heat. Many industrial processes require large quantities of steam and hot water, examples including breweries, paper mills, cement works and chemical processes. Similar situations are found in public institutions such as hospitals and universities. Heat at a relatively low temperature is required for heating buildings and operating air-conditioning systems. Chemical industries often need large quantities of hot gas containing a high proportion of free oxygen at a pressure sufficient to overcome the pressure loss in chemical reactors. The temperature limitation in the gas turbine cycle means that high air/fuel ratios must be used, resulting in a large proportion of unused oxygen in the exhaust, making the gas turbine exhaust highly suitable. The unit can be designed to meet the hot gas requirement, with or without shaft power (which may also be exported to a utility), and sometimes to burn a fuel which is a by-product of the chemical process. Figure 1.18 illustrates an early example of a cogeneration plant using Ruston gas turbines. It provides the whole of the electricity, process steam, heating steam and chilled water required for a factory. The use of eight gas turbines feeding four auxiliary fired waste heat boilers enables the changing power and heat demands during the day to be met by running the necessary number of units at substantially full power and therefore at peak efficiency. Newer plants are likely to use only one or two gas turbines, and it is common to export either much of the electricity or the steam produced to an external host. It is extremely difficult to balance heat and electrical loads through a wide variety of seasonal temperatures.

Another possibility being actively promoted at the start of the 21st century is the use of small-scale distributed cogeneration plant, using small gas turbines with ratings around 100 kW or even less. The idea is that these could be sold to relatively small users such as supermarkets or shopping malls. These small units are being referred to as *microturbines*; because of their small size they would be inherently simple machines, probably using a low pressure ratio regenerative cycle with a modest thermal efficiency, but making use of the low-grade heat in the exhaust for building heating, air-conditioning and hot water. It is clear that these applications would require a back-up connection to the local utility and also a back-up heating system. It is too early to say if this will prove viable but several major manufacturers are staking out a position along with a number of new entrants to the power generation field. Perhaps an analogy with computers should be considered; who in the early 1970s could have forecast the massive shift from centralized main-frame computers to the universal use of personal computers? It will probably take a minimum of five years before this trend becomes clear.

1.6   Marine and land transportation

In the early 1950s, as experience with aircraft gas turbines showed the promise of this new prime mover, gas turbines were seriously considered for applications such as merchant ships, trains, cars and trucks. Early gas turbines, however, had poor performance because of the low pressure ratios and cycle temperatures of the period, and this led to the development of complex cycles which inevitably proved difficult to perfect. Gas turbine merchant ships were built on an experimental basis on both sides of the Atlantic without success. Early attempts at building gas turbine locomotives were also a failure, as were proposals for road vehicle gas turbines. In the late 1990s, however, the aero-derivative gas turbine is beginning to penetrate niche markets in commercial shipping.

The gas turbine, however, has a long history of successful operation in navies where the advantages of compact size, high power density, low

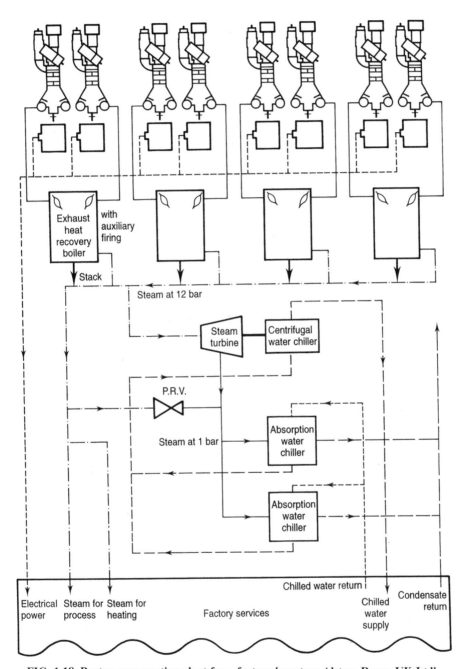

FIG. 1.18 Ruston cogeneration plant for a factory [courtesy Alstom Power UK Ltd]

noise and reduced manning have displaced the steam turbine in surface warships. Several navies, such as those of the UK, USA, Canada, the Netherlands and Japan, have accumulated many years of operating experience. A gas turbine was first used in a Motor Gun Boat in 1947 and aero-type engines (Rolls-Royce Proteus) were first used in fast patrol boats in 1958; the potential for using larger aero-derivative gas turbines for main propulsion of warships became clear.

A major disadvantage of the gas turbine in naval use is its poor specific fuel consumption at part load. If we consider a naval vessel having a maximum speed of, say, 36 knots, and a cruise speed of 18 knots, with the power required for hull-borne ships proportional to the cube of the speed, the cruise power will be only one-eighth of the maximum power; indeed, much time would be spent at speeds of less than 18 knots. To overcome this problem, and to ensure that gas turbines normally operate at high power and efficiency, combined power plants consisting of gas turbines in combination with steam turbines, diesels and gas turbines have been used. These go by names such as COSAG, CODOG, COGOG, COGAG, etc. CO stands for *combined*; S, D, G for *steam*, *diesel* and *gas turbine*. The final A or O stands for *and, or*. The 'or' requires explanation; with A the power output of the individual engines can be added; with O either plant can be used, but their outputs cannot be combined. COSAG was the earliest arrangement used by the Royal Navy where a ship's shaft was driven by both steam *and* gas turbines; the gearbox design permitted operation of the ship with either steam or gas turbine or both. The gas turbines were 6 MW industrial units, originally intended only for 'boost' purposes, or for rapid starts, but in practice they proved so popular that they became used for much longer periods. Another alternative is to combine a boost gas turbine with a cruise diesel; in this case the diesel power is small relative to the gas turbine power and there is little advantage in adding the powers. The vessel therefore operates in *either* the diesel *or* the gas turbine mode, i.e. CODOG. For naval use the diesel engine has the advantage of very good cruise fuel combustion, with the disadvantage of large bulk and a louder underwater noise level. The COGOG arrangement, with a small cruise gas turbine (4–5 MW) and a large boost gas turbine (20–25 MW), has been widely used. The small cruise gas turbines, however, are not competitive with diesels regarding specific fuel consumption and there appears to be a trend away from COGOG to CODOG. The COGAG arrangement, using gas turbines of the same size, has been used in the US Navy with four LM 2500 engines in large destroyers; this arrangement is also found with four Olympus engines in the Invincible class aircraft carriers of the Royal Navy.

The first all-gas turbine warships in the western world were the Canadian DDH-280 class in 1970, using a combination of Pratt and Whitney FT-4s for boost power and FT-12s for cruise; in the late 1980s the cruise engines were replaced by Allison 570s, providing both higher power and better fuel economy. The Royal Navy selected the Olympus as boost engine and the Rolls-Royce Tyne as cruise engine, and this machinery arrangement was also selected by the Royal Netherlands Navy and others. The Olympus has now been superseded by the more modern Rolls-Royce Spey. The Olympus and Tyne engines operated with great success in the Falklands War and the redundancy provided by the COGOG system proved to be very valuable. The US Navy has a large fleet powered by the LM 2500, which is used in both combatant and replenishment ships. The most recent Royal Navy ships (Type 23) use a combined gas turbine/diesel/electric propulsion system, where the propellers are driven by an electric motor; for low-speed towing of a sonar array power is obtained from the diesel. This system is known as CODLAG. The Canadian Patrol Frigate (CPF) uses a CODOG arrangement with two LM 2500s and a single cruise diesel of about 7 MW. With the increasing electrical needs of modern warships, and the absence of steam for use in turbogenerators, gas turbine-driven generators also offer a very compact source of electricity; Allison 501 units with about 3 MW capacity are used on US ships, but most other navies have opted for diesel generators.

With the availability of compact, high-power aero-derivatives fast container ships were brought into service in the early 1970s, with a service speed of 25 knots. The rapid increase in fuel prices in the mid 1970s made fuel costs prohibitive and the four ships in service were re-engined with diesels; the converted ships suffered a major loss in both speed and cargo capacity, but high speed could no longer be economically justified. In the 1990s, however, the gas turbine was again being used for fast merchant ships. The first successful application was in fast ferries, operating on routes such as Britain–Ireland, Mediterranean and Scandinavian links. These ships are typically catamarans powered by *water jets*, capable of operating at speeds in excess of 40 knots. Smaller ferries, of about 500 tonnes, have been built with two Solar Taurus engines of about 5 MW, while larger ferries, capable of carrying 1000 passengers and 300 cars, use LM 2500 or LM 1600 engines. The ABB GT35 is also used; this is somewhat heavier than the aero-derivative engines but is capable of using lower-grade fuels because of its conservative turbine inlet temperature.

Two major breakthroughs at the end of the 20th century were the selection of the LM 2500 for cruise ships and the Trent for container ships with a speed of 40 knots. In the cruise ship application the LM 2500 is used in a combined cycle, providing steam for both power and hotel services; the propellers are driven by electric motors and the machinery arrangement is referred to as COGES. Although high speed is not necessary, the compact nature of the power plant gives a valuable increase in high revenue passenger accommodation. The gas turbine also provides considerable improvement in emissions compared with the diesel, and this is becoming more important at the present time. The container ships, to be built by Fastship Lines, will use five Trents with a total installed power of 250 MW driving water jets.

The gas turbine has to date not made a significant contribution to land transportation. Union Pacific successfully operated large freight trains with gas turbine power from 1955 for 15–20 years, but these eventually gave way to diesels. Several high-speed passenger trains were built using helicopter-type gas turbines, but only with limited success; the most successful of these were built by the French and operated on the Amtrak system in the eastern USA. It appears, however, that electric traction using overhead pantographs is more suitable for high-speed passenger trains; the French TGV (Train à Grande Vitesse) uses electric traction with great success. In the late 1990s there was again interest in using aerotype gas turbines in areas that do not have the traffic density to justify the cost of electrification, but current proposals would use the gas turbines to generate electricity with separate electric motors on the driving wheels.

A considerable amount of work was done on gas turbines for long-haul trucks and buses, with engines of 200–300 kW being developed. The basic configuration for these units used a low pressure ratio cycle with a centrifugal compressor, free turbine and rotary heat-exchanger, but none of these proved competitive with the diesel. Similar efforts were expended on gas turbines for cars, with much of this sponsored by the US government, but the automobile gas turbine still appears to be far over the horizon and may never appear. There is no doubt that the cost of gas turbines could be significantly reduced if they were built in numbers approaching those of piston engines. The major problem to be overcome is still that of part-load fuel consumption, as cars spend most of their operating life at low power levels other than for acceleration and hill climbing. It is possible that small gas turbines could appear in hybrid propulsion systems, combined with electric power; these would be single-shaft units operating at constant speed and power. The gas turbine would be used for maximum power and also for charging the vehicle's propulsion batteries. In the late 1990s city buses were being tested on a prototype basis.

One major breakthrough achieved by the gas turbine was its choice for propulsion of the M1 tank built for the US Army, with over 10 000 engines supplied. This was a regenerative cycle, as the tank application requires a high percentage of operating time at low power. The M1 tanks obtained considerable battlefield experience in desert conditions in the Gulf War with considerable success. It has still not been proved that the gas turbine is superior to the diesel in this application and, to date, no other countries have followed the US lead.

1.7   Environmental issues

The first major application of the gas turbine was for jet propulsion for military aircraft, towards the end of World War II. The jet engine led to much higher aircraft speeds, and this was sufficiently important that serious deficiencies in fuel consumption and engine life could be ignored; the jet exhaust was noisy but this mattered little in military applications. When jet engines were considered for use in civil transport aircraft both fuel consumption and longer overhaul lives became of great importance, although noise was not then an issue. The appearance of significant numbers of jet-propelled aircraft at civil airports in the late 1950s rapidly resulted in noise becoming a problem that would severely inhibit the growth of air transport. The need to reduce engine noise was originally met by adding silencers, which were not very effective and caused serious loss of performance; it was clear that engine noise had to be properly understood and that engine design had to cater for noise reduction from the start of an aircraft design project. Mathematicians deduced that jet noise was proportional to (jet velocity)^8, so the basic requirement was immediately recognized to be provision of the necessary thrust at a reduced jet velocity with a resultant increase in airflow. This was precisely what the turbofan did to obtain high propulsive efficiency. It was fortunate indeed that the search for higher efficiency also resulted in lower noise. The bypass ratio on early turbofans was restricted by both lack of knowledge of three-dimensional flow effects in the longer fan blades and installation problems, particularly for engines buried in the wing; it was the advent of the pod-mounted engine that permitted the move to steadily increasing the bypass ratio. As bypass ratios increased, however, it was soon found that the high fluid velocities at the fan blade tips were another source of noise that was particularly troublesome during the landing approach, with the noise spread over a very wide area. This problem was attacked by the use of sound-absorbing materials in the intake duct and careful choice of spacing between rotor and stator blades. Aircraft noise reduction has required a vast amount of research and capital expenditure, but it can safely be said that current and future designs of aero engine will result in airport noise being greatly reduced. The exhaust of industrial gas turbines leaves at a low velocity and is discharged through a chimney, so the main source of noise occurring in the jet engine is absent. Owing to their compact nature and ease of installation, however, gas turbines may be located close to industrial areas and a low noise level is commonly specified; the requirement may be met by acoustic treatment of the intake system and baffles in the exhaust ducting.

When gas turbines were first considered for non-aircraft applications, the combination of rotating machinery with steady flow combustion and large amounts of excess air seemed to offer a relatively clean-burning power plant. In the late 1960s it was discovered that the notorious 'smog' occurring in Los Angeles was caused by a photo-chemical reaction between sunlight and oxides of nitrogen produced by automobile exhausts. This led to major research programmes to reduce oxides of nitrogen (referred to as NO_x) and unburned hydrocarbons (UHC) for piston engines. When gas turbines began to enter the market in applications such as pipelines, electricity generation and mechanical drives, they soon became subject to regulations limiting emissions and these have become steadily more stringent. The same is now true of emissions from aircraft engines. It will be apparent from section 6.7, where methods of reducing emissions are described, that different approaches are used for industrial plant and aircraft engines because of the variation in operational requirements. Oxides of nitrogen occur at very high combustion temperatures, and also increase with combustion inlet temperature; in other words, the very factors needed for high efficiency cause increased NO_x formation. In the early 1990s the design of low NO_x combustion systems was one of the key drivers in producing competitive gas turbines. The easiest approach for industrial plant was to use water or steam injection to reduce the peak combustion temperature, but this introduced a host of other problems and costs related to engine durability and provision of treated water. Emphasis was placed on the development of *dry* low NO_x systems, and the principal manufacturers have developed various solutions which entered service in the mid 1990s. Owing to the large amount of excess air used in combustion the production of UHC was less critical, but still subject to stringent restrictions. The main combustion product of any hydrocarbon fuel is carbon dioxide, which is believed to contribute significantly to global warming due to its 'greenhouse effect'. CO_2 emissions can be reduced only by improving engine efficiencies so that less fuel is burned, or by developing sources of power that do not entail the combustion of fossil fuels. To encourage this some countries, notably in Scandinavia, have introduced a CO_2 tax.

With the successful tackling of the noise and emission problems, the gas turbine once again offers the same promise of an environmentally benign power plant. Figure 1.19 shows an installation of a 70 MW combined cycle base-load plant located between two hospitals and immediately adjacent to a prime residential area.

1.8   Some future possibilities

The increasing scarcity and therefore cost of high-grade fossil fuels will necessitate the wider use of poor-quality coal and heavy fuel oil with a high sulphur content. Such fuel can be burnt in steam power stations, but only with expensive boiler maintenance and the costly cleaning of stack gases to meet ever more stringent anti-pollution regulations. Two other quite distinct approaches exist, both involving the use of gas turbines. The first makes use of the idea of *fluidized bed* combustion, and the second involves the transformation of the low-quality solid or liquid fuel into a clean gaseous fuel.

FIG. 1.19 70 MW combined cycle plant [courtesy Stewart and Stevenson]

A fluidized bed combustor consists essentially of a refractory-brick lined cylinder containing sand-sized refractory particles kept in suspension by an upward flow of air. When used in conjunction with a gas turbine, the required air can be bled from the compressor. If coal is being burnt, the oxides of sulphur formed are trapped in the ash, and if oil is the fuel they can be trapped by particles of limestone or dolomite in the bed. Figure 1.20 shows one possible scheme. It makes use of the fact that heat is transferred between a fluidized bed and any solid surface immersed in it with very high heat transfer coefficients. In this scheme, most of the compressor air is heated in a tubular heat-exchanger in the bed, and only the small amount of air bled for fluidization need be cleaned of dust in cyclone separators before being passed to the turbine. Corrosion and erosion problems are holding up development, but if they can be overcome the fluidized bed combustor opens up the possibility of burning coal mined by remote-controlled methods, or even the material in colliery spoil heaps. In the latter case not only would useful power be developed from hitherto unusable fuel, but valuable land would be reclaimed.

A prototype combined cycle plant using fluidized bed combustion was brought into service in Sweden in 1991. It was built to provide both power

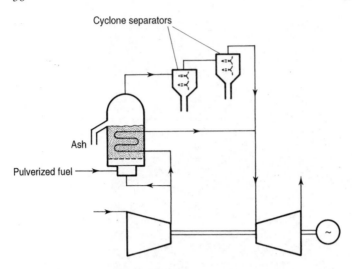

FIG. 1.20 Fluidized bed combustion

and heat, with a capacity of 135 MW of power and a district heating load
of 224 MW. Two gas turbines generated 34 MW of power, the balance
being provided by the steam turbine. This design did not use a heat-
exchanger in the bed as shown in Fig. 1.20, and all the compressor air
passed through the combustion system before being cleaned in cyclone
separators prior to entry to the turbine.

Before leaving the subject of fluidized bed combustion it is worth
mentioning another possible application: the incineration of municipal
waste. The waste is shredded and useful recoverable materials (steel, tin,
aluminium, etc.) are separated out using magnetic and flotation
techniques, and vibrating screens. The remainder, about 85 per cent, is
burnt in a fluidized bed combustor. Combustion of the waste maintains
the temperature somewhere between 700 °C and 800 °C, which is high
enough to consume the material without causing the bed to agglomerate.
Supplementary oil burners are provided for starting the unit. The hot
gases pass through several stages of cleaning to prevent erosion of the
turbine and to satisfy air pollution standards. Income from the electricity
generated and sale of recyclable materials is expected to reduce
substantially the cost of waste disposal compared with the conventional
land-fill method. It must be emphasized that the maximum temperature
that can be used in a fluidized bed is unlikely to be very high so that the
gas turbine efficiency will be low. Fluidized bed combustors are likely to be
used only for burning cheap or otherwise unusable fuels.

The second approach to the problem of using poor-quality coal or heavy
oil is its transformation into a clean gaseous fuel, this process being known
as *gasification*. The gasification process removes troublesome vanadium
and sodium impurities which cause corrosion in the turbine, and also the

sulphur which produces polluting oxides in the stack gases. There are a variety of processes for gasification, but most require large quantities of steam and the gasification plant can be integrated with a combined cycle; this is referred to as an IGCC (Integrated Gasification Combined Cycle) plant and a possible arrangement is shown in Fig. 1.21. Compressed air required for the process is bled from the gas turbine compressor. To overcome the pressure loss in the gasification plant the pressure is boosted in a separate compressor driven by a steam turbine. This would use some of the steam from the waste heat boiler, the major fraction supplying the power steam turbine. The gas produced by such a plant would have a very low calorific value—perhaps only 5000 kJ/m^3 compared with about 39 000 kJ/m^3 for natural gas. This is because of dilution by the nitrogen in the air supplied by the gasifier. The low calorific value carries little penalty, however, because all gas turbines operate with a weak mixture to limit the turbine inlet temperature. It simply means that the nitrogen normally fed directly to the combustion chamber in the compressor delivery air, now enters as part of the fuel already at the pressure required. Further discussion of this type of plant is given in section 6.8.

One other possible future application of the gas turbine should be mentioned: its use as an energy storage device. The overall efficiency of a country's electricity generating system can be improved if sufficient energy storage capacity is provided to enable the most efficient base-load stations to run night and day under conditions yielding peak efficiency. So far hydro-electric pumped storage plants have been built to meet the need, but suitable sites in Great Britain have virtually all been used. A possible alternative is illustrated in Fig. 1.22. Here a reversible motor/generator is coupled to either the compressor or the turbine. During the night, off-peak

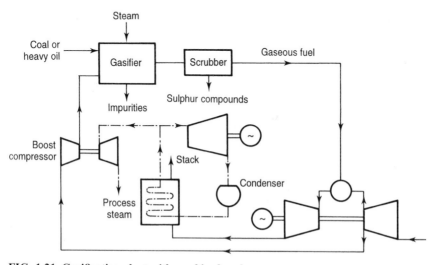

FIG. 1.21 Gasification plant with combined cycle

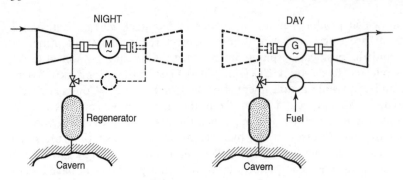

FIG. 1.22 Energy storage scheme

power is used to drive the compressor which delivers air to an underground cavern via a 'pebble bed' regenerator. The regenerator stores the heat in pebbles of alumina or silica. During the day the compressed air is released through the regenerator, picking up stored energy on its way to the turbine. To satisfy peak demand it may prove desirable also to burn some fuel in a combustion chamber to make up for heat losses in the regenerator. If the cavern is to be sufficiently small to make such a scheme economic, the pressure must be high—perhaps as high as 100 bar. This implies a high compressor delivery temperature of about 900 °C. By cooling the air in the regenerator the volume is further reduced, and at the same time the walls of the cavern are protected from the high temperature. Salt caverns excavated by washing have been proposed, and disused mine workings are another possibility if economic means can be found of sealing them adequately.

The first air-storage gas turbine plant was built by Brown Boveri and commissioned in Germany in 1978. It has no regenerator, but it incorporates a heat-exchanger, two-stage compression with intercooling and sequential (or reheat) combustion. An after-cooler protects the salt cavern walls from high temperature. The plant is for peak-load generation and produces up to 290 MW for periods of 1·5 hours three times per day, using about 12 hours for pumping the reservoir up to pressure.

1.9   Gas turbine design procedure

Although this book is primarily concerned with providing an introduction to the theory of the gas turbine, it also provides some material on the mechanical design and operational aspects of gas turbines. To place the contents of the book in proper perspective, a schematic diagram representing the overall design procedure is shown in Fig. 1.23. This gives some idea of the interrelationships between thermodynamic, aerodynamic, mechanical and control system design and emphasizes the need for

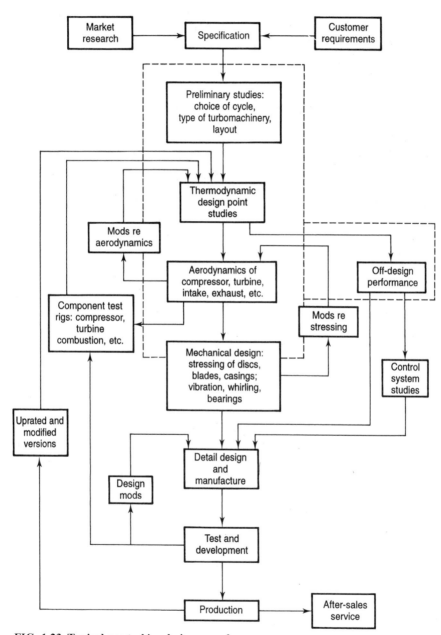

FIG. 1.23 Typical gas turbine design procedure

feedback between the various specialists. Manufacturing feasibility must be considered at the same time and the overall flow of information between the various groups is often referred to as *concurrent engineering*. The dotted lines enclose the topics covered in subsequent chapters.

The design process is shown as starting from a specification, resulting from either market research or a customer requirement; the process is similar for both aero and industrial engines. There are very few customers who are powerful enough to have an engine built to their specific requirement, and the specification normally results from market research. The development of a high-performance gas turbine is extremely costly, and may be shared by several companies; in the case of aero engines this may require multi-national consortia. The V 2500 turbofan, for example, was designed by companies in five nations: Rolls-Royce in the UK, Pratt and Whitney in the USA, Fiat in Italy, MTU in Germany, and Japan Aero Engines.

Successful engines are those which find a variety of applications, and their life-cycle from start of design to final service use may be in excess of 50 years. When the first edition of this book was written in 1950 the Rolls-Royce Dart was in the design stage, remaining in production until 1986; in late 1999 there were still nearly 1500 Darts in airline service, primarily in the Fokker F-27 and BAe 748. The design of the PT-6 started in 1958 and it is still in production for a wide variety of aircraft and helicopters; the US T-6 Texan trainer aircraft, for example, is scheduled to remain in production until 2015 and can be expected to continue in service until about 2030. The Lockheed P-3 Orion is scheduled to remain in service until at least 2030, and the original design of the Allison 501 dates back to the late 1940s. Further examples will be given later for both turbofans and industrial engines.

The specification is rarely a simple statement of required power and efficiency. Other factors of major importance, which vary with the application, include weight, cost, volume, life and noise, and many of these criteria act in opposition. For example, high efficiency inevitably incurs high capital cost but reduces operating costs; a simple engine of lower efficiency may be perfectly adequate for an application such as emergency power generation, where the engine may run for less than 50 hours per year. An important decision facing the designer is the choice of cycle, and this aspect will be covered in Chapters 2 and 3. It is essential to consider at an early stage what type of turbomachinery and combustor to use, and this will in large measure depend on the size of the engine; turbomachinery and combustor design will be dealt with in Chapters 4–7. The layout of the engine must also be considered, for example, whether a single or multi-shaft design should be used, and the behaviour of these different types of engine will be covered in Chapters 8 and 9.

The first major design step is to carry out *thermodynamic* design point studies. These are detailed calculations taking into account all important factors such as expected component efficiencies, air-bleeds, variable fluid properties and pressure losses, and would be carried out over a range of pressure ratio and turbine inlet temperature. A value for the specific output (i.e. power per unit mass flow of air) and specific fuel consumption will be determined for various values of the cycle parameters listed above. In industry these calculations would be carried out using sophisticated software, such as the commercially available program GASTURB, but it should be clearly understood that there is no mathematically defined optimum. For example, at a specified turbine inlet temperature a large increase in pressure ratio may give a minimal improvement in specific fuel consumption, and the resulting engine would be too complex and expensive to be practical; other considerations such as mechanical and aerodynamic design must be taken into account. Once the designer has selected a suitable choice of parameters, the specific output can be used to determine the airflow required to give the specified power.

It is important to realize that the choice of cycle parameters is strongly influenced by the engine *size*, and in particular by the airflow required. The turbine of a 500 kW engine, for example, would have very small blades which could not be cooled for reasons of manufacturing complexity and cost, limiting the maximum cycle temperature to about 1200 K. The pressure ratio would be restricted to allow blading of a reasonable size, and it would probably be necessary to use a centrifugal compressor of somewhat reduced efficiency; a heat-exchanger could be considered if high thermal efficiency were important. A 150 MW unit, however, could use sophisticated blading and could operate at a turbine inlet temperature of over 1600 K, some 400 K higher than that of the small engine. The large unit would use axial turbomachinery and the pressure ratio could be as high as 30:1.

The thermodynamic design point calculations having determined the airflow, pressure ratio and turbine inlet temperature, attention can be turned to the *aerodynamic* design of the turbomachinery. It is now possible to determine annulus dimensions, rotational speeds and number of stages. At this point it may well be found that difficulties arise which may cause the aerodynamicists to consult with the thermodynamic and mechanical designers to see if a change in the design point could be considered, perhaps a slight increase in turbine inlet temperature, a decrease in pressure ratio or an increase in rotational speed. The aerodynamic design of the turbomachinery must take into account *manufacturing* feasibility from the outset. When designing the centrifugal compressor for a small engine, for example, the space required for milling cutters between adjacent passages is of prime importance. In large turbofans, in contrast, the weight of the fan blades is dependent on the bypass ratio and the fan diameter, and the imbalance caused by the loss of a blade plays an important role in the design of the bearing support system and the structure required for fan containment.

The *mechanical* design can start only after the thermodynamic and aerodynamic design teams have established the key dimensions of the engine. It will then probably be found that stress or vibration problems may lead to further changes, the requirements of the stress and aerodynamics groups often being in opposition. At the same time as these studies are proceeding, *off-design* performance and *control system* design must be considered; off-design operation will include the effects of both varying ambient and flight conditions and reduced power operation. When designing a control system to ensure the safe and automatic operation of the engine, it is necessary to be able to predict parameters such as temperatures and pressure levels throughout the engine and to select some of these as control parameters.

The design of the engine must be carried out with consideration for future *growth*. Once the engine has entered service, there will be demands from customers for more powerful or more efficient versions, leading to the development of *uprated* engines. Indeed, such demands may often arise *before* the design process has been completed, particularly for aero engines. When engines have to be uprated, the designer must consider such methods as increasing the mass flow, turbine inlet temperature and component efficiencies, while maintaining the same basic engine design. A successful engine may have its power tripled during a long development cycle. Eventually, however, the engine will become dated and no longer competitive with newer technology designs. The timing of a decision to start a new design is of critical importance for the economic well-being of the manufacturer. References (1) and (2) describe the choice of design for a small Alstom industrial gas turbine and a Pratt and Whitney Canada turboprop. References (3) and (4) trace the design evolution of large industrial gas turbines by GE and Westinghouse over an extended period, showing how the power and efficiency are continuously improved by increasing pressure ratio, turbine inlet temperature and airflow. Reference (5) describes the evolution of a new gas turbine based on a previous successful family of engines, with the engine designed from the outset to be capable of driving both compressors and electric generators. The evolution of the Westinghouse 501 gas turbine from 1968 to 1993 is described in Ref. (4), and the table below shows how the power increased from 42 to 160 MW. Aerodynamic development allowed the pressure ratio to be raised from 7·5 to 14·6, while improvements in materials and blade cooling permitted a significant increase in cycle temperature; the net result was an improvement in thermal efficiency from 27·1 to 35·6 per cent. The 1968 engine had only the first nozzle row cooled, whereas the 1993 version required the cooling of six rows. The steady increase in exhaust gas temperature should be noted, this being an important factor in obtaining high thermal efficiency in a combined cycle application. The W501G, introduced in 1999 at 253 MW and 39 per cent efficiency, is shown in Fig. 1.24.

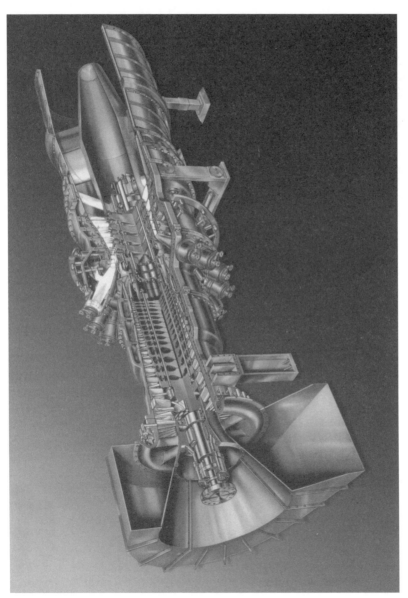

FIG. 1.24 W501G gas turbine [courtesy Siemens–Westinghouse]

Year	1968	1971	1973	1975	1981	1993
Power (MW)	42	60	80	95	107	160
Thermal efficiency (%)	27·1	29·4	30·5	31·2	33·2	35·6
Pressure ratio	7·5	10·5	11·2	12·6	14·0	14·6
Turbine inlet temp. (K)	1153	1161	1266	1369	1406	1533
Airflow (kg/s)	249	337	338	354	354	435
Exhaust gas temp. (°C)	474	426	486	528	531	584
No. of comp. stages	17	17	17	19	19	16
No. of turbine stages	4	4	4	4	4	4
No. of cooled rows	1	1	3	4	4	6

Reference (6) discusses the design of a high bypass ratio turbofan, with particular emphasis on the effect of the aircraft mission on the design. The original RB-211 was designed for use in long-haul aircraft and the RB211-535 derivative was developed for short-haul aircraft with typical flight times of 1–1·5 hours, resulting in a higher frequency of take-offs and climbs. Particular attention had to be paid to maintaining the same turbine life under these more arduous operating conditions. Reference (7) provides information on technological developments over an extended period and on innovative technologies that are required to produce major gains in engine performance and weight in future engines.

The foregoing should give the reader an overall, if superficial, view of the design process and may lead to the realization that the gas turbine industry can provide an exciting and rewarding technical career for a wide variety of highly skilled engineers.

2

Shaft power cycles

Enough has been said in the foregoing chapter for the reader to realize how great is the number of possible varieties of gas turbine cycle when multi-stage compression and expansion, heat-exchange, reheat and intercooling are incorporated. A comprehensive study of the performance of all such cycles, allowing for the inefficiencies of the various components, would result in a very large number of performance curves. We shall here concentrate mainly on describing methods of calculating cycle performance. For convenience the cycles are treated in two groups—shaft power cycles (this chapter) and aircraft propulsion cycles (Chapter 3). An important distinction between the two groups arises from the fact that the performance of aircraft propulsion cycles depends very significantly upon forward speed and altitude. These two variables do not enter into performance calculations for marine and land-based power plant to which this chapter is confined.

Before proceeding with the main task, it will be useful to review the performance of ideal gas turbine cycles in which perfection of the individual components is assumed. The specific work output and cycle efficiency then depend only on the pressure ratio and maximum cycle temperature. The limited number of performance curves so obtained enables the major effects of various additions to the simple cycle to be seen clearly. Such curves also show the upper limit of performance which can be expected of real cycles as the efficiency of gas turbine components is improved.

2.1 Ideal cycles

Analyses of ideal gas turbine cycles can be found in texts on engineering thermodynamics [e.g. Ref. (1)] and only a brief résumé will be given here. The assumption of ideal conditions will be taken to imply the following:

(a) Compression and expansion processes are reversible and adiabatic, i.e. isentropic.
(b) The change of kinetic energy of the working fluid between inlet and outlet of each component is negligible.

(c)   There are no pressure losses in the inlet ducting, combustion
      chambers, heat-exchangers, intercoolers, exhaust ducting, and ducts
      connecting the components.
(d)   The working fluid has the same composition throughout the cycle and
      is a perfect gas with constant specific heats.
(e)   The mass flow of gas is constant throughout the cycle.
(f)   Heat transfer in a heat-exchanger (assumed counterflow) is
      'complete', so that in conjunction with (d) and (e) the temperature rise
      on the cold side is the maximum possible and exactly equal to the
      temperature drop on the hot side.

Assumptions (d) and (e) imply that the combustion chamber, in which fuel
is introduced and burned, is considered as being replaced by a heater with
an external heat source. For this reason, as far as the calculation of
performance of ideal cycles is concerned, it makes no difference whether
one is thinking of them as 'open' or 'closed' cycles. The diagrammatic
sketches of plant will be drawn for the normal case of the open cycle.

Simple gas turbine cycle

The ideal cycle for the simple gas turbine is the Joule (or Brayton) cycle,
i.e. cycle 1234 in Fig. 2.1. The relevant steady flow energy equation is

$$Q = (h_2 - h_1) + \tfrac{1}{2}(C_2^2 - C_1^2) + W$$

where Q and W are the heat and work transfers per unit mass flow. Applying
this to each component, bearing in mind assumption (b), we have

$$W_{12} = -(h_2 - h_1) = -c_p(T_2 - T_1)$$
$$Q_{23} = (h_3 - h_2) \quad = c_p(T_3 - T_2)$$
$$W_{34} = (h_3 - h_4) \quad = c_p(T_3 - T_4)$$

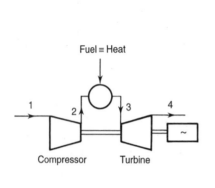

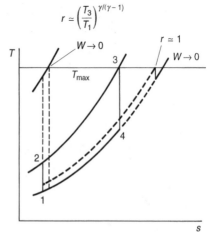

FIG. 2.1 Simple cycle

The cycle efficiency is

$$\eta = \frac{\text{net work output}}{\text{heat supplied}} = \frac{c_p(T_3 - T_4) - c_p(T_2 - T_1)}{c_p(T_3 - T_2)}$$

Making use of the isentropic p–T relation, we have

$$T_2/T_1 = r^{(\gamma-1)/\gamma} = T_3/T_4$$

where r is the pressure ratio $p_2/p_1 = r = p_3/p_4$. The cycle efficiency is then readily shown to be given by

$$\eta = 1 - \left(\frac{1}{r}\right)^{(\gamma-1)/\gamma} \tag{2.1}$$

The efficiency thus depends only on the pressure ratio and the nature of the gas. Figure 2.2(a) shows the relation between η and r when the working fluid is air ($\gamma = 1.4$), or a monatomic gas such as argon ($\gamma = 1.66$). For the remaining curves in this section air will be assumed to be the working fluid. The effect of using helium instead of air in a closed cycle is studied in section 2.6, where it is shown that the theoretical advantage indicated in Fig. 2.2(a) is not realized when component losses are taken into account.

The specific work output W, upon which the size of plant for a given power depends, is found to be a function not only of pressure ratio but also of maximum cycle temperature T_3. Thus

$$W = c_p(T_3 - T_4) - c_p(T_2 - T_1)$$

which can be expressed as

$$\frac{W}{c_p T_1} = t\left(1 - \frac{1}{r^{(\gamma-1)/\gamma}}\right) - (r^{(\gamma-1)/\gamma} - 1) \tag{2.2}$$

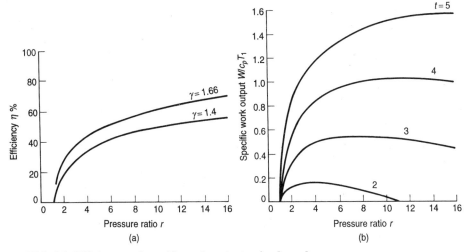

FIG. 2.2 **Efficiency and specific work output—simple cycle**

where $t = T_3/T_1$; T_1 is normally atmospheric temperature and is not a major independent variable. It is therefore convenient to plot the specific work output in non-dimensional form $(W/c_p T_1)$ as a function of r and t as in Fig. 2.2(b). The value of T_3, and hence t, that can be used in practice depends upon the maximum temperature which the highly stressed parts of the turbine can stand for the required working life: it is often called the 'metallurgical limit'. Early gas turbines used values of t between 3·5 and 4, but the introduction of air-cooled turbine blades allowed t to be raised to between 5 and 6.

A glance at the T–s diagram of Fig. 2.1 will show why a constant t curve exhibits a maximum at a certain pressure ratio: W is zero at $r = 1$ and also at the value of r for which the compression and expansion processes coincide, namely $r = t^{\gamma/(\gamma-1)}$. For any given value of t the optimum pressure ratio for maximum specific work output can be found by differentiating equation (2.2) with respect to $r^{(\gamma-1)/\gamma}$ and equating to zero: the result is

$$r_{\text{opt}}^{(\gamma-1)/\gamma} = \sqrt{t}$$

Since $r^{(\gamma-1)/\gamma} = T_2/T_1 = T_3/T_4$, this is equivalent to writing

$$\frac{T_2}{T_1} \times \frac{T_3}{T_4} = t$$

But $t = T_3/T_1$ and consequently it follows that $T_2 = T_4$. Thus the specific work output is a maximum when the pressure ratio is such that the compressor and turbine outlet temperatures are equal. For all values of r between 1 and $t^{\gamma/2(\gamma-1)}$, T_4 will be greater than T_2 and a heat-exchanger can be incorporated to reduce the heat transfer from the external source and so increase the efficiency.

Heat-exchange cycle

Using the nomenclature of Fig. 2.3, the cycle efficiency now becomes

$$\eta = \frac{c_p(T_3 - T_4) - c_p(T_2 - T_1)}{c_p(T_3 - T_5)}$$

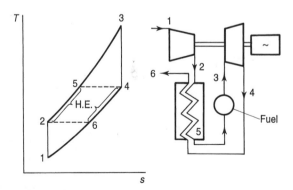

FIG. 2.3 Simple cycle with heat-exchange

With ideal heat-exchange $T_5 = T_4$, and on substituting the isentropic p–T relations the expression reduces to

$$\eta = 1 - \frac{r^{(\gamma-1)/\gamma}}{t} \qquad (2.3)$$

Thus the efficiency of the heat-exchange cycle is not independent of the maximum cycle temperature, and clearly it increases as t is increased. Furthermore it is evident that, for a given value of t, the efficiency increases with decrease in pressure ratio and not with increase in pressure ratio as for the simple cycle. The full lines in Fig. 2.4 represent the equation, each constant t curve starting at $r = 1$ with a value of $\eta = 1 - 1/t$, i.e. the Carnot efficiency. This is to be expected because in this limiting case the Carnot requirement of complete external heat reception and rejection at the upper and lower cycle temperature is satisfied.

The curves fall with increasing pressure ratio until a value corresponding to $r^{(\gamma-1)/\gamma} = \sqrt{t}$ is reached, and at this point equation (2.3) reduces to (2.1). This is the pressure ratio for which the specific work output curves of Fig. 2.2(b) reach a maximum and for which it was shown that $T_4 = T_2$. For higher values of r a heat-exchanger would *cool* the air leaving the compressor and so reduce the efficiency, and therefore the constant t lines have not been extended beyond the point where they meet the efficiency curve for the simple cycle which is shown dotted in Fig. 2.4.

The specific work output is unchanged by the addition of a heat-exchanger and the curves of Fig. 2.2(b) are still applicable. From these curves and those in Fig. 2.4 it can be concluded that to obtain an appreciable improvement in efficiency by heat-exchange, (a) a value of r appreciably less than the optimum for maximum specific work output should be used and (b) it is not necessary to use a higher cycle pressure ratio as the maximum cycle temperature is increased. Later it will be shown that for real cycles conclusion (a) remains true but conclusion (b) requires modification.

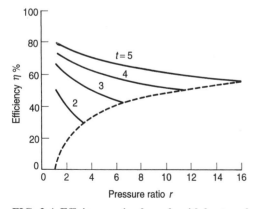

FIG. 2.4 Efficiency—simple cycle with heat-exchange

Reheat cycle

A substantial increase in specific work output can be obtained by splitting the expansion and reheating the gas between the high-pressure and low-pressure turbines. Figure 2.5(a) shows the relevant portion of the reheat cycle on the T–s diagram. That the turbine work is increased is obvious when one remembers that the vertical distance between any pair of constant pressure lines increases as the entropy increases: thus $(T_3 - T_4) + (T_5 - T_6) > (T_3 - T'_4)$.

Assuming that the gas is reheated to a temperature equal to T_3, differentiation of the expression for specific work output shows that the optimum point in the expansion at which to reheat is when the pressure ratios (and hence temperature drops and work transfers) for the HP and LP turbines are equal. With this optimum division, it is then possible to derive expressions for the specific output and efficiency in terms of r and t as before. Writing $c = r^{(\gamma-1)/\gamma}$ they become

$$\frac{W}{c_p T_1} = 2t - c + 1 - \frac{2t}{\sqrt{c}} \tag{2.4}$$

$$\eta = \frac{2t - c + 1 - 2t/\sqrt{c}}{2t - c - t/\sqrt{c}} \tag{2.5}$$

Comparison of the $W/c_p T_1$ curves of Fig. 2.6 with those of Fig. 2.2(b) shows that reheat markedly increases the specific output. Figure 2.5(b), however, indicates that this is achieved at the expense of efficiency. This is to be expected because one is adding a less efficient cycle (4′456 in Fig. 2.5(a)) to the simple cycle—less efficient because it operates over a smaller temperature range. Note that the reduction in efficiency becomes less severe as the maximum cycle temperature is increased.

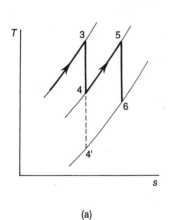

(a)

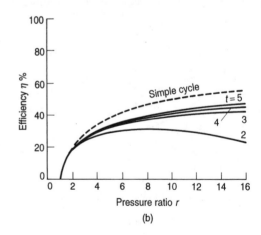

(b)

FIG. 2.5 Reheat cycle

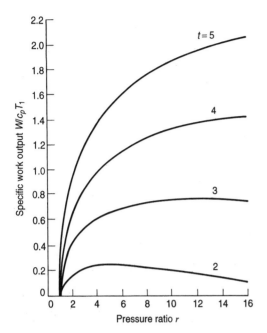

FIG. 2.6 Work output—reheat cycle

Cycle with reheat and heat-exchange

The reduction in efficiency due to reheat can be overcome by adding heat-exchange as in Fig. 2.7. The higher exhaust gas temperature is now fully utilized in the heat-exchanger and the increase in work output is no longer offset by the increase in heat supplied. In fact, when a heat-exchanger is employed, the efficiency is higher with reheat than without as shown by a comparison of Figs 2.8 and 2.4. The family of constant t lines exhibit the same features as those for the simple cycle with heat-exchange—each curve having the Carnot value at $r = 1$ and falling with increasing r to meet the

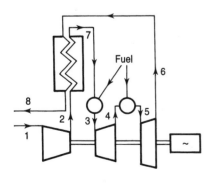

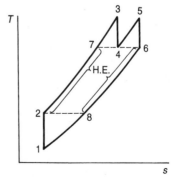

FIG. 2.7 Reheat cycle with heat-exchange

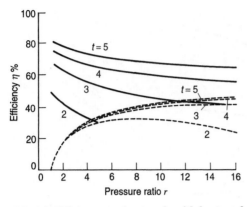

FIG. 2.8 Efficiency—reheat cycle with heat-exchange

corresponding efficiency curve of the reheat cycle without heat-exchange at the value of r corresponding to maximum specific work output.

Cycles with intercooled compression

A similar improvement in specific work output to that obtained by reheat can be achieved by splitting the compression and intercooling the gas between the LP and HP compressors; and, assuming that the air is intercooled to T_1, it can be shown that the specific work output is a maximum when the pressure ratios of the LP and HP compressors are equal. The use of intercoolers is seldom contemplated in practice because they are bulky and need large quantities of cooling water. The main advantage of the gas turbine, that it is compact and self-contained, is then lost. For this reason no performance curves for cycles with intercooling are included: suffice it to say that they are similar to Figs 2.5(b) and 2.6, although the increase in specific work output and reduction in efficiency with respect to the simple cycle are not so marked. (A modification to the low-temperature region of a cycle is normally less significant than a comparable modification to the high-temperature region.) As with reheat, intercooling increases the cycle efficiency only when a heat-exchanger is also incorporated, and then curves almost identical to those of Fig. 2.8 are obtained.

This discussion of ideal cycles should be sufficient to indicate the main effects of additions to the simple gas turbine. We have seen that the choice of pressure ratio will depend on whether it is high efficiency or high specific work output (i.e. small size) which is required; and that in cycles without heat-exchange a higher pressure ratio must be used to take advantage of a higher permissible turbine inlet temperature. It will be evident from what follows that these conclusions are also broadly true of practical cycles in which component losses are taken into account.

2.2 Methods of accounting for component losses

The performance of real cycles differs from that of ideal cycles for the following reasons.

(a) Because fluid velocities are high in turbomachinery the change in kinetic energy between inlet and outlet of each component cannot necessarily be ignored. A further consequence is that the compression and expansion processes are irreversible adiabatics and therefore involve an increase in entropy.

(b) Fluid friction results in pressure losses in combustion chambers and heat-exchangers, and also in the inlet and exhaust ducts. (Losses in the ducts connecting components are often included in the associated component losses.)

(c) If a heat-exchanger is to be of economic size, terminal temperature differences are inevitable, i.e. the compressed air cannot be heated to the temperature of the gas leaving the turbine.

(d) Slightly more work than that required for the compression process will be necessary to overcome bearing and 'windage' friction in the transmission between compressor and turbine, and to drive ancillary components such as fuel and oil pumps.

(e) The values of c_p and γ of the working fluid vary throughout the cycle due to changes of temperature and, with internal combustion, due to changes in chemical composition.

(f) The definition of the efficiency of an ideal cycle is unambiguous, but this is not the case for an open cycle with internal combustion. Knowing the compressor delivery temperature, composition of the fuel, and turbine inlet temperature required, a straightforward combustion calculation yields the fuel/air ratio necessary; and a combustion efficiency can also be included to allow for incomplete combustion. Thus it will be possible to express the cycle performance unambiguously in terms of fuel consumption per unit net work output, i.e. in terms of the *specific fuel consumption*. To convert this to a cycle efficiency it is necessary to adopt some convention for expressing the heating value of the fuel.

(g) With internal combustion, the mass flow through the turbine might be thought to be greater than that through the compressor by virtue of the fuel added. In practice, about 1–2 per cent of the compressed air is bled off for cooling turbine discs and blade roots, and we shall see later that the fuel/air ratio employed is in the region 0·01–0·02. For many cycle calculations it is sufficiently accurate to assume that the fuel added merely compensates for this loss. We will assume in the numerical examples which follow that the mass flows through the compressor and turbine are equal. When turbine inlet temperatures higher than about 1350 K are used, the turbine blading must be internally cooled as well as the disc and blade roots. We then have what is called an *air-cooled* turbine. Up to 15 per cent of the compressor delivery air might be bled for cooling purposes, and for an accurate estimate of cycle performance it is necessary to account explicitly for the variation of mass flow through the engine.

Methods of accounting for factors (*a*) to (*f*) must be discussed before giving examples of cycle calculations. We will also say a little more in general terms about cooling bleed flows.

Stagnation properties

The kinetic energy terms in the steady flow energy equation can be accounted for implicitly by making use of the concept of *stagnation* (or *total*) enthalpy. Physically, the stagnation enthalpy h_0 is the enthalpy which a gas stream of enthalpy h and velocity C would possess when brought to rest adiabatically and without work transfer. The energy equation then reduces to

$$(h_0 - h) + \tfrac{1}{2}(0 - C^2) = 0$$

and thus h_0 is defined by

$$h_0 = h + C^2/2 \tag{2.6}$$

When the fluid is a perfect gas, $c_p T$ can be substituted for h, and the corresponding concept of *stagnation* (or *total*) *temperature* T_0 is defined by

$$T_0 = T + C^2/2c_p \tag{2.7}$$

$C^2/2c_p$ is called the *dynamic temperature* and, when it is necessary to emphasize the distinction, T is referred to as the *static temperature*. An idea of the order of magnitude of the difference between T_0 and T is obtained by considering air at atmospheric temperature, for which $c_p = 1 \cdot 005 \, \text{kJ/kg K}$, flowing at $100 \, \text{m/s}$. Then

$$T_0 - T = \frac{100^2}{2 \times 1 \cdot 005 \times 10^3} \simeq 5 \, \text{K}$$

It follows from the energy equation that if there is no heat or work transfer T_0 will remain constant. If the duct is varying in cross-sectional area, or friction is degrading directed kinetic energy into random molecular energy, the static temperature will change—but T_0 will not. Applying the concept to an adiabatic compression, the energy equation becomes

$$W = -c_p(T_2 - T_1) - \tfrac{1}{2}(C_2^2 - C_1^2) = -c_p(T_{02} - T_{01})$$

Similarly for a heating process without work transfer,

$$Q = c_p(T_{02} - T_{01})$$

Thus if stagnation temperatures are employed there is no need to refer explicitly to the kinetic energy term. A practical advantage is that it is easier to measure the stagnation temperature of a high-velocity stream than the static temperature (see section 6.5).

When a gas is slowed down and the temperature rises there is a simultaneous rise in pressure. The *stagnation* (or *total*) *pressure* p_0 is defined in a similar way to T_0 but with the added restriction that the gas is imagined to be brought to rest not only adiabatically but also reversibly, i.e. isentropically. The stagnation pressure is thus defined by

$$\frac{p_0}{p} = \left(\frac{T_0}{T}\right)^{\gamma/(\gamma-1)} \tag{2.8}$$

Stagnation pressure, unlike stagnation temperature, is constant in a stream flowing without heat or work transfer only if friction is absent: the drop in stagnation pressure can be used as a measure of the fluid friction.

It is worth noting that p_0 is not identical with the usual pitot pressure p_0^* defined for incompressible flow by

$$p_0^* = p + \rho C^2/2$$

Substituting equation (2.7) in (2.8), and making use of $c_p = \gamma R/(\gamma - 1)$ and $p = \rho RT$, we have

$$p_0 = p\left(1 + \frac{\rho C^2}{2p} \times \frac{\gamma - 1}{\gamma}\right)^{\gamma/(\gamma-1)}$$

p_0^* is seen to be given by the first two terms of the binomial expansion. Thus p_0 approaches p_0^* as the velocity is decreased and compressibility effects become negligible. As an example of the difference at high velocities, for air moving with sonic velocity (Mach number $M = 1$), $p_0/p = 1.89$ whereas $p_0^*/p = 1.7$. Thus by assuming the flow to be incompressible the stagnation pressure would be underestimated by about 11 per cent.

Applying equation (2.8) to an isentropic compression between inlet 1 and outlet 2 we have the stagnation pressure ratio given by

$$\frac{p_{02}}{p_{01}} = \frac{p_{02}}{p_2} \times \frac{p_1}{p_{01}} \times \frac{p_2}{p_1} = \left(\frac{T_{02}}{T_2} \times \frac{T_1}{T_{01}} \times \frac{T_2}{T_1}\right)^{\gamma/(\gamma-1)} = \left(\frac{T_{02}}{T_{01}}\right)^{\gamma/(\gamma-1)}$$

Similarly, if required,

$$\frac{p_{02}}{p_1} = \left(\frac{T_{02}}{T_1}\right)^{\gamma/(\gamma-1)}$$

Thus p_0 and T_0 can be used in the same way as static values in isentropic p–T relations. Stagnation pressure and temperature are properties of the gas stream which can be used with static values to determine the combined thermodynamic and mechanical state of the stream. Such state points can be represented on the T–s diagram, as shown in Fig. 2.9 which depicts a

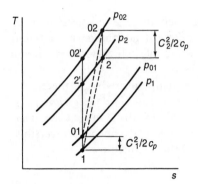

FIG. 2.9 Stagnation states

compression process between 'static' states 1 and 2; the differences between the constant p and p_0 lines have been exaggerated for clarity. The ideal stagnation state which would be reached after isentropic compression to the same, actual, outlet stagnation pressure is indicated by $02'$. *Primes will be attached to symbols to denote such ideal states throughout this book.*

Compressor and turbine efficiencies

(a) Isentropic efficiency
The efficiency of any machine, the object of which is the absorption or production of work, is normally expressed in terms of the ratio of actual and ideal work transfers. Because turbomachines are essentially adiabatic, the ideal process is isentropic and the efficiency is called an *isentropic efficiency*. Making use of the concept of stagnation enthalpy or temperature to take account of any change in kinetic energy of the fluid between inlet and outlet we have, for the compressor,

$$\eta_c = \frac{W'}{W} = \frac{\Delta h_0'}{\Delta h_0}$$

For a perfect gas $\Delta h_0 = c_p \Delta T_0$ and this relation is usually sufficiently accurate for real gases under conditions encountered in gas turbines if a mean c_p over the relevant range of temperature is used—see under heading 'Variation of specific heat' below for further discussion of this point. Furthermore, because the ideal and actual temperature changes are not very different, the mean c_p can be assumed the same for both so that the compressor isentropic efficiency is normally *defined* in terms of temperature as

$$\eta_c = \frac{T'_{02} - T_{01}}{T_{02} - T_{01}} \tag{2.9}$$

Similarly the turbine isentropic efficiency is defined as

$$\eta_t = \frac{W}{W'} = \frac{T_{03} - T_{04}}{T_{03} - T'_{04}} \tag{2.10}$$

When performing cycle calculations, values of η_c and η_t will be assumed and the temperature equivalents of the work transfers for a given pressure ratio are then found as follows:

$$T_{02} - T_{01} = \frac{1}{\eta_c}(T'_{02} - T_{01}) = \frac{T_{01}}{\eta_c}\left(\frac{T'_{02}}{T_{01}} - 1\right)$$

and finally

$$T_{02} - T_{01} = \frac{T_{01}}{\eta_c}\left[\left(\frac{p_{02}}{p_{01}}\right)^{(\gamma-1)/\gamma} - 1\right] \tag{2.11}$$

Similarly

$$T_{03} - T_{04} = \eta_t T_{03}\left[1 - \left(\frac{1}{p_{03}/p_{04}}\right)^{(\gamma-1)/\gamma}\right] \tag{2.12}$$

When the compressor is part of a stationary gas turbine, having a short intake fairing that can be regarded as part of the compressor, p_{01} and T_{01} in equation (2.11) will be equal to p_a and T_a respectively because the velocity of the ambient air is zero. This will be assumed to be the case throughout this chapter. Industrial gas turbine plant usually incorporates a long inlet duct and air filter, however, in which case an inlet pressure loss (Δp_i) must be deducted, i.e. p_{01} will be ($p_a - \Delta p_i$). Inlet losses vary with installation configurations and engine manufacturers often quote performance figures for zero inlet loss, with corrections for different levels of loss. The situation is rather different when the compressor is part of an aircraft propulsion unit because then there will be an intake duct of significant length in which ram compression takes place due to the forward speed of the aircraft. In this situation p_{01} and T_{01} would differ from p_a and T_a even if there were no friction losses, and it is always necessary to consider the intake and compressor as separate components. A discussion of how intake losses are then taken into account will be deferred until the next chapter.

In defining η_t according to equation (2.10) and thus taking the ideal work as proportional to ($T_{03} - T'_{04}$), we are implying that the kinetic energy in the exhaust gas is going to be utilized, e.g. in a subsequent turbine or in the propelling nozzle of a jet engine. But if the turbine is part of an industrial plant exhausting directly to atmosphere this kinetic energy is wasted. The ideal quantity of turbine work would then seem to be taken more appropriately as that produced by an isentropic expansion from p_{03} to the static outlet pressure p_4, with p_4 equal to the ambient pressure p_a. Thus η_t would be defined by

$$\eta_t = \frac{T_{03} - T_{04}}{T_{03}\left[1 - \left(\frac{1}{p_{03}/p_a}\right)^{(\gamma-1)/\gamma}\right]} \tag{2.13}$$

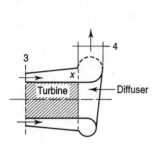

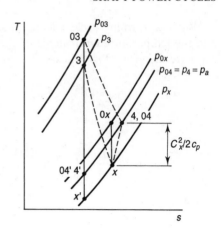

FIG. 2.10 Turbine with exhaust diffuser

In practice, even in such a case the kinetic energy of the gas immediately leaving the turbine is largely recovered in an exhaust diffuser which in effect increases the pressure ratio across the turbine: Fig. 2.10 indicates this for a diffuser which reduces the final velocity to a negligible value so that $p_{04} = p_4 = p_a$. The turbine pressure ratio is seen to be increased from p_{03}/p_a to p_{03}/p_x. The temperature equivalent of the turbine work $(T_{03} - T_{0x})$ is still given by $(T_{03} - T_{04})$, because no work is done in the diffuser and $T_{0x} = T_{04}$, but T_{04} is less than it would be if a diffuser were not fitted and p_x was equal to p_a. For ordinary cycle calculations there is no need to consider the turbine expansion $3 \rightarrow x$ and diffusion process $x \rightarrow 4$ separately. We may put $p_{04} = p_a$ in equation (2.12) and regard η_t as accounting also for the friction pressure loss in the diffuser $(p_{0x} - p_a)$. We then have equation (2.13), but must interpret it as applying to the turbine and exhaust diffuser combined rather than to the turbine alone. In this book, equation (2.12) with p_{04} put equal to p_a will be used for any turbine exhausting direct to the atmosphere: for any turbine delivering gas to a propelling nozzle, or to a second turbine in series, equation (2.12) will be employed as it stands.

(b) Polytropic efficiency
So far we have been referring to overall efficiencies applied to the compressor or turbine as a whole. When performing cycle calculations covering a range of pressure ratio, say to determine the optimum pressure ratio for a particular application, the question arises as to whether it is reasonable to assume fixed typical values of η_c and η_t. In fact it is found that η_c tends to decrease and η_t to increase as the pressure ratio for which the compressor and turbine are designed increases. The reason for this should be clear from the following argument based on Fig. 2.11. p and T are used instead of p_0 and T_0 to avoid a multiplicity of suffixes.

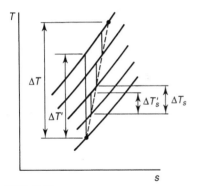

FIG. 2.11

Consider an axial flow compressor consisting of a number of successive stages. If the blade design is similar in successive blade rows it is reasonable to assume that the isentropic efficiency of a single stage, η_s, remains the same through the compressor. Then the overall temperature rise can be expressed by

$$\Delta T = \sum \frac{\Delta T_s'}{\eta_s} = \frac{1}{\eta_s} \sum \Delta T_s'$$

Also, $\Delta T = \Delta T'/\eta_c$ by definition of η_c, and thus

$$\frac{\eta_s}{\eta_c} = \frac{\sum \Delta T_s'}{\Delta T'}$$

But, because the vertical distance between a pair of constant pressure lines in the T–s diagram increases as the entropy increases, it is clear from Fig. 2.11 that $\sum \Delta T_s' > \Delta T'$. It follows that $\eta_c < \eta_s$ and that the difference will increase with the number of stages, i.e. with increase of pressure ratio. A physical explanation is that the increase in temperature due to friction in one stage results in more work being required in the next stage: it might be termed the 'preheat' effect. A similar argument can be used to show that for a turbine $\eta_t > \eta_s$. In this case frictional 'reheating' in one stage is partially recovered as work in the next.

These considerations have led to the concept of *polytropic* (or *small-stage*) *efficiency* η_∞, which is defined as the isentropic efficiency of an elemental stage in the process such that it is constant throughout the whole process. For a compression,

$$\eta_{\infty c} = \frac{\mathrm{d}T'}{\mathrm{d}T} = \text{constant}$$

But $T/p^{(\gamma-1)/\gamma} = \text{constant}$ for an isentropic process, which in differential form is

$$\frac{\mathrm{d}T'}{T} = \frac{\gamma-1}{\gamma} \frac{\mathrm{d}p}{p}$$

Substitution of dT' from the previous equation gives

$$\eta_{\infty c} \frac{dT}{T} = \frac{\gamma - 1}{\gamma} \frac{dp}{p}$$

Integrating between inlet 1 and outlet 2, with $\eta_{\infty c}$ constant by definition, we have

$$\eta_{\infty c} = \frac{\ln(p_2/p_1)^{(\gamma-1)/\gamma}}{\ln(T_2/T_1)} \tag{2.14}$$

This equation enables $\eta_{\infty c}$ to be calculated from measured values of p and T at inlet and outlet of a compressor. Equation (2.14) can also be written in the form

$$\frac{T_2}{T_1} = \left(\frac{p_2}{p_1}\right)^{(\gamma-1)/\gamma\eta_{\infty c}} \tag{2.15}$$

Finally, the relation between $\eta_{\infty c}$ and η_c is given by

$$\eta_c = \frac{T_2'/T_1 - 1}{T_2/T_1 - 1} = \frac{(p_2/p_1)^{(\gamma-1)/\gamma} - 1}{(p_2/p_1)^{(\gamma-1)/\gamma\eta_{\infty c}} - 1} \tag{2.16}$$

Note that if we write $(\gamma - 1)/\gamma\eta_{\infty c}$ as $(n - 1)/n$, equation (2.15) is the familiar relation between p and T for a polytropic process, and thus the definition of η_∞ implies that the non-isentropic process is polytropic. This is the origin of the term polytropic efficiency.

Similarly, since $\eta_{\infty t}$ is dT/dT', it can be shown that for an expansion between inlet 3 and outlet 4,

$$\frac{T_3}{T_4} = \left(\frac{p_3}{p_4}\right)^{\eta_{\infty t}(\gamma-1)/\gamma} \tag{2.17}$$

and

$$\eta_t = \frac{1 - \left(\dfrac{1}{p_3/p_4}\right)^{\eta_{\infty t}(\gamma-1)/\gamma}}{1 - \left(\dfrac{1}{p_3/p_4}\right)^{(\gamma-1)/\gamma}} \tag{2.18}$$

Making use of equations (2.16) and (2.18) with $\gamma = 1\cdot4$, Fig. 2.12 has been drawn to show how η_c and η_t vary with pressure ratio for a fixed value of polytropic efficiency of 85 per cent in each case.

In practice, as with η_c and η_t, it is normal to define the polytropic efficiencies in terms of stagnation temperatures and pressures. Furthermore, when employing them in cycle calculations, the most convenient equations to use will be shown to be those corresponding to (2.11) and (2.12), i.e. from equations (2.15) and (2.17),

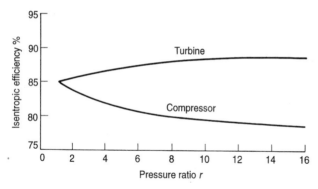

FIG. 2.12 Variation of turbine and compressor isentropic efficiency with pressure ratio for polytropic efficiency of 85%

$$T_{02} - T_{01} = T_{01}\left[\left(\frac{p_{02}}{p_{01}}\right)^{(n-1)/n} - 1\right] \tag{2.19}$$

where $(n-1)/n = (\gamma - 1)/\gamma\eta_{\infty c}$, and

$$T_{03} - T_{04} = T_{03}\left[1 - \left(\frac{1}{p_{03}/p_{04}}\right)^{(n-1)/n}\right] \tag{2.20}$$

where $(n-1)/n = \eta_{\infty t}(\gamma - 1)/\gamma$.

Again, for a compressor of an industrial gas turbine we shall take $p_{01} = p_a$ and $T_{01} = T_a$, while for turbines exhausting to atmosphere p_{04} will be put equal to p_a.

We end this sub-section with a reminder that the isentropic and polytropic efficiencies present the same information in different forms. When performing calculations over a range of pressure ratio, it is reasonable to assume constant polytropic efficiency; this automatically allows for a variation of isentropic efficiency with pressure ratio. In simple terms, the polytropic efficiency may be interpreted as representing the current state-of-the-art for a particular design organization. When determining the performance of a single cycle of interest, or analysing engine test data, it is more appropriate to use isentropic efficiencies.

Pressure losses

Pressure losses in the intake and exhaust ducting have been dealt with in the previous sub-section. In the combustion chamber a loss in stagnation pressure (Δp_b) occurs due to the aerodynamic resistance of flame-stabilizing and mixing devices, and also due to momentum changes produced by the exothermic reaction. These sources of loss are referred to in detail in Chapter 6. When a heat-exchanger is included in the plant there will also be frictional pressure losses in the passages on the air-side (Δp_{ha}) and gas-side (Δp_{hg}). As shown in Fig. 2.13, the pressure losses have the effect of

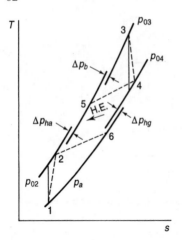

FIG. 2.13 Pressure losses

decreasing the turbine pressure ratio relative to the compressor pressure ratio and thus reduce the net work output from the plant. The gas turbine cycle is very sensitive to irreversibilities, because the net output is the difference of two large quantities (i.e. the 'work ratio' is low), so that the pressure losses have a significant effect on the cycle performance.

Fixed values of the losses can be fed into the cycle calculation directly. For example, for a simple cycle with heat-exchange we may determine the turbine pressure ratio p_{03}/p_{04} from

$$p_{03} = p_{02} - \Delta p_b - \Delta p_{ha} \quad \text{and} \quad p_{04} = p_a + \Delta p_{hg}$$

But again the question arises as to whether it is reasonable to assume constant values for the pressure drops when cycles of different pressure ratio are being compared. The frictional pressure losses will be roughly proportional to the local dynamic head of the flow ($\frac{1}{2}\rho C^2$ for incompressible flow), as with ordinary pipe flow. It might therefore be expected that the pressure drops Δp_{ha} and Δp_b will increase with cycle pressure ratio because the density of the fluid in the air-side of the heat-exchanger and in the combustion chamber is increased. Even though ρ is not proportional to p because T increases also, a better approximation might be to take Δp_{ha} and Δp_b as fixed proportions of the compressor delivery pressure. The turbine inlet pressure is then found from

$$p_{03} = p_{02}\left(1 - \frac{\Delta p_b}{p_{02}} - \frac{\Delta p_{ha}}{p_{02}}\right)$$

We shall express the pressure loss data in this way in subsequent numerical examples.

The pressure loss in the combustor can be minimized by using a large combustion chamber with consequent low velocities, which is feasible in an industrial unit where size is not critical. For an aero engine, where weight, volume and frontal area are all important, higher pressure losses are inevitable. The designer also has to consider pressure losses at a typical flight condition at high altitude, where the air density is low. Typical values of $\Delta p_b/p_{02}$ would be about 2–3 per cent for a large industrial unit and 6–8 per cent for an aero engine.

Heat-exchanger effectiveness

Heat-exchangers for gas turbines can take many forms, including counter-flow and cross-flow recuperators (where the hot and cold streams exchange heat through a separating wall) or regenerators (where the streams are brought cyclically into contact with a matrix which alternately absorbs and rejects heat). In all cases, using the notation of Fig. 2.13, the fundamental process is that the turbine exhaust gases reject heat at the rate of $m_t c_{p46}(T_{04} - T_{06})$ while the compressor delivery air receives heat at the rate of $m_c c_{p25}(T_{05} - T_{02})$. For conservation of energy, assuming that the mass flows m_t and m_c are equal,

$$c_{p46}(T_{04} - T_{06}) = c_{p25}(T_{05} - T_{02}) \tag{2.21}$$

But both T_{05} and T_{06} are unknown and a second equation is required for their evaluation. This is provided by the equation expressing the efficiency of the heat-exchanger.

Now the maximum possible value of T_{05} is when the 'cold' air attains the temperature of the incoming hot gas T_{04}, and one possible measure of performance is the ratio of actual energy received by the cold air to the maximum possible value, i.e.

$$\frac{m_c c_{p25}(T_{05} - T_{02})}{m_c c_{p24}(T_{04} - T_{02})}$$

The mean specific heat of air will not be very different over the two temperature ranges, and it is usual to define the efficiency in terms of temperature alone and to call it the *effectiveness* (or *thermal-ratio*) of the heat-exchanger. Thus

$$\text{effectiveness} = \frac{T_{05} - T_{02}}{T_{04} - T_{02}} \tag{2.22}$$

When a value of effectiveness is specified, equation (2.22) enables the temperature at inlet to the combustion chamber T_{05} to be determined. Equation (2.21) then yields T_{06} if required. Note that the mean specific heats c_{p46} and c_{p25} are not approximately equal and cannot be cancelled, because the former is for turbine exhaust gas and the latter for air.

In general, the larger the volume of the heat-exchanger the higher can be the effectiveness, but the increase in effectiveness becomes asymptotic beyond a certain value of heat transfer surface area; the cost of heat-exchangers is largely determined by their surface area, and with this in mind modern heat-exchangers are designed to have an effectiveness of about 0·90. The maximum permissible turbine *exit* temperature is fixed by the materials used for the construction of the heat-exchanger. With stainless steel, this temperature should not be much above 900 K. It follows that the turbine inlet temperature of a regenerative cycle is limited for this reason. Heat-exchangers are subjected to severe thermal stresses during start-up and are not used where frequent starts are required, e.g. for peak-load electricity generation. They are best suited for applications where the gas turbine operates for extended periods at a steady power, as may occur on a pipeline. It should be noted, however, that many pipeline stations are located in remote areas where transportation and installation of a heat-exchanger present major problems; for this reason, some modern heat-exchangers are built up from a series of identical modules. In recent years heat-exchangers have seldom been used, because they no longer offer an advantage relative to simple cycles of high pressure ratio or, in the case of base-load applications, relative to combined cycle plant. In the late 1990s, however, a high-efficiency small electric power unit of 4 MW, the Solar Mercury, was introduced. This engine was designed from the outset to be built with a heat-exchanger, and the design of the turbomachinery and combustor was configured to provide the most convenient flow paths to and from the heat-exchanger. The resulting layout is shown in Fig. 2.14, where it can be seen that the inlet to the compressor is located near the centre of the engine and

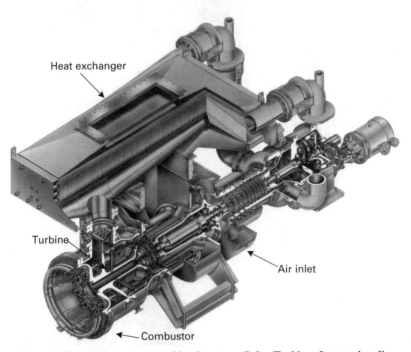

FIG. 2.14 Solar Mercury gas turbine [courtesy Solar Turbines International]

the combustor is located at one end; this enables combustor modifications to be easily incorporated, for example for use with different fuels or as new low-emission technologies are developed. If microturbines are successfully introduced to the market, they will undoubtedly require heat-exchangers to give a reasonable cycle efficiency because of their low pressure ratios; they would be expected to be used in cogeneration systems where a HRSG would produce steam or hot water for process use. A microturbine self-contained package is shown in Fig. 2.15, which clearly shows the very small size of the actual turbomachinery and the relatively large size of the heat-exchanger. The future search for cycle efficiencies exceeding 60 per cent could see the return of the heat-exchanger in complex cycles, and a survey of the possibilities is given in Ref. (2).

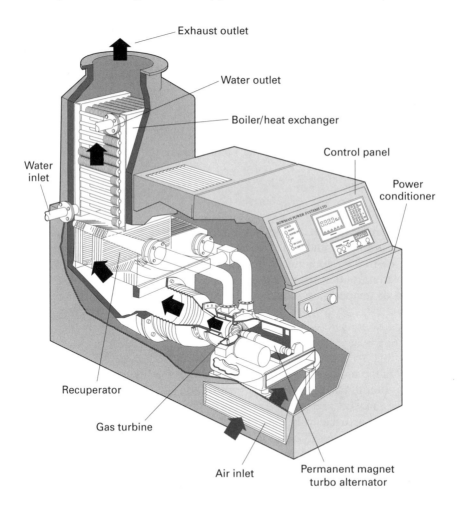

FIG. 2.15 Microturbine package [courtesy Bowman Power Systems]

Mechanical losses

In all gas turbines, the power necessary to drive the compressor is transmitted directly from the turbine without any intermediate gearing. Any loss that occurs is therefore due only to bearing friction and windage. This loss is very small and it is normal to assume that it amounts to about 1 per cent of the power necessary to drive the compressor. If the transmission efficiency is denoted by η_m, we have the work output required to drive the compressor given by

$$W = \frac{1}{\eta_m} c_{p12}(T_{02} - T_{01})$$

We shall take η_m to be 99 per cent for all numerical examples.

Any power used to drive ancillary components such as fuel and oil pumps can often be accounted for simply by subtracting it from the net output of the unit. Power absorbed in any gearing between the gas turbine and the load can be dealt with similarly. Except to say that such losses can be significant, especially for small gas turbines of low power, we shall not consider them further. Their consideration would not require the method of cycle performance estimation to be modified except when the gas turbine has a separate power turbine. Power for fuel and oil pumps will then be taken from the compressor turbine (because under some operating conditions the power turbine is stationary).

Variation of specific heat

The properties c_p and γ play an important part in the estimation of cycle performance, and it is necessary to take account of variations in values due to changing conditions through the cycle. In general, for real gases over normal working ranges of pressure and temperature, c_p is a function of temperature alone. The same is true of γ because it is related to c_p by

$$\frac{\gamma - 1}{\gamma} = \frac{\tilde{R}}{Mc_p} \tag{2.23}$$

where $\tilde{R}$ is the molar (universal) gas constant and M the molecular weight. The variation of c_p and γ with temperature for air is shown in Fig. 2.16 by the curves marked zero fuel/air ratio. Only the left-hand portion is of interest because even with a pressure ratio as high as 35 the compressor delivery temperature will be no more than about 800 K.

In the turbine of an open-cycle plant the working fluid will be a mixture of combustion gases. Most gas turbines run on kerosene which has a composition to which the formula C_nH_{2n} is a good approximation. If some such composition is assumed, the products analysis can be calculated for various fuel/air ratios. Knowing the specific heats and molecular weights

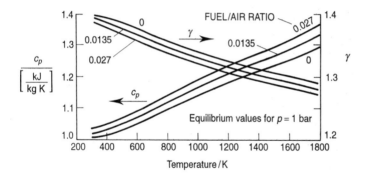

FIG. 2.16 c_p **and** γ **for air and typical combustion gases**

of the constituents it is then a simple matter to calculate the mean values of c_p and γ for the mixture. Figure 2.16 shows that c_p increases and γ decreases with increase in fuel/air ratio. It is worth noting that the mean molecular weight of the combustion products from typical hydrocarbon fuels is little different from that of air, and therefore c_p and γ are closely related by equation (2.23) with $\tilde{R}/M = R_{\mathrm{air}} = 0.287\,\mathrm{kJ/kg\,K}$.

The calculation of the product analyses is very lengthy when dissociation is taken into account and then, because pressure has a significant effect on the amount of dissociation, c_p and γ become a function of pressure as well as temperature. Accurate calculations of this kind have been made, and the results are tabulated in Ref. (3). Dissociation begins to have a significant effect on c_p and γ at a temperature of about 1500 K, and above this temperature the curves of Fig. 2.16 are strictly speaking applicable only to a pressure of 1 bar. In fact at 1800 K, both for air and products of combustion corresponding to low values of fuel/air ratio, a reduction of pressure to 0.01 bar increases c_p by only about 4 per cent and an increase to 100 bar decreases c_p by only about 1 per cent: the corresponding changes in γ are even smaller. In this book we will ignore any effect of pressure, although many aircraft and industrial gas turbines are designed to use turbine inlet temperatures in excess of 1500 K.

Now compressor temperature rises and turbine temperature drops will be calculated using equations such as (2.11) and (2.12), or (2.19) and (2.20). For accurate calculations a method of successive approximation would be required, i.e. it would be necessary to guess a value of γ, calculate the temperature change, take a more accurate mean value of γ and recalculate the temperature change. In fact, if this degree of accuracy is required it is better to use tables or curves of enthalpy and entropy as described, for example, in Ref. (1). For preliminary design calculations and comparative cycle calculations, however, it has been found to be sufficiently accurate to assume the following fixed values of c_p and γ for the compression and expansion processes respectively,

$$\text{air:} \quad c_{pa} = 1\cdot005 \text{ kJ/kg K}, \ \gamma_a = 1\cdot40 \text{ or } \left(\frac{\gamma}{\gamma-1}\right)_a = 3\cdot5$$

$$\text{combustion gases:} \quad c_{pg} = 1\cdot148 \text{ kJ/kg K}, \ \gamma_g = 1\cdot333 \text{ or } \left(\frac{\gamma}{\gamma-1}\right)_g = 4\cdot0$$

The reason why this does not lead to much inaccuracy is that c_p and γ vary in opposing senses with T. For cycle analysis we are interested in calculating compressor and turbine work from the product $c_p\Delta T$. Suppose that the temperature for which the above values of c_p and γ are the true values is lower than the actual mean temperature. γ is then higher than it should be and ΔT will be overestimated. This will be compensated in the product $c_p\Delta T$ by the fact that c_p will be lower than it should be. The actual temperatures at various points in the cycle will not be very accurate, however, and for the detailed design of the components it is necessary to know the exact conditions of the working fluid: the more accurate iterative approaches mentioned above must then be employed.

Fuel/air ratio, combustion efficiency and cycle efficiency

The performance of real cycles can be unambiguously expressed in terms of the specific fuel consumption, i.e. fuel mass flow per unit net power output. To obtain this the fuel/air ratio must be found. In the course of calculating the net output per unit mass flow of air the temperature at inlet to the combustion chamber (T_{02}) will have been obtained; and the temperature at outlet (T_{03}), which is the maximum cycle temperature, will normally be specified. The problem is therefore to calculate the fuel/air ratio f required to transform unit mass of air at T_{02} and f kg of fuel at the fuel temperature t_f to $(1 + f)$ kg of products at T_{03}.

Since the process is adiabatic with no work transfer, the energy equation is simply

$$\sum (m_i h_{i03}) - (h_{a02} + f h_f) = 0$$

where m_i is the mass of product i per unit mass of air and h_i its specific enthalpy. Making use of the enthalpy of reaction at a reference temperature of 25 °C, ΔH_{25}, the equation can be expanded in the usual way [see Ref. (1)] to become

$$(1 + f)c_{pg}(T_{03} - 298) + f\Delta H_{25} + c_{pa}(298 - T_{02}) + fc_{pf}(298 - T_f) = 0$$

where c_{pg} is the mean specific heat of the products over the temperature range 298 K to T_{03}. ΔH_{25} should be the enthalpy of reaction per unit mass of fuel with the H_2O in the products in the vapour phase, because T_{03} is high and above the dew point. For common fuels ΔH_{25} may be taken from tables, or alternatively it may be evaluated from the enthalpies of formation of the reactants. It is usual to assume that the fuel temperature is the same as the reference temperature, so that the fourth term on the LHS of the equation is zero. The term will certainly be small because f is low (≈ 0.02) and c_{pf} for liquid hydrocarbon fuels is only about $2\,\text{kJ/kg K}$. Finally, we have already discussed in the previous section, 'Variation of specific heat', the calculation of the mean specific heat of the products c_{pg} as a function of f and T, so we are left with an equation from which f can be obtained for any given values of T_{02} and T_{03}.

Such calculations are too lengthy to be undertaken for every individual cycle calculation—particularly if dissociation is significant because then the $f\Delta H_{25}$ term must be modified to allow for the incompletely burnt carbon and hydrogen arising from the dissociated CO_2 and H_2O. It is usually sufficiently accurate to use tables or charts which have been compiled for a typical fuel composition. Figure 2.17 shows the combustion temperature rise $(T_{03} - T_{02})$ plotted against fuel/air ratio for various values of inlet temperature (T_{02}), and these curves will be used for all numerical examples in this book. It is a small-scale version of larger and more accurate graphs given in Ref. (4). The reference fuel for which the data have been calculated is a hypothetical liquid hydrocarbon containing 13.92 per cent H and 86.08 per cent C, for which the stoichiometric fuel/air ratio is 0.068 and ΔH_{25} is $-43\,100\,\text{kJ/kg}$. The curves are certainly adequate for any kerosene burnt in dry air. Methods are given in Ref. (4) whereby the data can be used for hydrocarbon fuels which differ widely in composition from the reference fuel, or where the fuel is burnt in a reheat chamber, i.e. not in air but in the products of combustion from a previous chamber in the cycle.

The data for Fig. 2.17 have been calculated on the assumption that the fuel is completely burnt, and thus the abscissa could be labelled 'theoretical fuel/air ratio'. The most convenient method of allowing for combustion loss is by introducing a *combustion efficiency* defined by

$$\eta_b = \frac{\text{theoretical } f \text{ for given } \Delta T}{\text{actual } f \text{ for given } \Delta T}$$

This is the definition used in this book. An alternative method is to regard the ordinate as the theoretical ΔT for a given f and define the efficiency in terms of the ratio: actual ΔT/theoretical ΔT. Neither definition is quite the same as the fundamental definition based on the ratio of actual energy released to the theoretical quantity obtainable. (η_b differs from it because of the small additional heat capacity of the products arising from the increase in fuel needed to produce the given temperature.) But in practice combustion is so nearly complete—98–99 per cent—that the efficiency is difficult to measure accurately and the three definitions of efficiency yield virtually the same result.

Once the fuel/air ratio is known, the fuel consumption m_f is simply $f \times m$ where m is the air mass flow, and the specific fuel consumption can be found directly from

$$SFC = \frac{f}{W_N}$$

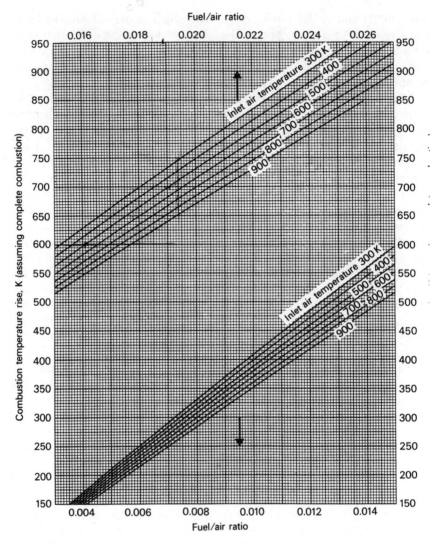

FIG. 2.17 Combustion temperature rise v. fuel/air ratio

Since the fuel consumption is normally measured in kg/h, while W_N is in kW per kg/s of airflow, the *SFC* in kg/kW h is given by the following numerical equation:†

$$\frac{SFC}{[\text{kg/kW h}]} = \frac{f}{W_N/[\text{kW s/kg}]} \times \frac{[\text{s}]}{[\text{h}]} = \frac{3600\,f}{W_N/[\text{kW s/kg}]}$$

† The fundamental principle employed here is that a physical quantity, or the symbol representing it, is equal to (pure number × unit) and *never* the number alone. Consequently an equation relating numbers only will, in addition to pure numbers and dimensional ratios, contain only quotients of symbols and units, e.g. $W_N/[\text{kW s/kg}]$.

When dealing with a specific engine rather than cycle calculations it is more covenient to calculate the *SFC* from

$$\frac{SFC}{[\text{kg/kW h}]} = \frac{m_f}{\text{power}}$$

If the thermal efficiency of the cycle is required, it must be defined in the form 'work output/heat supplied' even though the combustion process is adiabatic and in the thermodynamic sense no heat is supplied. We know that if the fuel is burnt under ideal conditions, such that the products and reactants are virtually at the same temperature (the reference temperature 25 °C), the rate of energy release in the form of heat will be

$$m_f Q_{\text{gr},p} = f m Q_{\text{gr},p}$$

where m_f is the fuel flow and $Q_{\text{gr},p}$ is the gross (or higher) calorific value at constant pressure. In the gas turbine it is not possible to utilize the latent heat of the H_2O vapour in the products and the convention of using the net calorific value has been adopted in most countries. Thus the cycle efficiency may be defined as

$$\eta = \frac{W_N}{f Q_{\text{net},p}}$$

With the units used here, the equivalent numerical equation becomes

$$\eta = \frac{W_N/[\text{kW s/kg}]}{f \times Q_{\text{net},p}/[\text{kJ or kW s/kg}]}$$

or, using the above numerical equation for *SFC*,

$$\eta = \frac{3600}{SFC/[\text{kg/kW h}] \times Q_{\text{net},p}/[\text{kJ/kg}]}$$

$Q_{\text{net},p}$ is sensibly equal in magnitude but opposite in sign to the enthalpy of reaction ΔH_{25} referred to earlier, and the value of 43 100 kJ/kg will be used for all numerical examples.

When referring to the thermal efficiency of actual gas turbines, manufacturers often prefer to use the concept of *heat rate* rather than efficiency. The reason is that fuel prices are normally quoted in terms of pounds sterling (or dollars) per megajoule, and the heat rate can be used to evaluate fuel cost directly. Heat rate is defined as ($SFC \times Q_{\text{net},p}$), and thus expresses the heat input required to produce a unit quantity of power. It is normally expressed in kJ/kW h, in which case the corresponding thermal efficiency can be found from 3600/(heat rate).

Bleed flows

In section 2.1 the importance of high turbine inlet temperature was demonstrated. Turbine inlet temperatures, however, are limited by metallurgical considerations and many modern engines make use of air-cooled blades to permit operation at elevated temperatures. It is possible to operate with uncooled blades up to about 1350–1400 K; and then it can be assumed that the mass flow remains constant throughout as explained earlier. At higher temperatures it is necessary to extract air to cool both stator and rotor blades. The required bleeds may amount to 15 per cent or more of the compressor delivery flow in an advanced engine, and must be properly accounted for in accurate calculations. The overall air cooling system for such an engine is complex, but the methods of dealing with bleeds can be illustrated using a simplified example. A single-stage turbine with cooling of both the stator and rotor blades will be considered, as shown schematically in Fig. 2.18; β_D, β_S and β_R denote bleeds for cooling the disc, stator and rotor respectively. Bleeds are normally specified as a percentage of the compressor delivery flow.

Note that bleed β_D acts to prevent the flow of hot gases down the face of the turbine disc, but it does pass through the rotor. The stator bleed also passes through the rotor and both contribute to the power developed. The rotor bleed, however, does not contribute to the work output and the reduction in mass flow results in an increase in both temperature drop and pressure ratio for the specified power. Useful work would be done by the rotor bleed only if there was another turbine downstream.

If the airflow at compressor delivery is m_a, then the flow available to the rotor, m_R, is given by

$$m_R = m_a(1 - \beta_R) + m_f$$

The fuel flow is found from the fuel/air ratio required for the given combustion temperature rise and combustion inlet temperature, and the air available for combustion, i.e.

$$m_a(1 - \beta_D - \beta_S - \beta_R)$$

Hence the fuel flow is given by

$$m_f = m_a(1 - \beta_D - \beta_S - \beta_R) \cdot f$$

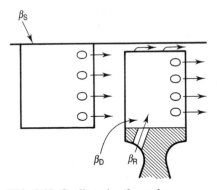

FIG. 2.18 Cooling air schematic

It should be noted that the cooling flows from the stator and disc, at the compressor delivery temperature, will cause some reduction in the effective temperature at entry to the rotor. This effect can be estimated by carrying out an enthalpy balance assuming complete mixing of the flows. Such perfect mixing does not occur in practice, but accurate cycle calculations would require a good estimate to be made of this cooling of the main flow.

In a typical cooled stator with a β_S of 6 per cent, the stator outlet temperature may be reduced by about 100 K. Since the rotor inlet temperature is equal to the stator outlet temperature this implies a reduction in turbine power output. In addition there will be a small drop in efficiency due to the mixing of the bleeds with the main stream.

The distribution of cooling air in a hypothetical multi-stage turbine is shown in Fig. 2.19, which gives some idea of the complexity and the need to understand fully the location of the various bleeds.

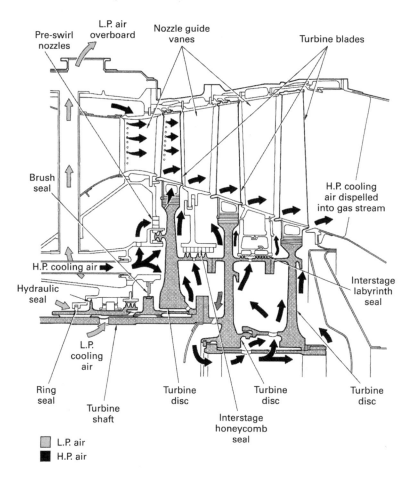

FIG. 2.19 Cooling air distribution [courtesy Rolls-Royce plc]

2.3   Design point performance calculations

Before assessing the effect of component losses on the general performance
curves for the various cycles considered in section 2.1, it is necessary to
outline the method of calculating the performance in any particular case
for specified values of the design parameters. These parameters will include
the compressor pressure ratio, turbine inlet temperature, component
efficiencies and pressure losses.

Several examples will be used to show how these effects are incorporated
in realistic cycle calculations. For the first example we consider a single-
shaft unit with a heat-exchanger, using modest values of pressure ratio and
turbine inlet temperature, and typical isentropic efficiences. A second
example shows how to calculate the performance of a simple cycle of high
pressure ratio and turbine temperature, which incorporates a free power
turbine. Finally, we will evaluate the performance of an advanced cycle
using reheat, suitable for use in both simple and combined cycle
applications: this example will illustrate the use of polytropic efficiencies.

EXAMPLE 2.1

Determine the specific work output, specific fuel consumption and cycle
efficiency for a heat-exchange cycle, Fig. 2.20, having the following specification:

Compressor pressure ratio	4·0
Turbine inlet temperature	1100 K
Isentropic efficiency of compressor, η_c	0·85
Isentropic efficiency of turbine, η_t	0·87
Mechanical transmission efficiency, η_b	0·99
Combustion efficiency, η_b	0·98
Heat-exchanger effectiveness	0·80
Pressure losses—	
Combustion chamber, Δp_b	2% comp. deliv. press.
Heat-exchanger air-side, Δp_{ha}	3% comp. deliv. press.
Heat-exchanger gas-side, Δp_{hg}	0·04 bar
Ambient conditions, p_a, T_a	1 bar, 288 K

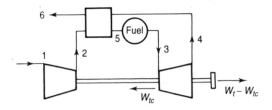

FIG. 2.20 Heat-exchange cycle

Since $T_{01} = T_a$ and $p_{01} = p_a$, and $\gamma = 1.4$, the temperature equivalent of the compressor work from equation (2.11) is

$$T_{02} - T_a = \frac{T_a}{\eta_c} \left[\left(\frac{p_{02}}{p_a} \right)^{(\gamma-1)/\gamma} - 1 \right]$$

$$= \frac{288}{0.85} [4^{1/3.5} - 1] = 164.7 \text{ K}$$

Turbine work required to drive compressor per unit mass flow is

$$W_{tc} = \frac{c_{pa}(T_{02} - T_a)}{\eta_m} = \frac{1.005 \times 164.7}{0.99} = 167.2 \text{ kJ/kg}$$

$$p_{03} = p_{02} \left(1 - \frac{\Delta p_b}{p_{02}} - \frac{\Delta p_{ha}}{p_{02}} \right) = 4.0(1 - 0.02 - 0.03) = 3.8 \text{ bar}$$

$$p_{04} = p_a + \Delta p_{hg} = 1.04 \text{ bar, and hence } p_{03}/p_{04} = 3.654$$

Since $\gamma = 1.333$ for the expanding gases, the temperature equivalent of the total turbine work from equation (2.13) is

$$T_{03} - T_{04} = \eta_t T_{03} \left[1 - \left(\frac{1}{p_{03}/p_{04}} \right)^{(\gamma-1)/\gamma} \right]$$

$$= 0.87 \times 1100 \left[1 - \left(\frac{1}{3.654} \right)^{1/4} \right] = 264.8 \text{ K}$$

Total turbine work per unit mass flow is

$$W_t = c_{pg}(T_{03} - T_{04}) = 1.148 \times 264.8 = 304.0 \text{ kJ/kg}$$

Remembering that the mass flow is to be assumed the same throughout the unit, the specific work output is simply

$$W_t - W_{tc} = 304 - 167.2 = 136.8 \text{ kJ/kg (or kW s/kg)}†$$

(It follows that for a 1000 kW plant an air mass flow of 7.3 kg/s would be required.) To find the fuel/air ratio we must first calculate the combustion temperature rise $(T_{03} - T_{05})$.

From equation (2.22),

$$\text{heat-exchanger effectiveness} = 0.80 = \frac{T_{05} - T_{02}}{T_{04} - T_{02}}$$

† Note: 1 kW s/kg $\equiv$ 0.6083 hp s/lb.

$$T_{02} = 164\cdot7 + 288 = 452\cdot7\,\text{K, and } T_{04} = 1100 - 264\cdot8 = 835\cdot2\,\text{K}$$

Hence

$$T_{05} = 0\cdot80 \times 382\cdot5 + 452\cdot7 = 758\cdot7\,\text{K}$$

From Fig. 2.17, for a combustion chamber inlet air temperature of 759 K and a combustion temperature rise of $(1100 - 759) = 341\,\text{K}$, the theoretical fuel/air ratio required is 0·0094 and thus

$$f = \frac{\text{theoretical } f}{\eta_b} = \frac{0\cdot0094}{0\cdot98} = 0\cdot0096$$

The specific fuel consumption is therefore

$$SFC = \frac{f}{W_t - W_{tc}} = \frac{3600 \times 0\cdot0096}{136\cdot8} = 0\cdot253\,\text{kg/kW h}\dagger$$

Finally, the cycle efficiency is

$$\eta = \frac{3600}{SFC \times Q_{net,p}} = \frac{3600}{0\cdot253 \times 43\,100} = 0\cdot331$$

EXAMPLE 2.2

Determine the specific work output, specific fuel consumption and cycle efficiency for a simple cycle gas turbine with a free power turbine (Fig. 2.21) given the following specification:

Compressor pressure ratio	12·0
Turbine inlet temperature	1350 K
Isentropic efficiency of compressor, η_c	0·86
Isentropic efficiency of each turbine, η_t	0·89
Mechanical efficiency of each shaft, η_m	0·99
Combustion efficiency	0·99
Combustion chamber pressure loss	6% comp. deliv. press.
Exhaust pressure loss	0·03 bar
Ambient conditions, p_a, T_a	1 bar, 288 K

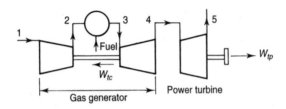

FIG. 2.21 Free turbine unit

† Note: 1 kg/kW h ≡ 1·644 lb/hp h.

Proceeding as in the previous example,

$$T_{02} - T_{01} = \frac{288}{0.86}[12^{1/3.5} - 1] = 346.3 \text{ K}$$

$$W_{tc} = \frac{1.005 \times 346.3}{0.99} = 351.5 \text{ kJ/kg}$$

$$p_{02} = 12.0(1 - 0.06) = 11.28 \text{ bar}$$

The intermediate pressure between the two turbines, p_{04}, is unknown, but can be determined from the fact that the compressor turbine produces just sufficient work to drive the compressor. The temperature equivalent of the compressor turbine work is therefore

$$T_{03} - T_{04} = \frac{W_{tc}}{c_{pg}} = \frac{351.5}{1.148} = 306.2 \text{ K}$$

The corresponding pressure ratio can be found using equation (2.12)

$$T_{03} - T_{04} = \eta_t T_{03}\left[1 - \left(\frac{1}{p_{03}/p_{04}}\right)^{\gamma-1/\gamma}\right]$$

$$306.2 = 0.89 \times 1350\left[1 - \left(\frac{1}{p_{03}/p_{04}}\right)^{0.25}\right]$$

$$\frac{p_{03}}{p_{04}} = 3.243$$

$$T_{04} = 1350 - 306.2 = 1043.8 \text{ K}$$

The pressure at entry to the power turbine, p_{04}, is then found to be 11.28/3.243 = 3.478 bar and the power turbine pressure ratio is 3.478/(1 + 0.03) = 3.377.

The temperature drop in the power turbine can now be obtained from equation (2.12),

$$T_{04} - T_{05} = 0.89 \times 1043.8\left[1 - \left(\frac{1}{3.377}\right)^{0.25}\right] = 243.7 \text{ K}$$

and the specific work output, i.e. power turbine work per unit air mass flow, is

$$W_{tp} = c_{pg}(T_{04} - T_{05})\eta_m$$
$$W_{tp} = 1.148(243.7)0.99 = 277.0 \text{ kJ/kg (or kW s/kg)}$$

The compressor delivery temperature is $288 + 346.3 = 634.3 \text{ K}$ and the combustion temperature rise is $1350 - 634.3 = 715.7 \text{ K}$; from Fig. 2.17, the theoretical fuel/air ratio required is 0.0202 giving an actual fuel/air ratio of 0.0202/0.99 = 0.0204.

The *SFC* and cycle efficiency, η, are then given by

$$SFC = \frac{f}{W_{tp}} = \frac{3600 \times 0 \cdot 0204}{277 \cdot 9} = 0 \cdot 265 \, \text{kg/kW h}$$

$$\eta = \frac{3600}{0 \cdot 265 \times 43100} = 0 \cdot 315$$

It should be noted that the cycle calculations have been carried out as above to determine the *overall* performance. It is important to realize, however, that they also provide information that is needed by other groups such as the aerodynamic and control design groups, as discussed in section 1.9. The temperature at entry to the power turbine, T_{04}, for example, may be required as a control parameter to prevent operation above the metallurgical limiting temperature of the compressor turbine. The exhaust gas temperature (*EGT*), T_{05}, would be important if the gas turbine were to be considered for combined cycle or cogeneration plant, which were first mentioned in Chapter 1, and will also be discussed in section 2.5. For the cycle of Example 2.2, $T_{05} = 1043 \cdot 8 - 243 \cdot 7 = 800 \cdot 1 \, \text{K}$ or 527 °C, which is suitable for use with a waste heat boiler. When thinking of combined cycle plant, a higher *TIT* might be desirable because there would be a consequential increase in *EGT*, permitting the use of a higher steam temperature and a more efficient steam cycle. If the cycle pressure ratio were increased to increase the efficiency of the gas cycle, however, the *EGT* would be decreased resulting in a lower steam cycle efficiency. The next example will illustrate how the need to make a gas turbine suitable for more than one application can affect the choice of cycle parameters.

EXAMPLE 2.3

Consider the design of a high pressure ratio, single-shaft cycle with reheat at some point in the expansion when used *either* as a separate unit, *or* as part of a combined cycle. The power required is 240 MW at 288 K and 1·01 bar.

Compressor pressure ratio	30
Polytropic efficiency (compressor and turbines)	0·89
Turbine inlet temperature (both turbines)	1525 K
$\Delta p/p_{02}$ (1st combustor)	0·02
$\Delta p/p_{04}$ (2nd combustor)	0·04
Exhaust pressure	1·02 bar

The plant is shown in Fig. 2.22. A heat-exchanger is not used because it would result in an exhaust temperature that would be too low for use with a high-efficiency steam cycle.

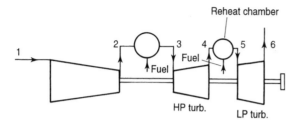

FIG. 2.22 Reheat cycle

To simplify the calculation it will be assumed that the mass flow is constant throughout, ignoring the effect of the substantial cooling bleeds that would be required with the high turbine inlet temperatures specified. The reheat pressure is not specified, but as a starting point it is reasonable to use the value giving the same pressure ratio in each turbine. (As shown in section 2.1, this division of the expansion leads to equal work in each turbine and a maximum net work output for the *ideal* reheat cycle.)

It is convenient to start by evaluating $(n-1)/n$ for the polytropic compression and expansion. Referring to equations (2.19) and (2.20), we have:

$$\text{for compression,} \quad \frac{n-1}{n} = \frac{1}{\eta_{\infty c}}\left(\frac{\gamma-1}{\gamma}\right) = \frac{1}{0\cdot89}\left(\frac{0\cdot4}{1\cdot4}\right) = 0\cdot3210$$

$$\text{for expansion,} \quad \frac{n-1}{n} = \eta_{\infty t}\left(\frac{\gamma-1}{\gamma}\right) = 0\cdot89\left(\frac{0\cdot333}{1\cdot333}\right) = 0\cdot2223$$

Making the usual assumption that $p_{01} = p_a$ and $T_{01} = T_a$, we have

$$\frac{T_{02}}{T_{01}} = (30)^{0\cdot3210} \qquad T_{02} = 858\cdot1\text{ K}$$

$$T_{02} - T_{01} = 570\cdot1\text{ K}$$
$$p_{02} = 30 \times 1\cdot01 = 30\cdot3\text{ bar}$$
$$p_{03} = 30\cdot3(1\cdot00 - 0\cdot02) = 29\cdot69\text{ bar}$$
$$p_{06} = 1\cdot02\text{ bar, so} \quad \frac{p_{03}}{p_{06}} = 29\cdot11$$

Theoretically, the optimum pressure ratio for each turbine would be $\sqrt{(29\cdot11)} = 5\cdot395$. The 4 per cent pressure loss in the reheat combustor has to be considered, so a value of 5.3 for p_{03}/p_{04} could be assumed. Then

$$\frac{T_{03}}{T_{04}} = (5\cdot3)^{0\cdot2223} \qquad T_{04} = 1052\cdot6\text{ K}$$

$$p_{04} = 29\cdot69/5\cdot3 = 5\cdot602\text{ bar}$$
$$p_{05} = 5\cdot602(1\cdot00 - 0\cdot04) = 5\cdot378\text{ bar}$$
$$p_{05}/p_{06} = 5\cdot378/1\cdot02 = 5\cdot272$$

$$\frac{T_{05}}{T_{06}} = (5\cdot272)^{0\cdot2223} \qquad T_{06} = 1053\cdot8\text{ K}$$

Assuming unit flow of $1.0\,\text{kg/s}$, and a mechanical efficiency of 0.99

$$\text{Turbine output, } W_t = 1.0 \times 1.148\{(1525 - 1052.6)$$
$$+ (1525 - 1053.8)\} \times 0.99$$
$$= 1072.3\,\text{kJ/kg}$$

$$\text{Compressor input, } W_c = 1.0 \times 1.005 \times 570.1$$
$$= 573.0\,\text{kJ/kg}$$

$$\text{Net work output, } W_N = 1072.3 - 573.0 = 499.3\,\text{kJ/kg}$$

Flow required for 240 MW is given by

$$m = \frac{240\,000}{499.3} = 480.6\,\text{kg/s}$$

The combustion temperature rise in the first combustor is $(1525 - 858) = 667\,\text{K}$, and with an inlet temperature of $858\,\text{K}$ the fuel/air ratio is found from Fig. 2.17 to be 0.0197. For the second combustor the temperature rise is $(1525 - 1052.6) = 472.4\,\text{K}$ giving a fuel/air ratio of 0.0142. The total fuel/air ratio is then

$$f = \frac{0.0197 + 0.0142}{0.99} = 0.0342$$

and the thermal efficiency is given by

$$\eta = \frac{499.3}{0.0342 \times 43\,100} = 33.9\%$$

This is a reasonable efficiency for simple cycle operation, and the specific output is excellent. Examination of the temperature at turbine exit, however, shows that $T_{06} = 1053.8\,\text{K}$ or $780.8\,°\text{C}$. This temperature is too high for efficient use in a combined cycle plant. A reheat steam cycle using conventional steam temperatures of about 550–575 $°\text{C}$ would require a turbine exit temperature of about 600 $°\text{C}$.

The turbine exit temperature could be reduced by increasing the reheat pressure, and if the calculations are repeated for a range of reheat pressure the results shown in Fig. 2.23 are obtained. It can be seen that a reheat pressure of 13 bar gives an exhaust gas temperature (*EGT*) of 605 $°\text{C}$; the specific output is about 10 per cent lower than the optimum value, but the thermal efficiency is substantially improved to 37.7 per cent. Further increases in reheat pressure would give slightly higher efficiencies, but the *EGT* would be reduced below 600 $°\text{C}$ resulting in a less efficient steam cycle. With a reheat pressure of 13 bar, the first turbine has a pressure ratio of 2.284 while the second has a pressure ratio of 12.23, differing markedly from the equal pressure ratios we assumed at the outset. This example illustrates some of the problems that arise when a gas turbine must be designed for more than one application.

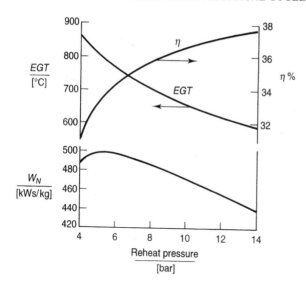

FIG. 2.23 Effect of varying reheat pressure

Figure 2.24 compares the cross-section of the ABB GT26 reheat (sequential combustion) gas turbine with the conventional ABB GT13E2 gas turbine. It can be seen that the use of reheat provides an increase from 165 to 241 MW for the same footprint, primarily due to the increase in specific output. This increase in power density has proved valuable in re-powering applications where the existing site is cramped for space.

2.4 Comparative performance of practical cycles

The large number of variables involved make it impracticable to derive algebraic expressions for the specific output and efficiency of real cycles. On the other hand, the type of step-by-step calculation illustrated in the previous section is ideally suited for computer programming, each of the design parameters being given a set of values in turn to elicit their effect upon the performance.

Some performance curves will now be presented to show the main differences between practical and ideal cycles, and the relative importance of some of the parameters. The curves are definitely *not* a comprehensive set from which designers can make a choice of cycle for a particular application. To emphasize that too much importance should not be attached to the values of specific output and efficiency the full specification of the parameters has not been given: it is sufficient to note that those parameters which are not specified on the curves are kept constant. All the curves use compressor pressure ratio r_c as abscissa, the turbine pressure ratio being less than r_c by virtue of the pressure losses. The cycle efficiency

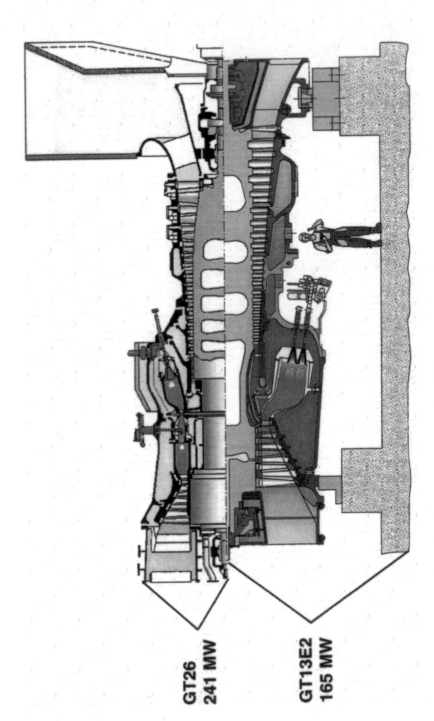

FIG 2.24 Reheat and conventional gas turbines [courtesy Alstom]

has been evaluated to facilitate comparison with the ideal curves of section 2.1. In practice it is usual to quote *SFC* rather than efficiency, not only because its definition is unambiguous, but also because it provides both a direct indication of fuel consumption and a measure of cycle efficiency to which it is inversely proportional.

Simple gas turbine cycle

When component losses are taken into account the efficiency of the simple cycle becomes dependent upon the maximum cycle temperature T_{03} as well as pressure ratio, Fig. 2.25. Furthermore, for each temperature the efficiency has a peak value at a particular pressure ratio. The fall in efficiency at higher pressure ratios is due to the fact that the reduction in fuel supply to give the fixed turbine inlet temperature, resulting from the higher compressor delivery temperature, is outweighed by the increased work necessary to drive the compressor. Although the optimum pressure ratio for maximum efficiency differs from that for maximum specific output, the curves are fairly flat near the peak and a pressure ratio between the two optima can be used without much loss in efficiency. It is perhaps worth pointing out that the lowest pressure ratio which will give an acceptable performance is always chosen: it might even be slightly lower than either optimum value. Mechanical design considerations beyond the scope of this book may affect the choice: such considerations include the number of compressor and turbine stages required, the avoidance of excessively small blades at the high-pressure end of the compressor, and whirling speed and bearing problems associated with the length of the compressor–turbine combination.

The advantage of using as high a value of T_{03} as possible, and the need to use a higher pressure ratio to take advantage of a higher permissible temperature, is evident from the curves. The efficiency increases with T_{03} because the component losses become relatively less important as the ratio

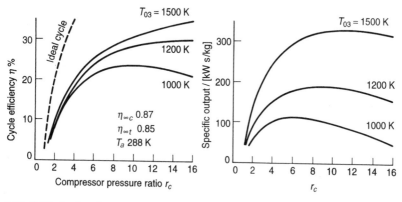

FIG. 2.25 Cycle efficiency and specific output of simple gas turbine

of positive turbine work to negative compressor work increases, although the gain in efficiency becomes marginal as T_{03} is increased beyond 1200 K (particularly if a higher temperature requires a complex turbine blade cooling system which incurs additional losses). There is nothing marginal, however, about the gain in specific work output with increase in T_{03}. The consequent reduction in size of plant for a given power is very marked, and this is particularly important for aircraft gas turbines as will be emphasized in the next chapter.

The following figures illustrate the relative importance of some of the other parameters. Changes in efficiency are quoted as simple *differences* in percentages. With a T_{03} of 1500 K, and a pressure ratio near the optimum value, an increase of 5 per cent in the polytropic efficiency of either the compressor or turbine would increase the cycle efficiency by about 4 per cent and the specific output by about 65 kW s/kg. (If isentropic efficiencies had been used, the turbine loss would have been seen to be more important than the compressor loss but the use of polytropic efficiencies obscures this fact.) A reduction in combustion chamber pressure loss from 5 per cent of the compressor delivery pressure to zero would increase the cycle efficiency by about 1·5 per cent and the specific output by about 12 kW s/kg. The remaining parameter of importance is the ambient temperature, to which the performance of gas turbines is particularly sensitive.

The ambient temperature affects both the compressor work (proportional to T_a) and the fuel consumption (a function of $T_{03} - T_{02}$). An increase in T_a reduces both specific output and cycle efficiency, although the latter is less affected than the former because for a given T_{03} the combustion temperature rise is reduced. Considering again the case of $T_{03} = 1500$ K and a pressure ratio near the optimum, an increase in T_a from 15 to 40 °C reduces the efficiency by about 25 per cent and the specific output by about 62 kW s/kg. The latter is nearly 20 per cent of the output, which emphasizes the importance of designing a gas turbine to give the required power output at the highest ambient temperature likely to be encountered.

Heat-exchange (or regenerative) cycle

As far as the specific work output is concerned, the addition of a heat-exchanger merely causes a slight reduction due to the additional pressure losses: the curves retain essentially the same form as those in Fig. 2.25. The efficiency curves are very different, however, as shown in Fig. 2.26. Heat-exchange increases the efficiency substantially and markedly reduces the optimum pressure ratio for maximum efficiency. Unlike the corresponding curves for the ideal cycle, they do not rise to the Carnot value at $r_c = 1$ but fall to zero at the pressure ratio at which the turbine provides just sufficient work to drive the compressor: at this point there will be positive heat input with zero net work output. The spacing of the

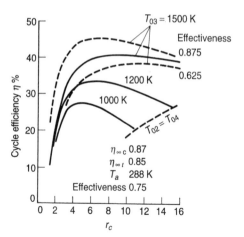

FIG. 2.26 Heat-exchange cycle

constant T_{03} curves in Fig. 2.26 indicates that when a heat-exchanger is used the gain in efficiency is no longer merely marginal as T_{03} is raised above 1200 K. This is a most important feature of the heat-exchange cycle, because progress in materials science and blade cooling techniques has enabled the permissible temperature to be increased at an average rate of about 10 K per year and there is every hope that this will continue. It should also be noted that the optimum pressure ratio for maximum efficiency increases as T_{03} is increased. Our study of the ideal heat-exchange cycle suggested that no such increase in pressure ratio was required: this was the conclusion under the heading 'Heat-exchange cycle' in section 2.1 which it was stated would have to be modified. Nevertheless, it remains true to say that no very high pressure ratio is ever required for a heat-exchange cycle, and the increase in weight and cost due to a heat-exchanger is partially offset by the reduction in size of the compressor.

The dotted curves have been added to show the effect of heat-exchanger effectiveness. Not only does an increase in effectiveness raise the cycle efficiency appreciably, but it also reduces the value of the optimum pressure ratio still further. Since the optimum r_c for maximum efficiency is below that for maximum specific output, it is inevitable that a plant designed for high efficiency will suffer a space and weight penalty. For example, with a T_{03} of 1500 K and an effectivenes of 0·75, the optimum r_c for maximum efficiency is about 10 at which (from Fig. 2.25) the specific output is not far short of the peak value. But with the effectiveness increased to 0·875 the optimum r_c is reduced to about 6, at which the specific output is only about 90 per cent of the peak value. The position is similar at the lower (and at present more realistic) values of T_{03}.

An alternative method of presenting performance characteristics is to plot the variation of specific fuel consumption and specific output on a single figure for a range of values of pressure ratio and turbine inlet temperature.

Examples for realistic values of polytropic efficiency, pressure losses and heat-exchanger effectiveness are shown in Figs 2.27(a) and (b) for simple and heat-exchange cycles respectively. The marked effect of increasing T_{03} on specific output in both cases is clearly evident. Such plots also contrast clearly the small effect of T_{03} on *SFC* for the simple cycle with the much greater effect when a heat-exchanger is used.

Heat-exchange (regenerative) cycle with reheat or intercooling

Our study of ideal cycles suggested that there is no virtue in employing reheat without heat-exchange because of the deleterious effect on efficiency. This is generally true of practical cycles also, and so we will not include performance curves for this case here. (Exceptions to the rule will be discussed at the end of this section.) With heat-exchange, addition of reheat improves the specific output considerably without loss of efficiency (cf. Figs 2.28 and 2.25/26). The curves of Fig. 2.28 are based on the assumption that the gas is reheated to the maximum cycle temperature at the point in the expansion giving equal pressure ratios for the two turbines. The gain in efficiency due to reheat obtained with the ideal cycle is not realized in practice, partly because of the additional pressure loss in the reheat chamber and the inefficiency of the expansion process, but primarily because the effectiveness of the heat-exchanger is well short of unity and the additional energy in the exhaust gas is not wholly recovered. It is important to use a pressure ratio not less than the optimum value for maximum efficiency, because at lower pressure ratios the addition of reheat can actually reduce the efficiency as indicated by the curves.

Reheat has not been widely used in practice because the additional combustion chamber, and the associated control problems, can offset the advantage gained from the decrease in size of the main components consequent upon the increase in specific output. With the exception of the application mentioned at the end of this section, reheat would certainly be considered only (a) if the expansion had to be split between two turbines for other reasons and (b) if the additional flexibility of control provided by the reheat fuel supply was thought to be desirable. With regard to (a), it must be noted that the natural division of expansion between a compressor turbine and power turbine may not be the optimum point at which to reheat and if so the full advantage of reheat will not be realized. Finally, readers familiar with steam turbine design will understand that reheat also introduces additional mechanical problems arising from the decrease in gas density, and hence the need for longer blading, in the low-pressure stages.

Intercooling, which has a similar effect upon the performance of the ideal heat-exchange cycle as reheat, does not suffer from the same defects. When incorporated in a practical cycle, even allowing for the additional pressure loss there is, in addition to the marked increase in specific output, a worthwhile improvement in efficiency. Nevertheless, as pointed out in

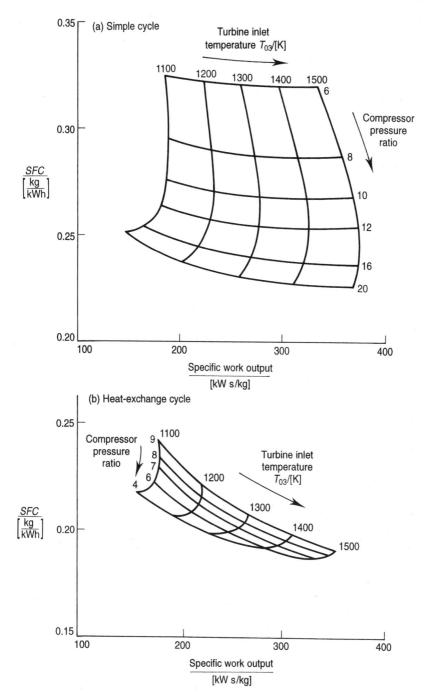

FIG. 2.27 Cycle performance curves

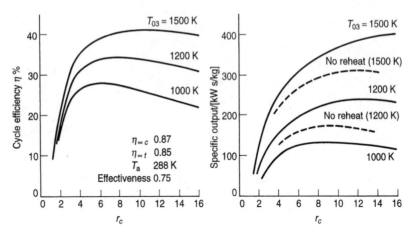

FIG. 2.28 Cycle with heat-exchange and reheat

section 2.1, intercoolers tend to be bulky, and if they require cooling water the self-contained nature of the gas turbine is lost. The cycle is attractive for naval applications, however, because cooling water is readily available—the ocean. Not only does the intercooled cycle with heat-exchange yield a thermal efficiency in excess of 40 per cent, but it also has a good part-load efficiency. The latter feature means that it might be possible to use a single engine rather than a COGOG arrangement, thus offsetting the greater bulk and cost of the engine.

We have already met one exception to the rule that there is little advantage in incorporating reheat without heat-exchange—in Example 2.3. In general, when a gas turbine is part of a combined cycle plant, or cogeneration scheme, a heat-exchanger would be superfluous. The increase in the gas turbine exhaust temperature due to reheat is utilized by the bottoming steam cycle or waste-heat boiler. The addition of reheat to the gas cycle can then result in an increase in efficiency as well as specific output. Because of the need to reduce the environmental impact of burning fossil fuels, there is increasing pressure to seek the highest possible efficiency from our large base-load power stations. The reheat gas turbine, reintroduced by ABB in the late 1990s, has proved to be a highly successful solution and many of these units entered service in the early part of the 21st century. Studies suggest that efficiencies of over 60 per cent may be achieved by using complex cycles involving intercooling, heat-exchange and reheat [Ref. (5)].

We may conclude this section by emphasizing that in practice most gas turbines utilize a high-pressure ratio simple cycle. The only other widely used cycle is the low-pressure ratio heat-exchange variety, but even so the number of engines built with heat-exchangers is only a small fraction of the output of the gas turbine industry. The other modifications mentioned do not normally show sufficient advantage to offset the increased complexity and cost.

2.5   Combined cycles and cogeneration schemes

In the gas turbine, practically all the energy not converted to shaft power is available in the exhaust gases for other uses. The only limitation is that the final exhaust temperature, i.e. the stack temperature, should not be reduced below the dewpoint to avoid corrosion problems arising from sulphur in the fuel. The exhaust heat may be used in a variety of ways. If it is wholly used to produce steam in a waste heat boiler or heat recovery steam generator for a steam turbine, with the object of augmenting the shaft power produced, the system is referred to as a *combined gas/steam cycle power plant* or simply a *combined cycle plant* (Fig. 1.3). Alternatively, the exhaust heat may be used to produce hot water or steam for district or factory heating, hot gas or steam for some chemical process, hot gas for distillation plant, or steam for operating an absorption refrigerator in water chilling or air-conditioning plant. The shaft power will normally be used to produce electricity. In such circumstances the system is referred to as a *cogeneration* or CHP plant (Fig. 1.18). We shall consider briefly the main characteristics of these two types of system in turn.

Combined cycle plant

The optimization of binary cycles is too complex to be discussed in detail in this book, but the more important decisions that have to be taken when designing combined cycle plant can be described in broad terms.

Figure 2.29(a) shows the gas and steam conditions in the boiler on a T–H diagram. The enthalpy rise between feed water inlet and steam outlet must equal the enthalpy drop of the exhaust gases in the heat recovery steam generator (HRSG), and the pinch point and terminal temperature differences cannot be less than about 20 °C if the boiler is to be of

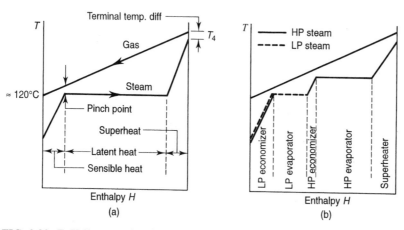

FIG. 2.29 *T–H* diagrams for single- and dual-pressure combined cycle plant

economic size. It follows that a reduction in gas turbine exhaust temperature will lead to a reduction in the steam pressure that can be used for the steam cycle. In the combined cycle, therefore, selection of a higher compressor pressure ratio to improve the gas turbine efficiency may lead to a fall in steam cycle efficiency and a reduction in overall thermal efficiency. In practice, heavy industrial gas turbines must be acceptable in both simple and combined cycle markets and it appears that a pressure ratio of around 15, giving an exhaust gas temperature (EGT) of between 550 and 600 °C, is suitable for both. Aircraft derivative engines with design pressure ratios of 25–35 yield lower values of EGT of about 450 °C which result in the use of less efficient steam cycles.

Combined cycle plant is used for large base-load generating stations and the overall thermal efficiency, while very important, is not the ultimate criterion, which is the cost of a unit of electricity sent out. This will depend on both the efficiency and the capital cost of the plant. For example, if a dual-pressure steam cycle was used, as in the plant shown in Fig. 1.3, a higher efficiency would be obtained because the average temperature at which heat is transferred to the steam is increased. Figure 2.29(b) will make this clear. The additional complication would certainly increase the cost of the boiler and steam turbine, and a detailed study would be required to see if this would lead to a reduction in the cost of electricity produced. The results of such a study can be found in Ref. (6). Dual-pressure cycles have, in fact, been widely used; with modern gas turbines yielding exhaust gas temperatures close to 600 °C, triple-pressure cycles with reheat have been found to be economic and are now commonly used.

Two other decisions have to be made at an early stage in the design process. There is the possibility of associating more than one gas turbine with a single HRSG and steam turbine. Such an arrangement would permit each gas turbine to be fitted with its own diverter valve and stack so that any particular gas turbine could be shut down for maintenance. A second option requiring a decision is whether or not to arrange for extra fuel to be burnt in the HRSG, referred to as *supplementary firing*. This could be used for increasing the peak load for short periods, but would increase the capital cost of the HRSG substantially. The exhaust gases from the supplementary firing do not pass through a turbine, which opens the possibility of burning heavy oil or coal in the HRSG. This is usually ruled out by the logistic difficulty and cost of supplying a large plant with more than one fuel. It may become economic in the future, however, if the prices of oil and gas increase sharply relative to coal.

We will conclude this sub-section by showing how to estimate the improvement in performance resulting from the addition of an exhaust-heated Rankine cycle to a gas cycle. Only the simple case of a single-pressure superheat cycle (Fig. 2.29(a)) will be dealt with. The heat transferred from the exhaust gas is $m_g c_{pg}(T_4 - T_{stack})$ where T_4 is the temperature at exit from the turbine. This is equal to $m_s(h - h_w)$ where m_s is the steam flow and h and h_w are the specific enthalpies at steam turbine inlet and HRSG inlet respectively. The superheat temperature is fixed by T_4 and the terminal temperature difference, and the enthalpy will depend on the pressure. The pinch point temperature, T_p, is fixed by the pinch point temperature difference and the saturation temperature of the steam. The mass flow of steam is obtained from

$$m_s(h - h_f) = m_g c_{pg}(T_4 - T_p)$$

where h_f is the saturated liquid enthalpy. The stack temperature can then be determined from

$$m_g c_{pg}(T_p - T_{stack}) = m_s(h_f - h_w)$$

The total power from the combined cycle is $W_{gt} + W_{st}$ but the heat input (from the fuel burned in the gas turbine) is unchanged. The overall efficiency is then given by

$$\eta = \frac{W_{gt} + W_{st}}{Q}$$

It is useful to express this in terms of the gas turbine and steam turbine efficiencies to illustrate their individual effects. $W_{st} = \eta_{st} Q_{st}$, where Q_{st} is the notional heat supplied to the steam turbine:

$$Q_{st} = m_g c_{pg}(T_4 - T_{stack})$$

This, in turn, can be expressed in terms of the heat rejection from the gas turbine assuming that the turbine exhaust gas at T_4 is cooled without useful energy extraction from T_{stack} to the ambient temperature T_a, i.e. $m_g c_{pg}(T_4 - T_a)$ or $Q(1 - \eta_{gt})$. Thus

$$Q_{st} = m_g c_{pg}(T_4 - T_a)\frac{(T_4 - T_{stack})}{(T_4 - T_a)} = Q(1 - \eta_{gt})\frac{(T_4 - T_{stack})}{(T_4 - T_a)}$$

The overall efficiency is then given by

$$\eta = \frac{W_{gt}}{Q} + \frac{W_{st}}{Q} = \frac{W_{gt}}{Q} + \frac{\eta_{st} Q_{st}}{Q}$$

or

$$\eta = \eta_{gt} + \eta_{st}(1 - \eta_{gt})\frac{(T_4 - T_{sack})}{(T_4 - T_a)}$$

It can now be seen that the overall efficiency is influenced by both the gas turbine exhaust temperature and the stack temperature. A typical modern gas turbine used in combined cycle applications would have an exhaust gas temperature of around 600 °C and a thermal efficiency of about 34 per cent. The stack temperature could be around 140 °C if liquid fuel were being burned, or 120 °C when the fuel was natural gas which has a very low sulphur content. A single-pressure steam cycle might give around 32 per cent thermal efficiency. Hence

$$\eta = 0.34 + 0.32(1 - 0.34)\left[\frac{600 - 120}{600 - 15}\right]$$

$$= 0.513$$

Using a more complex steam cycle with $\eta_{st} = 0.36$, the overall efficiency would be raised to 53·5 per cent, which is typical of modern combined cycles.

Cogeneration plant

In a cogeneration plant, such as that in Fig. 1.18, the steam generated in the HRSGs is used for several different purposes such as a steam turbine drive for a centrifugal water chiller, a heat source for absorption chillers, and to meet process and heating requirements. Each user, such as a hospital or paper making factory, will have different requirements for steam, and even though the same gas turbine may be suitable for a variety of cogeneration plant, the HRSGs would be tailored to the specific requirements of the user.

A major problem in the thermodynamic design of a cogeneration plant is the balancing of the electric power and steam loads. The steam requirement, for example, will depend on the heating and cooling loads which may be subject to major seasonal variations: the electrical demand may peak in either winter or summer depending on the geographical location. In some cases, the system may be designed to meet the primary heating load and the gas turbine selected may provide significantly more power than demanded by the plant; in this case, surplus electricity would be exported to the local utility. Another possibility is for electricity generation to be the prime requirement, with steam sold as a by-product to an industrial complex for process use. The economics and feasibility of a cogeneration plant are closely related to the commercial arrangements that can be made with the local electric utility or the steam host. It would be necessary, for example, to negotiate rates at which the utility would purchase power and also to ensure that back-up power would be available during periods when the gas turbine was out of service; these could be planned outages, for maintenance, or forced outages resulting from system failure. Such considerations are crucial when the cogeneration plant forms the basis of a district heating system.

When designing a cogeneration plant care must be taken in specifying the ambient conditions defining the *site rating*. In cold climates, for example, the gas turbine exhaust temperature would be significantly reduced, resulting in a loss of steam generating capacity; this may require the introduction of supplementary firing in the HRSG. If the design were based on standard day conditions, it could be quite unsatisfactory on a very cold day. The site rating must take into account the seasonal variation of temperature and ensure that the critical requirements can be met. The off-design performance of gas turbines will be discussed in Chapter 8.

The overall efficiency of a cogeneration plant may be defined as the sum of the net work and useful heat output per unit mass flow, divided by the product of the fuel/air ratio and $Q_{net,p}$ of the fuel. For the purpose of preliminary cycle calculations it is sufficiently accurate to evaluate the useful heat output per unit mass flow as $c_{pg}(T_{in} - 393)$ bearing in mind the need to keep the stack temperature from falling much below 120 °C. T_{in} would be the temperature at entry to the HRSG and would normally be the exhaust gas temperature of the turbine. It is unlikely that a regenerative cycle would be used in cogeneration plant because the low exhaust temperature would limit steam production.

Repowering

Another way in which gas turbines can be used with steam turbines is the *repowering* of old power stations. Steam turbines have much longer lives than boilers, so it is possible to create an inexpensive combined cycle plant by replacing an existing fired boiler with a gas turbine and HRSG. The steam conditions for which power stations were designed in the 1950s enable a good match to be achieved between the steam and gas cycles. The power output can be trebled, while the thermal efficiency is increased from about 25 per cent to better than 40 per cent. The result is a vastly superior power plant installed on an existing site. The problems, and cost, of finding a suitable location for a new power station are avoided, and this contributes to the reduction in cost of electricity sent out. A typical repowering project is described in Ref. (7).

An interesting example of repowering is the conversion of an uncompleted nuclear plant to a base-load combined cycle plant, using the steam turbines and generators of the existing scheme but replacing the nuclear reactor as a heat source by twelve 85 MW gas turbines and HRSGs. The net result in this instance is a total output of 1380 MW, plus a substantial supply of steam to a chemical plant.

2.6   Closed-cycle gas turbines

The main features of the closed-cycle gas turbine have been described in section 1.3, where it was suggested that a monatomic gas would have advantages over air as the working fluid. Although, as we pointed out, closed-cycle gas turbines are unlikely to be widely used, it will be instructive to quantify some of these advantages. To this end, the performance of a particular closed cycle will be evaluated for air and helium.

Figure 2.30 is a sketch of the plant, annotated with some of the assumed operating conditions, viz. LP compressor inlet temperature and pressure, HP compressor inlet temperature, and turbine inlet temperature. The compressor inlet temperatures would be fixed by the temperature of the available cooling water and the required temperature difference between water and gas. A modest turbine inlet temperature of 1100 K has been chosen. (This might have been achievable in the core of the high-temperature nuclear reactor (HTR) discussed in section 1.3.) The higher we can make the pressure in the circuit the smaller the plant becomes, and we shall assume that a minimum pressure of 20 bar is practicable. Typical values of component efficiencies and pressure loss are: $\eta_{\infty c} = 0.89$, $\eta_{\infty t} = 0.88$, heat-exchanger effectiveness 0.7; pressure loss as percentage of component inlet pressure: in precooler and intercooler 1.0 per cent each, in hot and cold sides of heat-exchanger 2.5 per cent each, in gas heater 3 per cent.

The performance will be evaluated over a range of compressor pressure ratio p_{04}/p_{01}. It will be assumed that the compression is split between LP and HP compressors such that $p_{02}/p_{01} = (p_{04}/p_{01})^{0.5}$ which leads to an approximately equal division of work input. The usual values of $\gamma = 1.4$ and $c_p = 1.005\,\text{kJ/kg K}$ are assumed to hold throughout the cycle when air is the working fluid, and for helium $\gamma = 1.666$ and $c_p = 5.193\,\text{kJ/kg K}$. The cycle efficiency is given by

$$\eta = \frac{[(T_{06} - T_{07}) - (T_{04} - T_{03}) - (T_{02} - T_{01})]}{(T_{06} - T_{05})}$$

When comparing the specific work output of cycles using fluids of different density, it is more useful to express it in terms of output per unit volume flow than per unit mass flow, because the size of plant is determined by the

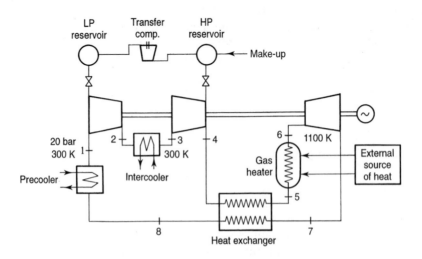

FIG. 2.30 Example of a closed-cycle plant

former. Here we shall evaluate the specific output from the product of the output per unit mass flow and the density at inlet to the compressor where $T = 300\,K$ and $p = 20\,bar$. The density of air at this state is $23 \cdot 23\,kg/m^3$ whereas for helium (molar mass $4\,kg/kmol$) it is only $3 \cdot 207\,kg/m^3$.

Figure 2.31 shows the results of the calculations and we will consider first the efficiency curves. They suggest that the helium cycle has a slightly lower efficiency. In fact more accurate calculations, allowing for the variation of c_p and γ with temperature in the case of air, lead to almost identical maximum efficiencies. (With helium, c_p and γ do not vary significantly with temperature over the range of interest.) As will be shown below, however, the heat transfer characteristics of helium are better than those of air so that it is probable that a higher heat-exchanger effectiveness can be used with helium without making the heat-exchanger excessively large. The portion of a dotted curve in Fig. 2.31 indicates the benefit obtained by raising the effectiveness from 0·7 to 0·8 for the helium cycle: the efficiency is then 39·5 per cent compared with about 38 per cent for the air cycle.

That the heat transfer characteristics are better for helium can be deduced quite simply from the accepted correlation for heat transfer in turbulent flow in tubes [Ref. (1)], viz.

$$Nu = 0 \cdot 023\,Re^{0 \cdot 8}\,Pr^{0 \cdot 4}$$

where Nu is the Nusselt number (hd/k). The Prandtl number $(c_p\mu/k)$ is approximately the same for air and helium, so that the heat transfer coefficients will be in the ratio

$$\frac{h_h}{h_a} = \left(\frac{Re_h}{Re_a}\right)^{0 \cdot 8}\left(\frac{k_h}{k_a}\right) \tag{2.24}$$

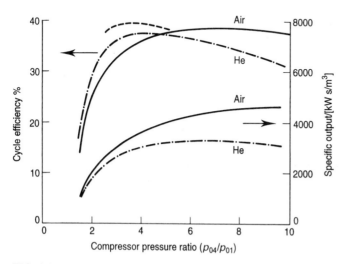

FIG. 2.31 Comparison of air and helium closed-cycle performance

Suffixes h and a refer to helium and air respectively. From tables of properties, e.g. Ref. (8), it may be seen that the thermal conductivities vary with temperature, but the ratio k_h/k_a is of the order of 5 over the relevant range of temperature. The Reynolds number $(\rho C d/\mu)$ is a function of velocity as well as properties of the fluid, and the flow velocity in the heat-exchanger (and coolers) is determined by the pressure drop entailed. As may be found in any text on fluid mechanics, or Ref. (1), the pressure loss for turbulent flow in tubes is given by

$$\Delta p = \left(\frac{4L}{d}\right)\tau = \left(\frac{4L}{d}\right)f\left(\frac{\rho C^2}{2}\right)$$

where τ is the wall shear stress and f the friction factor given by the Blasius law

$$f = \frac{0\cdot0791}{Re^{0\cdot25}} = \frac{0\cdot0791}{(\rho C d/\mu)^{0\cdot25}}$$

It follows that for equal pressure losses,

$$(\rho^{0\cdot75}C^{1\cdot75}\mu^{0\cdot25})_h = (\rho^{0\cdot75}C^{1\cdot75}\mu^{0\cdot25})_a$$

ρ_h/ρ_a is $0\cdot138$ and μ_h/μ_a is approximately $1\cdot10$, so that

$$\frac{C_h}{C_a} = \left(\frac{1}{0\cdot138^{0\cdot75} \times 1\cdot10^{0\cdot25}}\right)^{1/1\cdot75} = 2\cdot3$$

Thus for similar pressure losses the flow velocity of helium can be double that of air.

Using this result we have

$$\frac{Re_h}{Re_a} = \frac{\rho_h C_h \mu_h}{\rho_a C_a \mu_a} = \frac{0\cdot138 \times 2\cdot3}{1\cdot10} = 0\cdot29$$

Finally, from equation (2.24)

$$\frac{h_h}{h_a} = 0\cdot29^{0\cdot8} \times 5 = 1\cdot86$$

and the heat transfer coefficient for helium is therefore almost twice that for air. This implies that the heat-exchanger need have only half the surface area of tubing for the same temperature difference, or that a higher effectiveness can be used economically.

Turning now to the specific work output curves of Fig. 2.31, from which the comparative size of plant can be deduced, it will be appropriate to make the comparison at the pressure ratio yielding maximum efficiency in each case, i.e. around 4 for helium and 7 for air. Evidently about 45 per cent greater volume flow is required with helium than with air for a given power output. The reason is that under conditions of maximum efficiency, in each case the compressor temperature rises and turbine temperature drops are similar for helium and air, so that the specific outputs on a *mass* flow basis are in the ratio $c_{ph}/c_{pa} \approx 5$. The density of helium is only 0·138 times that of air, however, so that on a volume flow basis the specific output with helium is about 0·7 of that with air.

The increase in size of the heat-exchanger and coolers due to this higher volume flow will be more than offset by the reduction arising from the better heat transfer coefficients and the higher flow velocities that can be used with helium. Higher velocities will also reduce the diameter of the ducts connecting the components of the plant. A 45 per cent increase in volume flow with a doubling of the velocity would imply a 15 per cent reduction in diameter.

The effect of using helium instead of air on the size of the turbomachinery will only be fully appreciated after the chapters on compressors and turbines have been studied. We have already noted that the work done per unit mass flow in these components is about five times as large with helium as with air. One might expect that five times the number of stages of blading will be required. This is not so, however, because the Mach number limitation on flow velocity and peripheral speed (which determine the work that can be done in each stage) is virtually removed when helium is the working fluid. The reason is that the sonic velocity, $(\gamma R T)^{0.5}$, is much higher in helium. The gas constant is inversely proportional to the molar mass so that the ratio of sonic velocities at any given temperature becomes

$$\frac{a_h}{a_a} = \left(\frac{\gamma_h R_h}{\gamma_a R_a}\right)^{0.5} = \left(\frac{1 \cdot 66 \times 29}{1 \cdot 40 \times 4}\right)^{0.5} = 2 \cdot 94$$

The work per stage of blading can probably be quadrupled with helium, so that the number of stages is only increased in the ratio 5/4. On top of this we have a reduction in annulus area, and therefore height of blading, consequent upon the higher flow velocities through the turbomachinery. Only a detailed study would show whether the turbomachinery will in fact be much larger for the helium cycle.

Because helium is a relatively scarce resource, the decision to use it in large power plant will probably rest not so much on thermodynamic considerations as the satisfactory solution of the practical problem of sealing the system at the high pressures required. Experience with sealing high-pressure carbon dioxide in gas-cooled reactors has shown that leakage is substantial, and helium, being a lighter gas, will be even more difficult to contain. Although in section 1.7 it was stated that closed-cycle gas turbines are unlikely to be used in nuclear power plant, in-depth studies of the technological requirements for future applications continued until the mid 1990s [Ref. (9)].

NOMENCLATURE

The most widely used symbols in the book are introduced here and they
will not be repeated in the lists at the end of other chapters.

C	velocity
c_p	specific heat at constant pressure
f	fuel/air ratio by weight
h	specific enthalpy
ΔH	enthalpy of reaction
m	mass flow
M	molecular weight, Mach number
n	polytropic index
p	absolute pressure
Q	heat transfer per unit mass flow
$Q_{net,p}$	net calorific value at constant p
$R, \tilde{R}$	specific, molar (universal), gas constant
r	pressure ratio
s	specific entropy
T	absolute temperature
t	temperature ratio
W	specific work (power) output
γ	ratio of specific heats
η	efficiency
ρ	density

Suffixes

0	stagnation value
1, 2, 3, etc.	reference planes
∞	polytropic
a	ambient, air
b	combustion chamber
c	compressor
f	fuel
g	gas
h	heat-exchanger
i	intake, mixture constituent
m	mechanical
N	net
s	stage
t	turbine

3

Gas turbine cycles
for aircraft propulsion

Aircraft gas turbine cycles differ from shaft power cycles in that the useful power output is in the form of *thrust*: the whole of the thrust of the turbojet and turbofan is generated in propelling nozzles, whereas with the turboprop most is produced by a propeller with only a small contribution from the exhaust nozzle. A second distinguishing feature is the need to consider the effect of forward speed and altitude on the performance. It was the beneficial aspect of these parameters, together with a vastly superior power/weight ratio, that enabled the gas turbine to so rapidly supplant the reciprocating engine for aircraft propulsion except for low-powered light aircraft.

The designer of aircraft engines must recognize the differing requirements for take-off, climb, cruise and manoeuvring, the relative importance of these being different for civil and military applications and for long- and short-haul aircraft. In the early days it was common practice to focus on the take-off thrust, but this is no longer adequate. Engines for long-range civil aircraft, for example, require low *SFC* at cruise speed and altitude, while the thrust level may be determined either by take-off thrust on the hottest day likely to be encountered or by the thrust required at top of climb. Evidently the selection of design conditions is much more complex than for a land-based unit. As examples, 'design point' calculations will be shown for take-off (static) and cruise conditions.

The chapter opens with a discussion of the criteria appropriate for evaluating the performance of jet propulsion cycles, and of the additional parameters required to allow the losses in the intake and propelling nozzle to be taken into account. The cycle performance of turbojet, turbofan and turboprop are then discussed in turn. Methods for calculating the variation of actual performance with altitude and forward speed for any specific engine will be presented in Chapter 8. For the analysis of other forms of jet power plant, such as ramjets and rockets, the reader must turn to specialized texts on aircraft propulsion such as Ref. (1).

3.1 Criteria of performance

Consider the schematic diagram of a propulsive duct shown in Fig. 3.1. Relative to the engine, the air enters the intake with a velocity C_a equal and opposite to the forward speed of the aircraft, and the power unit accelerates the air so that it leaves with the jet velocity C_j. The 'power unit' may consist of a gas turbine in which the turbine merely drives the compressor, one in which part of the expansion is carried out in a power turbine driving a propeller, or simply a combustion chamber as in ramjet engines. For simplicity we shall assume here that the mass flow m is constant (i.e. that the fuel flow is negligible), and thus the *net thrust F* due to the rate of change of momentum is

$$F = m(C_j - C_a) \tag{3.1}$$

mC_j is called the *gross momentum thrust* and mC_a the *intake momentum drag*. When the exhaust gases are not expanded completely to p_a in the propulsive duct, the pressure p_j in the plane of the exit will be greater than p_a and there will be an additional pressure thrust exerted over the jet exit area A_j equal to $A_j(p_j - p_a)$ as indicated in Fig. 3.1. The net thrust is then the sum of the *momentum thrust* and the *pressure thrust*, namely

$$F = m(C_j - C_a) + A_j(p_j - p_a) \tag{3.2}$$

When the aircraft is flying at a uniform speed C_a in level flight the thrust must be equal and opposite to the drag of the aircraft at that speed.

In what follows we shall assume there is complete expansion to p_a in the propelling nozzle and therefore that equation (3.1) is applicable. From this equation it is clear that the required thrust can be obtained by designing the engine to produce either a high-velocity jet of small mass flow or a low-velocity jet of high mass flow. The question arises as to what is the most efficient combination of these two variables and a qualitative answer is provided by the following simple analysis.

The *propulsion efficiency* η_p can be defined as the ratio of the useful propulsive energy or thrust power (FC_a) to the sum of that energy and the unused kinetic energy of the jet. The latter is the kinetic energy of the jet *relative to the earth*, namely $m(C_j - C_a)^2/2$. Thus

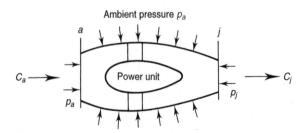

FIG. 3.1 Propulsive duct

$$\eta_p = \frac{mC_a(C_j - C_a)}{m[C_a(C_j - C_a) + (C_j - C_a)^2/2]} = \frac{2}{1 + (C_j/C_a)} \tag{3.3}$$

η_p is often called the *Froude efficiency*. Note that it is in no sense an overall power plant efficiency, because the unused enthalpy in the jet is ignored. From equations (3.1) and (3.3) it is evident that

(a) *F* is a maximum when $C_a = 0$, i.e. under static conditions, but η_p is then zero;

(b) η_p is a maximum when $C_j/C_a = 1$ but then the thrust is zero.

We may conclude that although C_j must be greater than C_a the difference should not be too great. This is the reason for the development of the family of propulsion units shown in Fig. 3.2. Taken in the order shown they provide propulsive jets of decreasing mass flow and increasing jet velocity, and therefore in that order they will be suitable for aircraft of increasing design cruising speed. In practice the choice of power plant can be made only when the specification of the aircraft is known: it will depend not only on the required cruise speed but also on such factors as the desired range of the aircraft and maximum rate of climb. Because the thrust and fuel consumption of a jet propulsion unit vary with both cruise speed and altitude (air density), the latter is also an important parameter. Figure 3.3 indicates

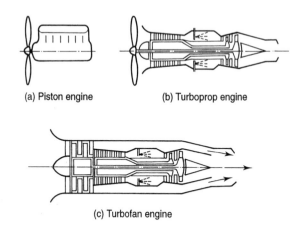

(a) Piston engine   (b) Turboprop engine

(c) Turbofan engine

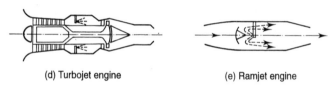

(d) Turbojet engine   (e) Ramjet engine

FIG. 3.2 Propulsion engines

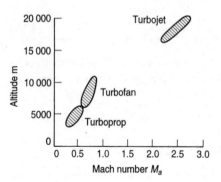

FIG. 3.3 Flight regimes

the flight regimes found to be suitable for the broad categories of power plant installed in civil aircraft.

The propulsion efficiency is a measure of the effectiveness with which the propulsive duct is being used for propelling the aircraft, and it is not the *efficiency of energy conversion* within the power plant itself that we will symbolize by η_e. The rate of energy supplied in the fuel may be regarded as $m_f Q_{net,p}$ where m_f is the fuel mass flow. This is converted into potentially useful kinetic energy for propulsion $m(C_j^2 - C_a^2)/2$, together with unusable enthalpy in the jet $mc_p(T_j - T_a)$. η_e is therefore defined by

$$\eta_e = \frac{m(C_j^2 - C_a^2)/2}{m_f Q_{net,p}} \tag{3.4}$$

The *overall efficiency* η_o is the ratio of the useful work done in overcoming drag to the energy in the fuel supplied, i.e.

$$\eta_o = \frac{mC_a(C_j - C_a)}{m_f Q_{net,p}} = \frac{FC_a}{m_f Q_{net,p}} \tag{3.5}$$

It is easy to see that the denominator of equation (3.3), namely $m[C_a(C_j - C_a) + (C_j - C_a)^2/2]$ is equal to the numerator of equation (3.4), and hence that

$$\eta_o = \eta_p \eta_e \tag{3.6}$$

The object of the simple analysis leading to equation (3.6) is to make the point that the efficiency of an aircraft power plant is inextricably linked to the aircraft speed. A crude comparison of different engines can be made, however, if the engine performance is quoted at two operating conditions: the sea-level static performance at maximum power (i.e. at maximum turbine inlet temperature) that must meet the aircraft take-off requirements, and the cruise performance at the optimum cruise speed and altitude of the aircraft for which it is intended. The ambiguous concept of efficiency is discarded in favour of the *specific fuel consumption* which, for aircraft engines, is usually defined as the fuel consumption *per unit thrust* (e.g. kg/h N). The overall efficiency given by equation (3.5) can be written in the form

$$\eta_o = \frac{C_a}{SFC} \times \frac{1}{Q_{net,p}} \tag{3.7}$$

With a given fuel, the value of $Q_{net,p}$ will be constant and it can be seen that the overall efficiency is proportional to C_a/SFC, rather than $1/SFC$ as for shaft power units.

Another important performance parameter is the *specific thrust F_s*, namely the thrust *per unit mass flow of air* (e.g. Ns/kg). This provides an indication of the relative size of engines producing the same thrust because the dimensions of the engine are primarily determined by the airflow. Size is important because of its association not only with weight but also with frontal area and the consequent drag. Note that the *SFC* and specific thrust are related by

$$SFC = \frac{f}{F_s} \tag{3.8}$$

where f is the fuel/air ratio.

When estimating the cycle performance at altitude we shall need to know the way in which ambient pressure and temperature vary with height above sea level. The variation depends to some extent upon the season and latitude, but it is usual to work with an average or 'standard' atmosphere. The *International Standard Atmosphere* (ISA) corresponds to average values at middling latitudes and yields a temperature decreasing by about 3·2 K per 500 m of altitude up to 11 000 m after which it is constant at 216·7 K until 20 000 m. Above this height the temperature starts to increase again slowly. Once the temperature is fixed, the variation of pressure follows according to the laws of hydrostatics. An abridged tabulated version of the ISA. is given at the end of this chapter, and this will be used for subsequent numerical examples; it should be appreciated, however, that actual ambient conditions can vary widely from ISA values both at sea level and high altitude. For high-subsonic and supersonic aircraft it is more appropriate to use Mach number rather than m/s for aircraft speed, because the drag is more a function of the former. It must be remembered that for a given speed in m/s the Mach number will rise with altitude up to 11 000 m because the temperature is falling. Figure 3.4 shows a plot of M_a versus C_a for sea level and 11 000 m obtained from ISA data and $M_a = C_a/a$, where a is the sonic velocity $(\gamma R T_a)^{1/2}$.

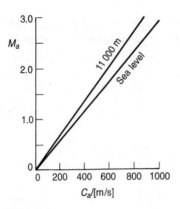

FIG. 3.4

3.2   Intake and propelling nozzle efficiencies

The nomenclature to be adopted is indicated in Fig. 3.5, which illustrates a simple turbojet engine and the ideal cycle upon which it operates. The turbine produces just sufficient work to drive the compressor, and the remaining part of the expansion is carried out in the propelling nozzle. Because of the significant effect of forward speed, the intake must be considered as a separate component: it cannot be regarded as part of the compressor as it often was in Chapter 2. Before proceeding to discuss the performance of aircraft propulsion cycles, it is necessary to describe how the losses in the two additional components—intake and propelling nozzle—are to be taken into account.

Intakes

The intake is a critical part of an aircraft engine installation, having a significant effect on both engine efficiency and aircraft safety. The prime requirement is to minimize the pressure loss up to the compressor face while ensuring that the flow enters the compressor with a uniform pressure and velocity, at all flight conditions. Non-uniform, or distorted, flow may

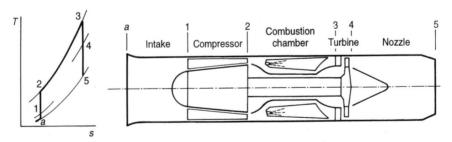

FIG. 3.5 Simple turbojet engine and ideal cycle

cause compressor surge which can result in either engine flame-out or severe mechanical damage due to blade vibration induced by unsteady aerodynamic effects. Even with a well-designed intake, it is difficult to avoid some flow distortion during rapid manoeuvring.

Successful engines will find applications in a variety of aircraft having widely different installations and inlet systems. Engines may be installed in the fuselage, in pods (wing or rear fuselage mounted), or buried in the wing root. Three-engined aircraft will require a different arrangement for the centre engine than for the wing- or tail-mounted engines. The L-1011 and DC-10 both have engines mounted at the rear, but use radically different forms of intake: the former has the engine mounted in the rear fuselage behind a long S-bend duct, while the latter has a straight-through duct with the engine mounted in the vertical fin as shown in Fig. 3.6. The design of the intake involves a compromise between aerodynamic and structural requirements.

Current designs of compressor require the flow to enter the first stage at an axial Mach number in the region of 0·4–0·5. Subsonic aircraft will typically cruise at M 0·8–0·85, while supersonic aircraft may operate at speeds from M 2 to M 2·5. At take-off, with zero forward speed, the engine will operate at maximum power and airflow. The intake must therefore satisfy a wide range of operating conditions and the design of the intake requires close collaboration between the aircraft and engine designers. The aerodynamic design of intakes is covered in Ref. (2).

Ways of accounting for the effect of friction in the intake, suitable for use in cycle calculations, can be found by treating the intake as an adiabatic duct. Since there is no heat or work transfer, the stagnation temperature is constant although there will be a loss of stagnation pressure due to friction and due to shock waves at supersonic flight speeds. Under static conditions or at very low forward speeds the intake acts as a nozzle in which the air accelerates from zero velocity or low C_a to C_1 at the compressor inlet. At normal forward speeds, however, the intake performs as a diffuser with the air decelerating from C_a to C_1 and the static pressure

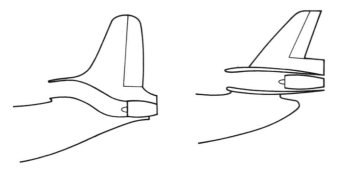

FIG. 3.6 S-bend and straight-through intakes

rising from p_a to p_1. Since it is the stagnation pressure at the compressor inlet which is required for cycle calculations, it is the pressure rise $(p_{01} - p_a)$ that is of interest and which is referred to as the *ram pressure rise*. At supersonic speeds it will comprise the pressure rise across a system of shock waves at the inlet (see Appendix A.7) followed by that due to subsonic diffusion in the remainder of the duct.

The intake efficiency can be expressed in a variety of ways, but the two most commonly used are the *isentropic efficiency* η_i (defined in terms of temperature rises) and the *ram efficiency* η_r (defined in terms of pressure rises). Referring to Fig. 3.7, we have

$$T_{01} = T_{0a} = T_a + \frac{C_a^2}{2c_p}$$

and

$$\frac{p_{01}}{p_a} = \left(\frac{T'_{01}}{T_a}\right)^{\gamma/(\gamma-1)}$$

where T'_{01} is the temperature which would have been reached after an isentropic ram compression to p_{01}. T'_{01} can be related to T_{01} by introducing an isentropic efficiency η_i defined by

$$\eta_i = \frac{T'_{01} - T_a}{T_{01} - T_a} \tag{3.9}$$

It follows that

$$T'_{01} - T_a = \eta_i \frac{C_a^2}{2c_p}$$

so that η_i can be regarded as the fraction of the inlet dynamic temperature which is made available for isentropic compression in the intake. The intake pressure ratio can then be found from

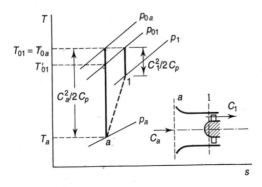

FIG. 3.7 Intake loss

$$\frac{p_{01}}{p_a} = \left[1 + \frac{T'_{01} - T_a}{T_a}\right]^{\gamma/(\gamma-1)} = \left[1 + \eta_i \frac{C_a^2}{2c_p T_a}\right]^{\gamma/(\gamma-1)} \tag{3.10a}$$

Remembering that $M = C/(\gamma R T)^{1/2}$ and $\gamma R = c_p(\gamma - 1)$, this equation can be written as

$$\frac{p_{01}}{p_a} = \left[1 + \eta_i \frac{\gamma - 1}{2} M_a^2\right]^{\gamma/(\gamma-1)} \tag{3.10b}$$

The stagnation temperature can also be expressed in terms of M_a as

$$\frac{T_{01}}{T_a} = \left[1 + \frac{\gamma - 1}{2} M_a^2\right] \tag{3.11}$$

The ram efficiency η_r is defined by the ratio of the ram pressure rise to the inlet dynamic head, namely

$$\eta_r = \frac{p_{01} - p_a}{p_{0a} - p_a}$$

η_r can be shown to be almost identical in magnitude to η_i and the two quantities are interchangeable: curves relating them can be found in Ref. (2). Apart from the fact that η_r is easier to measure experimentally, it has no advantage over η_i and we shall use the latter in this book. For subsonic intakes, both η_i and η_r are found to be independent of inlet Mach number up to a value of about 0·8 and thence their suitability for cycle calculations. They both suffer equally from the drawback of implying zero stagnation pressure loss when C_a is zero, because then $p_{01}/p_a = 1$ and $p_a = p_{0a}$. This is not serious because under these conditions the average velocity in the intake is low, and the flow is accelerating, so that the effect of friction is very small.

The intake efficiency will depend upon the location of the engine in the aircraft (in wing, pod or fuselage), but we shall assume a value of 0·93 for η_i in the numerical examples relating to subsonic aircraft that follow. It would be less than this for supersonic intakes, the value decreasing with increase in inlet Mach number. In practice, neither η_i nor η_r is used for supersonic intakes and it is more usual to quote values of stagnation pressure ratio p_{01}/p_{0a} as a function of Mach number. p_{01}/p_{0a} is called the *pressure recovery factor* of the intake. Knowing the pressure recovery factor, the pressure ratio p_{01}/p_a can be found from

$$\frac{p_{01}}{p_a} = \frac{p_{01}}{p_{0a}} \times \frac{p_{0a}}{p_a}$$

where p_{0a}/p_a is given in terms of M_a by the isentropic relation (8) of Appendix A, namely

$$\frac{p_{0a}}{p_a} = \left[1 + \frac{\gamma - 1}{2} M_a^2\right]^{\gamma/(\gamma-1)}$$

Some data on the performance of supersonic intakes can be found in Ref. (3); their design is the province of highly specialized aerodynamicists and much of the information is classified. A rough working rule adopted by the American Department of Defense for the pressure recovery factor *relating to the shock system itself* is

$$\left(\frac{p_{01}}{p_{0a}}\right)_{\text{shock}} = 1\cdot0 - 0\cdot075(M_a - 1)^{1\cdot35}$$

which is valid when $1 < M_a < 5$. To obtain the overall pressure recovery factor, $(p_{01}/p_{0a})_{\text{shock}}$ must be multiplied by the pressure recovery factor for the subsonic part of the intake.

Propelling nozzles

We shall here use the term 'propelling nozzle' to refer to the component in which the working fluid is expanded to give a high-velocity jet. With a simple jet engine, as in Fig. 3.5, there will be a single nozzle downstream of the turbine. The turbofan may have two separate nozzles for the hot and cold streams, as in Fig. 3.15, or the flows may be mixed and leave from a single nozzle. Between the turbine exit and propelling nozzle there will be a *jet pipe* of a length determined by the location of the engine in the aircraft. In the transition from the turbine annulus to circular jet pipe some increase in area is provided to reduce the velocity, and hence friction loss, in the jet pipe. When thrust boosting is required, an *afterburner* may be incorporated in the jet pipe as indicated in Fig. 3.8. (Afterburning in a jet engine is sometimes referred to as 'reheating', although it is not equivalent to the reheating between turbines sometimes proposed for industrial gas turbines which would be in constant operation.) Depending on the location of the engine in the aircraft, and on whether reheat is to be incorporated for thrust boosting, the 'propelling nozzle' will comprise some or all of the items shown in Fig. 3.8.

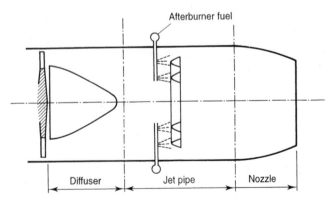

FIG. 3.8 'Propelling nozzle' system

The question immediately arises as to whether a simple convergent nozzle is adequate or whether a convergent–divergent nozzle should be employed. As we shall see from subsequent cycle calculations, even with moderate cycle pressure ratios the pressure ratio p_{04}/p_a will be greater than the critical pressure ratio† over at least part of the operational range of forward speed and altitude. Although a convergent–divergent nozzle might therefore appear to be necessary, it must be remembered that it is *thrust* which is required and not maximum possible jet velocity. Certainly it can be shown that for an isentropic expansion the thrust produced is a maximum when complete expansion to p_a occurs in the nozzle: the pressure thrust $A_5(p_5 - p_a)$ arising from incomplete expansion does not entirely compensate for the loss of momentum thrust due to a smaller jet velocity. But this is no longer true when friction is taken into account because the theoretical jet velocity is not achieved. Furthermore, the use of a convergent–divergent nozzle would result in significant increases in engine weight, length and diameter which would in turn result in major installation difficulties and a penalty in aircraft weight.

For values of p_{04}/p_a up to about 3 there is no doubt, from experiments described in Ref. (4), that the thrust produced by a convergent nozzle is as large as that from a convergent–divergent nozzle even when the latter has an exit/throat area ratio suited precisely to the pressure ratio. At operating pressure ratios less than the design value, a convergent–divergent nozzle of fixed proportions would certainly be *less* efficient because of the loss incurred by the formation of a shock wave in the divergent portion. For these reasons aircraft gas turbines normally employ a convergent propelling nozzle. A secondary advantage of this type is the relative ease with which the following desirable features can be incorporated:

(a) *Variable area*, which is essential when an afterburner is incorporated, for reasons discussed in section 3.8. Earlier engines sometimes used variable area to improve starting performance, but this is seldom necessary today. Figure 3.9(a) illustrates the 'iris' and 'central plug' methods of achieving variation of propelling nozzle area.

(b) *Thrust reverser* to reduce the length of runway required for landing, used almost universally in civil transport aircraft.

(c) *Noise suppression*. Most of the jet noise is due to the mixing of the high-velocity hot stream with the cold atmosphere, and the intensity decreases as the jet velocity is reduced. For this reason the jet noise of the turbofan is less than that of the simple turbojet. In any given case, the noise level can be reduced by accelerating the mixing process and

† An estimate of the value of the critical pressure ratio can be obtained by assuming isentropic flow, i.e. by putting $\gamma = 1\cdot333$ in equation (12) of Appendix A,

$$\frac{p_{04}}{p_c} = \left(\frac{\gamma + 1}{2}\right)^{\gamma/(\gamma-1)} = 1\cdot853$$

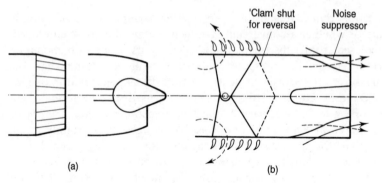

FIG. 3.9 **Variable area, thrust reversal and noise suppression**

this is normally achieved by increasing the surface area of the jet stream as shown in Fig. 3.9(b).

It should not be thought that convergent–divergent nozzles are never used. At high supersonic speeds the large ram pressure rise in the intake results in a very high nozzle pressure ratio. The value of p_{04}/p_a is then many times larger than the critical pressure ratio and may be as high as 10–20 for flight Mach numbers in the range 2–3. Variable exit/throat area is essential to avoid shock losses over as much of the operating range as possible, and the additional mechanical complexity must be accepted. The main limitations on the design are:

(*a*) the exit diameter must be within the overall diameter of the engine—otherwise the additional thrust is offset by the increased external drag;

(*b*) in spite of the weight penalty, the included angle of divergence must be kept below about 30° because the loss in thrust associated with divergence (non-axiality) of the jet increases sharply at greater angles.

The design of the convergent–divergent nozzle is especially critical for supersonic transport aircraft, such as the Concorde, which spend extended periods at high supersonic speeds. The range and payload are so strongly affected by the performance of the propulsion system as a whole, that it is necessary to use both a continuously variable intake, and a nozzle capable of continuous variation of both the throat and exit areas, as discussed in Ref. (5). Military aircraft normally operate subsonically with a supersonic 'dash' capability for only short periods, and supersonic fuel consumption is not as critical.

After these opening remarks, we may now turn our attention to methods of allowing for propelling nozzle loss in cycle calculations. We shall restrict our attention to the almost universally used convergent nozzle. Two approaches are commonly used: one via an isentropic efficiency η_j and the other via a specific thrust coefficient K_F. The latter is defined as the ratio of the actual specific gross thrust, namely $[mC_5 + A_5(p_5 - p_a)]/m$, to that which would have resulted from isentropic flow. When the expansion to p_a is completed in the nozzle, i.e. when $p_{04}/p_a < p_{04}/p_c$, K_F becomes simply the ratio of actual to isentropic jet velocity which is the 'velocity coefficient' often used by steam turbine designers. Under these conditions it will be easy to see from the following that $\eta_j = K_F^2$. Although K_F is the easier to measure on nozzle test rigs, it is not so useful for our present purpose as η_j.

Figure 3.10 illustrates the real and isentropic processes on the T–s diagram; and η_j is defined by

$$\eta_j = \frac{T_{04} - T_5}{T_{04} - T'_5}$$

It follows that for given inlet conditions (p_{04}, T_{04}) and an assumed value of η_j, T_5 is given by

$$T_{04} - T_5 = \eta_j T_{04}\left[1 - \left(\frac{1}{p_{04}/p_5}\right)^{(\gamma-1)/\gamma}\right] \tag{3.12}$$

This is also the temperature equivalent of the jet velocity ($C_5^2/2c_p$) because $T_{05} = T_{04}$. For pressure ratios up to the critical value, p_5 will be put equal to p_a in equation (3.12), and the pressure thrust is zero. Above the critical pressure ratio the nozzle is choked, p_5 remains at the critical value p_c, and C_5 remains at the sonic value $(\gamma R T_5)^{1/2}$. The outstanding question is how the critical pressure should be evaluated for non-isentropic flow.

The critical pressure ratio p_{04}/p_c is the pressure ratio p_{04}/p_5 which yields $M_5 = 1$. Consider the corresponding critical temperature ratio T_{04}/T_c: it is the same for both isentropic and irreversible adiabatic flow. This follows from the fact that $T_{04} = T_{05}$ in all cases of adiabatic flow with no work transfer, and thus

$$\frac{T_{04}}{T_5} = \frac{T_{05}}{T_5} = 1 + \frac{C_5^2}{2c_p T_5} = 1 + \frac{\gamma - 1}{2}M_5^2$$

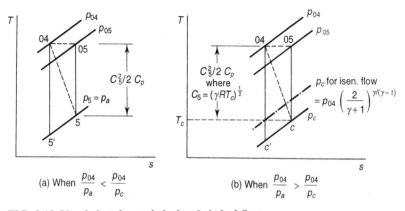

(a) When $\dfrac{p_{04}}{p_a} < \dfrac{p_{04}}{p_c}$

(b) When $\dfrac{p_{04}}{p_a} > \dfrac{p_{04}}{p_c}$

FIG. 3.10 Nozzle loss for unchoked and choked flows

Putting $M_5 = 1$ we have the familiar expression

$$\frac{T_{04}}{T_c} = \frac{\gamma + 1}{2} \tag{3.13}$$

Having found T_c from equation (3.13), we see from Fig. 3.10(b) that the value of η_j yields T'_c which is the temperature reached after an isentropic expansion to the real critical pressure p_c, namely

$$T'_c = T_{04} - \frac{1}{\eta_j}(T_{04} - T_c)$$

p_c is then given by

$$p_c = p_{04}\left(\frac{T'_c}{T_{04}}\right)^{\gamma/(\gamma-1)} = p_{04}\left[1 - \frac{1}{\eta_j}\left(1 - \frac{T_c}{T_{04}}\right)\right]^{\gamma/(\gamma-1)}$$

On substituting for T_c/T_{04} from equation (3.13), we have the critical pressure ratio in the convenient form

$$\frac{p_{04}}{p_c} = \frac{1}{\left[1 - \frac{1}{\eta_j}\left(\frac{\gamma - 1}{\gamma + 1}\right)\right]^{\gamma/(\gamma-1)}} \tag{3.14}$$

This method of using η_j to determine the critical pressure ratio yields results consistent with a more detailed analysis involving the momentum equation given in Ref. (6). In particular, it shows that the effect of friction is to increase the pressure drop required to achieve a Mach number of unity.

The remaining quantity necessary for evaluating the pressure thrust $A_5(p_c - p_a)$ is the nozzle area A_5. For a given mass flow m, this is given *approximately* by

$$A_5 = \frac{m}{\rho_c C_c} \tag{3.15}$$

with ρ_c obtained from p_c/RT_c and C_c from $[2c_p(T_{04} - T_c)]^{\frac{1}{2}}$ or $(\gamma RT_c)^{\frac{1}{2}}$. It is only an approximate value of the exit area because allowance must be made for the thickness of the boundary layer in the real flow. Furthermore, for conical nozzles, the more complete analysis in Ref. (6) shows that the condition $M = 1$ is reached just downstream of the plane of the exit when the flow is irreversible. In practice, the exit area necessary to give the required engine operating conditions is found by trial and error during development tests. Furthermore, it is found that individual engines of the same type will require slightly different nozzle areas, due to tolerance build-ups and small changes in component efficiencies. Minor changes to the nozzle area can be made using *nozzle trimmers*, which are small tabs used to block a portion of the nozzle area.

The value of η_j is obviously dependent on a wide variety of factors, such as the length of the jet pipe, and whether the various auxiliary features mentioned earlier are incorporated because they inevitably introduce additional frictional losses. Another factor is the amount of swirl in the gases leaving the turbine, which should be as low as possible (see Chapter 7). We shall assume a value of 0·95 for η_j in the following cycle calculations.

3.3 Simple turbojet cycle

Figure 3.11 shows the real turbojet cycle on the T–s diagram for comparison with the ideal cycle of Fig. 3.5. The method of calculating the design point performance at any given aircraft speed and altitude is illustrated in the following example.

EXAMPLE 3.1

Determination of the specific thrust and *SFC* for a simple turbojet engine, having the following component performance at the design point at which the cruise speed and altitude are M 0·8 and 10 000 m.

Compressor pressure ratio	8·0
Turbine inlet temperature	1200 K
Isentropic efficiency:	
of compressor, η_c	0·87
of turbine, η_t	0·90
of intake, η_i	0·93
of propelling nozzle, η_j	0·95
Mechanical transmission efficiency η_m	0·99
Combustion efficiency η_b	0·98
Combustion pressure loss Δp_b	4% comp. deliv. press.

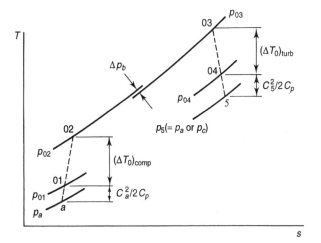

FIG. 3.11 Turbojet cycle with losses

From the ISA table, at $10\,000\,\mathrm{m}$

$$p_a = 0\cdot2650\,\text{bar}, \quad T_a = 223\cdot3\,\text{K} \quad \text{and} \quad a = 299\cdot5\,\text{m/s}$$

The stagnation conditions after the intake may be obtained as follows:

$$\frac{C_a^2}{2c_p} = \frac{(0\cdot8 \times 299\cdot5)^2}{2 \times 1\cdot005 \times 1000} = 28\cdot6\,\text{K}$$

$$T_{01} = T_a + \frac{C_a^2}{2c_p} = 223\cdot3 + 28\cdot6 = 251\cdot9\,\text{K}$$

$$\frac{p_{01}}{p_a} = \left[1 + \eta_i \frac{C_a^2}{2c_p T_a}\right]^{\gamma/(\gamma-1)} = \left[1 + \frac{0\cdot93 \times 28\cdot6}{223\cdot3}\right]^{3\cdot5} = 1\cdot482$$

$$p_{01} = 0\cdot2650 \times 1\cdot482 = 0\cdot393\,\text{bar}$$

At outlet from the compressor,

$$p_{02} = \left(\frac{p_{02}}{p_{01}}\right)p_{01} = 8\cdot0 \times 0\cdot393 = 3\cdot144\,\text{bar}$$

$$T_{02} - T_{01} = \frac{T_{01}}{\eta_c}\left[\left(\frac{p_{02}}{p_{01}}\right)^{(\gamma-1)/\gamma} - 1\right] = \frac{251\cdot9}{0\cdot87}\left[8\cdot0^{1/3\cdot5} - 1\right] = 234\cdot9\,\text{K}$$

$$T_{02} = 251\cdot9 + 234\cdot9 = 486\cdot8\,\text{K}$$

$W_t = W_c/\eta_m$ and hence

$$T_{03} - T_{04} = \frac{c_{pa}(T_{02} - T_{01})}{c_{pg}\eta_m} = \frac{1\cdot005 \times 234\cdot9}{1\cdot148 \times 0\cdot99} = 207\cdot7\,\text{K}$$

$$T_{04} = 1200 - 207\cdot7 = 992\cdot3\,\text{K}$$

$$p_{03} = p_{02}\left(1 - \frac{\Delta p_b}{p_{02}}\right) = 3\cdot144(1 - 0\cdot04) = 3\cdot018\,\text{bar}$$

$$T'_{04} = T_{03} - \frac{1}{\eta_t}(T_{03} - T_{04}) = 1200 - \frac{207\cdot7}{0\cdot90} = 969\cdot2\,\text{K}$$

$$p_{04} = p_{03}\left(\frac{T'_{04}}{T_{03}}\right)^{\gamma/(\gamma-1)} = 3\cdot018\left(\frac{969\cdot2}{1200}\right)^4 = 1\cdot284\,\text{bar}$$

The nozzle pressure ratio is therefore

$$\frac{p_{04}}{p_a} = \frac{1\cdot284}{0\cdot265} = 4\cdot845$$

The critical pressure ratio, from equation (3.14), is

$$\frac{p_{04}}{p_c} = \frac{1}{\left[1 - \frac{1}{\eta_j}\left(\frac{\gamma-1}{\gamma+1}\right)\right]^{\gamma/(\gamma-1)}} = \frac{1}{\left[1 - \frac{1}{0\cdot95}\left(\frac{0\cdot333}{2\cdot333}\right)\right]^4} = 1\cdot914$$

Since $p_{04}/p_a > p_{04}/p_c$ the nozzle is choking.†

$$T_5 = T_c = \left(\frac{2}{\gamma+1}\right) T_{04} = \frac{2 \times 992 \cdot 3}{2 \cdot 333} = 850 \cdot 7 \text{ K}$$

$$p_5 = p_c = p_{04}\left(\frac{1}{p_{04}/p_c}\right) = \frac{1 \cdot 284}{1 \cdot 914} = 0 \cdot 671 \text{ bar}$$

$$\rho_5 = \frac{p_c}{RT_c} = \frac{100 \times 0 \cdot 671}{0 \cdot 287 \times 850 \cdot 7} = 0 \cdot 275 \text{ kg/m}^3$$

$$C_5 = (\gamma R T_c)^{\frac{1}{2}} = (1 \cdot 333 \times 0 \cdot 287 \times 850 \cdot 7 \times 1000)^{\frac{1}{2}} = 570 \cdot 5 \text{ m/s}$$

$$\frac{A_5}{m} = \frac{1}{\rho_5 C_5} = \frac{1}{0 \cdot 275 \times 570 \cdot 5} = 0 \cdot 006\,374 \text{ m}^2 \text{ s/kg}$$

The specific thrust is

$$F_s = (C_5 - C_a) + \frac{A_5}{m}(p_c - p_a)$$
$$= (570 \cdot 5 - 239 \cdot 6) + 0 \cdot 006\,374(0 \cdot 671 - 0 \cdot 265)10^5$$
$$= 330 \cdot 9 + 258 \cdot 8 = 589 \cdot 7 \text{ N s/kg}$$

From Fig. 2.17, with $T_{02} = 486 \cdot 8$ K and $T_{03} - T_{02} = 1200 - 486.8 = 713.2$ K, we find that the theoretical fuel/air ratio required is $0 \cdot 0194$. Thus the actual fuel/air ratio is

$$f = \frac{0 \cdot 0194}{0 \cdot 98} = 0 \cdot 0198$$

The specific fuel consumption is therefore

$$SFC = \frac{f}{F_s} = \frac{0 \cdot 0198 \times 3600}{589 \cdot 7} = 0 \cdot 121 \text{ kg/h N}$$

For cycle *optimization*, calculations would normally be done on the basis of specific thrust and *SFC*. A common problem, however, is the determination of actual engine performance to meet a particular aircraft thrust requirement. The engine designer needs to know the airflow, fuel flow and nozzle area; the airflow and nozzle area are also important to the aircraft designer who must determine the installation dimensions.

If, for example, the cycle conditions in the example were selected to meet a thrust requirement of 6000 N, then

$$m = \frac{F}{F_s} = \frac{6000}{589 \cdot 7} = 10 \cdot 17 \text{ kg/s}$$

The fuel flow is given by

$$m_f = fm = 0 \cdot 0198 \times 10 \cdot 17 \times 3600 = 725 \cdot 2 \text{ kg/h}$$

---

† The next example in this chapter shows how the calculation is performed when the nozzle is unchoked.

(It should be noted that fuel flow is normally measured and indicated in kg/h rather than kg/s.)

The nozzle area follows from the continuity equation:

$$A_5 = 0{\cdot}006\,374 \times 10{\cdot}17 = 0{\cdot}0648\,\mathrm{m}^2$$

Optimization of the turbojet cycle

When considering the design of a turbojet the basic thermodynamic parameters at the disposal of the designer are the turbine inlet temperature and the compressor pressure ratio. It is common practice to carry out a series of design point calculations covering a suitable range of these two variables, using fixed polytropic efficiencies for the compressor and turbine, and to plot *SFC* versus specific thrust with turbine inlet temperature T_{03} and compressor pressure ratio r_c as parameters. Such calculations may be made for several appropriate flight conditions of forward speed and altitude. Typical results applying to a subsonic cruise condition are shown in Fig. 3.12. The effects of turbine inlet temperature and compressor pressure ratio will be considered in turn.

It can be seen that specific thrust is strongly dependent on the value of T_{03}, and utilization of the highest possible temperature is desirable in order to keep the engine as small as possible for a given thrust. At a constant pressure ratio, however, an increase in T_{03} will cause some increase in

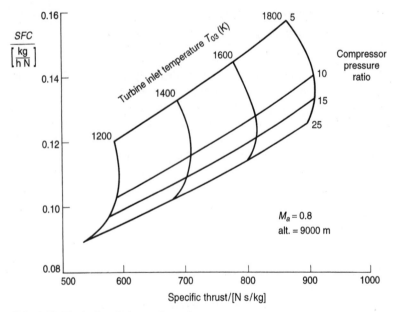

FIG. 3.12 Typical turbojet cycle performance

SFC. This is in contrast to the effect of T_{03} on shaft power cycle performance, where increasing T_{03} improves both specific power and *SFC* as discussed in section 2.4. Nevertheless the gain in specific thrust with increasing temperature is invariably more important than the penalty in increased *SFC*, particularly at high flight speeds where small engine size is essential to reduce both weight and drag.

The effect of increasing the pressure ratio r_c is clearly to reduce the *SFC*. At a fixed value of T_{03}, increasing the pressure ratio initially results in an increase in specific thrust but eventually leads to a decrease; and the optimum pressure ratio for maximum specific thrust increases as the value of T_{03} is increased. Evidently the effects of pressure ratio follow the pattern already observed to hold for shaft power cycles and need no further comment.

Figure 3.12 applies to a particular subsonic cruise condition. When such calculations are repeated for a higher cruising speed at the same altitude it is found that in general, for any given values of r_c and T_{03}, the *SFC* is increased and the specific thrust is reduced. These effects are due to the combination of an increase in inlet momentum drag and an increase in compressor work consequent upon the rise in inlet temperature. Corresponding curves for different altitudes show an increase in specific thrust and a decrease in *SFC* with increasing altitude, due to the fall in temperature and the resulting reduction in compressor work. Perhaps the most notable effect of an increase in the design cruise speed is that the optimum compressor pressure ratio for maximum specific thrust is reduced. This is because of the larger ram compression in the intake. The higher temperature at the compressor inlet and the need for a higher jet velocity make the use of a high turbine inlet temperature desirable—and indeed essential for economic operation of supersonic aircraft.

The thermodynamic optimization of the turbojet cycle cannot be isolated from mechanical design considerations, and the choice of cycle parameters depends very much on the type of aircraft. While high turbine temperatures are thermodynamically desirable they mean the use of expensive alloys and cooled turbine blades leading to an increase in complexity and cost, or to the acceptance of a decrease in engine life. The thermodynamic gains of increased pressure ratio must be considered in the light of increased weight, complexity and cost due to the need for more compressor and turbine stages and perhaps the need for a multi-spool configuration. Figure 3.13 illustrates the relation between performance and design considerations. A small business jet or trainer, for example, needs a simple, reliable engine of low cost; *SFC* is not critical because of the relatively small amount of flying done, and low pressure ratio turbojets of modest turbine inlet temperature were satisfactory for early aircraft. The Rolls-Royce Viper and General Electric J85, with values of around 5 to 6 and 1100 K, were widely used and remain in service at the turn of the century; indeed, the J85 is expected to remain in service in the T-38

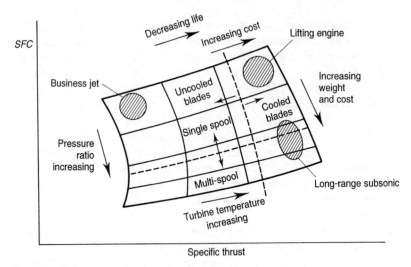

FIG. 3.13 Performance and design considerations

advanced trainer up to 2020. In recent years, however, noise restrictions have led to the displacement of the turbojet by the turbofan. For the business jet, this change has also been driven by the need for longer range, and the improved *SFC* of the turbofan permits an increased range for a given amount of fuel. Another example of interest was the development of specialized lifting engines for Vertical Take-Off and Landing (VTOL), where the prime requirement was for maximum thrust per unit weight and volume, with *SFC* less critical because of very low running times; these requirements were met using low pressure ratio units with a very high turbine inlet temperature (permissible because of the short life required). The compressor pressure ratio was governed by the maximum that could be handled by a single-stage turbine. This type of engine was not widely used because of the increased aircraft complexity, but engines of exceptionally high thrust/weight ratio were built in the early 1960s. Lastly, turbojets of high pressure ratio were used in early commercial aircraft and bombers because of the need for long range and hence low *SFC*. The Boeing B-52 entered service in 1955 with jet engines, and converted to low bypass ratio turbofans in 1961; the B-52 is scheduled to remain in service up to 2030. Turbojets have now been totally superseded by turbofans for commercial subsonic aircraft. The last remaining application of turbojets for commercial aircraft is in the Concorde, but they would not be considered for a new supersonic transport because of the take-off noise. In future, such aircraft would need an engine with take-off noise comparable to that of conventional turbofans. This will require the design and development of *variable cycle* engines capable of operating as turbofans during take-off and as turbojets (or very low bypass turbofans) at supersonic cruise. The subsonic fuel consumption of a supersonic transport is important because a considerable portion of any journey is flown at subsonic speeds, and it follows that the optimization procedure cannot be carried out around a single cruise condition.

Variation of thrust and SFC with flight conditions for a given engine

The reader is reminded that we have been discussing the results of design point cycle calculations. Curves such as those of Fig. 3.12 do not represent what happens to the performance of a particular engine when the turbine inlet temperature, forward speed or altitude differ from the design values. The method of arriving at such data is described in Chapter 8: here we will merely note some of the more important aspects of the behaviour of a turbojet.

At different flight conditions both the thrust and SFC will vary, due to the change in air mass flow with density and the variation of momentum drag with forward speed. Furthermore, even if the engine were run at a fixed rotational speed, the compressor pressure ratio and turbine inlet temperature would change with intake conditions. Typical variations of thrust and SFC with change in altitude and Mach number, for a simple turbojet operating at its maximum rotational speed, are shown in Fig. 3.14. It can be seen that thrust decreases significantly with increasing altitude, due to the decrease in ambient pressure and density, even though the specific thrust increases with altitude due to the favourable effect of the lower intake temperature. Specific fuel consumption, however, shows some improvement with increasing altitude. It will be shown in Chapter 8 that SFC is dependent upon ambient temperature, but not pressure, and hence its change with altitude is not so marked as that of thrust. It is obvious from the variation in thrust and SFC that the fuel consumption will be

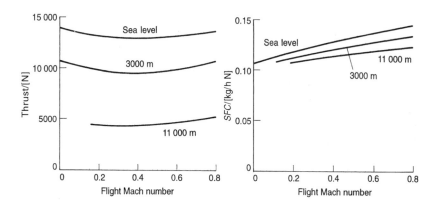

FIG. 3.14 Variation of thrust and SFC with Mach number and altitude for typical turbojet engine

greatly reduced at high altitudes. Reference to Fig. 3.14 shows that with increase of Mach number at a fixed altitude the thrust initially decreases, due to increasing momentum drag, and then starts to increase due to the beneficial effects of the ram pressure ratio; at supersonic Mach numbers this increase in thrust is substantial.

Indication of thrust

The thrust of a jet engine can be measured directly in a test facility using a strain-gauged load cell. It is obviously important that a pilot has an accurate indication of available thrust, particularly for take-off in critical conditions of runway length, aircraft weight and ambient conditions. Surprisingly, after 50 years of jet engine operations, there is still no satisfactory method of measuring thrust on an aircraft, and as a result it is necessary to use indirect methods of *indicating* the thrust from existing measurements of engine performance. Thrust variation is controlled by fuel flow, and will be limited by factors such as the maximum permissible values of rotational speed, turbine inlet temperature and fuel flow.

For a simple jet engine with a fixed area convergent nozzle, it can be shown that the thrust is related to the pressure ratio across the nozzle and this has been widely used for setting thrust on aircraft installations. Using the terminology of the previous worked example, the turbine exit conditions p_{04} and T_{04} will normally be measured, the ambient pressure p_a is known and the nozzle area A_5 is fixed. At take-off conditions, with the aircraft stationary, the nozzle can be assumed to be choked and the thrust is given by

$$F = mC_5 + A_5(p_5 - p_a)$$

From the continuity equation, equation of state and steady flow energy equation,

$$m = \rho_5 A_5 C_5; \quad \rho_5 = \frac{p_5}{RT_5}; \quad C_5^2 = 2C_p(T_{04} - T_5)$$

Rewriting and expressing in terms of pressure and temperature ratios,

$$F = \rho_5 A_5 C_5^2 + A_5(p_5 - p_a)$$

$$= \frac{p_5}{RT_5} A_5 \times 2C_p(T_{04} - T_5) + A_5(p_5 - p_a)$$

$$= \frac{2C_p}{RT_5} A_5 p_5 (T_{04} - T_5) + A_5(p_5 - p_a)$$

$$= 2\frac{\gamma}{\gamma - 1} A_5 \left(\frac{p_5}{p_{04}}\right) \left(\frac{p_{04}}{p_a}\right) p_a \left(\frac{T_{04}}{T_5} - 1\right) + A_5 p_a \left(\frac{p_5}{p_{04}} \cdot \frac{p_{04}}{p_a} - 1\right)$$

Recognizing that p_{04}/p_5 is the critical pressure ratio, this can be expressed as

$$\frac{F}{A_5 p_a} = \left[2\frac{\gamma}{\gamma-1}\cdot\left(\frac{2}{\gamma+1}\right)^{\frac{\gamma}{\gamma-1}}\left(\frac{p_{04}}{p_a}\right)\left(\frac{\gamma+1}{2}-1\right)\right] + \left\{\left(\frac{2}{\gamma+1}\right)^{\frac{\gamma}{\gamma-1}}\frac{p_{04}}{p_a}-1\right\}$$

$$= \frac{p_{04}}{p_a}\left\{\left(\frac{2}{\gamma+1}\right)^{\frac{\gamma}{\gamma-1}}\left[\left(\frac{2\gamma}{\gamma-1}\right)\left(\frac{\gamma-1}{2}\right)+1\right]\right\} - 1$$

Thus we can express $F/A_5 p_a$ as a function of p_{04}/p_a and γ, giving

$$\frac{F}{A_5 p_a} = \frac{p_{04}}{p_a}\left\{\left(\frac{2}{\gamma+1}\right)^{\frac{\gamma-1}{\gamma}}(\gamma+1)\right\} - 1$$

This is a very simple relationship of the form

$$\frac{F}{A_5 p_a} = K\left[\frac{p_{04}}{p_a}\right] - 1 \quad \text{where } K = f(\gamma)$$

This relationship is suitable for indicating the thrust at static conditions, but there is also a requirement for thrust indication in flight. The ratio of the turbine exit pressure to the compressor inlet pressure (p_{04}/p_{01}) is referred to as the *engine pressure ratio* (*EPR*). The ratio of the compressor inlet pressure to the ambient pressure (p_{01}/p_a) is referred to as the *ram pressure ratio* (*RPR*). The nozzle pressure ratio can be expressed as

$$\frac{p_{04}}{p_a} = \frac{p_{04}}{p_{01}}\cdot\frac{p_{01}}{p_a} = EPR \times RPR$$

It then follows that

$$\left(\frac{F}{A_5 p_a}+1\right)\bigg/ RPR = K(EPR)$$

where $K = 1{\cdot}2594$ for $\gamma = 1{\cdot}333$.

EPR has been widely used as the parameter for setting thrust on jet engines, being easily measured and reliable.

3.4 The turbofan engine

The turbofan engine was originally conceived as a method of improving the propulsion efficiency of the jet engine by reducing the mean jet velocity, particularly for operation at high subsonic speeds. It was soon apparent, however, that a lower jet velocity also resulted in less jet noise, an important factor when large numbers of jet-propelled aircraft entered commercial service. In the turbofan a portion of the total flow *bypasses* part of the compressor, combustion chamber, turbine and nozzle before being ejected through a separate nozzle as shown in Fig. 3.15. Thus the thrust is made up of two components, the cold stream (or *fan*) thrust and the hot stream thrust. Figure 3.15 shows an engine with separate exhausts, but it is sometimes desirable to mix the two streams and eject them as a single jet of reduced velocity.

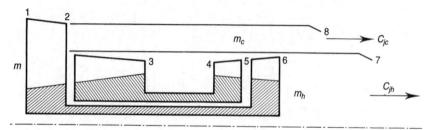

FIG. 3.15 Twin-spool turbofan engine

Turbofan engines are usually described in terms of *bypass ratio*, defined as the ratio of the flow through the bypass duct (cold stream) to the flow at entry to the high-pressure compressor (hot stream).† With the notation of Fig. 3.15 the bypass ratio B is given by

$$B = \frac{m_c}{m_h}$$

It immediately follows that

$$m_c = \frac{mB}{B+1}, \quad m_h = \frac{m}{B+1} \quad \text{and} \quad m = m_c + m_h$$

For the particular case where both streams are expanded to atmospheric pressure in the propelling nozzles, the net thrust is given by

$$F = (m_c C_{jc} + m_h C_{jh}) - mC_a$$

The design point calculations for the turbofan are similar to those for the turbojet; in view of this, only the differences in calculation will be outlined.

(a) Overall pressure ratio and turbine inlet temperature are specified as before, but it is also necessary to specify the bypass ratio (B) and the fan pressure ratio (*FPR*).

(b) From the inlet conditions and *FPR*, the pressure and temperature of the flow leaving the fan and entering the bypass duct can be calculated. The mass flow down the bypass duct can be established from the total flow and the bypass ratio. The cold stream thrust can then be calculated as for the jet engine, noting that air is the working fluid. It is necessary to check whether the fan nozzle is choked or unchoked; if choked the pressure thrust must be calculated.

(c) In the two-spool configuration shown in Fig. 3.15 the fan is driven by the LP turbine. Calculations for the HP compressor and turbine are

---

† The terms turbofan and bypass engine may both be encountered, often referring to the same engine. Early engines with a small portion of the flow bypassing the combustion chamber (low value of bypass ratio) were initially referred to as bypass engines. As the bypass ratio is increased, the optimum pressure ratio for the bypass stream is reduced and can eventually be provided by a single compressor stage. The term turbofan was originally used for engines of high bypass ratio but is increasingly employed for all bypass engines.

quite standard, and conditions at entry to the LP turbine can then be found. Considering the work requirement of the low pressure rotor,

$$mc_{pa}\Delta T_{012} = \eta_m m_h c_{pg}\Delta T_{056}$$

and hence

$$\Delta T_{056} = \frac{m}{m_h} \times \frac{c_{pa}}{\eta_m c_{pg}} \times \Delta T_{012} = (B+1) \times \frac{c_{pa}}{\eta_m c_{pg}} \times \Delta T_{012}$$

The value of B may range from 0·3 to 8 or more, and its value has a major effect on the temperature drop and pressure ratio required from the LP turbine. Knowing T_{05}, η_t and ΔT_{056}, the LP turbine pressure ratio can be found and conditions at entry to the hot stream nozzle can be established; the calculation of the hot stream thrust is then quite straightforward.

(d) If the two streams are mixed it is necessary to find the conditions after mixing by means of an enthalpy and momentum balance; this will be considered following an example on the performance of an engine similar to that shown in Fig. 3.15.

EXAMPLE 3.2

The following data apply to a twin-spool turbofan engine, with the fan driven by the LP turbine and the compressor by the HP turbine. Separate cold and hot nozzles are used.

Overall pressure ratio	25·0
Fan pressure ratio	1·65
Bypass ratio m_c/m_h	5·0
Turbine inlet temperature	1550 K
Fan, compressor and turbine *polytropic* efficiency	0·90
Isentropic efficiency of each propelling nozzle	0·95
Mechanical efficiency of each spool	0·99
Combustion pressure loss	1·50 bar
Total air mass flow	215 kg/s

It is required to find the thrust and SFC under sea-level static conditions where the ambient pressure and temperature are 1·0 bar and 288 K.

The values of $(n-1)/n$ for the polytropic compression and expansion are:

for compression, $\dfrac{n-1}{n} = \dfrac{1}{\eta_{\infty c}}\left(\dfrac{\gamma-1}{\gamma}\right)_a = \dfrac{1}{0\cdot9 \times 3\cdot5} = 0\cdot3175$

for expansion, $\dfrac{n-1}{n} = \eta_{\infty t}\left(\dfrac{\gamma-1}{\gamma}\right)_g = \dfrac{0\cdot9}{4} = 0\cdot225$

Under static conditions $T_{01} = T_a$ and $p_{01} = p_a$ so that, using the nomenclature of Fig. 3.15,

$$\frac{T_{02}}{T_{01}} = \left(\frac{p_{02}}{p_{01}}\right)^{(n-1)/n} \quad \text{yields } T_{02} = 288 \times 1.65^{0.3175} = 337.6 \text{ K}$$

$$T_{02} - T_{01} = 337.6 - 288 = 49.6 \text{ K}$$

$$\frac{p_{03}}{p_{02}} = \frac{25.0}{1.65} = 15.15$$

$$T_{03} = T_{02}\left(\frac{p_{03}}{p_{02}}\right)^{(n-1)/n} = 337.6 \times 15.15^{0.3175} = 800.1 \text{ K}$$

$$T_{03} - T_{02} = 800.1 - 337.6 = 462.5 \text{ K}$$

The cold nozzle pressure ratio is

$$\frac{p_{02}}{p_a} = FPR = 1.65$$

and the critical pressure ratio for this nozzle is

$$\frac{p_{02}}{p_c} = \frac{1}{\left[1 - \frac{1}{\eta_j}\left(\frac{\gamma-1}{\gamma+1}\right)\right]^{\gamma/(\gamma-1)}} = \frac{1}{\left[1 - \frac{1}{0.95}\left(\frac{0.4}{2.4}\right)\right]^{3.5}} = 1.965$$

Thus the cold nozzle is not choking, so that $p_8 = p_a$ and the cold thrust F_c is given simply by

$$F_c = m_c C_8$$

The nozzle temperature drop, from equation (3.12), is

$$T_{02} - T_8 = \eta_j T_{02}\left[1 - \left(\frac{1}{p_{02}/p_a}\right)^{(\gamma-1)/\gamma}\right]$$

$$= 0.95 \times 337.6\left[1 - \left(\frac{1}{1.65}\right)^{1/3.5}\right] = 42.8 \text{ K}$$

and hence

$$C_8 = \left[2c_p(T_{02} - T_8)\right]^{\frac{1}{2}} = (2 \times 1.005 \times 42.8 \times 1000)^{\frac{1}{2}} = 293.2 \text{ m/s}$$

Since the bypass ratio B is 5.0,

$$m_c = \frac{mB}{B+1} = \frac{215 \times 5.0}{6.0} = 179.2 \text{ kg/s}$$

$$F_c = 179.2 \times 293.2 = 52\,532 \text{ N}$$

Considering the work requirement of the HP rotor,

$$T_{04} - T_{05} = \frac{c_{pa}}{\eta_m c_{pg}}(T_{03} - T_{02}) = \frac{1 \cdot 005 \times 462 \cdot 5}{0 \cdot 99 \times 1 \cdot 148} = 409 \cdot 0 \text{ K}$$

and for the LP rotor

$$T_{05} - T_{06} = (B + 1)\frac{c_{pa}}{\eta_m c_{pg}}(T_{02} - T_{01}) = \frac{6 \cdot 0 \times 1 \cdot 005 \times 49 \cdot 6}{0 \cdot 99 \times 1 \cdot 148} = 263 \cdot 2 \text{ K}$$

Hence

$$T_{05} = T_{04} - (T_{04} - T_{05}) = 1550 - 409 \cdot 0 = 1141 \cdot 0$$
$$T_{06} = T_{05} - (T_{05} - T_{06}) = 1141 \cdot 0 - 263 \cdot 2 = 877 \cdot 8$$

p_{06} may then be found as follows.

$$\frac{p_{04}}{p_{05}} = \left(\frac{T_{04}}{T_{05}}\right)^{n/(n-1)} = \left(\frac{1550}{1141 \cdot 0}\right)^{1/0 \cdot 225} = 3 \cdot 902$$

$$\frac{p_{05}}{p_{06}} = \left(\frac{T_{05}}{T_{06}}\right)^{n/(n-1)} = \left(\frac{1141 \cdot 0}{877 \cdot 8}\right)^{1/0 \cdot 225} = 3 \cdot 208$$

$$p_{04} = p_{03} - \Delta p_b = 25 \cdot 0 \times 1 \cdot 0 - 1 \cdot 50 = 23 \cdot 5 \text{ bar}$$

$$p_{06} = \frac{p_{04}}{(p_{04}/p_{05})(p_{05}/p_{06})} = \frac{23 \cdot 5}{3 \cdot 902 \times 3 \cdot 208} = 1 \cdot 878 \text{ bar}$$

Thus the hot nozzle pressure ratio is

$$\frac{p_{06}}{p_a} = 1 \cdot 878$$

while the critical pressure ratio is

$$\frac{p_{06}}{p_c} = \frac{1}{\left[1 - \frac{1}{0 \cdot 95}\left(\frac{0 \cdot 333}{2 \cdot 333}\right)\right]^4} = 1 \cdot 914$$

This nozzle is also unchoked, and hence $p_7 = p_a$.

$$T_{06} - T_7 = \eta_j T_{06}\left[1 - \left(\frac{1}{p_{06}/p_a}\right)^{(\gamma-1)/\gamma}\right]$$

$$= 0 \cdot 95 \times 877 \cdot 8\left[1 - \left(\frac{1}{1 \cdot 878}\right)^{\frac{1}{4}}\right] = 121 \cdot 6 \text{ K}$$

$$C_7 = [2c_p(T_{06} - T_7)]^{\frac{1}{2}} = [2 \times 1 \cdot 148 \times 121 \cdot 6 \times 1000]^{\frac{1}{2}}$$
$$= 528 \cdot 3 \text{ m/s}$$

$$m_h = \frac{m}{B + 1} = \frac{215}{6 \cdot 0} = 35 \cdot 83 \text{ kg/s}$$

$$F_h = 35 \cdot 83 \times 528 \cdot 3 = 18 \, 931 \text{ N}$$

Thus the total thrust is

$$F_c + F_h = 52\,532 + 18\,931 = 71\,463\,\text{N or } 71.5\,\text{kN}$$

The fuel flow can readily be calculated from the known temperatures in the combustor and the airflow through the combustor, i.e. m_h. The combustion temperature rise is $(1550 - 800) = 750\,\text{K}$ and the combustion inlet temperature is 800 K. From Fig. 2.17 the ideal fuel/air ratio is found to be 0.0221 and the actual fuel/air ratio is then $(0.0221/0.99) = 0.0223$. Hence the fuel flow is given by

$$m_f = 0.0223 \times 35.83 \times 3600 = 2876.4\,\text{kg/h}$$

and

$$SFC = \frac{2876.4}{71\,463} = 0.0403\,\text{kg/h N}$$

Because both nozzles were unchoked, the thrust could be evaluated without calculating the nozzle areas. It is always good practice, however, to calculate key pieces of information which may be required for other purposes. In both cases the area can be calculated from continuity, i.e. $m = \rho AC$. The density is obtained from $\rho = p/RT$, where p and T are the static values in the plane of the nozzle; for both nozzles p will equal p_a. The following results are obtained for the two streams:

	Cold	Hot
Static pressure (bar)	1.0	1.0
Static temperature (K)	192.6	749.8
Density (kg/m^3)	1.191	0.4647
Mass flow (kg/s)	179.2	35.83
Velocity (m/s)	293.2	528.3
Nozzle area (m^2)	0.5132	0.1459

The cold nozzle area is much larger than the hot one, and Fig. 1.12(b) shows the physical appearance of an engine of similar cycle and bypass ratio; Fig. 1.12(a) shows an engine of lower bypass ratio, around 2.5.

This example illustrated the method followed when a propelling nozzle is unchoked, while the previous example showed how a choked nozzle may be dealt with.

Note that at static conditions the bypass stream contributes approximately 74 per cent of the total thrust. At a forward speed of 60 m/s, which is approaching a normal take-off speed, the momentum drag mC_a will be 215×60 or 12 900 N; the ram pressure ratio and temperature rise will be negligible and thus the net thrust is reduced to 58 563 N. The drop in thrust during take-off is even more marked for engines of higher bypass ratio and for this reason it is preferable to quote turbofan thrusts at a typical take-off speed rather than at static conditions.

Mixing of hot and cold streams

Mixing is essential for an afterburning turbofan when maximum thrust boosting is required, to avoid the need for two reheat combustion systems. In certain cases mixing may also be advantageous in subsonic transport applications, giving a small but significant gain in *SFC*. We shall present a simple method of dealing with mixing in a constant area duct, with no losses and assuming adiabatic flow. The duct is shown schematically in Fig. 3.16, with the hot and cold flows beginning to mix at plane A and with complete mixing achieved by plane B.

Starting from the enthalpy balance, with suffix *m* denoting the properties of the mixed stream,

$$m_c c_{pc} T_{02} + m_h c_{ph} T_{06} = m c_{pm} T_{07}$$

where $m = m_c + m_h$.

We also have the following equations that relate the properties of a mixture of gases to those of its constituents:

$$c_{pm} = \frac{m_c c_{pc} + m_h c_{ph}}{(m_c + m_h)}$$

$$R_m = \frac{m_c R_c + m_h R_h}{(m_c + m_h)}$$

$$\left(\frac{\gamma}{\gamma - 1}\right)_m = \frac{R_m}{c_{pm}}$$

From the momentum balance,

$$(m_c C_c + p_2 A_2) + (m_h C_h + p_6 A_6) = m C_7 + p_7 A_7$$

If there is no swirl in the jet pipe downstream of plane A, the *static* pressure will be uniform across the duct, and so $p_2 = p_6$.

From continuity,

$$m = \rho_7 C_7 A_7$$

It is the pressure after mixing, p_{07}, that is required for the cycle calculation, because this is the stagnation pressure at entry to the propelling nozzle. When this has been found, the calculation for the thrust is the same as described previously. The calculation of p_{07} is simplified if

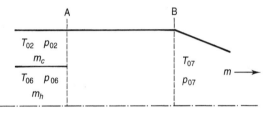

FIG. 3.16 Mixing in a constant area duct

we work in terms of the Mach numbers in the hot and cold streams: the hot stream Mach number is fixed by the turbine design, and typically M_6 will be about 0·5. Having selected a value of M_6, the procedure is as follows. We will make use of the standard relations between M and static and stagnation p and T, equations (8) in Appendix A.2.

(a) Knowing M_6, T_{06} and p_{06}, we can determine p_6 and C_h: p_6 and T_6 give ρ_6, hence A_6 follows from continuity. We now know $(m_h C_h + p_6 A_6)$.

(b) With $p_6 = p_2$, p_2/p_{02} yields the value of M_2. With M_2, p_{02} and T_{02} known we can now find C_c and A_2 so that $(m_c C_c + p_2 A_2)$ is known.

(c) $(mC_7 + A_7 p_7)$ is now obtained from the momentum balance.

(d) $A_7 = A_6 + A_2$, and $m = \rho C_7 A_7 = \dfrac{p_7}{R_m T_7} C_7 A_7$

(e) T_{07} is known from the enthalpy balance, but neither p_7 nor p_{07} is known. Guess a value of M_7, and then find T_7 and C_7; continuity then gives p_7.

(f) It is now necessary to check that $(mC_7 + A_7 p_7)$ is equal to the value obtained from the momentum balance in (c).

(g) Iterate on M_7 until correct p_7 is found.

(h) p_{07} is then obtained from p_7, M_7.

The fan outlet pressure (p_{02}) should be only slightly higher than the turbine outlet pressure (p_{06}) to keep mixing losses to a minimum; typically, p_{02}/p_{06} should be about 1·05–1·07. In practice, quite small changes in cycle parameters can cause significant changes in the ratio p_{02}/p_{06} and negate the benefits of mixing. No hard and fast rules can be given and the decision on whether to use mixing or not will also be influenced by installation and engine weight considerations, combined with a detailed investigation of the pressure losses caused by the mixer; Ref. (7) describes an experimental investigation of mixing losses.

Turbofans for use in fighter aircraft use mixed exhausts; typically they have a low bypass ratio, say 0·3 to 0·5, and are much closer in configuration to the straight jet engine than commercial high bypass engines. Fighter engines require *afterburning*† for take-off and combat manoeuvring and the bypass and core flows are mixed before entering the afterburner. The GE F404 is shown in Fig. 3.17. Noting that the fan exit and turbine exit pressures must be similar, it follows that the fan pressure ratio (*FPR*) and engine pressure ratio (*EPR*) are almost equal. It was shown in a previous section that the thrust of a simple jet engine is related directly to *EPR* and it follows that for a low bypass military engine thrust will basically be determined by the *FPR*. Modern military engines typically use a three-stage fan with a pressure ratio of 3·5–4·0. Calculated values of specific thrust as a function of *FPR* are shown in Fig. 3.18, with the main

† Afterburning will be covered later in this chapter.

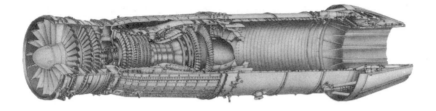

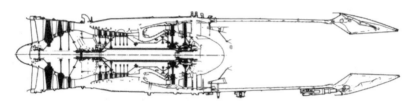

FIG. 3.17 F404 military turbofan [courtesy GE Aircraft Engines]

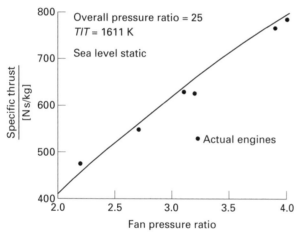

FIG. 3.18 Specific thrust v. fan pressure ratio

parameters as shown; the bypass ratio is selected to give matching pressures in the bypass and core streams. The full curve is from calculations and the six symbols are data from actual engines obtained from published data. Useful information on selection of cycles for military engines is given in Ref. (8).

Optimization of the turbofan cycle

Designers of turbofans have four thermodynamic parameters at their disposal: overall pressure ratio and turbine inlet temperature (as for the simple turbojet), and also bypass ratio and fan pressure ratio.

Optimization of the cycle is somewhat complex but the basic principles are easily understood.

Let us consider an engine with the overall pressure ratio and bypass ratio both specified. If we select a value of turbine inlet temperature the energy input is fixed because the combustion chamber airflow and entry temperature are determined by the chosen operating conditions. The remaining variable is the fan pressure ratio and as a first step it is necessary to consider the variation of specific thrust and specific fuel consumption with *FPR*. If we start with a low value of *FPR*, the fan thrust will be small and the work extracted from the LP turbine will also be small; thus little energy will be extracted from the hot stream and a large value of hot thrust will result. As the *FPR* is raised it is evident that the fan thrust will increase and the hot thrust will decrease. A typical variation of specific thrust and *SFC* with *FPR*, for a range of turbine temperature, is shown in Fig. 3.19. It can be seen that for any value of turbine inlet temperature there will be an optimum value of *FPR*; optimum values of *FPR* for minimum *SFC* and maximum specific thrust coincide because of the fixed energy input. Taking the values of *SFC* and specific thrust for each of these values of *FPR* in turn, a curve of *SFC* against specific thrust may be plotted as shown in Fig. 3.20(a). Note that each point on this curve is the result of a previous optimization, and is associated with a particular value of *FPR* and turbine inlet temperature.

The foregoing calculations may be repeated for a series of bypass ratios, still at the same overall pressure ratio, to give a family of curves as shown in Fig. 3.20(b). This plot yields the optimum variation of *SFC* with specific thrust for the selected overall pressure ratio as shown by the dotted envelope curve. The procedure can then be repeated for a range of overall pressure ratio. It will be clear that the optimization procedure is lengthy and that a large amount of detailed calculation is necessary. The

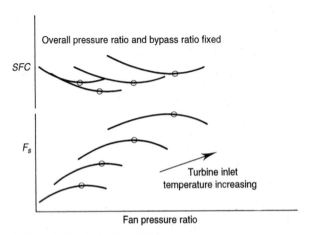

FIG. 3.19 Optimization of fan pressure ratio

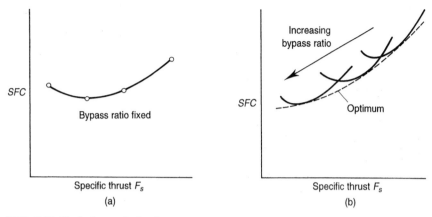

FIG. 3.20 Turbofan optimization

qualitative results of such a series of calculations can be summarized as follows:

(a) Increasing bypass ratio improves *SFC* at the expense of a significant reduction in specific thrust.
(b) The optimum fan pressure ratio increases with turbine inlet temperature.
(c) The optimum fan pressure ratio decreases with the increase of bypass ratio.
(d) The optimum *SFC* requires *low* specific thrust, this being particularly important for high bypass turbofans.

Civil engines will always use a single-stage fan to minimise noise and the pressure ratio will be about 1·5 to 1·8, being dependent on the overall cycle parameters and flight condition. Examples are shown in Figs 1.12(a) and (b) and 1.16. Military engines may have two- or three-stage fans with pressure ratios as high as 4·0.

The choice of cycle parameters is dependent on the aircraft application, and both high and low bypass ratios have their place. Specific fuel consumption is of major importance for long-range subsonic transport aircraft and the requirement can best be met by using a bypass ratio of 4–6 and a high overall pressure ratio, combined with a high turbine inlet temperature. A higher bypass ratio of 8–9 was chosen for the GE90, which entered service in 1995, representing the first major change in this parameter for many years. Military aircraft with a supersonic dash capability and a requirement for good subsonic *SFC* would use a much lower bypass ratio, perhaps 0·5–1, to keep the frontal area down, and afterburning would be used for supersonic operation. Engines currently under development, with significantly higher values of *TIT*, will permit operation up to speeds of about *M* 1·4 without the use of afterburning. Short-haul commercial aircraft are not as critical as long-haul aircraft regarding *SFC* and for many years bypass ratios of around 1 were used; modern designs, however, use higher bypass ratios similar to those used in long-haul aircraft. The prime reason for this change is the significant decrease in engine noise resulting from increased bypass ratio. Another reason for concentrating on engines of low *SFC*, so that they are suitable for a wide range of aircraft, is the escalating cost of developing new engines.

Figure 3.20(b) showed that optimizing the *SFC* required the use of high values of *BPR* resulting in engines of *low* specific thrust. These curves apply only to *uninstalled* engine performance, and installation effects must be considered for evaluation in a specific aircraft. High *BPR* engines would have a large-diameter fan and both diameter and weight would increase with *BPR*; ground clearance effects would cause increases in undercarriage length and weight, increasing aircraft weight and thrust required. A mock-up of the nacelle for a Rolls-Royce Trent is shown in Fig. 3.21. The flattened portion on the intake is to ease the ground clearance problem, and the thrust reverser doors are shown in the open position; the thrust reversers only operate on the fan stream. The nacelle must be designed for low drag and also easy maintenance access to accessories, such as gearboxes, oil tanks and the engine control system.

The combination of the nacelle and its support system is normally referred to as the 'pod' and it is instructive to do a simple analysis on the effect of pod drag. The propulsion efficiency will be modified to give

$$\eta_p = \frac{2}{1 + \dfrac{C_j}{C_a}} \left[\frac{\text{net thrust} - \text{pod drag}}{\text{net thrust}} \right]$$

The ideal propulsion efficiency can be expressed in a different form as

$$\eta_p = \frac{2}{1 + \dfrac{\text{gross thrust}}{\text{momentum drag}}} = \frac{2 \times \text{momentum drag}}{\text{momentum drag} + \text{gross thrust}}$$

$$= \frac{2 \times \text{momentum drag}}{2 \times \text{momentum drag} + \text{net thrust}} = \frac{2}{2 + \dfrac{\text{net thrust}}{\text{momentum drag}}}$$

Thus, including the effect of pod drag

$$\eta_p = \frac{2}{2 + \dfrac{\text{net thrust}}{\text{momentum drag}}} \left[\frac{\text{net thrust} - \text{pod drag}}{\text{net thrust}} \right]$$

$$= \frac{2}{2 + \dfrac{\text{net thrust}}{\text{momentum drag}}} \left[1 - \frac{\left(\dfrac{\text{pod drag}}{\text{momentum drag}} \right)}{\left(\dfrac{\text{net thrust}}{\text{momentum drag}} \right)} \right]$$

FIG. 3.21 Nacelle for high bypass turbofan [courtesy Rolls-Royce plc]

The ratio of (net thrust/momentum drag) is directly related to the bypass ratio, decreasing with increase in *BPR*. Using α to denote (pod drag/ momentum drag), the effect of pod drag on η_p can be readily evaluated giving the results shown in Fig. 3.22. It can be seen that at high values of *BPR* the effect of pod drag is significant and the installation effects must be carefully evaluated by the aircraft manufacturer, working in conjunction with the engine manufacturer.

It was mentioned earlier that, because of their reduced mean jet velocity, turbofans produce less exhaust noise than turbojets. At first sight it would appear that noise considerations would demand the highest possible bypass ratio, resulting in a low jet velocity. Unfortunately, however, as bypass ratio is increased the resulting high tip speed of the fan leads to a large increase in fan noise. Indeed at approach conditions, with the engine operating at a low thrust setting, the fan noise predominates; fan noise is essentially produced at discrete frequencies which can be much more irritating than the broad-band jet noise. The problem can be alleviated by

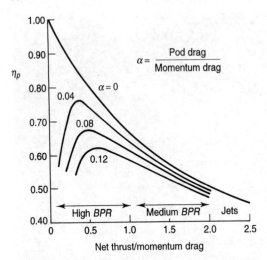

FIG. 3.22 **Effect of pod drag on propulsion efficiency**

acoustic treatment of the intake duct, avoiding the use of inlet guide vanes, and careful choice of axial spacing between the fan rotor and stator blades.

Indication of thrust

It was shown earlier that *EPR* was a convenient measurement for indicating thrust on simple turbojets. Most large turbofans, however, use separate exhausts and this complicates the issue of thrust indication. In Example 3.2, it was shown that at take-off the cold stream contributed approximately 74 per cent of the total thrust, with only 26 per cent from the hot stream. The use of the conventionally defined *EPR*, i.e. turbine exit pressure divided by compressor inlet pressure, provides an indication of *hot stream* thrust only. *EPR* was widely used on the early low bypass turbofans, such as the Pratt and Whitney JT-8D, which was built in very large numbers.

The advent of the high bypass turbofan led to different approaches by the various engine manufacturers. Pratt and Whitney continued to use *EPR* on their large engines, but Rolls-Royce selected Integrated Engine Pressure Ratio (*IEPR*) for the RB-211, and GE selected fan speed (N_1) for the CF-6; *IEPR* is a weighted mean value given by

$$IEPR = \frac{\text{fan pressure} \times \text{fan outlet area} + \text{nozzle pressure} \times \text{hot nozzle area}}{\text{fan outlet area} + \text{hot nozzle area}}$$

These different methods give comparable results and all have their advantages and disadvantages. *IEPR*, for example, is the most rigorous method but requires extensive instrumentation of the fan flow. *EPR* is familiar to many operators, but has the disadvantage of measuring only a small component of the thrust. Fan speed is probably the simplest to install; although it does not actually give a direct indication of thrust, this can be calibrated on a test bed. These three methods are compared in Ref. (9); it should be recognized that whatever thrust indication system is used, it should be capable of showing in-service degradation of thrust.

Turbofan configurations

The cycle parameters for a turbofan have a much greater effect on the mechanical design of the engine than in the case of the turbojet. This is because variation in bypass ratio implies variation in component diameters and rotational speeds, and the configuration of engines of low and high bypass ratio may be completely different.

Some early turbofans were directly developed from existing turbojets, and this led to the 'aft–fan' configuration shown in Fig. 3.23. A combined turbine-fan was mounted downstream of the gas-generator turbine. Two major problems arise with this configuration. Firstly, the blading of the turbine–fan unit must be designed to give turbine blade sections for the hot stream and compressor blade sections for the cold stream. This obviously leads to blading of high cost and, because the entire blade must be made from the turbine material, of high weight. The other problem is that of sealing between the two streams. The aft-fan configuration has not been used in a new design for many years, but is a possible contender for ultra high bypass (UHB) engines. In this case, however, the turbine and fan sections would probably be connected by a gearbox and the complex blades used in earlier engines would not be required.

For moderate bypass and overall pressure ratios the simple two-spool arrangement of Fig. 3.15 is adequate. At very high bypass ratios, especially when combined with high overall pressure ratios, design problems arise because the fan rotational speed must be much lower than that of the high-pressure rotor; the important limitations on blade tip speed will be discussed in Chapter 5.

Four different configurations which may be used to obtain high bypass ratio and high overall pressure ratio are shown in Fig. 3.24. The configuration of Fig. 3.24(a) suffers from the fact that the later stages on

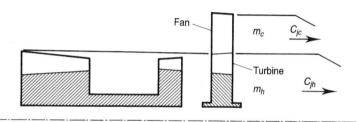

FIG. 3.23 Aft-fan configuration

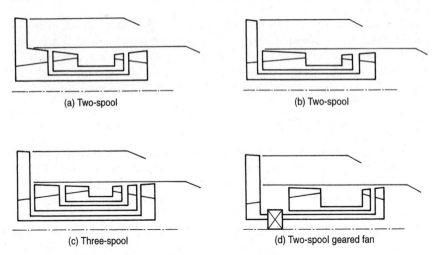

(a) Two-spool

(b) Two-spool

(c) Three-spool

(d) Two-spool geared fan

FIG. 3.24 Configurations for high bypass ratio turbofans

the LP rotor, usually called 'booster stages', contribute little because of their low blade speed. Scheme (b) is more attractive, but requires a very high pressure ratio from the HP compressor which leads to the instability problems referred to in section 1.2. The three-spool arrangement of Fig. 3.24(c) is in many ways the most attractive concept, with a modest pressure ratio from each compressor. All of these have been used for large turbofans; the V 2500 (Fig. 1.12(b)) is an example of (a) while the Rolls-Royce Trent (Fig. 1.16) is an example of (c). A geared-fan arrangement as in (d) is possible for smaller engines, and units of this sort have been developed from a turboprop background. The power requirements for the fan of a large turbofan may be about 60 MW and the design of a lightweight gearbox to handle this power would be difficult; design studies, however, are being carried out for ultra high bypass ratio (UHB) engines with variable pitch fans, where the low rotational speed of the large-diameter fan requires a gearbox to avoid using a large number of turbine stages.

3.5 The turboprop engine

The turboprop engine differs from the shaft power units dealt with in Chapter 2 in that some of the useful output appears as jet thrust. In this case, therefore, it is necessary to combine shaft power and jet thrust. This can be done in a number of ways, but in all cases a knowledge of the aircraft speed is involved.

Power must eventually be delivered to the aircraft in the form of thrust power, just as it is with a piston engine driving a propeller. The *thrust power* (*TP*) can be expressed in terms of shaft power (*SP*), propeller efficiency η_{pr} and jet thrust F by

$$TP = (SP)\eta_{pr} + FC_a$$

In practice the shaft power will account for a large proportion of the enthalpy drop available at the gas-generator exit, and thrust power is therefore largely dependent on the propeller efficiency which may vary significantly with flight conditions.

It is desirable to find some way of expressing the power so that it can be readily compared with that of a piston engine, and so that it is not quite so dependent on propeller efficiency. The same basic engine may be used in conjunction with a variety of propellers for different applications, and it is the performance of the engine itself which is then of most interest to us. A more suitable way of expressing the power is to quote the *equivalent* (or *effective*) *power EP* defined as

$$EP = \frac{TP}{\eta_{pr}} = SP + \frac{FC_a}{\eta_{pr}}$$

η_{pr} now affects only the smaller term. The equivalent power is an arbitrarily defined quantity and it should not be quoted without reference to the flight speed. Note that by definition the *EP* and *SP* are equal at static conditions, although there is some beneficial jet thrust. Allowance must be made for this when comparing engines under take-off conditions. Experiments have shown that an average propeller produces a thrust of about 8·5 N per kW of power input under static conditions, so that the *take-off EP* is conventionally taken as $SP + (F/8·5)$ with *SP* in kW and *F* in newtons.

Turboprops are usually rated on the basis of equivalent power at take-off conditions, and the specific fuel consumption and specific power are often expressed in terms of that equivalent power. It is nevertheless desirable to quote both the shaft power and jet thrust available at any condition of interest and it should be recognized that the equivalent power is merely a useful, but artificial, concept. In view of the similarity between the turboprop and the shaft power units discussed in Chapter 2 it is not necessary to elaborate on cycle requirements. The only basic difference is that the designer can choose the proportions of the available enthalpy drop used to produce shaft power and jet thrust. It can be shown that there is an optimum division for any given flight speed and altitude; a simple rule of thumb is to design so that the turbine exit pressure is equal to the compressor inlet pressure. The equivalent power is not particularly sensitive to turbine exit pressure in this range. Turboprops generally operate with the nozzle unchoked and use a simple straight-through tailpipe rather than a convergent nozzle.

The combined efficiency of the power turbine, propeller, and the necessary reduction gear, is well below that of an equivalent propelling nozzle. It follows that η_e for a turboprop engine is lower than that of a turbojet or turbofan engine. The turboprop has held its position for speeds up to M 0·6 because the propulsion efficiency is so much higher than that of the turbojet; the turboprop is widely used in business aircraft and regional airliners, mostly at power levels of 500–2000 kW. The propeller efficiency, using conventional propeller design methods, decreases drastically at flight speeds above M 0·6 and for this reason the turboprop did not become widely used for longer haul aircraft, being superseded by turbofans of equivalent propulsion efficiency; an exception was long-endurance patrol aircraft, which fly to the search area at M 0·6 and then loiter at much lower flight speeds. In the early 1980s, however, considerable attention was focused on the design of propellers with the goal of obtaining propeller efficiencies of 0·80 at flight Mach numbers of 0·8. The successful development of such propellers would give large savings in fuel compared to existing turbofans. The design of the propeller was markedly different from conventional designs, using highly swept supersonic blading and 8–10 blades; these devices were called 'propfans' to differentiate them from conventional propellers. Studies showed that propfan-powered aircraft would require much higher powers than previously experienced with turboprops, and at power levels in excess of 8000 kW the gearbox design becomes difficult. A further major problem is the transmission of propeller noise to the passenger cabin, and it was widely recognized that a substantial increase in noise over existing turbofan levels would not be acceptable to passengers; it is probable that this can be overcome only by the use of 'pusher' configurations with the propellers mounted behind the passenger cabin. Turboprop engines are surveyed in Ref. (10).

An alternative configuration, a compromise between the turboprop and the turbofan, was the 'unducted fan' (UDF), developed by General Electric in the 1980s. In this approach, two counter-rotating variable pitch fans were directly coupled to counter-rotating turbines with no stators, eliminating the need for a gearbox. The arrangement shows a similarity to the aft-fan configuration described earlier. This revolutionary approach demonstrated a significant reduction in fuel consumption in flight tests, but encountered problems with noise and cabin vibration; airlines were also reluctant to pioneer this new technology, and not all engine manufacturers were convinced that the gearbox could be eliminated. The UDF concept, however, was also considered to be a promising technology for the development of long-range cruise missiles, where its excellent *SFC* would improve range for a fixed quantity of fuel on board.

The status of advanced turboprops, propfans and unducted fans in the mid 1980s was very similar to that of turbofans in the mid 1950s. At that time there was a heated controversy regarding the relative merits of turboprops and turbofans, with the latter being the clear winner. In the mid 1990s the turbofan once again emerged as the winner; fuel prices, however, have been stable over a long period, but if this were to change significantly engines having ultra low fuel consumption would be essential.

The conventional turboprop dominated the market for smaller aircraft of 10–60 seats with flight speeds in the range of 400–600 km/h until the mid 1990s. These aircraft were primarily used on flights with a duration of 60–90 minutes, with a range of perhaps 400–500 km; a widespread system of commuter airlines was built up, providing both short-range flights between smaller centres and feed to major airline hubs. In the late 1990s, however, the advent of turbofan-powered *regional jets* with capacities of 30–70 seats resulted in a major shift away from turboprops; aircraft such as the Canadair RJ and Embraer 145 proved to have much higher passenger appeal than the existing turboprops, and manufacturers such as BAe, Fokker, Saab and Dornier abandoned the turboprop field. While there should be a continuous small demand for turboprops, it seems likely that the trend to the regional jets will gather momentum. In recent years, high-power turboprops such as the PW 150 (3800 kW) and the Rolls-Royce AE 2100 (4500 kW) have been developed, but they have yet to achieve significant market penetration.

3.6   The turboshaft engine

The *turboshaft* engine, in which the output power drives a helicopter rotor, is of great importance and is virtually universally used because of its low weight and high power. In the helicopter application, free turbine configurations are always used. Ideally, the helicopter rotor should operate at constant speed by changing the pitch, and the power is varied by changing the gas-generator speed. In practice, during transient operation, there will be a slight change in the rotor speed and this must be minimized by careful attention to the engine transient response and control system. While at first sight the design requirements of turboprops and turboshafts appear to be the same, the former may be optimized for cruise at an altitude of 6000–8000 m, while the latter is optimized for operation at very low altitudes. Helicopter speeds are limited to about 160 knots (due to aerodynamic limitations on the rotor blades) so jet thrust is not critical and turboshafts are designed to produce the maximum available shaft power.

For many years, military requirements dictated helicopter design and, particularly, production. Civil helicopters have not made significant inroads into passenger transportation, and fill niche markets, such as transportation to off-shore rigs, search and rescue, logging operations, air ambulances and police operations. In many of these applications, range is not critical, and low engine weight is generally more important than engine fuel consumption; in the case of North Sea off-shore operations, however, range is becoming steadily more important, as new rigs are drilled further from land. The power required is typically from 400 to 2000 kW, so the relatively small size requires a fairly simple design. The dominant engines in recent years have been the GE T-700 for military engines and the Rolls-Royce Allison 250 for commercial engines.

A new application for the turboshaft engine, which emerged at the turn of the century, is the development of *tilt-rotor* aircraft, which aims at combining the vertical take-off and landing capability of the helicopter with the higher flight speed of the turboprop. The V-22 Osprey, developed for the US military, is scheduled to enter service in the early part of the 21st century, following an extended development programme. The two large-diameter rotors are operated horizontally for take-off and landing, with the engines rotated to the vertical position; following successful take-off, the engines are gradually rotated to the horizontal position, with the rotors in the vertical orientation, acting as propellers and providing a speed capability of about 300 knots. The mechanical arrangement is complex, because of the need to provide safe operation in the event of an engine failure. It would clearly be catastrophic if an engine shutdown caused the rotor to stop; for this reason, each engine drives both rotors by means of a system of cross-shafting traversing the entire wing. The engines used are the Rolls-Royce AE 1107 of 4600 kW, which is basically similar to the Rolls-Royce AE 2100 turboprop; considerable modifications are required to permit operation in the vertical mode, particularly with regard to lubrication systems. Bell are proposing to use this technology in the Bell 609 civil tilt-rotor, which will be powered by PT-6 turboshafts and is scheduled to enter service around 2005.

3.7 Auxiliary Power Units

All aircraft have substantial requirements for on-board power supply, ranging from electrical and hydraulic power to compressed air. During normal flight, these requirements are met by extracting power from the engines using an accessory gearbox. Civil aircraft, however, require power for considerable periods with the engines shut down; aircraft servicing and the loading of passengers, baggage and freight must be done before starting the engines. The passenger compartment must be maintained at a comfortable level of temperature and relative humidity in ambient temperatures ranging from −40 to 40 °C. The power needed for aircraft systems on the ground may be obtained either from a ground cart, providing a mobile source of electric power and conditioned air, or from an *Auxiliary Power Unit* (APU) installed in the aircraft. The APU is a small gas turbine specifically designed to meet the on-board demands of the aircraft; the obvious need is for ground power before take-off and after landing, but for ETOPS qualified aircraft a prime requirement is the capability of in-flight starting after hours of cold soaking at high altitudes.

This results from the need for increased electrical and hydraulic power extraction from the remaining engine following an in-flight shutdown, and the need to maintain a redundant power supply for safety.

The first commercial jets, such as the Comet, 707 and DC-8, did not have APUs installed. These were the premium transports of their day, and only operated from major airports where ground supplies were readily available. When smaller jets such as the DC-9, 727 and BAC 1-11 were introduced, they began to use less sophisticated airports where ground power was not always available, and they all incorporated APUs. The next generation of large aircraft (747, DC-10, L-1011, A300) all used APUs and these are now standard on civil transports. Most small turboprops and regional jets use APUs, as their operations require rapid turn-around often at small airports without adequate ground facilities. Military aircraft use APUs to allow for operations from a wide variety of bases, making them self-sufficient and capable of rapid take-offs in a combat situation.

In its earliest form, the APU was a simple single-shaft unit in which the compressor was over-sized relative to the turbine, allowing up to 30 per cent of the inlet flow to be bled off at compressor delivery for aircraft services. The useful power output could be split between compressed air and electric power by means of a bleed valve. With the bleed valve fully open, maximum compressed airflow could be achieved; this would be needed for engine starting. With the valve closed, the electrical power could be maximized, and modulation of the bleed valve position could balance the supply of compressed air and electrical power. Air starting makes use of a small air turbine driving the HP rotor through an accessory gearbox and typically requires compressed air at about 4 atmospheres. This immediately gives rise to a problem with the thermal efficiency of the APU. With the simple arrangement, the cycle pressure ratio of the APU is limited to 4 and this obviously results in a very inefficient cycle. This problem can be overcome at the expense of an increase in complexity of the APU.

The standard approach used in modern APUs is to use a separate *load compressor* driven by a gas turbine, with a common air intake for both the load compressor and the gas turbine compressor. This allows the load compressor to be designed for a pressure ratio of about 4, while the gas turbine pressure ratio can be significantly higher, typically in the range of 8–12. Figure 3.25 shows a schematic of the Honeywell 131 APU used in the MD80 family of aircraft. The load compressor is a centrifugal compressor with variable inlet guide vanes, which are necessary to accommodate airflow variation while running at constant speed, which is required to maintain constant frequency electrical output.

Good thermal efficiency is a desirable but not dominant requirement. Other factors of prime importance include low volume, low noise, low weight, reliability, ease of maintenance and overall cost of ownership. In civil aircraft, the APU is invariably installed in the tail cone, where volume

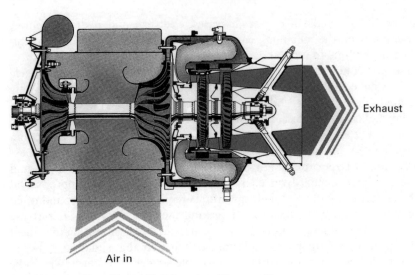

Exhaust

Air in

FIG. 3.25 Auxiliary Power Unit [courtesy Honeywell]

is at a premium and critical hydraulic and control lines must be shielded from debris in the event of a mechanical failure.

3.8   Thrust augmentation

If the thrust of an engine has to be increased above the original design value several alternatives are available. Increase of turbine inlet temperature, for example, will increase the specific thrust and hence the thrust for a given engine size. Alternatively the mass flow through the engine could be increased without altering the cycle parameters. Both of these methods imply some redesign of the engine, and either or both may be used to uprate an existing engine.

Frequently, however, there will be a requirement for a temporary increase in thrust, e.g. for take-off, for acceleration from subsonic to supersonic speed or during combat manoeuvres; the problem then becomes one of *thrust augmentation*. Numerous schemes for thrust augmentation have been proposed, but the two methods most widely used are *liquid injection* and *afterburning* (or *reheat*).

Liquid injection is primarily useful for increasing take-off thrust. Substantial quantities of liquid are required, but if the liquid is consumed during take-off and initial climb the weight penalty is not significant. Spraying water into the compressor inlet causes evaporation of the water droplets, resulting in extraction of heat from the air; the effect of this is equivalent to a drop in compressor inlet temperature. Chapter 8 will show that reducing the temperature at entry to a turbojet will increase the thrust, due to the increase in pressure ratio and mass flow resulting from the effective increase in rotational speed. In practice a mixture of water and methanol is used; the methanol lowers the freezing point of water, and in addition it will burn when it reaches the combustion chamber. Liquid is sometimes injected directly into the combustion chamber. The resulting 'blockage' forces the compressor to operate at a higher pressure ratio causing the thrust to increase. In both cases the mass of liquid injected adds to the useful mass flow, but this is a secondary effect. Liquid injection is now seldom used in aircraft engines.

Afterburning, as the name implies, involves burning additional fuel in the jet pipe as shown in Fig. 3.8. In the absence of highly stressed rotating blades the temperature allowable following afterburning is much higher than the turbine inlet temperature. Stoichiometric combustion is desirable for maximum thrust augmentation and final temperatures of around 2000 K are possible. Figure 3.26 shows the T–s diagram for a simple turbojet with the addition of afterburning to 2000 K. The large increase in fuel flow required is evident from the relative temperature rises in the combustion chamber and afterburner, and the penalty in increased *SFC* is heavy. Assuming that a choked convergent nozzle is used, the jet velocity will correspond to the sonic velocity at the appropriate temperature in the plane of the nozzle, i.e. T_7 or T_5 depending on whether the engine is operated with or without afterburning. Thus the jet velocity can be found from $(\gamma R T_c)^{1/2}$, with T_c given by either $T_{06}/T_c = (\gamma + 1)/2$ or $T_{04}/T_c = (\gamma + 1)/2$. It follows that the jet velocity is proportional to $\sqrt{T_0}$ at inlet to the propelling nozzle, and that the gross momentum thrust, relative to that of the simple turbojet, will be increased in the ratio $\sqrt{(T_{06}/T_{04})}$. For the temperatures shown in Fig. 3.26 this amounts to $\sqrt{(2000/959)}$ or 1·44. As an approximation, the increase in fuel would be in the ratio $(2000 - 959) + (1200 - 565)$ with afterburning, to $(1200 - 565)$

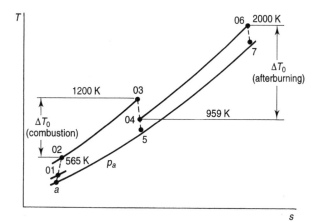

FIG. 3.26 Cycle of turbojet with afterburning

without afterburning, i.e. 2·64. Thus a 44 per cent increase in thrust is obtained at the expense of a 164 per cent increase in fuel flow, and clearly afterburning should be used only for short periods. This might be the thrust augmentation under take-off conditions where the gross thrust is equal to the net thrust. At high forward speeds, however, the gain is much greater and is often well over 100 per cent. This is because for a fixed momentum drag an increase in gross thrust represents a considerably greater increase in net thrust. The Concorde makes use of afterburning for transonic acceleration from M 0·9 to M 1·4; the significant increase in net thrust provides faster acceleration through the high drag regime near M 1·0, resulting in a reduction in fuel consumption in spite of the short-term increase in fuel flow. Afterburning offers even greater gains for low bypass ratio turbofans, because of the relatively low temperature after mixing of the hot and cold streams and the larger quantities of excess air available for combustion; military turbofans use afterburning for take-off and combat manoeuvring.

It is essential for engines fitted with an afterburner to incorporate a variable area nozzle because of the large change in density of the flow approaching the nozzle resulting from the large change in temperature. Afterburning will normally be brought into operation when the engine is running at its maximum rotational speed, corresponding to its maximum unaugmented thrust. The afterburner should be designed so that the engine will continue to operate at the same speed when it is in use, and hence the nozzle must pass the same mass flow at a much reduced density. This can be achieved only if a variable nozzle is fitted permitting a significant increase in nozzle area. Note that the pressure thrust will also be increased owing to the enlarged nozzle area.

The pressure loss in the afterburner can be significant. Combustion pressure losses are discussed in Chapter 6, where it is shown that the pressure loss is due to both fluid friction and momentum changes resulting from heat addition. In combustion chambers the former predominates, but in afterburners the losses due to momentum changes are much more important. The temperature rise is determined by the turbine outlet temperature and the fuel/gas ratio in the afterburner, and the pressure loss due to momentum changes can be determined using the Rayleigh functions and the method outlined in Appendix A.4. This pressure loss is found to be a function of both the temperature ratio across the afterburner and the Mach number at inlet to the duct. If the inlet Mach number is too high, heat release can result in the downstream Mach number reaching 1·0 and this places an upper limit on the allowable heat release: the phenomenon is referred to as *thermal choking*. Figure 3.27 shows values of the pressure loss due to momentum changes and emphasizes the need for a low Mach number. Typically, the exit Mach number from the turbine of a jet engine will be about 0·5 and it is necessary to introduce a diffuser between the turbine and afterburner to reduce the Mach number to about 0·25–0·30 before introducing the afterburner fuel.

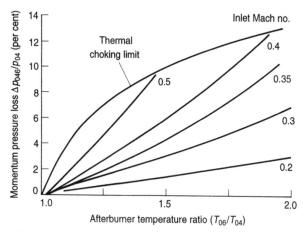

FIG. 3.27 Momentum pressure loss

Even when not in use, an afterburner incurs some penalty in pressure loss due to the presence of the burners and flame stabilizing devices. Another disadvantage of this method of thrust augmentation is that the very high jet velocities resulting from a large degree of afterburning result in a noisy exhaust. The afterburning Olympus engines used on Concorde provide about 15–20 per cent increase in take-off thrust, resulting in an exhaust temperature of around 1400 K. The increased noise level, though a serious concern, is significantly less than might be expected from experience with military aircraft. It is most unlikely, however, that any future supersonic transport will use afterburning for take-off, as a prime requirement will be take-off noise levels comparable to current subsonic aircraft; as mentioned earlier, this will necessitate the development of variable cycle engines.

3.9 Miscellaneous topics

Common cores

The *core* of an engine, consisting of the HP compressor, combustor, HP turbine and associated shafting, poses the most demanding and costly development problems because of both the high temperatures and the stress levels resulting from the high tip speeds. With a multi-spool engine it is possible to match the developed core with different LP systems, to reduce the development cost of a new engine. Using the common core approach, for example, a manufacturer could develop both a high bypass ratio turbofan for civil applications and a low bypass ratio turbofan for military use. The F101 low bypass engine was developed by GE for use in the B-1B bomber, and the core was later mated to a high bypass ratio fan and a new turbine designed by SNECMA to produce the CFM56 turbofan used in the A320, 737 and A340;

in recent years, this has been the most widely sold civil engine, clearly demonstrating the validity of using a common core. Rolls-Royce Allison used the core of the AE 2100 turboprop to provide power to drive a fan instead of a propellor, resulting in the AE 3007 turbofan used in the Embraer regional jet family. A final example is the use of a common core in the PW 530 and 545 turbofans, covering different thrust ranges for the business jet market; the 530 has a take-off thrust of 13·9 kN, compared with 19·6 kN for the 545. The 530 uses the simplest configuration possible, while the 545 adds a booster stage and incorporates a larger diameter fan to increase the airflow. Figure 3.28 shows the two engines drawn on a common centre line.

Engines for VTOL

An aircraft designed for VTOL requires a take-off thrust of about 120 per cent of the take-off weight, because all the lift has to be produced by the propulsion system. This inevitably resulted in engines that were much bigger than were required for cruise, so they had to be throttled back significantly with severe penalties on fuel burn; off-design performance will be discussed in Chapters 8 and 9, where it will be seen that gas turbines are inefficient at low power settings. Many approaches were tried, for both civil and military applications. It eventually became clear that VTOL was not economic for civil applications and there were major problems with engine noise in urban areas. Development then focused on military applications, primarily for fighters; the VTOL capability offered the chance to operate from unprepared areas or from battle-damaged airfields. As mentioned earlier, special lift engines of very high thrust to weight ratio were developed over 30 years ago for use in multi-engine applications, but these installations proved to be too complex to be practical.

The only really successful system to date is the single-engined *vectored thrust* installation of the Rolls-Royce Pegasus in the Harrier/AV-8B. The Pegasus is basically a conventional turbofan with separate exhausts, with vectoring nozzles for both hot and cold streams; two nozzles are used for each stream resulting in a 'four poster' arrangement, shown in Fig. 3.29. This system requires installation close to the centre of gravity of the aircraft and the cycle must be selected to give approximately equal thrust from the hot and cold streams. The vectoring capability means that the thrust can be aligned vertically for vertical take-off and horizontally for normal flight; a particular advantage is that the thrust may be vectored to give components of vertical and horizontal thrust, or even some reverse thrust. This flexibility permits the aircraft to operate either as VTOL, Conventional Take-Off and Landing (CTOL) or Reduced Take-Off and Landing (RTOL). It has been shown that even a short take-off run, of say 100 m, can give a significant increase in useful payload.

It has been found in practice, and well documented in the Falklands War, that the best solution is to operate military aircraft in the STOVL (Short

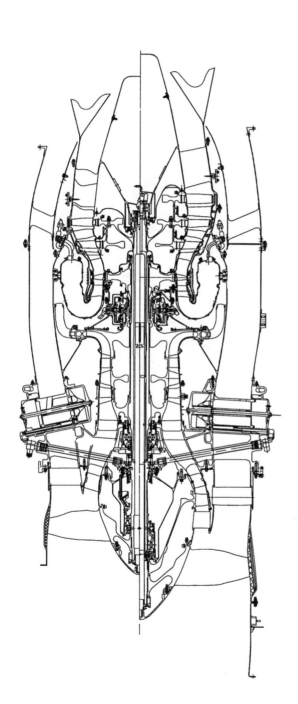

FIG. 3.28 Common core, PW 530 (top) and PW 545 [courtesy Pratt and Whitney Canada]

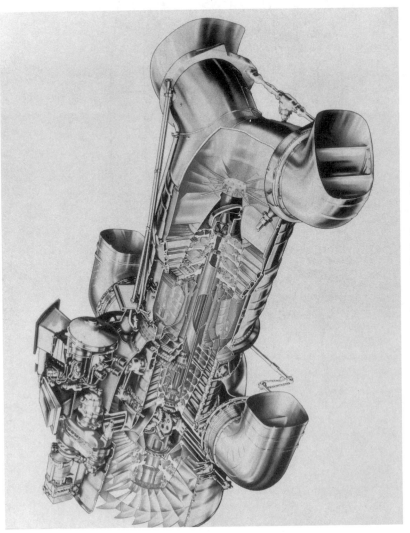

FIG. 3.29 Pegasus vectored thrust engine [courtesy Rolls-Royce plc]

Take-Off and Vertical Landing) mode, allowing the aircraft to operate from simple bases but with the capability of taking off or landing vertically when required.

The major problem facing the engine designer is reconciling the very different thrust requirements of take-off and normal flight. The problems are magnified for an aircraft with a supersonic requirement, which would only be required for a short duration with most of the time spent at high subsonic speeds. The Joint Strike Fighter (JSF), currently under development with work being done jointly in the USA and the UK, requires a considerably more complex propulsion system than the Harrier. There are two competitors, both employing vectored thrust, but with very different configurations. One uses an afterburning turbofan with a vectoring final nozzle and a forward-mounted lift fan used only for take-off; the lift fan is shaft driven by the LP turbine. Its shaft is perpendicular to that of the drive shaft and a mechanical clutch is used to permit coupling and decoupling. The other configuration uses a direct lift system, similar to the Harrier, without the forward nozzles; the two rear nozzles are vectored as in the Harrier and the afterburner is diverted through a vectoring nozzle, giving a 'three poster' arrangement. It will be appreciated that vectoring a variable area nozzle at temperatures of 2000 K is a formidable mechanical design problem, and this is also the case for the clutch.

NOMENCLATURE

a	sonic velocity
A	cross-sectional area
B	bypass ratio (m_c/m_h)
F	net thrust
F_s	specific thrust
K_F	specific thrust coefficient
M	Mach number
η_e	efficiency of energy conversion
η_i	intake efficiency
η_j	nozzle efficiency
η_m	mechanical efficiency
η_o	overall efficiency
η_p	propulsion (Froude) efficiency
η_{pr}	propeller efficiency
η_r	ram efficiency
η_∞	polytropic efficiency

Suffixes

c	critical condition, cold stream
h	hot stream
j	jet
m	mixed

International Standard Atmosphere

$\dfrac{z}{[\text{m}]}$	$\dfrac{p}{[\text{bar}]}$	$\dfrac{T}{[\text{K}]}$	ρ/ρ_0	$\dfrac{a}{[\text{m/s}]}$
0	1·01325	288·15	1·0000	340·3
500	0·9546	284·9	0·9529	338·4
1 000	0·8988	281·7	0·9075	336·4
1 500	0·8456	278·4	0·8638	334·5
2 000	0·7950	275·2	0·8217	332·5
2 500	0·7469	271·9	0·7812	330·6
3 000	0·7012	268·7	0·7423	328·6
3 500	0·6578	265·4	0·7048	326·6
4 000	0·6166	262·2	0·6689	324·6
4 500	0·5775	258·9	0·6343	322·6
5 000	0·5405	255·7	0·6012	320·5
5 500	0·5054	252·4	0·5694	318·5
6 000	0·4722	249·2	0·5389	316·5
6 500	0·4408	245·9	0·5096	314·4
7 000	0·4111	242·7	0·4817	312·3
7 500	0·3830	239·5	0·4549	310·2
8 000	0·3565	236·2	0·4292	308·1
8 500	0·3315	233·0	0·4047	306·0
9 000	0·3080	229·7	0·3813	303·8
9 500	0·2858	226·5	0·3589	301·7
10 000	0·2650	223·3	0·3376	299·5
10 500	0·2454	220·0	0·3172	297·4
11 000	0·2270	216·8	0·2978	295·2
11 500	0·2098	216·7	0·2755	295·1
12 000	0·1940	216·7	0·2546	295·1
12 500	0·1793	216·7	0·2354	295·1
13 000	0·1658	216·7	0·2176	295·1
13 500	0·1533	216·7	0·2012	295·1
14 000	0·1417	216·7	0·1860	295·1
14 500	0·1310	216·7	0·1720	295·1
15 000	0·1211	216·7	0·1590	295·1
15 500	0·1120	216·7	0·1470	295·1
16 000	0·1035	216·7	0·1359	295·1
16 500	0·09572	216·7	0·1256	295·1
17 000	0·08850	216·7	0·1162	295·1
17 500	0·08182	216·7	0·1074	295·1
18 000	0·07565	216·7	0·09930	295·1
18 500	0·06995	216·7	0·09182	295·1
19 000	0·06467	216·7	0·08489	295·1
19 500	0·05980	216·7	0·07850	295·1
20 000	0·05529	216·7	0·07258	295·1

Density at sea level $\rho_0 = 1\cdot2250\,\text{kg/m}^3$.
Extracted from: ROGERS G F C and MAYHEW Y R
Thermodynamic and Transport Properties of Fluids
(Blackwell 1995)

4

Centrifugal compressors

Very rapid progress in the development of gas turbines was made during the Second World War, where attention was focused on the simple turbojet unit. German efforts were based on the axial flow compressor, but British developments used the centrifugal compressor—Refs (1) and (2). It was recognized in Britain that development time was critical and much experience had already been gained on the design of small high-speed centrifugal compressors for supercharging reciprocating engines. Centrifugal compressors were used in early British and American fighter aircraft and also in the original Comet airliners which were the first gas turbine powered civil aircraft in regular service. As power requirements grew, however, it became clear that the axial flow compressor was more suitable for large engines. The result was that a very high proportion of development funding was diverted to the axial type, leading to the availability of axial compressors with an appreciably higher isentropic efficiency than could be achieved by their centrifugal counterparts.

By the late 1950s, however, it became clear that smaller gas turbines would have to use centrifugal compressors, and serious research and development work started again. Small turboprops, turboshafts and auxiliary power units (APUs) have been made in very large numbers and have nearly all used centrifugal compressors; notable examples include the Pratt and Whitney Canada PT-6, the Honeywell 331 and the large stable of APUs built by the latter organization. They are also used for the high-pressure spools in small turbofans, see Fig. 1.12(a). Centrifugals were used primarily for their suitability for handling small volume flows, but other advantages include a shorter length than an equivalent axial compressor, better resistance to Foreign Object Damage (FOD), less susceptibility to loss of performance by build-up of deposits on the blade surfaces and the ability to operate over a wider range of mass flow at a particular rotational speed. The importance of the latter feature, in alleviating problems of matching operating conditions with those of the associated turbine, will be made clear in Chapter 8.

A pressure ratio of around 4:1 can readily be obtained from a single-stage compressor made of aluminium alloys, and in section 2.4 it was shown that this is adequate for a heat-exchange cycle when the turbine inlet temperature is in the region of 1000–1200 K. Many proposals for vehicular gas turbines were based on this arrangement and manufacturers such as Leyland, Ford, General Motors and Chrysler built development engines which never went into production. The advent of titanium alloys, permitting much higher tip speeds, combined with advances in aerodynamics now permit pressure ratios of greater than 8:1 to be achieved in a single stage. When higher pressure ratios are required, the centrifugal compressor may be used in conjunction with an axial-flow compressor (Fig. 1.11), or as a two-stage centrifugal (Fig. 1.10). Even though the latter arrangement involves rather complex ducting between stages, it is still regarded as a practical proposition. Reference (2) of Chapter 1 describes the design process leading to the choice of a twin-spool all-centrifugal compressor for the Pratt and Whitney Canada PW 100 turboprop which entered service in 1984.

Centrifugal compressors are widely used on natural gas pipelines, directly driven by the free power turbine of the prime mover. The same design methods are applicable but these machines would normally operate at low pressure ratios (1·2–1·4) and at very high inlet pressures. Multi-stage centrifugal compressors may also be used in high pressure ratio processes with up to five stages, with intercooling between stages; these may be found in applications such as air separation plants, and the compressors may be driven by steam turbines or electric motors via a speed-increasing gearbox. Centrifugal compressors are also used in very large-scale refrigeration plants, with the compressors run at constant speed driven by large single-shaft industrial gas turbines. This chapter, however, will deal with the design of the type of centrifugal compressor used in gas turbines.

4.1 Principle of operation

The centrifugal compressor consists essentially of a stationary casing containing a rotating *impeller* which imparts a high velocity to the air, and a number of fixed diverging passages in which the air is decelerated with a consequent rise in static pressure. The latter process is one of diffusion, and consequently the part of the compressor containing the diverging passages is known as the *diffuser*. Figure 4.1(a) is a diagrammatic sketch of a centrifugal compressor. The impeller may be single- or double-sided as in 1(b) or 1(c), but the fundamental theory is the same for both. The double-sided impeller was required in early aero-engines because of the relatively small flow capacity of the centrifugal compressor for a given overall diameter.

Air is sucked into the impeller eye and whirled round at high speed by the vanes on the impeller disc. At any point in the flow of air through the impeller, the centripetal acceleration is obtained by a pressure head, so that the static pressure of the air increases from the eye to the tip of the

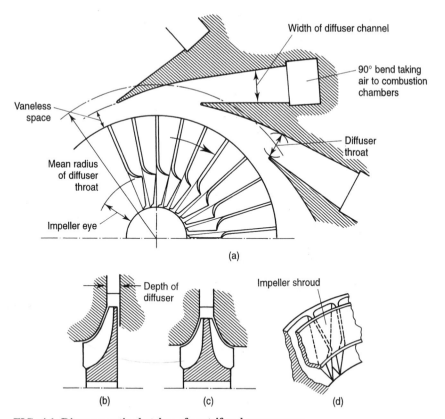

FIG. 4.1 Diagrammatic sketches of centrifugal compressors

impeller. The remainder of the static pressure rise is obtained in the diffuser, where the very high velocity of the air leaving the impeller tip is reduced to somewhere in the region of the velocity with which the air enters the impeller eye; it should be appreciated that friction in the diffuser will cause some loss in *stagnation* pressure. The normal practice is to design the compressor so that about half the pressure rise occurs in the impeller and half in the diffuser.

It will be appreciated that owing to the action of the vanes in carrying the air around with the impeller, there will be a slightly higher static pressure on the forward face of a vane than on the trailing face. The air will thus tend to flow round the edges of the vanes in the clearance space between the impeller and the casing. This naturally results in a loss of efficiency, and the clearance must be kept as small as possible. A shroud attached to the vanes, Fig. 4.1(d), would eliminate such a loss, but the manufacturing difficulties are vastly increased and there would be a disc friction or 'windage' loss associated with the shroud. Although shrouds have been used on superchargers and process compressors, they are not used on impellers for gas turbines.

The impellers of modern centrifugal compressors operate with very high tip speeds resulting in very high stress levels. It will be shown in the next section that backswept curved vanes are desirable for compressors of high pressure ratio, but for many years designers were forced to use radial vanes because of the tendency for curved vanes to straighten out under the action of the considerable centrifugal force involved, setting up undesirable bending stresses in the vanes. Modern methods of stress analysis combined with stronger materials, however, now permit backswept vanes to be used in high-performance compressors.

4.2 Work done and pressure rise

Since no work is done on the air in the diffuser, the energy absorbed by the compressor will be determined by the conditions of the air at the inlet and outlet of the impeller. Figure 4.2 shows the nomenclature employed.

In the first instance it will be assumed that the air enters the impeller eye in the axial direction, so that the initial angular momentum of the air is zero. The axial portion of the vanes must be curved so that the air can pass smoothly into the eye. The angle which the leading edge of a vane makes with the tangential direction α will be given by the direction of the relative velocity of the air at inlet, V_1, as shown in Fig. 4.2.

If the air leaves the impeller tip with an absolute velocity C_2, it will have a tangential or whirl component C_{w2}, and a comparatively small radial component C_{r2}. Under ideal conditions C_2 would be such that the whirl component is equal to the impeller tip speed U, as shown by the velocity triangle at the top of Fig. 4.2. Due to its inertia, the air trapped between

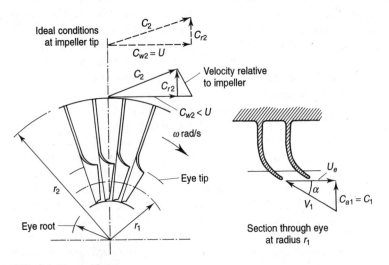

FIG. 4.2 Nomenclature

the impeller vanes is reluctant to move round with the impeller, and we have already noted that this results in a higher static pressure on the leading face of a vane than on the trailing face. It also prevents the air from acquiring a whirl velocity equal to the impeller speed. This effect is known as *slip*. How far the whirl velocity at the impeller tip falls short of the tip speed depends largely upon the number of vanes on the impeller. The greater the number of vanes, the smaller the slip, i.e. the more nearly C_{w2} approaches U. It is necessary in design to assume a value for the *slip factor* σ, where σ is defined as the ratio C_{w2}/U. Various approximate analyses of the flow in an impeller channel have led to formulae for σ: the one appropriate to radial-vaned impellers which seems to agree best with experiment is that due to Stanitz, Ref. (4):

$$\sigma = 1 - \frac{0 \cdot 63\pi}{n}$$

where n is the number of vanes.

As explained in any elementary text on applied thermodynamics, the theoretical torque which must be applied to the impeller will be equal to the rate of change of angular momentum experienced by the air. Considering unit mass flow of air, this torque is given by

$$\text{theoretical torque} = C_{w2}r_2 \tag{4.1}$$

If ω is the angular velocity, the work done on the air will be

$$\text{theoretical work done} = C_{w2}r_2\omega = C_{w2}U$$

Or, introducing the slip factor,

$$\text{theoretical work done} = \sigma U^2 \tag{4.2}$$

For convenience, in both the chapters on compressors we shall treat the work done on the air as a positive quantity.

Due to friction between the casing and the air carried round by the vanes, and other losses which have a braking effect such as disc friction or 'windage', the applied torque and therefore the actual work input is greater than this theoretical value. A *power input factor* ψ can be introduced to take account of this, so that the actual work done on the air becomes

$$\text{work done} = \psi\sigma U^2 \tag{4.3}$$

If $(T_{03} - T_{01})$ is the stagnation temperature rise across the whole compressor then, since no energy is added in the diffuser, this must be equal to the stagnation temperature rise $(T_{02} - T_{01})$ across the impeller alone. It will therefore be equal to the temperature equivalent of the work done on the air given by equation (4.3), namely

$$T_{03} - T_{01} = \frac{\psi\sigma U^2}{c_p} \tag{4.4}$$

where c_p is the mean specific heat over this temperature range. Typical values for the power input factor lie in the region of $1\cdot035$–$1\cdot04$.

So far we have merely considered the work which must be put into the compressor. If a value for the *overall* isentropic efficiency η_c is assumed, then it is known how much of the work is usefully employed in raising the pressure of the air. The *overall* stagnation pressure ratio follows as

$$\frac{p_{03}}{p_{01}} = \left(\frac{T'_{03}}{T_{01}}\right)^{\gamma/(\gamma-1)} = \left[1 + \frac{\eta_c(T_{03} - T_{01})}{T_{01}}\right]^{\gamma/(\gamma-1)} = \left[1 + \frac{\eta_c\psi\sigma U^2}{c_p T_{01}}\right]^{\gamma/(\gamma-1)}$$

(4.5)

The distinction between the power input factor and the slip factor should be clearly understood: they are neither independent of one another nor of η_c. The power input factor represents an increase in the work input, the whole of which is absorbed in overcoming frictional loss and therefore degraded into thermal energy. The fact that the outlet temperature is raised by this loss, and incidentally by other frictional losses as well, enables the maximum cycle temperature to be reached without burning so much fuel, so that as far as the efficiency of the whole gas turbine unit is concerned these losses are not entirely wasteful. Nevertheless, this effect is outweighed by the fact that more turbine work is used in driving the compressor and isentropic (i.e. frictionless adiabatic) compression is the ideal at which to aim. It follows that the power input factor should be as low as possible, a low value of ψ implying simultaneously a high value of η_c. It should be appreciated that η_c depends also upon the friction loss in the diffuser which does not affect the argument up to equation (4.4). For this reason it is not helpful to consider ψ implicitly as part of η_c.

The slip factor, on the other hand, is a factor limiting the work capacity of the compressor even under isentropic conditions, and this quantity should be as great as possible. Clearly the more nearly C_{w2} approaches U, the greater becomes the rate at which work can usefully be put into a compressor of given size. Unfortunately an increase in the number of vanes, which would increase σ, entails an increase in the solidity of the impeller eye, i.e. a decrease in the effective flow area. Additional friction losses arise because, for the same mass flow or 'throughput', the inlet velocity must be increased. Thus the additional work input that can be employed by increasing the number of vanes may not result in an increase in that portion which is usefully employed in raising the pressure of the air; it may only increase the thermal energy produced by friction resulting in an increase in ψ and reduction in η_c. A suitable compromise must be found, and present-day practice is to use the number of vanes which give a slip factor of about $0\cdot9$, i.e. about 19 or 21 vanes (see also under heading 'Mach number in the diffuser', in section 4.4).

From equation (4.5) it will be seen that the remaining factors influencing the pressure ratio for a given working fluid are the impeller tip speed U, and the inlet temperature T_{01}. Any lowering of the inlet temperature T_{01} will clearly increase the pressure ratio of the compressor for a given work input, but it is not a variable under the control of the designer. Reference to texts on strength of materials will show that the centrifugal stresses in a rotating disc are proportional to the square of the rim speed. For single-sided impellers of light alloy, U is limited to about 460 m/s by the maximum allowable centrifugal stresses in the impeller: such a speed yields a pressure ratio of about 4:1. Higher speeds can be used with more expensive materials such as titanium and pressure ratios of over 8:1 are now possible. Because of the additional disc loading, lower speeds must be used for double-sided impellers.

EXAMPLE 4.1

The following data are suggested as a basis for the design of a single-sided centrifugal compressor:

Power input factor ψ	1·04
Slip factor σ	0·9
Rotational speed N	290 rev/s
Overall diameter of impeller	0·5 m
Eye tip diameter	0·3 m
Eye root diameter	0·15 m
Air mass flow m	9 kg/s
Inlet stagnation temperature T_{01}	295 K
Inlet stagnation pressure p_{01}	1·1 bar
Isentropic efficiency η_c	0·78

Requirements are (a) to determine the pressure ratio of the compressor and the power required to drive it assuming that the velocity of the air at inlet is axial; (b) to calculate the inlet angle of the impeller vanes at the root and tip radii of the eye, assuming that the axial inlet velocity is constant across the eye annulus; and (c) to estimate the axial depth of the impeller channels at the periphery of the impeller.

(a) Impeller tip speed $U = \pi \times 0·5 \times 290 = 455·5$ m/s.
Temperature equivalent of the work done on unit mass flow of air is

$$T_{03} - T_{01} = \frac{\psi\sigma U^2}{c_p} = \frac{1·04 \times 0·9 \times 455·5^2}{1·005 \times 10^3} = 193 \text{ K}$$

$$\frac{p_{03}}{p_{01}} = \left[1 + \frac{\eta_c(T_{03} - T_{01})}{T_{01}}\right]^{\gamma/(\gamma-1)} = \left(1 + \frac{0·78 \times 193}{295}\right)^{3·5} = 4·23$$

Power required $= mc_p(T_{03} - T_{01}) = 9 \times 1·005 \times 193 = 1746$ kW

(b) To find the inlet angle of the vanes it is necessary to determine the inlet velocity which in this case is axial, i.e. $C_{a1} = C_1$. C_{a1} must satisfy the continuity equation $m = \rho_1 A_1 C_{a1}$, where A_1 is the flow area at inlet. Since the density ρ_1 depends upon C_1, and both are unknown, a trial and error process is required.

The iterative procedure is not critically dependent on the initial value assumed for the axial velocity, but clearly it is desirable to have some rational basis for obtaining an estimated value for starting the iteration. The simplest way of obtaining a reasonable estimate of the axial velocity is to calculate the density on the basis of the known *stagnation* temperature and pressure; in practice this will give a density that is too high and a velocity that is too low. Having obtained an initial estimate of the axial velocity, the density can be recalculated and thence the actual velocity from the continuity equation; if the assumed and calculated velocities do not agree it is necessary to iterate until agreement is reached (only the final trial is shown below). Note that it is normal to design for an axial velocity of about 150 m/s, this providing a suitable compromise between high flow per unit frontal area and low frictional losses in the intake.

$$\text{Annulus area of impeller eye, } A_1 = \frac{\pi(0{\cdot}3^2 - 0{\cdot}15^2)}{4} = 0{\cdot}053\,\text{m}^2$$

Based on stagnation conditions:

$$\rho_1 \simeq \frac{p_{01}}{RT_{01}} = \frac{1{\cdot}1 \times 100}{0{\cdot}287 \times 295} = 1{\cdot}30\,\text{kg/m}^3$$

$$C_{a1} = \frac{m}{\rho_1 A_1} = \frac{9}{1{\cdot}30 \times 0{\cdot}053} = 131\,\text{m/s}$$

Since $C_1 = C_{a1}$, the equivalent dynamic temperature is

$$\frac{C_1^2}{2c_p} = \frac{131^2}{2 \times 1{\cdot}005 \times 10^3} = \frac{1{\cdot}31^2}{0{\cdot}201} = 8{\cdot}5\,\text{K}$$

$$T_1 = T_{01} - \frac{C_1^2}{2c_p} = 295 - 8{\cdot}5 = 286{\cdot}5\,\text{K}$$

$$p_1 = \frac{p_{01}}{(T_{01}/T_1)^{\gamma/(\gamma-1)}} = \frac{1{\cdot}1}{(295/286{\cdot}5)^{3{\cdot}5}} = 0{\cdot}992\,\text{bar}$$

$$\rho_1 = \frac{p_1}{RT_1} = \frac{0{\cdot}992 \times 100}{0{\cdot}287 \times 286{\cdot}5} = 1{\cdot}21\,\text{kg/m}^3$$

Check on C_{a1}:

$$C_{a1} = \frac{m}{\rho_1 A_1} = \frac{9}{1{\cdot}21 \times 0{\cdot}053} = 140\,\text{m/s}$$

Final trial:

>   Try $C_{a1} = C_1 = 145 \, \text{m/s}$

>   Equivalent dynamic temperature is

$$\frac{C_1^2}{2c_p} = \frac{145^2}{2 \times 1 \cdot 005 \times 10^3} = \frac{1 \cdot 45^2}{0 \cdot 201} = 10 \cdot 5 \, \text{K}$$

$$T_1 = T_{01} - \frac{C_1^2}{2c_p} = 295 - 10 \cdot 5 = 284 \cdot 5 \, \text{K}$$

$$p_1 = \frac{p_{01}}{(T_{01}/T_1)^{\gamma/(\gamma-1)}} = \frac{1 \cdot 1}{(295/284 \cdot 5)^{3 \cdot 5}} = 0 \cdot 968 \, \text{bar}$$

$$\rho_1 = \frac{p_1}{RT_1} = \frac{0 \cdot 968 \times 100}{0 \cdot 287 \times 284 \cdot 5} = 1 \cdot 185 \, \text{kg/m}^3$$

Check on C_{a1}:

$$C_{a1} = \frac{m}{\rho_1 A_1} = \frac{9}{1 \cdot 185 \times 0 \cdot 053} = 143 \, \text{m/s}$$

This is a good agreement and a further trial using $C_{a1} = 143 \, \text{m/s}$ is unnecessary because a small change in C has little effect upon ρ. For this reason it is more accurate to use the final value $143 \, \text{m/s}$, rather than the mean of $145 \, \text{m/s}$ (the trial value) and $143 \, \text{m/s}$. The vane angles can now be calculated as follows:

>   Peripheral speed at the impeller eye tip radius
> $$= \pi \times 0 \cdot 3 \times 290 = 273 \, \text{m/s}$$
>   and at eye root radius $= 136 \cdot 5 \, \text{m/s}$
> α at root $= \tan^{-1} 143/136 \cdot 5 = 46 \cdot 33°$
> α at tip $= \tan^{-1} 143/273 = 27 \cdot 65°$

(c) The shape of the impeller channel between eye and tip is very much a matter of trial and error. The aim is to obtain as uniform a change of flow velocity up the channel as possible, avoiding local decelerations up the trailing face of the vane which might lead to flow separation. Only tests on the machine can show whether this has been achieved: the flow analyses already referred to [Ref. (4)] are for inviscid flow and are not sufficiently realistic to be of direct use in design. To calculate the required depth of the impeller channel at the periphery we must make some assumptions regarding both the radial component of velocity at the tip, and the division of losses between the impeller and the diffuser so that the density can be evaluated. The radial component of velocity will be relatively small and can be chosen by the designer; a suitable value is obtained by making it approximately equal to the axial velocity at inlet to the eye.

To estimate the density at the impeller tip, the static pressure and temperature are found by calculating the absolute velocity at this point and using it in conjunction with the stagnation pressure which is calculated from the assumed loss up to this point. Figure 4.3 may help the reader to follow the calculation.

Making the choice $C_{r2} = C_{a1}$, we have $C_{r2} = 143$ m/s

$$C_{w2} = \sigma U = 0.9 \times 455.5 = 410 \text{ m/s}$$

$$\frac{C_2^2}{2c_p} = \frac{C_{r2}^2 + C_{w2}^2}{2c_p} = \frac{1.43^2 + 4.10^2}{0.201} = 93.8 \text{ K}$$

Assuming that half the total loss, i.e. $0.5(1 - \eta_c) = 0.11$, occurs in the impeller, the effective efficiency of compression from p_{01} to p_{02} will be 0.89 so that

$$\frac{p_{02}}{p_{01}} = \left(1 + \frac{0.89 \times 193}{295}\right)^{3.5} = 1.582^{3.5}$$

Now $(p_2/p_{02}) = (T_2/T_{02})^{3.5}$, and $T_{02} = T_{03} = 193 + 295 = 488$ K, so that

$$T_2 = T_{02} - \frac{C_2^2}{2c_p} = 488 - 93.8 = 394.2 \text{ K}$$

$$\frac{p_2}{p_{02}} = \left(\frac{394.2}{488}\right)^{3.5}$$

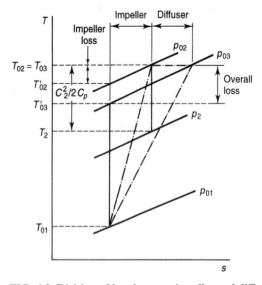

FIG. 4.3 Division of loss between impeller and diffuser

Hence, since $(p_2/p_{01}) = (p_2/p_{02})(p_{02}/p_{01})$,

$$\frac{p_2}{p_{01}} = \left(1 \cdot 582 \times \frac{394 \cdot 2}{488}\right)^{3 \cdot 5} = 2 \cdot 35$$

$$p_2 = 2 \cdot 35 \times 1 \cdot 1 = 2 \cdot 58 \text{ bar}$$

$$\rho_2 = \frac{p_2}{RT_2} = \frac{2 \cdot 58 \times 100}{0 \cdot 287 \times 394 \cdot 2} = 2 \cdot 28 \text{ kg/m}^3$$

The required area of cross-section of flow in the radial direction at the impeller tip is

$$A = \frac{m}{\rho_2 C_{r2}} = \frac{9}{2 \cdot 28 \times 143} = 0 \cdot 0276 \text{ m}^2$$

Hence the depth of impeller channel

$$= \frac{0 \cdot 0276}{\pi \times 0 \cdot 5} = 0 \cdot 0176 \text{ m or } 1 \cdot 76 \text{ cm}$$

This result will be used when discussing the design of the diffuser in the next section.

Before leaving the subject of the impeller, it is worth noting the effect of using backswept curved vanes, which we said at the end of section 4.1 are increasingly being used for high-performance compressors. The velocity triangle at the tip section for a backswept impeller is shown in Fig. 4.4,

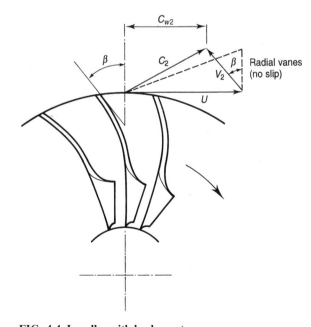

FIG. 4.4 Impeller with backswept vanes

drawn for the ideal case of zero slip for ease of understanding. The corresponding triangle for the radial-vaned impeller is shown by dotted lines. Assuming the radial component of velocity to be the same, implying the same mass flow, it can be seen that the velocity relative to the tip, V_2, is increased while the absolute velocity of the fluid, C_2, is reduced. These changes imply less stringent diffusion requirements in both the impeller and diffuser, tending to increase the efficiency of both components. The backsweep angle β may be in the region of 30–40°. The work-absorbing capacity of the rotor is reduced, however, because C_{w2} is lower and the temperature rise will be less than would be obtained with the radial-vaned impeller. This effect is countered by the increased efficiency and it must be remembered that the ultimate goal is high pressure ratio and efficiency rather than high temperature rise. The use of backswept vanes also gives the compressor a wider operating range of airflow at a given rotational speed, which is important for matching the compressor to its driving turbine; this will be discussed in Chapter 8.

4.3   The diffuser

In Chapter 6 it will be seen that the problem of designing an efficient combustion system is eased if the velocity of the air entering the combustion chamber is as low as possible. It is necessary, therefore, to design the diffuser so that only a small part of the stagnation temperature at the compressor outlet corresponds to kinetic energy. Usually the velocity of the air at the compressor outlet is in the region of 90 m/s.

As will be emphasized throughout this book, it is much more difficult to arrange for an efficient deceleration of flow than it is to obtain an efficient acceleration. There is a natural tendency in a diffusing process for the air to break away from the walls of the diverging passage, reverse its direction, and flow back in the direction of the pressure gradient—see Fig. 4.5(a). If the divergence is too rapid, this may result in the formation of eddies with consequent transfer of some kinetic energy into internal energy and a reduction in useful pressure rise. A small angle of divergence,

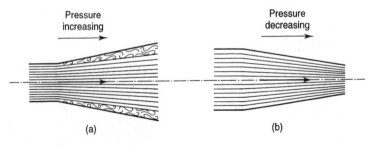

FIG. 4.5 Diffusing and accelerating flow

however, implies a long diffuser and a high value of skin friction loss. Experiments have shown that the optimum included angle of divergence is about $7°$, although angles of up to twice this value can be used with diffusers of low length/width (or radius) ratio without incurring a serious increase in stagnation pressure loss. Empirical design curves for diffusers of both circular and rectangular cross-section can be found in Ref. (8). During acceleration in a converging passage, on the other hand, the gas naturally tends to fill the passage and follow the boundary walls closely, as in Fig. 4.5(b), however rapid the rate of convergence. Only the normal frictional losses will be incurred in this case.

In order to control the flow of air effectively and carry out the diffusion process in as short a length as possible, the air leaving the impeller is divided into a number of separate streams by fixed diffuser vanes. Usually the passages formed by the vanes are of constant depth, the width diverging in accordance with the shape of the vanes, as shown in Fig. 4.1. The angle of the diffuser vanes at the leading edge must be designed to suit the direction of the absolute velocity of the air at the radius of the leading edges, so that the air will flow smoothly over the vanes. As there is always a radial gap between the impeller tip and the leading edges of the vanes, this direction will not be that with which the air leaves the impeller tip. The reason for the vaneless space after the impeller will be explained in section 4.4, when the effects of compressibility in this region are discussed.

To find the correct inlet angle for the diffuser vanes, the flow in the vaneless space must be considered. No further energy is supplied to the air after it leaves the impeller so that, neglecting the effect of friction, the angular momentum $C_w r$ must be constant. Hence C_w decreases from impeller tip to diffuser vane ideally in inverse proportion to the radius. For a channel of constant depth, the area of flow in the radial direction is directly proportional to the radius. The radial velocity C_r will therefore also decrease from impeller tip to diffuser vane, in accordance with the equation of continuity. If both C_r and C_w decrease, then the resultant velocity C will decrease from the impeller tip, and some diffusion evidently takes place in the vaneless space. The consequent increase in density means that C_r will not decrease in inverse proportion to the radius as does C_w, and the way in which C_r varies must be found from the continuity equation. An example at the end of this section will show how this may be done. When C_r and C_w have been calculated at the radius of the leading edges of the diffuser vanes, then the direction of the resultant velocity can be found and hence the inlet angle of the vanes.

It will be apparent that the direction of the airflow in the vaneless space will vary with mass flow and pressure ratio, so that when the compressor is operating under conditions other than those of the design point the air may not flow smoothly into the diffuser passages in which event some loss of efficiency will result. In gas turbines where weight and complexity are not so important as high part-load efficiency, it is possible to incorporate adjustable diffuser vanes so that the inlet angle is correct over a wide range of operating conditions.

For a given pressure and temperature at the leading edge of the diffuser vanes, the mass flow passed will depend upon the total throat area of the diffuser passages (see Fig. 4.1). Once the number of vanes and the depth of passage have been decided upon, the throat width can be calculated to suit the mass flow required under given conditions of temperature and pressure. For reasons given later in section 4.6, the number of diffuser vanes is appreciably less than the number of impeller vanes. The length of the diffuser passages will of course be determined by the maximum angle of divergence permissible, and the amount of diffusion required, and use would be made of the data in Ref. (8) to arrive at an optimum design. Although up to the throat the vanes must be curved to suit the changing direction of flow, after the throat the airflow is fully controlled and the walls of the passage may be straight. Note that diffusion can be carried out in a much shorter flow path once the air is controlled than it can be in a vaneless space where the air follows an approximately logarithmic spiral path (for an incompressible fluid $\tan^{-1}(C_r/C_w) = $ constant).

After leaving the diffuser vanes, the air may be passed into a volute (or scroll) and thence to a single combustion chamber (via a heat-exchanger if one is employed). This would be done only in an industrial gas turbine unit: indeed, in some small industrial units the diffuser vanes are omitted and the volute alone is used. For aircraft gas turbines, where volume and frontal area are important, the individual streams of air may be retained, each diffuser passage being connected to a separate combustion chamber. Alternatively the streams can be fed into an annular combustion chamber surrounding the shaft connecting the compressor and turbine.

EXAMPLE 4.2

Consider the design of a diffuser for the compressor dealt with in the previous example. The following additional data will be assumed:

Radial width of vaneless space	5 cm
Approximate mean radius of diffuser throat	0·33 m
Depth of diffuser passages	1·76 cm
Number of diffuser vanes	12

Required are (a) the inlet angle of the diffuser vanes, and (b) the throat width of the diffuser passages which are assumed to be of constant depth. For simplicity it will be assumed that the additional friction loss in the short distance between impeller tip and diffuser throat is small and therefore that 50 per cent of the overall loss can be considered to have occurred up to the diffuser throat. *For convenience suffix 2 will be used to denote any plane in the flow after the impeller tip, the context making it clear which plane is under consideration.*

(a) Consider conditions at the radius of the diffuser vane leading edges, i.e. at $r_2 = 0.25 + 0.05 = 0.3$ m. Since in the vaneless space $C_w r = $ constant for constant angular momentum,

$$C_{w2} = 410 \times \frac{0.25}{0.30} = 342 \, \text{m/s}$$

The radial component of velocity can be found by trial and error. The iteration may be started by assuming that the temperature equivalent of the resultant velocity is that corresponding to the whirl velocity, but only the final trial is given here.

Try $C_{r2} = 97 \, \text{m/s}$

$$\frac{C_2^2}{2c_p} = \frac{3.42^2 + 0.97^2}{0.201} = 62.9 \, \text{K}$$

Ignoring any additional loss between the impeller tip and the diffuser vane leading edges at 0.3 m radius, the stagnation pressure will be that calculated for the impeller tip, namely it will be that given by

$$\frac{p_{02}}{p_{01}} = 1.582^{3.5}$$

Proceeding as before we have

$$T_2 = 488 - 62.9 = 425.1 \, \text{K}$$

$$\frac{p_2}{p_{02}} = \left(\frac{425.1}{488} \right)^{3.5}$$

$$\frac{p_2}{p_{01}} = \left(1.582 \times \frac{425.1}{488} \right)^{3.5} = 3.07$$

$$p_2 = 3.07 \times 1.1 = 3.38 \, \text{bar}$$

$$\rho_2 = \frac{3.38 \times 100}{0.287 \times 425.1} = 2.77 \, \text{kg/m}^3$$

Area of cross-section of flow in radial direction

$$= 2\pi \times 0.3 \times 0.0176 = 0.0332 \, \text{m}^2$$

Check on C_{r2}:

$$C_{r2} = \frac{9}{2.77 \times 0.0332} = 97.9 \, \text{m/s}$$

Taking C_r as 97.9 m/s, the angle of the diffuser vane leading edge for zero incidence should be

$$\tan^{-1}(C_{r2}/C_{w2}) = \tan^{-1} \frac{97.9}{342} = 16°$$

(b)   The throat width of the diffuser channels may be found by a similar calculation for the flow at the assumed throat radius of 0·33 m.

$$C_{w2} = 410 \times \frac{0·25}{0·33} = 311 \text{ m/s}$$

Try $C_{r2} = 83$ m/s

$$\frac{C_2^2}{2c_p} = \frac{3·11^2 + 0·83^2}{0·201} = 51·5 \text{ K}$$

$$T_2 = 488 - 51·5 = 436·5 \text{ K}$$

$$\frac{p_2}{p_{01}} = \left(1·582 \times \frac{436·5}{488}\right)^{3·5} = 3·37$$

$$p_2 = 3·37 \times 1·1 = 3·71 \text{ bar}$$

$$\rho_2 = \frac{3·71 \times 100}{0·287 \times 436·5} = 2·96 \text{ kg/m}^3$$

As a first approximation we may neglect the thickness of the diffuser vanes, so that the area of flow in the radial direction

$$= 2\pi \times 0·33 \times 0·0176 = 0·0365 \text{ m}^2$$

Check on C_{r2}:

$$C_{r2} = \frac{9}{2·96 \times 0·0365} = 83·3 \text{ m/s}$$

$$\text{Direction of flow} = \tan^{-1} \frac{83·3}{311} = 15°$$

Now $m = \rho_2 A_{r2} C_{r2} = \rho_2 A_2 C_2$, or $A_2 = A_{r2} C_{r2}/C_2$, and hence area of flow in the direction of resultant velocity, i.e. total throat area of the diffuser passages, is

$$0·0365 \sin 15° = 0·009\,45 \text{ m}^2$$

With 12 diffuser vanes, the width of the throat in each passage of depth 0·0176 m is therefore

$$\frac{0·009\,45}{12 \times 0·0176} = 0·0440 \text{ m or } 4·40 \text{ cm}$$

For any required outlet velocity, knowing the total loss for the whole compressor, it is a simple matter to calculate the final density and hence the flow area required at outlet: the length of the passage after the throat then depends on the chosen angle of divergence.

On the basis of such calculations as these it is possible to produce a preliminary layout of the diffuser. Some idea of the thickness of the vanes at various radii can then be obtained enabling more accurate estimates of flow areas and velocities to be made. When arriving at a final value of the

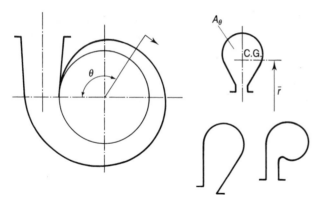

FIG. 4.6 Volutes

throat area, an empirical contraction coefficient will also be introduced to allow for the blockage effect of the boundary layer. The design of the diffuser is essentially a process of successive approximation: sufficient information has been given to indicate the method of approach.

It has been shown that a diffuser comprising channels of circular cross-section can be more efficient than the conventional type with rectangular channels. The passages are formed by drilling holes tangentially through a ring surrounding the impeller, and after the throats the passages are conical. Overall isentropic efficiencies of over 85 per cent are claimed for compressors using this 'pipe' diffuser—see Ref. (5).

Earlier it was stated that the air leaving the diffuser passages might be collected in a volute. The simplest method of volute design assumes that the angular momentum of the flow remains constant, i.e. friction in the volute is neglected. Using the nomenclature of Fig. 4.6, the cross-sectional area A_θ at any angle θ can then be shown to be given by

$$\frac{A_\theta}{\bar{r}} = V \frac{\theta}{2\pi K}$$

where K is the constant angular momentum $C_w r$, V is the volume flow, and $\bar{r}$ is the radius of the centroid of the cross-section of the volute. The shape of the cross-section has been shown to have little effect upon the friction loss: the various shapes shown in Fig. 4.6 all yield volutes of similar efficiency. For further information and useful references the reader may turn to Ref. (6).

4.4 Compressibility effects

It is well known (see Appendix A) that breakdown of flow and excessive pressure loss can be incurred if the velocity of a compressible fluid, relative to the surface over which it is moving, reaches the speed of sound in the fluid. This is a most important phenomenon in a diffusing process where there is always a tendency for the flow to break away from the boundary even at low speeds. When an effort is made to obtain the maximum possible mass flow from the smallest possible compressor, as is done most particularly in the design of aircraft gas turbines, the air speeds are very high. It is of the utmost importance that the Mach numbers at certain points in the flow do not exceed the value beyond which the losses increase rapidly due to the formation of shock waves. The critical Mach number is usually less than unity when calculated on the basis of the mean velocity of the fluid relative to the boundary, because the actual relative velocity near the surface of a curved boundary may be in excess of the mean velocity. As a general rule, unless actual tests indicate otherwise, the Mach numbers are restricted to about 0·8.

Consider now the Mach numbers at vital points in the compressor, beginning with the intake.

Inlet Mach number for impeller and diffuser

At the intake, the air is deflected through a certain angle before it passes into the radial channels on the impeller. There is always a tendency for the air to break away from the convex face of the curved part of the impeller vane. Here then is a point at which the Mach number will be extremely important; a shock wave might occur as shown in Fig. 4.7(a).

The inlet velocity triangle for the impeller eye is shown in Fig. 4.7(b). The full lines represent the case of axial inlet velocity which we have considered up to now. It has also been assumed that the axial velocity is uniform from the root to the tip of the eye. In this case, the velocity of the air

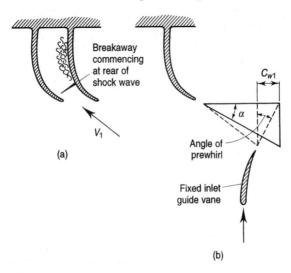

FIG. 4.7 **Effect of prewhirl**

relative to the vane, V_1, will reach a maximum at the eye tip where the vane speed is greatest. The inlet Mach number will be given by

$$M = \frac{V_1}{\sqrt{(\gamma R T_1)}}$$

where T_1 is the static temperature at the inlet.

Even though this Mach number is satisfactory under ground-level atmospheric conditions, if the compressor is part of an aircraft gas turbine the Mach number may well be too high at altitude. It is possible to reduce the relative velocity, and hence the Mach number, by introducing *prewhirl* at the intake. This is achieved by allowing the air to be drawn into the impeller eye over curved *inlet guide vanes* attached to the compressor casing. For the same axial velocity, and hence approximately the same mass flow, the relative velocity is reduced as shown by the dotted triangle in Fig. 4.7(b). An additional advantage of using prewhirl is that the curvature of the impeller vanes at the inlet is reduced, i.e. the inlet angle α is increased.

This method of reducing the Mach number unfortunately reduces the work capacity of the compressor. The air now has an initial whirl component C_{w1}, so that the rate of change of angular momentum per unit mass flow of air is

$$C_{w2} r_2 - C_{w1} r_1$$

If C_{w1} is constant over the eye of the impeller, then the initial angular momentum will increase from root to tip of the eye. The amount of work done on each kilogram of air will therefore depend upon the radius at which it enters the eye of the impeller. A mean value of the work done per kg can be found by using the initial angular momentum of the air at the mean radius of the eye. There is, however, no point in reducing the work capacity of the compressor unnecessarily, and the Mach number is only high at the tip. It is clearly preferable to vary the prewhirl, gradually reducing it from a maximum at the tip to zero at the root of the eye. This may be done if the inlet guide vanes are suitably twisted.

We will now check the relative Mach numbers in the compressor dealt with in Examples 4.1 and 4.2. Consider first the inlet Mach number for the tip radius of the impeller eye.

Inlet velocity	$= 143 \, \text{m/s}$ and is axial
Eye tip speed	$= 273 \, \text{m/s}$
Relative velocity at tip	$= \sqrt{(143^2 + 273^2)} = 308 \, \text{m/s}$
Velocity of sound	$= \sqrt{(1 \cdot 4 \times 0 \cdot 287 \times 284 \cdot 5 \times 10^3)}$
	$= 338 \, \text{m/s}$
Maximum Mach number at inlet	$= 308/338 = 0 \cdot 91$

This would not be considered satisfactory even if it were actually the maximum value likely to be reached. But if the compressor is part of an aircraft engine required to operate at an altitude of 11 000 m, where the atmospheric temperature is only about 217 K, we must calculate the Mach number under these conditions. Since there will be a temperature rise due to the ram effect in the intake when the aircraft has a forward speed, the effect of drop in atmospheric temperature will not be quite so great as might be expected. We will take 90 m/s as being the minimum speed likely to be reached at high altitude.

Temperature equivalent of
    forward speed $= 4\,\mathrm{K}$
Inlet stagnation temperature $= 217 + 4 = 221\,\mathrm{K}$
Temperature equivalent of axial
    inlet velocity from the first
    example† $= 10\cdot5\,\mathrm{K}$
Inlet static temperature at altitude $= 210\cdot5\,\mathrm{K}$

$$\text{Inlet Mach number at altitude} \quad = 0\cdot91\left(\frac{284\cdot5}{210\cdot5}\right)^{\frac{1}{2}} = 1\cdot06$$

This is clearly too high and we will find the Mach number when 30 degrees of prewhirl are introduced. In this case the absolute inlet velocity will be slightly higher than before, so that the inlet static temperature will be slightly lower. A new value for the axial velocity must be found by the usual trial and error process. Reverting to the original sea-level static case:

Try $C_{a1} = 150\,\mathrm{m/s}$

$$C_1 = \frac{150}{\cos 30} = 173\cdot2\,\mathrm{m/s}$$

Temperature equivalent of $C_1 = 14\cdot9\,\mathrm{K}$

$$T_1 = 295 - 14\cdot9 = 280\cdot1\,\mathrm{K}$$

$$p_1 = 0\cdot918\,\text{bar and } \rho_1 = 1\cdot14\,\mathrm{kg/m^3}$$

$$\text{Check on } C_{a1} = \frac{9}{1\cdot14 \times 0\cdot053} = 149\,\mathrm{m/s}$$

Whirl velocity at inlet, $C_{w1} = 149\tan 30$

$$= 86\,\mathrm{m/s}$$

$$\text{Maximum relative velocity} = \sqrt{\left[149^2 + (273 - 86)^2\right]}$$

$$= 239\ \mathrm{m/s}$$

Hence maximum inlet Mach number when $T_{01} = 295\,\mathrm{K}$ is

$$\frac{239}{\sqrt{(1\cdot4 \times 0\cdot287 \times 280\cdot1 \times 10^3)}} = 0\cdot71$$

---

† This assumes that the mass flow through the compressor at altitude will be such as to give correct flow direction into the eye. Then if the rotational speed is unaltered the axial velocity must be the same.

Under altitude conditions this would rise to little more than 0·8 and 30 degrees of prewhirl can be regarded as adequate.

To show the effect of the 30 degrees of prewhirl on the pressure ratio, we will take the worst case and assume that the prewhirl is constant over the eye of the impeller.

$$\text{Speed of impeller eye at mean radius, } U_e = \frac{273 + 136\cdot5}{2}$$

$$= 204\cdot8 \text{ m/s}$$

$$\text{Actual temperature rise } = \frac{\psi}{c_p}(\sigma U^2 - C_{w1}U_e)$$

$$= \frac{1\cdot04(0\cdot9 \times 455\cdot5^2 - 86 \times 204\cdot8)}{1\cdot005 \times 10^3} = 175 \text{ K}$$

$$\frac{p_{03}}{p_{01}} = \left[1 + \frac{0\cdot78 \times 175}{295}\right]^{3\cdot5} = 3\cdot79$$

This pressure ratio may be compared with the figure of 4·23 obtained with no prewhirl. It is sometimes advantageous to use adjustable inlet guide vanes to improve the performance under off-design conditions.

We next consider the relevant Mach numbers in the *diffuser*. The maximum value will occur at the entry to the diffuser: that is, at the impeller tip. Once again the values calculated in the previous examples of this chapter will be used.

In Example 4.1 the temperature equivalent of the resultant velocity of the air leaving the impeller was found to be

$$\frac{C_2^2}{2c_p} = 93\cdot8 \text{ K}$$

and hence

$$C_2 = 434 \text{ m/s}$$

T_2 was found to be 394·2 K, and thus the Mach number at the impeller tip equals

$$\frac{434}{\sqrt{(1\cdot4 \times 0\cdot287 \times 394\cdot2 \times 10^3)}} = 1\cdot09$$

Now consider the leading edge of the diffuser vanes. In Example 4.2 the whirl velocity was found to be 342 m/s and the radial component 97·9 m/s. The resultant velocity at this radius is therefore 356 m/s. The static temperature was 425·1 K at this radius so that the Mach number is

$$\frac{356}{\sqrt{(1\cdot4 \times 0\cdot287 \times 425\cdot1 \times 10^3)}} = 0\cdot86$$

In the particular design under consideration, the Mach number is 1·09 at the impeller tip and 0·86 at the leading edges of the diffuser vanes. It has been found that as long as the radial velocity component is subsonic, Mach numbers greater than unity can be used at the impeller tip without loss of efficiency: it appears that supersonic diffusion can occur without the formation of shock waves if it is carried out at constant angular momentum with vortex motion in the vaneless space, Ref. (7). But the Mach number at the leading edge of the diffuser vanes is rather high and it would probably be advisable to increase the radial width of the vaneless space or the depth of the diffuser to reduce the velocity at this radius.

High Mach numbers at the leading edges of the diffuser vanes are undesirable, not only because of the danger of shock losses, but because they imply high air speeds and comparatively large pressures at the stagnation points where the air is brought to rest locally at the leading edges of the vanes. This causes a circumferential variation in static pressure, which is transmitted upstream in a radial direction through the vaneless space to the impeller tip. Although the variation will have been considerably reduced by the time it reaches the impeller, it may well be large enough to excite the impeller vanes and cause mechanical failure due to vibrational fatigue cracks in the vanes. This will occur when the exciting frequency, which depends on the rotational speed and relative number of impeller and diffuser vanes, is of the same order as one of the natural frequencies of the impeller vanes. To reduce the likelihood of this, care is taken to see that the number of vanes in the impeller is not an exact multiple of the number in the diffuser; it is common practice to use a prime number for the impeller vanes and an even number for the diffuser vanes.

The reason for the vaneless space will now be apparent: both the dangers of shock losses and excessive circumferential variation in static pressure would be considerably increased if the leading edges of the diffuser vanes were too near the impeller tip where the Mach numbers are very high.

4.5   Non-dimensional quantities for plotting compressor characteristics

The performance of a compressor may be specified by curves of delivery pressure and temperature plotted against mass flow for various fixed values of rotational speed. These characteristics, however, are dependent on other variables such as the conditions of pressure and temperature at entry to the compressor and the physical properties of the working fluid. Any attempt to allow for full variations of all these quantities over the working range would involve an excessive number of experiments and make a concise presentation of the results impossible. Much of this complication may be eliminated by using the technique of dimensional analysis, by which the variables involved may be combined to form a smaller and more manageable number of dimensionless groups. As will be shown, the complete characteristics of any particular compressor may then be specified by only two sets of curves.

Before embarking on the dimensional analysis of the behaviour of a compressor, the following special points should be noted.

(a) When considering the dimensions of temperature, it is convenient always to associate with it the gas constant R so that the combined variable RT, being equal to p/ρ, has the dimensions $\mathbf{ML^{-1}T^{-2}/ML^{-3}} = \mathbf{L^2T^{-2}}$. Thus they are the same as those of (velocity)^2. When the same gas, e.g. air, is being employed during both the testing and subsequent use of the compressor, R can finally be eliminated, but if for any reason there is a change from one gas to another it must be retained in the final expressions.

(b) A physical property of the gas which undoubtedly influences the behaviour of the compressor is its density ρ but, if pressure p and the RT product are also cited, its inclusion is superfluous because $\rho = p/RT$.

(c) A further physical property of the gas which in theory would also influence the problem would be its viscosity. The presence of this variable would ultimately result in the emergence of a dimensionless group having the character of a Reynolds number. In the highly turbulent conditions which generally prevail in machines of this type, it is found from experience that the influence of this group is negligibly small over the normal operating range† and it is customary to exclude it from turbomachinery analysis.

Bearing the above points in mind, it is now possible to consider the various quantities which will both influence the behaviour of the compressor and depend upon it. The solution of the problem may then be stated in the form of an equation in which a function of all these variables is equated to zero. Thus in the present instance, it would be reasonable to state that

$$\text{Function}(D, N, m, p_{01}, p_{02}, RT_{01}, RT_{02}) = 0 \tag{4.6}$$

where D is a characteristic linear dimension of the machine (usually taken as the impeller diameter) and N is the rotational speed. Note that now we are simply considering the performance of the compressor as a whole; *suffix 2 will be used to refer to conditions at the compressor outlet for the remainder of this chapter.*

By the principle of dimensional analysis, often referred to as the Pi theorem, it is known that the function of seven variables expressed by

---

† To reduce the power requirements when testing large compressors, it is often necessary to throttle the flow at inlet, giving a drop in inlet pressure and hence mass flow. The resultant drop in density lowers the Reynolds number, and it is known that the performance of compressors drops off with a substantial reduction in Reynolds number.

equation (4.6) is reducible to a different function of $7 - 3 = 4$ non-dimensional groups formed from these variables. The reduction by 3 is due to the presence of the three fundamental units, **M**, **L**, **T**, in the dimensions of the original variables. Various techniques exist for the formation of these non-dimensional groups and it is possible in theory to obtain an infinite variety of self-consistent sets of these groups. Generally there is little difficulty in deriving them by inspection, having decided on the most suitable quantities to 'non-dimensionalize' by using the others. In this case, it is most useful to have the non-dimensional forms of p_{02}, T_{02}, m and N and the groups emerge as

$$\frac{p_{02}}{p_{01}}, \frac{T_{02}}{T_{01}}, \frac{m\sqrt{(RT_{01})}}{D^2 p_{01}}, \frac{ND}{\sqrt{(RT_{01})}}$$

This now means that the performance of the machine in regard to the variations of delivery pressure and temperature with mass flow, rotational speed and inlet conditions can be expressed in the form of a function of these groups. When we are concerned with the performance of a machine of fixed size compressing a specified gas, R and D may be omitted from the groups so that

$$\text{Function}\left(\frac{p_{02}}{p_{01}}, \frac{T_{02}}{T_{01}}, \frac{m\sqrt{T_{01}}}{p_{01}}, \frac{N}{\sqrt{T_{01}}}\right) = 0 \tag{4.7}$$

The quantities $m\sqrt{T_{01}}/p_{01}$ and $N/\sqrt{T_{01}}$ are usually termed the 'non-dimensional' mass flow and rotational speed respectively, although they are not truly dimensionless.

A function of this form can be expressed graphically by plotting one group against another for various fixed values of a third. In this particular case, experience has shown that the most useful plots are those of the pressure and temperature ratios p_{02}/p_{01} and T_{02}/T_{01} against the non-dimensional mass flow $m\sqrt{T_{01}}/p_{01}$ using the non-dimensional rotational speed $N/\sqrt{T_{01}}$ as a parameter. As will be shown in the next section, it is also possible to replace the temperature ratio by a derivative of it, namely the isentropic efficiency.

Finally, it is worth noting one particular physical interpretation of the non-dimensional mass flow and rotational speed parameters. The former can be written as

$$\frac{m\sqrt{(RT)}}{D^2 p} = \frac{\rho A C \sqrt{(RT)}}{D^2 p} = \frac{\rho A C \sqrt{(RT)}}{RTD^2 p} \propto \frac{C}{\sqrt{(RT)}} \propto M_F$$

and the latter as

$$\frac{ND}{\sqrt{(RT)}} = \frac{U}{\sqrt{(RT)}} \propto M_R$$

Thus the parameters can be regarded as a flow Mach number M_F and a rotational speed Mach number M_R. All operating conditions covered by a pair of values of $m\sqrt{T}/p$ and $N/\sqrt{T}$ should give rise to similar velocity triangles, so that the vane angles and airflow directions will match and the compressor will yield the same performance in terms of pressure ratio, temperature ratio and isentropic efficiency. This is what the non-dimensional method of plotting the characteristics implies.

4.6 Compressor characteristics

The value of the dimensional analysis is now evident. It is only necessary to plot two sets of curves in order to describe the performance of a compressor completely. The stagnation pressure and temperature ratios are plotted separately against 'non-dimensional' mass flow in the form of a family of curves, each curve being drawn for a fixed value of the 'non-dimensional' rotational speed. From these two sets of curves it is possible to construct constant speed curves of isentropic efficiency plotted against 'non-dimensional' mass flow because this efficiency is given by

$$\eta_c = \frac{T'_{02} - T_{01}}{T_{02} - T_{01}} = \frac{(p_{02}/p_{01})^{(\gamma-1)/\gamma} - 1}{(T_{02}/T_{01}) - 1}$$

Before describing a typical set of characteristics, it will be as well to consider what might be expected to occur when a valve, placed in the delivery line of a compressor running at constant speed, is slowly opened. The variation in pressure ratio is shown in Fig. 4.8. When the valve is shut and the mass flow is zero, the pressure ratio will have some value A, corresponding to the centrifugal pressure head produced by the action of the impeller on the air trapped between the vanes. As the valve is opened and flow commences, the diffuser begins to contribute its quota of pressure rise, and the pressure ratio increases. At some point B, where the efficiency approaches its maximum value, the pressure ratio will reach a maximum,

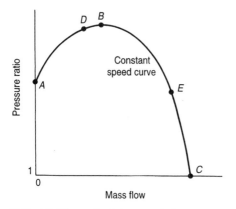

FIG. 4.8 Theoretical characteristic

and any further increase in mass flow will result in a fall of pressure ratio. For mass flows greatly in excess of that corresponding to the design mass flow, the air angles will be widely different from the vane angles, breakaway of the air will occur, and the efficiency will fall off rapidly. In this hypothetical case the pressure ratio drops to unity at *C*, when the valve is fully opened and all the power is absorbed in overcoming internal frictional resistance.

In actual practice, though point *A* could be obtained if desired, most of the curve between *A* and *B* could not be obtained owing to the phenomenon of *surging*. Surging is associated with a sudden drop in delivery pressure, and with violent aerodynamic pulsation which is transmitted throughout the whole machine. It may be explained as follows. If we suppose that the compressor is operating at some point *D* on the part of the characteristic having positive slope, then a decrease in mass flow will be accompanied by a fall of delivery pressure. If the pressure of the air downstream of the compressor does not fall quickly enough, the air will tend to reverse its direction and flow back in the direction of the resulting pressure gradient. When this occurs, the pressure ratio drops rapidly. Meanwhile, the pressure downstream of the compressor has fallen also, so that the compressor will now be able to pick up again to repeat the cycle of events which occurs at high frequency.

This surging of the air may not happen immediately the operating point moves to the left of *B* in Fig. 4.8, because the pressure downstream of the compressor may first of all fall at a greater rate than the delivery pressure. Sooner or later, as the mass flow is reduced, the reverse will apply and the conditions are inherently unstable between *A* and *B*. As long as the operating point is on the part of the characteristic having a negative slope, however, decrease of mass flow is accompanied by a rise of delivery pressure and stability of operation is assured. In a gas turbine, the actual point at which surging occurs depends upon the swallowing capacity of the components downstream of the compressor, e.g. the turbine, and the way in which this swallowing capacity varies over the range of operating conditions.

Surging probably starts to occur in the diffuser passages, where the flow is retarded by frictional forces near the vanes: certainly the tendency to surge appears to increase with the number of diffuser vanes. This is due to the fact that it is very difficult to split the flow of air so that the mass flow is the same in each passage. When there are several diffuser channels to every impeller channel, and these deliver into a common outlet pipe, there is a tendency for the air to flow up one channel and down another when the conditions are conducive to surging. If this occurs in only one pair of channels the delivery pressure will fall, and thus increase the likelihood of surging. For this reason, the number of diffuser vanes is usually less than the number of impeller vanes. The conditions of flow are then approximately the same in each diffuser passage, because if each is supplied with air from several impeller channels the variations in pressure and velocity between the channels will be evened out by the time the air reaches the diffuser. Surging is therefore not likely to occur until the instability has reached a point at which reversal of flow will occur in most of the diffuser passages simultaneously.

There is one other important cause of instability and poor performance, which may contribute to surge but can exist in the nominally stable operating range: this is the *rotating stall*. When there is any non-uniformity in the flow or geometry of the channels between vanes or blades, breakdown in the flow in one channel, say B in Fig. 4.9, causes the air to be deflected in such a way that channel C receives fluid at a reduced angle of incidence and channel A at an increased incidence. Channel A then stalls, resulting in a reduction of incidence to channel B enabling the flow in that channel to recover. Thus the stall passes from channel to channel: at the impeller eye it would rotate in a direction opposite to the direction of rotation of the impeller. Rotating stall may lead to aerodynamically induced vibrations resulting in fatigue failures in other parts of the gas turbine.

Returning now to consider the hypothetical constant speed curve ABC in Fig. 4.8, there is an additional limitation to the operating range, between B and C. As the mass flow increases and the pressure decreases, the density is reduced and the radial component of velocity must increase. At constant rotational speed this must mean an increase in resultant velocity and hence in angle of incidence at the diffuser vane leading edge. Sooner or later, at some point E say, the position is reached where no further increase in mass flow can be obtained and choking is said to have occurred. This point represents the maximum delivery obtainable at the particular rotational speed for which the curve is drawn. Other curves may be obtained for different speeds, so that the actual variation of pressure ratio over the complete range of mass flow and rotational speed will be shown by curves such as those in Fig. 4.10(a). The left-hand extremities of the constant speed curves may be joined up to form what is known as the *surge line*, while the right-hand extremities represent the points where choking occurs.

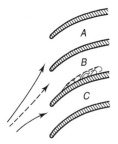

FIG. 4.9 Rotating stall

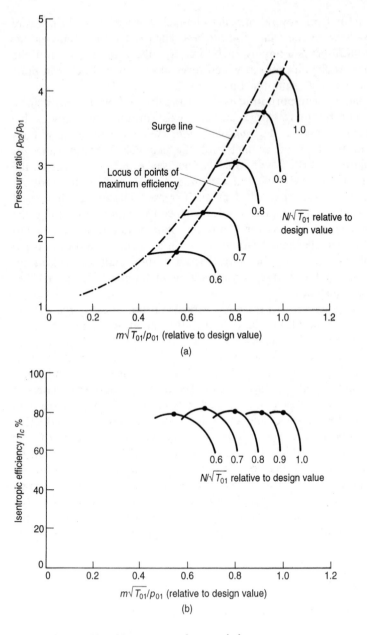

FIG. 4.10 Centrifugal compressor characteristics

The temperature ratio is a simple function of the pressure ratio and isentropic efficiency, so that the form of the curves for temperature ratio plotted on the same basis will be similar to Fig. 4.10(a); there is no need to give a separate diagram here. From these two sets of curves the isentropic efficiency may be plotted as in Fig. 4.10(b) or, alternatively, contour lines for various values of the efficiency may be superimposed upon Fig. 4.10(a). The efficiency varies with mass flow at a given speed in a similar manner to the pressure ratio, but the maximum value is approximately the same at all speeds. A curve representing the locus of operating points for maximum efficiency can be obtained as shown by the dotted curve in Fig. 4.10(a). Ideally, the gas turbine should be so designed that the compressor will always be operating on this curve. Chapter 8 will describe a method for estimating the position of the operating line on the compressor characteristics.

In conclusion, mention must be made of two other parameters frequently used in preference to $m\sqrt{T_{01}}/p_{01}$ and $N/\sqrt{T_{01}}$ when plotting compressor characteristics. These are the *equivalent flow* $m\sqrt{\theta}/\delta$ and *equivalent speed* $N/\sqrt{\theta}$, where $\theta = T_{01}/T_{ref}$ and $\delta = p_{01}/p_{ref}$. The reference ambient state is normally that corresponding to the ISA at sea level, namely 288 K and 1·013 bar. When the compressor is operating with the reference intake condition, $m\sqrt{\theta}/\delta$ and $N/\sqrt{\theta}$ are equal to the actual mass flow and rotational speed respectively. With this method of plotting the characteristics, the numbers on the axes are recognizable quantities.

4.7 Computerized design procedures

Although the approach to the design of centrifugal compressors given here is adequate as an introduction to the subject, the reader should be aware that more sophisticated methods are available. For example, Ref. (9) outlines a computerized design procedure developed at the National Gas Turbine Establishment (now Royal Aircraft Establishment). This makes use of the Marsh 'matrix throughflow' aerodynamic analysis for determining the shape of the impeller channels. It includes a program for checking the stresses in the impeller, and concludes with one which predicts the performance characteristics of the compressor. The procedure leads to a numerical output of co-ordinates for defining the shape of the impeller and diffuser vanes, in such a form that it can be fed directly to numerically controlled machine tools used for manufacturing these components. Tests have suggested that compressors designed in this way have an improved performance.

Reference (10) describes a method of performance prediction which has been compared with test results from seven different compressors, including one designed on the basis of Ref. (9). The form of the pressure ratio v. mass flow characteristics, and the choking flow, were predicted satisfactorily. The efficiency was within ±1–2 per cent of the experimental value at the design speed, with somewhat larger discrepancies at low speeds. No general means have yet been devised for predicting the surge line.

Reference (11) reviews centrifugal compressor design methods that are commonly used in industry. The design process, starting with the preliminary design, and its reliance on empirical rules through the state-of-the-art design using computational fluid dynamics (CFD), is presented. The basics of the mechanical design of the impeller are also introduced.

NOMENCLATURE

n	number of vanes
N	rotational speed
r	radius
U	impeller speed at tip
U_e	impeller speed at mean radius of eye
V	relative velocity, volume flow
α	vane angle
σ	slip factor
ψ	power input factor
ω	angular velocity

Suffixes

a	axial component, ambient
r	radial component
w	whirl component

5
Axial flow compressors

The importance of a high overall pressure ratio in reducing specific fuel consumption was referred to in sections 2.4 and 3.3, and the difficulties of obtaining a high pressure ratio with the centrifugal compressor were pointed out in Chapter 4. From an early stage in the history of the gas turbine, it was recognized that the axial flow compressor had the potential for both higher pressure ratio and higher efficiency than the centrifugal compressor. Another major advantage, especially for jet engines, was the much larger flow rate possible for a given frontal area. These potential gains have now been fully realized as the result of intensive research into the aerodynamics of axial compressors: the axial flow machine dominates the field for large powers and the centrifugal compressor is restricted to the lower end of the power spectrum where the flow is too small to be handled efficiently by axial blading.

Early axial flow units had pressure ratios of around 5:1 and required about 10 stages. Over the years the overall pressure ratios available have risen dramatically, and some turbofan engines have pressure ratios exceeding 40:1. Continued aerodynamic development has resulted in a steady increase in stage pressure ratio, with the result that the number of stages for a given overall pressure ratio has been greatly reduced. There has been in consequence a reduction in engine weight for a specified level of performance, which is particularly important for aircraft engines. It should be realized, however, that high stage pressure ratios imply high Mach numbers and large gas deflections in the blading which would not generally be justifiable in an industrial gas turbine where weight is not critical; industrial units, built on a much more restricted budget than an aircraft engine, will inevitably use more conservative design techniques resulting in more stages.

In accordance with the introductory nature of this book, attention will be focused on the design of subsonic compressors. True supersonic compressors, i.e. those for which the velocity at entry is everywhere supersonic, have not proceeded beyond the experimental stage. *Transonic* compressors, however, in which the velocity relative to a moving row of blades is supersonic over *part* of the blade height, are now successfully used in both aircraft and industrial gas turbines. The reader must turn to more advanced texts for a full discussion of these topics.

5.1 Basic operation

The axial flow compressor consists of a series of stages, each stage comprising a row of rotor blades followed by a row of stator blades: the individual stages can be seen in Fig. 5.1. The working fluid is initially accelerated by the rotor blades, and then decelerated in the stator blade passages wherein the kinetic energy transferred in the rotor is converted to static pressure. The process is repeated in as many stages as are necessary to yield the required overall pressure ratio.

The flow is always subject to an adverse pressure gradient, and the higher the pressure ratio the more difficult becomes the design of the compressor. The process consists of a series of diffusions, both in the rotor and stator blade passages: although the *absolute* velocity of the fluid is increased in the rotor, it will be shown that the fluid velocity *relative* to the rotor is decreased, i.e. there is diffusion within the rotor passages. The need for moderate rates of change of cross-sectional area in a diffusing flow has already been emphasized in the previous chapter. This limit on the diffusion in each stage means that a single compressor stage can provide only a relatively small pressure ratio, and very much less than can be used by a turbine with its advantageous pressure gradient, converging blade passages, and accelerating flow (Fig. 5.2). This is why a single turbine stage can drive a large number of compressor stages.

Careful design of compressor blading based on both aerodynamic theory and experiment is necessary, not only to prevent wasteful losses, but also to ensure a minimum of stalling troubles which are all too prevalent in axial compressors, especially if the pressure ratio is high. Stalling, as in the case of isolated aerofoils, arises when the difference between the flow direction and the blade angle (i.e. the angle of incidence) becomes excessive. The fact that the pressure gradient is acting against the flow

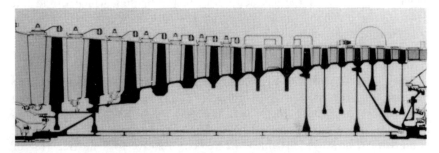

FIG. 5.1 16-stage high pressure ratio compressor [courtesy General Electric]

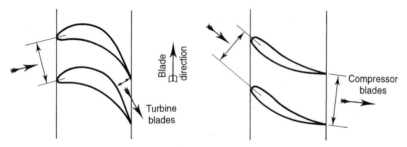

Blade direction

Turbine blades

Compressor blades

FIG. 5.2 Comparison of typical forms of turbine and compressor rotor blades

direction is always a danger to the stability of the flow, and flow reversals may easily occur at conditions of mass flow and rotational speed which are different from those for which the blades were designed.

The compressor shown in Fig. 5.1 makes use of inlet guide vanes (IGVs), which guide the flow into the first stage. Many industrial units have variable IGVs, permitting the flow angle entering the first stage to vary with rotational speed to improve the off-design performance. Most aircraft engines have now dispensed with IGVs, however, mainly to obtain the maximum possible flow per unit area and minimum engine weight. Other benefits include an easing of noise and icing problems.

Figure 5.1 shows the marked change in blade size from front to rear in a high pressure ratio compressor. For reasons which will appear later, it is desirable to keep the axial velocity approximately constant throughout the compressor. With the density increasing as the flow progresses through the machine, it is therefore necessary to reduce the flow area and hence the blade height. When the machine is running at a lower speed than design, the density in the rear stages will be far from the design value, resulting in incorrect axial velocities which will cause blade stalling and compressor surge. Several methods may be used to overcome this problem, all of which entail increased mechanical complexity. The approach of Rolls-Royce and Pratt and Whitney has been to use multi-spool configurations, whereas General Electric has favoured the use of variable stator blades; the IGVs and first six stator rows of a GE compressor can be seen to be pivoted in Fig. 5.1. For turbofan engines the large difference in diameter between the fan and the rest of the compressor requires the use of multi-spool units; Pratt and Whitney and General Electric have used two spools, but Rolls-Royce have used three spools on the RB-211. Another possibility is the use of blow-off valves, and on advanced engines it is sometimes necessary to include all these schemes. It is important to realize that the designer must consider at the outset the performance of the compressor at conditions far from design, although detailed discussion of these matters is left to the end of the chapter.

In early axial compressors the flow was entirely subsonic, and it was found necessary to use aerofoil section blading to obtain a high efficiency. The need to pass higher flow rates at high pressure ratios increased the Mach numbers, which became especially critical at the tip of the first row of rotor blades. Eventually it became necessary to design blading for transonic compressors, where the flow over part of the blade is supersonic. It was found that the most effective blading for transonic stages consisted of sections based on circular arcs, often referred to as biconvex blading. As Mach numbers were further increased it was found that blade sections based on parabolas became more effective, and most high-performance stages no longer use aerofoil sections.

Long-term improvements in gas turbine power and efficiency have resulted from continuous development of compressors, combined with increases in cycle temperature. The number of stages required for a given pressure ratio is important both for aircraft engines, to minimize weight, and for industrial engines, to minimize cost. Openly published data from two manufacturers, Alstom and Rolls-Royce, will be used to illustrate trends; similar data could be shown for all major manufacturers. The dates are approximate, but quite adequate to show the continuous progress. The Alstom line of small industrial engines originated with the Ruston organization and relevant data are tabulated below.

Engine	Date	Power [MW]	Pressure ratio	Stages
TB 5000	1970	3·9	7·8	12
Tornado	1981	6·75	12·3	15
Typhoon	1989	4·7	14·1	10
Tempest	1995	7·7	13·9	10
Cyclone	2000	12·9	16·9	11

Rolls-Royce data for civil aero engines are given in a similar format; the Avon was a simple single-spool turbojet used in the Comet 4, the Spey was a twin-spool low bypass ratio turbofan and the RB 211/Trent are large high bypass ratio turbofans.

Engine	Date	Thrust [kN]	Pressure ratio	Stages
Avon	1958	44	10	17
Spey	1963	56	21	17
RB-211	1972	225	29	14
Trent	1995	356	41	15

The Typhoon compressor can be seen in Fig. 1.13 and the Trent in Fig. 1.16.

Before looking at the basic theory of the axial flow compressor it should be emphasized that successful compressor design is very much an art, and all the major engine manufacturers have developed a body of knowledge which is kept proprietary for competitive reasons. Bearing in mind the introductory nature of this text, the aim of this chapter will be to present only the fundamentals of compressor design.

5.2   Elementary theory

The working fluid in an axial flow compressor is normally air, but for closed-cycle gas turbines other gases such as helium or carbon dioxide might be used. The analysis which follows is applicable to any gas, but unless otherwise noted it will be assumed that the working fluid is air.

A sketch of a typical stage is shown in Fig. 5.3. Applying the steady flow energy equation to the rotor, and recognizing that the process can be assumed to be adiabatic, it can readily be seen that the power input is given by

$$W = mc_p(T_{02} - T_{01}) \tag{5.1}$$

Repeating with the stator, where the process can again be assumed adiabatic and there is zero work input, it follows that $T_{02} = T_{03}$. All the power is absorbed in the rotor, and the stator merely transforms kinetic energy to an increase in static pressure with the stagnation temperature remaining constant. The increase in stagnation pressure is accomplished wholly within the rotor and, in practice, there will be some decrease in stagnation pressure in the stator due to fluid friction. Losses will also occur in the rotor and the stagnation pressure rise will be less than would

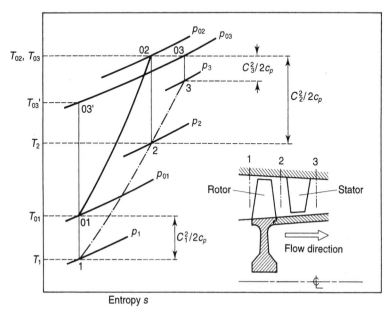

FIG. 5.3 Compressor stage and T–s diagram

be obtained with an isentropic compression and the same power input. A T–s diagram for the stage, showing the effect of losses in both rotor and stator, is also shown in Fig. 5.3.

Obtaining the power input to the stage from simple thermodynamics is no help in designing the blading. For this purpose we need to relate the power input to the stage velocity triangles. Initially attention will be focused on a simple analysis of the flow at the mean height of a blade where the peripheral speed is U, assuming the flow to occur in a tangential plane at the mean radius. This two-dimensional approach means that in general the flow velocity will have two components, one axial (denoted by subscript a) and one tangential (denoted by subscript w, implying a whirl velocity). This simplified analysis is reasonable for the later stages of an axial flow compressor where the blade height is small and the blade speeds at root and tip are similar. At the front end of the compressor, however, the blades are much longer, there are marked variations in blade speed from root to tip, and it becomes essential to consider three-dimensional effects in analysing the flow: these will be treated in a later section.

The velocity vectors and associated velocity diagram for a typical stage are shown in Fig. 5.4. The air approaches the rotor with a velocity C_1 at an angle α_1 from the axial direction; combining C_1 vectorially with the blade speed U gives the velocity relative to the blade, V_1, at an angle β_1 from the axial direction. After passing through the rotor, which increases the absolute velocity of the air, the fluid leaves the rotor with a relative velocity V_2 at an angle β_2 determined by the rotor blade outlet angle.

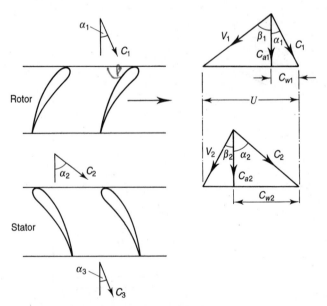

FIG. 5.4 Velocity triangles for one stage

Assuming that the design is such that the axial velocity C_a is kept constant, the value of V_2 can be obtained and the outlet velocity triangle constructed by combining V_2 and U vectorially to give C_2 at angle α_2. The air leaving the rotor at α_2 then passes to the stator where it is diffused to a velocity C_3 at angle α_3; typically the design is such that $C_3 \approx C_1$ and $\alpha_3 \approx \alpha_1$ so that the air is prepared for entry to another similar stage.

Assuming that $C_a = C_{a1} = C_{a2}$, two basic equations follow immediately from the geometry of the velocity triangles. These are

$$\frac{U}{C_a} = \tan \alpha_1 + \tan \beta_1 \tag{5.2}$$

$$\frac{U}{C_a} = \tan \alpha_2 + \tan \beta_2 \tag{5.3}$$

By considering the change in angular momentum of the air in passing through the rotor, the following expression for the power input to the stage can be deduced:

$$W = mU(C_{w2} - C_{w1})$$

where C_{w1} and C_{w2} are the tangential components of fluid velocity before and after the rotor. This expression can be put in terms of the axial velocity and air angles to give

$$W = mUC_a(\tan \alpha_2 - \tan \alpha_1)$$

It is more useful, however, to express the power in terms of the rotor blade air angles, β_1 and β_2. It can readily be seen from equations (5.2) and (5.3) that $(\tan \alpha_2 - \tan \alpha_1) = (\tan \beta_1 - \tan \beta_2)$. Thus the power input is given by

$$W = mUC_a(\tan \beta_1 - \tan \beta_2) \tag{5.4}$$

This input energy will be absorbed usefully in raising the pressure of the air and wastefully in overcoming various frictional losses. But regardless of the losses, or in other words of the efficiency of compression, the whole of this input will reveal itself as a rise in stagnation temperature of the air. Equating (5.1) and (5.4), the stagnation temperature rise in the stage, ΔT_{0S}, is given by

$$\Delta T_{0S} = T_{03} - T_{01} = T_{02} - T_{01} = \frac{UC_a}{c_p}(\tan \beta_1 - \tan \beta_2) \tag{5.5}$$

The pressure rise obtained will be strongly dependent on the efficiency of the compression process, as should be clear from Fig. 5.3. Denoting the isentropic efficiency of the stage by η_S, where $\eta_S = (T'_{03} - T_{01})/(T_{03} - T_{01})$, the stage pressure ratio is then given by

$$R_S = \frac{p_{03}}{p_{01}} = \left[1 + \frac{\eta_S \Delta T_{0S}}{T_{01}}\right]^{\gamma/(\gamma-1)} \tag{5.6}$$

It can now be seen that to obtain a high temperature rise in a stage, which is desirable to minimize the number of stages for a given overall pressure ratio, the designer must combine

(i)   high blade speed
(ii)   high axial velocity
(iii) high fluid deflection ($\beta_1 - \beta_2$) in the rotor blades.

Blade stresses will obviously limit the blade speed, and it will be seen in the next section that aerodynamic considerations and the previously mentioned adverse pressure gradient combine to limit (ii) and (iii).

5.3   Factors affecting stage pressure ratio

Tip speed

The centrifugal stress in the rotor blades depends on the rotational speed, the blade material and the length of the blade. The maximum centrifugal tensile stress, which occurs at the blade root, can be seen to be given by

$$(\sigma_{ct})_{max} = \frac{\rho_b \omega^2}{a_r} \int_r^t ar \, dr$$

where ρ_b is the density of the blade material, ω is the angular velocity, a is the cross-sectional area of the blade at any radius, and suffixes r and t refer to root and tip of the blade. If, for simplicity, the blade cross-section is assumed to be constant from root to tip

$$(\sigma_{ct})_{max} = \frac{\rho_b}{2}(2\pi N)^2(r_t^2 - r_r^2) = 2\pi N^2 \rho_b A$$

where N is the rotational speed and A the annulus area. The tip speed, U_t, is given by $2\pi N r_t$ so the equation for centrifugal tensile stress can be written also as

$$(\sigma_{ct})_{max} = \frac{\rho_b}{2} U_t^2 \left[1 - \left(\frac{r_r}{r_t}\right)^2\right]$$

The ratio r_r/r_t is normally referred to as the *hub–tip ratio*. It is immediately apparent that the centrifugal stress is proportional to the square of the tip speed, and that a reduction of hub–tip ratio increases the blade stress. It will be seen later that the hub–tip ratio is also very important with regard to aerodynamic considerations.

In practice, the blade sectional area will be decreased with radius to relieve the blade root stress and the loading on the disc carrying the blades, so that the integral would be evaluated numerically or graphically. A simple analytical expression can be deduced, however, for a linear variation of cross-sectional area from root to tip. If the hub–tip ratio is b and the ratio of the cross-sectional area at the tip to that at the root is d, the stress in a tapered blade is given by

$$(\sigma_{ct})_{max} = \frac{\rho_b}{2} U_t^2 (1 - b^2) K$$

where $K = 1 - [(1 - d)(2 - b - b^2)/3(1 - b^2)]$.

Typical values of K would range from 0·55 to 0·65 for tapered blades.

Direct centrifugal tensile stresses are not often of major concern in compressor blades. The first-stage blades, being the longest, are the most highly stressed, but the later stages often appear to be very moderately stressed. The designer should not be lulled into a false sense of security by this, because the blades are also subject to fluctuating gas bending stresses which may cause fatigue failure. Although the problem is significantly more difficult for compressors than turbines, because of the high probability of aerodynamic vibration resulting from flow instability when some stages are operating in a stalled condition, a discussion of methods of predicting gas bending stresses is left to Chapter 7.

For tip speeds of around 350 m/s, stress problems are not usually critical in the sizing of the annulus. Tip speeds of 450 m/s, however, are common in the fans of high bypass ratio turbofans, and with their low hub–tip ratios the design of the disc to retain the long heavy blades becomes critical. The inner radius of the annulus may then be dictated by disc stressing constraints.

Axial velocity

The expression for stage temperature rise, which in conjunction with an isentropic efficiency determines the stage pressure ratio, showed the desirability of using a high value of axial velocity. A high axial velocity is also required to provide a high flow rate per unit frontal area, which is important for turbojet and turbofan engines.

The axial velocity at inlet, however, must be limited for aerodynamic reasons. Considering a first stage with no IGVs, the entry velocity will be purely axial in direction and the velocity diagram at entry to the rotor will be as shown by the right-angled triangle in Fig. 5.5. The velocity relative to the rotor is given by $V_1^2 = C_1^2 + U^2$ and, assuming the axial velocity to be constant over the blade height, the maximum relative velocity will occur at the tip, i.e. V_{1t}. The static temperature at rotor entry, T_1, is given by $T_{01} - (C_1^2/2c_p)$ and the local acoustic velocity by $a = \sqrt{(\gamma R T_1)}$. The Mach number relative to the rotor tip is V_{1t}/a, and for a given speed U_t it is therefore determined by the axial velocity at entry to the stage as indicated by the curves in Fig. 5.5. Axial velocities for an industrial gas turbine will usually be of the order of 150 m/s whereas for advanced aero engines they could be up to 200 m/s.

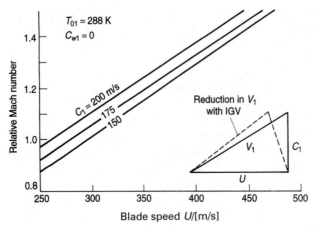

FIG. 5.5 Relative Mach number at rotor entry

On early compressors the design had to be such that the Mach number at the rotor tip was subsonic, but in the early 1950s it became possible to use transonic Mach numbers up to about 1·1 without introducing excessive losses. With fans of large bypass ratio the Mach number at the rotor tip may be of the order of 1·5. The dotted velocity triangle in Fig. 5.5 shows how the Mach number at entry may be slightly reduced by means of IGVs, and IGVs were aerodynamically necessary until the ability to operate at transonic Mach numbers was demonstrated. As will be explained in section 5.12, the twin-spool compressor goes some way towards alleviating the tip Mach number problem, because the LP compressor runs at a lower speed than the HP compressor, thereby reducing the tip speed at entry. The acoustic velocity increases in successive stages because of the progressive increase in static temperature, so the Mach number problem diminishes for the later stages. We have already seen that the later stages are also not so critical from the mechanical point of view, short blades implying low stresses.

For a given frontal diameter, the flow area can be increased by decreasing the hub–tip ratio. As this ratio is reduced to very small values, however, the incremental gain in flow area becomes less and less, and as mentioned previously the mechanical design of the first-stage disc becomes difficult; it will be seen later that the Mach number at the tip of the first-stage stator blades also becomes important. For these reasons, hub–tip ratios much below 0·4 are not used for aero engines and significantly higher values are normal for industrial gas turbines.

When air is the working fluid it is generally found that compressibility effects become critical before stress considerations, i.e. U_t and C_a are limited by the need to keep the relative gas velocity to an acceptable level. Closed-cycle gas turbines, with external combustion, are not restricted to using air as the working fluid and both helium and carbon dioxide are possible working fluids. With a light gas such as helium, the gas constant is much higher than for air and the acoustic velocity is correspondingly greater; in this case, it is found that the Mach numbers are low and stress limitations predominate. For a heavier gas such as carbon dioxide, the opposite is true.

High fluid deflections in the rotor blades

Recalling equation (5.5) for convenience, the stage temperature rise was given by $\Delta T_{0S} = UC_a(\tan \beta_1 - \tan \beta_2)/c_p$. For most compressor stages it can be assumed with little error that the value of U is the same at inlet and outlet of a rotor blade for any particular streamline, and the velocity triangles can conveniently be drawn on a common base as in Fig. 5.6. The amount of deflection required in the rotor is shown by the directions of the relative velocity vectors V_1 and V_2, and the change in whirl velocity is ΔC_w. Considering a fixed value of β_1, it is obvious that increasing the deflection by reducing β_2 entails a reduction in V_2. In other words, high fluid deflection implies a high rate of diffusion. The designer must have some method for assessing the allowable diffusion, and one of the earliest criteria used was the *de Haller number*, defined as V_2/V_1; a limit of $V_2/V_1 \not< 0.72$ was set, lower values leading to excessive losses. Because of its extreme simplicity, the de Haller number is still used in preliminary design work, but for final design calculations a criterion called the *diffusion factor* is preferred. The latter concept was developed by NACA (the precursor of NASA), and it is widely used on both sides of the Atlantic.

To explain the diffusion factor it is necessary to make a small foray into cascade testing, which will be discussed more fully in a later section. Figure 5.7 shows a pair of typical blades which have a pitch s and a chord c. The air passing over an aerofoil will accelerate to a higher velocity on the convex surface, and in a stationary row this will give rise to a drop in static pressure; for this reason the convex surface is known as the suction

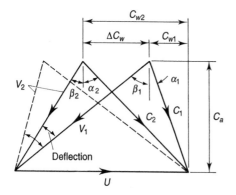

FIG. 5.6 Effect of increasing fluid deflection

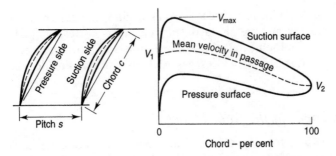

FIG. 5.7 Blade spacing and velocity distribution through passage

side of the blade. On the concave surface, the pressure side, the fluid will be decelerated. The velocity distribution through the blade passage will be of the form shown: the maximum velocity on the suction surface will occur at around 10–15 per cent of the chord from the leading edge and will then fall steadily until the outlet velocity is reached. The losses in a blade row arise primarily from the growth of boundary layers on the suction and pressure sides of the blade. These surface boundary layers come together at the blade trailing edge to form a wake, giving rise to a local drop in stagnation pressure. Relatively thick surface boundary layers, resulting in high losses, have been found to occur in regions where rapid changes of velocity are occurring, i.e. in regions of high velocity gradient. Figure 5.7 would suggest that these would be most likely to occur on the suction surface. The derivation of the NACA diffusion factor is based on the establishment of the velocity gradient on the suction surface in terms of V_1, V_2 and V_{max} in conjunction with results from cascade tests. From the cascade tests it was deduced that the maximum velocity $V_{max} \approx V_1 + 0.5(\Delta C_w s/c)$. In simplified form,† the diffusion factor, D, can be expressed as

$$D \approx \frac{V_{max} - V_2}{V_1} \approx \frac{V_1 + \dfrac{\Delta C_w}{2} \cdot \dfrac{s}{c} - V_2}{V_1}$$

$$\approx 1 - \frac{V_2}{V_1} + \frac{\Delta C_w}{2V_1} \cdot \frac{s}{c} \tag{5.7}$$

The variation of friction loss with D obtained from a large number of NACA tests [Ref. (6)] is shown in Fig. 5.8. These tests were carried out over a wide range of cascade geometries for a particular aerofoil section, and were found to be generally applicable as long as the maximum local Mach numbers were subsonic or only slightly supersonic. It can be seen that for the rotor hub region and stators the losses are unaffected by variation in D up to 0·6; in the rotor tip region, however, the losses

---

† The full derivation of the diffusion factor is given in Ref. (4).

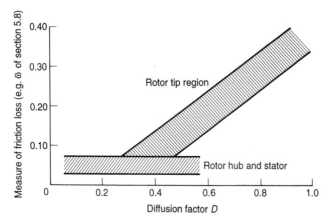

FIG. 5.8 Variation of friction loss with diffusion factor

increase rapidly at values of D above 0·4. The great merit of the diffusion factor, as a criterion for placing a limit on the permissible gas deflection, is its relative simplicity and the fact that it can be established as soon as the velocity diagram has been constructed and a value of the pitch/chord ratio has been chosen. In American practice the term *solidity*, the inverse of pitch/chord ratio, is used.

5.4   Blockage in the compressor annulus

Because of the adverse pressure gradient in compressors, the boundary layers along the annulus walls thicken as the flow progresses. The main effect is to reduce the area available for flow below the geometric area of the annulus. This will have a considerable effect on the axial velocity through the compressor and must be allowed for in the design process. The flow is extremely complex, with successive accelerations and decelerations combined with changes in tangential flow direction; the effects of tip clearance are also significant, making the calculation of boundary layer growth extremely difficult. For this reason, compressor designers normally make use of empirical correction factors based on experimental data from compressor tests.

Early British experiments revealed that the stage temperature rise was always less than would be given by equation (5.5). The explanation of this is based on the fact that the radial distribution of axial velocity is not constant across the annulus, but becomes increasingly 'peaky' as the flow proceeds, settling down to a fixed profile at about the fourth stage. This is illustrated in Fig. 5.9 which shows typical axial velocity profiles in the first and fourth stages. To show how the change in axial velocity affects the work-absorbing capacity of the stage, equation (5.4) can be recast using equation (5.2) as

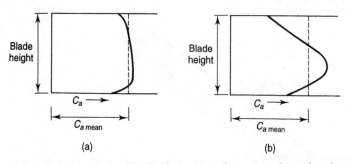

FIG. 5.9 Axial velocity distributions: (a) at first stage, (b) at fourth stage

$$W = mU[(U - C_a \tan \alpha_1) - C_a \tan \beta_2]$$
$$= mU[U - C_a(\tan \alpha_1 + \tan \beta_2)] \tag{5.8}$$

The outlet angles of the stator and rotor blades determine the values of α_1 and β_2 which can therefore be regarded as fixed (unlike α_2 which varies with C_a). Equation (5.8) shows that an increase in C_a will result in a decrease of W. If it is assumed that the compressor has been designed for a constant radial distribution of axial velocity as shown by the dotted line in Fig. 5.9, the effect of an increase in C_a in the central region of the annulus will be to reduce the work capacity of the blading in that area. The reduction in C_a at the root and tip might be expected to compensate for this effect by increasing the work capacity of the blading close to the annulus walls. Unfortunately the influence of both the boundary layers on the annulus walls and the blade tip clearance has an adverse effect on this compensation and the net result is a decrease in total work capacity. This effect becomes more pronounced as the number of stages is increased.

The reduction in work capacity can be accounted for by use of the *work-done factor* λ, which is a number less than unity. The actual stage temperature rise is then given by

$$\Delta T_{0S} = \frac{\lambda}{c_p} UC_a(\tan \beta_1 - \tan \beta_2) \tag{5.9}$$

Because of the variation of axial velocity profile through the compressor, the mean work-done factor will vary as shown in Fig. 5.10.

Care should be taken to avoid confusion of the work-done factor with the idea of an efficiency. If W is the value of the work input calculated from equation (5.4), then λW is the measure of the actual work which can be supplied to the stage. Having established the actual temperature rise, application of the isentropic efficiency of the stage yields the pressure ratio in accordance with equation (5.6).

It was mentioned earlier that a high axial velocity was required for a high stage temperature rise, and at first sight the explanation of work-done

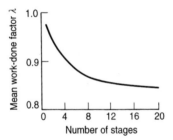

FIG. 5.10 Variation of mean work-done factor with number of stages

factor, on the basis of a lower temperature rise in regions of high axial velocity, appears to be contradictory. This is not so for the following reason. At the design stage, where the values of β_1 and β_2 can be selected to give a satisfactory ΔT_{0S} within the limits set by the de Haller number or diffusion factor, a high axial velocity is required. Once the design has been specified, however, and the angles α_1, β_1 and β_2 are fixed, then equation (5.8) correctly shows that an increase in axial velocity will reduce the stage temperature rise.

An alternative approach is to assign 'blockage factors' to reduce the effective annulus area to allow for the growth in boundary layer thickness, and this is the American design practice. Both the 'work-done factor' and 'blockage factor' represent empirical corrections based on a particular organization's experience of compressor development.

5.5 Degree of reaction

It was shown in an earlier section that in an axial compressor stage, diffusion takes place in both rotor and stator, and there will be an increase in static pressure through both rows. The *degree of reaction* Λ provides a measure of the extent to which the rotor contributes to the overall static pressure rise in the stage. It is normally defined in terms of enthalpy rise as follows,

$$\Lambda = \frac{\text{static enthalpy rise in the rotor}}{\text{static enthalpy rise in the stage}}$$

but, because the variation of c_p over the relevant temperature ranges is negligible, the degree of reaction can be expressed more conveniently in terms of temperature rises.

The degree of reaction is a useful concept in compressor design, and it is possible to obtain a formula for it in terms of the various velocities and air angles associated with the stage. This will be done for the most common case in which it is assumed that (*a*) C_a is constant through the stage, and (*b*) the air leaves the stage with the same absolute velocity with which it enters, i.e. $C_3 = C_1$ and hence $\Delta T_S = \Delta T_{0S}$. If ΔT_A and ΔT_B denote the static temperature rises in the rotor and stator respectively, then equation (5.5) gives, per unit mass flow,

$$W = c_p(\Delta T_A + \Delta T_B) = c_p \Delta T_S = UC_a(\tan \beta_1 - \tan \beta_2)$$
$$= UC_a(\tan \alpha_2 - \tan \alpha_1) \qquad (5.10)$$

Since all the work input to the stage takes place in the rotor, the steady flow energy equation yields

$$W = c_p \Delta T_A + \tfrac{1}{2}(C_2^2 - C_1^2)$$

so that with equation (5.10)

$$c_p \Delta T_A = UC_a(\tan \alpha_2 - \tan \alpha_1) - \tfrac{1}{2}(C_2^2 - C_1^2)$$

But $C_2 = C_a \sec \alpha_2$ and $C_1 = C_a \sec \alpha_1$, and hence

$$c_p \Delta T_A = UC_a(\tan \alpha_2 - \tan \alpha_1) - \tfrac{1}{2}C_a^2(\sec^2 \alpha_2 - \sec^2 \alpha_1)$$
$$= UC_a(\tan \alpha_2 - \tan \alpha_1) - \tfrac{1}{2}C_a^2(\tan^2 \alpha_2 - \tan^2 \alpha_1)$$

From the definition of Λ,

$$\Lambda = \frac{\Delta T_A}{\Delta T_A + \Delta T_B}$$

$$= \frac{UC_a(\tan \alpha_2 - \tan \alpha_1) - \tfrac{1}{2}C_a(\tan^2 \alpha_2 - \tan^2 \alpha_1)}{UC_a(\tan \alpha_2 - \tan \alpha_1)}$$

$$= 1 - \frac{C_a}{2U}(\tan \alpha_2 + \tan \alpha_1)$$

By the addition of equations (5.2) and (5.3),

$$\frac{2U}{C_a} = \tan \alpha_1 + \tan \beta_1 + \tan \alpha_2 + \tan \beta_2$$

Hence

$$\Lambda = \frac{C_a}{2U}\left[\frac{2U}{C_a} - \frac{2U}{C_a} + \tan \beta_1 + \tan \beta_2\right]$$

$$= \frac{C_a}{2U}(\tan \beta_1 + \tan \beta_2) \qquad (5.11)$$

Because the case of 50 per cent reaction is important in design, it is of interest to see the consequences of putting $\Lambda = 0\cdot5$. In this case, from equation (5.11)

$$\tan \beta_1 + \tan \beta_2 = \frac{U}{C_a}$$

and it immediately follows from equations (5.2) and (5.3) that

$$\tan \alpha_1 = \tan \beta_2, \quad \text{i.e. } \alpha_1 = \beta_2$$
$$\tan \beta_1 = \tan \alpha_2, \quad \text{i.e. } \alpha_2 = \beta_1$$

Furthermore, since C_a is constant through the stage,

$$C_a = C_1 \cos \alpha_1 = C_3 \cos \alpha_3$$

It was initially assumed that $C_1 = C_3$ and hence $\alpha_1 = \alpha_3$. Because of this equality of angles, namely $\alpha_1 = \beta_2 = \alpha_3$ and $\beta_1 = \alpha_2$, the velocity diagram of Fig. 5.6 becomes symmetrical and blading designed on this basis is sometimes referred to as *symmetrical blading*. From the symmetry of the velocity diagram, it follows that $C_1 = V_2$ and $V_1 = C_2$.

It must be pointed out that in deriving equation (5.11) for Λ, a work-done factor λ of unity has implicitly been assumed in making use of equation (5.10). A stage designed with symmetrical blading will always be referred to as a 50 per cent reaction stage, although the value of Λ actually achieved will differ slightly from 0·5 because of the influence of λ.

It is of interest to consider the significance of the limiting values of Λ, viz. 0 and 1·0. The degree of reaction was defined in terms of enthalpy rises but could also have been expressed in terms of changes in static pressure. Making use of the thermodynamic relation $T \, ds = dh - v \, dp$ and assuming incompressible isentropic flow

$$0 = dh - dp/\rho$$

which on integration from state 1 to state 2 becomes

$$h_2 - h_1 = (p_2 - p_1)/\rho$$

showing that enthalpy and static pressure changes are related. By putting $\Lambda = 0$ in equation (5.11) it can readily be seen that $\beta_1 = -\beta_2$; the rotor blades are then of impulse type (i.e. passage area the same at inlet and outlet) and all the static pressure rise occurs in the stator. Conversely, for $\Lambda = 1·0$ the stators are of impulse type. In the interests of obtaining the most efficient overall diffusion, it is desirable to share the diffusion between both components of the stage, and for this reason the use of 50 per cent reaction is attractive. We are not suggesting, however, that the efficiency is very sensitive to the precise degree of reaction chosen. In practice, as will appear later, the degree of reaction may show considerable variation across the annulus (i.e. along the blade span), especially for stages of low hub–tip ratio, and the designer will primarily be concerned to satisfy more stringent conditions set by a limiting diffusion factor and Mach number.

At this point the reader may wish to see an example of a preliminary mean diameter design to consolidate the previous material. If so, he or she can turn to section 5.7. To enable the full design procedure to be illustrated, however, we need to consider three-dimensional effects, and these will be introduced in the next section.

5.6   Three-dimensional flow

The elementary theory presented in section 5.2 assumed that the flow in the compressor annulus is two-dimensional, meaning that any effect due to radial movement of the fluid is ignored. The assumption of two-dimensional flow is quite reasonable for stages in which the blade height is small relative to the mean diameter of the annulus, i.e. those of hub–tip ratio greater than about 0·8, which would be typical of the later stages of a compressor. The front stages, however, have lower values of hub–tip ratio, and values as low as 0·4 are used for the first stage of aero-engine compressors so that a high mass flow can be passed through a machine of low frontal area. When the compressor has a low hub–tip ratio in the first stage and a high hub–tip ratio in the later stages, the annulus will have a substantial taper (see Fig. 5.1) and the streamlines will not lie on a surface of revolution parallel to the axis of the rotor as previously assumed. Under these conditions, the flow must have a radial component of velocity, although it will generally be small compared with the axial and whirl components.

A second cause of radial movement occurs as follows. Because the flow has a whirl component, the pressure must increase with radius, i.e. up the blade height, to provide the force associated with the centripetal acceleration of the fluid. In the course of adjusting itself to provide a balance between the pressure forces and the inertia forces, the flow will undergo some movement in the radial direction.

With a low hub–tip ratio, the variation in blade speed from root to tip is large and this will have a major effect on the shape of the velocity triangles and the resulting air angles. Furthermore, the aforementioned change of pressure, and hence density, with radius will cause the fluid velocity vectors to change in magnitude and these too affect the shape of the velocity triangles. It follows that the air angles at the mean diameter will be far from representative of those at the root and tip of a blade row. For high efficiency it is essential that the blade angles match the air angles closely at all radii, and the blade must therefore be twisted from root to tip to suit the changing air angles.

The basic equation expressing the balance between pressure forces and inertia forces can be derived by considering the forces acting on the fluid element shown in Fig. 5.11. The whirl component of velocity is shown in Fig. 5.11(a), and the axial component and much smaller radial component resulting from the curvature of the streamline in Fig. 5.11(b). The inertia forces in the radial direction arise from

(i)   the centripetal force associated with circumferential flow;
(ii)   the radial component of the centripetal force associated with the flow along the curved streamline;
(iii) the radial component of the force required to produce the linear acceleration along the streamline.

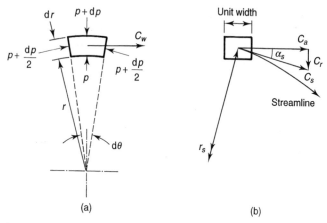

FIG. 5.11 Radial equilibrium of fluid element

The total inertia force, F_1, must be produced by the pressure forces acting on the element in the radial direction. (The acceleration in the radial direction may amount to several thousand times the acceleration due to gravity so that gravitational forces can be neglected.)

Considering a fluid element of unit width, having a density ρ, the three terms of the inertia force can be expressed as follows. The centripetal force associated with the circumferential flow is

$$F_{(i)} = \frac{mC_w^2}{r} = (\rho r \, dr \, d\theta)\frac{C_w^2}{r}$$

For the flow along the curved streamline, the radial force is given by

$$F_{(ii)} = \frac{mC_S^2}{r_S} \cos \alpha_S = (\rho r \, dr \, d\theta)\frac{C_S^2}{r_S} \cos \alpha_S$$

where the suffix S refers to the component along the streamline and r_S is the radius of curvature of the streamline. For the acceleration along the streamline, the radial component of force is

$$F_{(iii)} = m\frac{dC_S}{dt} \sin \alpha_S = (\rho r \, dr \, d\theta)\frac{dC_S}{dt} \sin \alpha_S$$

With the curvature shown in Fig. 5.11(b) the forces $F_{(ii)}$ and $F_{(iii)}$ are in the same direction as $F_{(i)}$. Thus the total inertia force, F_I, is given by

$$F_I = \rho r \, dr \, d\theta \left[\frac{C_w^2}{r} + \frac{C_S^2}{r_S} \cos \alpha_S + \frac{dC_S}{dt} \sin \alpha_S \right]$$

The pressure force, F_P, producing this inertia force is obtained by resolving in the radial direction to give

$$F_P = (p + dp)(r + dr)d\theta - pr \, d\theta - 2\left(p + \frac{dp}{2}\right)dr\frac{d\theta}{2}$$

The third term in the equation results from the resolution of the pressure forces on the two sides of the element in the radial–axial plane, on which it is assumed that the pressure is the average of the two extremes, namely $p + (dp/2)$. Equating the forces F_I and F_P, and neglecting second-order terms, we are left with

$$\frac{1}{\rho}\frac{dp}{dr} = \frac{C_w^2}{r} + \frac{C_S^2}{r_S}\sin\alpha_S + \frac{dC_S}{dt}\sin\alpha_S \qquad (5.12)$$

This is the complete radial equilibrium equation which includes all contributory factors.

For most design purposes it can be assumed that r_S is so large, and α_S so small, that the last two terms of equation (5.12) can be ignored. Thus, finally, we have

$$\frac{1}{\rho}\frac{dp}{dr} = \frac{C_w^2}{r} \qquad (5.13)$$

and this is what will be referred to as the *radial equilibrium equation*. In effect, the radial component of velocity is being neglected. Certainly in the spaces between the blade rows C_r is very much smaller than either C_a or C_w and can safely be assumed to be negligible.

Equation (5.13) enables us to deduce an energy equation which expresses the variation of enthalpy with radius. The stagnation enthalpy h_0 at any radius r where the absolute velocity is C is given by

$$h_0 = h + \frac{C^2}{2} = h + \tfrac{1}{2}(C_a^2 + C_w^2)$$

and the variation of enthalpy with radius is therefore

$$\frac{dh_0}{dr} = \frac{dh}{dr} + C_a\frac{dC_a}{dr} + C_w\frac{dC_w}{dr} \qquad (5.14)$$

From the thermodynamic relation $T\,ds = dh - dp/\rho$,

$$\frac{dh}{dr} = T\frac{ds}{dr} + ds\frac{dT}{dr} + \frac{1}{\rho}\frac{dp}{dr} - \frac{1}{\rho^2}\frac{d\rho}{dr}dp$$

Dropping second-order terms

$$\frac{dh}{dr} = T\frac{ds}{dr} + \frac{1}{\rho}\frac{dp}{dr}$$

Substituting for dh/dr in equation (5.14)

$$\frac{dh_0}{dr} = T\frac{ds}{dr} + \frac{1}{\rho}\frac{dp}{dr} + C_a\frac{dC_a}{dr} + C_w\frac{dC_w}{dr}$$

Using the radial equilibrium equation (5.13), the second term on the right-hand side of the above equation can be replaced by C_w^2/r, leaving the basic equation for the analysis of flow in the compressor annulus as

$$\frac{dh_0}{dr} = T\frac{ds}{dr} + C_a\frac{dC_a}{dr} + C_w\frac{dC_w}{dr} + \frac{C_w^2}{r}$$

The term $T\,ds/dr$ represents the radial variation of loss across the annulus, and may be significant in detailed design calculations; this is especially true if Mach numbers relative to the blade are supersonic and shock losses occur. For our purposes, however, it will be assumed that the entropy gradient term can be ignored and the final form of the equation is given by

$$\frac{dh_0}{dr} = C_a\frac{dC_a}{dr} + C_w\frac{dC_w}{dr} + \frac{C_w^2}{r} \qquad (5.15)$$

This will be referred to as the *vortex energy equation*.

Apart from regions near the walls of the annulus, the stagnation enthalpy (and temperature) will be uniform across the annulus at entry to the compressor. If the frequently used design condition of *constant specific work at all radii* is applied, then although h_0 will increase progressively through the compressor in the axial direction, its radial distribution will remain uniform. Thus $dh_0/dr = 0$ in any plane between a pair of blade rows. Equation (5.15) then reduces to

$$C_a\frac{dC_a}{dr} + C_w\frac{dC_w}{dr} + \frac{C_w^2}{r} = 0 \qquad (5.16)$$

A special case may now be considered in which C_a is maintained constant across the annulus, so that $dC_a/dr = 0$. Equation (5.16) then reduces to

$$\frac{dC_w}{dr} = -\frac{C_w}{r}, \quad \text{or} \quad \frac{dC_w}{C_w} = -\frac{dr}{r}$$

which on integration gives

$$C_w r = \text{constant} \qquad (5.17)$$

Thus the whirl velocity varies inversely with radius, this being known as the *free vortex condition*.

It can therefore be seen that the three conditions of (*a*) constant specific work, (*b*) constant axial velocity, and (*c*) free vortex variation of whirl velocity, naturally satisfy the radial equilibrium equation (5.13) and are therefore conducive to the design flow conditions being achieved. It would at first appear that they constitute an ideal basis for design. Unfortunately, there are certain disadvantages associated with the resultant 'free vortex blading', described in later sections, which influence the designer in considering other combinations of basic conditions. Free vortex designs, however, are widely used in axial flow turbines and will be discussed further in Chapter 7.

One disadvantage of free vortex blading will be dealt with here: the marked variation of degree of reaction with radius. In section 5.5 it was shown that for the case where $C_3 = C_1$ and $C_{a1} = C_{a2} = C_a$, the degree of reaction can be expressed by

$$\Lambda = 1 - \frac{C_a}{2U}(\tan \alpha_2 + \tan \alpha_1)$$

which can readily be written in terms of whirl velocities as

$$\Lambda = 1 - \frac{C_{w2} + C_{w1}}{2U} \tag{5.18}$$

Remembering that $U = U_m r / r_m$, where U_m is the blade speed at the mean radius of the annulus r_m, equation (5.18) can be written

$$\Lambda = 1 - \frac{C_{w2}r + C_{w1}r}{2U_m r^2 / r_m}$$

For a free vortex design $C_w r = $ constant, so that

$$\Lambda = 1 - \frac{\text{constant}}{r^2} \tag{5.19}$$

Evidently the degree of reaction increases markedly from root to tip of the blade. Even if the stage has the desirable value of 50 per cent at the mean radius, it may well be too low at the root and too high at the tip for good efficiency. Because of the lower blade speed at the root section, more fluid deflection is required for a given work input, i.e. a greater rate of diffusion is required at the root section. It is, therefore, particularly undesirable to have a low degree of reaction in this region, and the problem is aggravated as the hub–tip ratio is reduced.

When considering other possible sets of design conditions, it is usually desirable to retain the constant specific work-input condition to provide a constant stage pressure ratio up the blade height. It would be possible, however, to choose a variation of one of the other variables, say C_w, and determine the variation of C_a from equation (5.16). The radial equilibrium requirement would still be satisfied. (It should be clear that in general a design can be based on arbitrarily chosen radial variations of any two variables and the appropriate variation of the third can be determined by inserting them into equation (5.15).)

As an illustration we shall use the normal design condition (a) constant specific work input at all radii, together with (b) an arbitrary whirl velocity distribution which is compatible with (a). To obtain constant work input, $U(C_{w2} - C_{w1})$ must remain constant across the annulus. Let us consider distributions of whirl velocity at inlet to and outlet from the rotor blade given by

$$C_{w1} = aR^n - \frac{b}{R} \quad \text{and} \quad C_{w2} = aR^n + \frac{b}{R} \tag{5.20}$$

where a, b and n are constants and R is the radius ratio r/r_m. At any radius r the blade speed is given by $U = U_m R$. It is immediately seen that $(C_{w2} - C_{w1}) = 2b/R$ and hence that

$$U(C_{w2} - C_{w1}) = 2bU_m$$

which is independent of radius. The two design conditions (a) and (b) are therefore compatible. We shall consider three special cases: where $n = -1$, 1 and 0 respectively. (It will be shown later that a and b are not arbitrary constants but depend upon the chosen values of degree of reaction and stage temperature rise.)

When $n = -1$

The whirl distributions become

$$C_{w1} = \frac{a}{R} - \frac{b}{R} \quad \text{and} \quad C_{w2} = \frac{a}{R} + \frac{b}{R}$$

which are of free vortex form, $C_w r = \text{constant}$. From what has gone before it should be clear that $C_a = \text{constant}$ must be the third design condition required to ensure radial equilibrium, i.e. to satisfy equation (5.16). It also follows that the variation of Λ is given by equation (5.19), which, since $(C_{w2}r + C_{w1}r) = 2ar_m$, can be written

$$\Lambda = 1 - \frac{2ar_m}{2U_m Rr} = 1 - \frac{a}{U_m R^2} \tag{5.21}$$

When $n = 1$

$$C_{w1} = aR - \frac{b}{R} \quad \text{and} \quad C_{w2} = aR + \frac{b}{R}$$

Rewriting equation (5.16) in terms of the dimensionless R we have

$$C_a \, dC_a + C_w \, dC_w + \frac{C_w^2}{R} \, dR = 0$$

Integrating from the mean radius ($R = 1$) to any other radius,

$$-\tfrac{1}{2} \left[C_a^2 \right]_1^R = \tfrac{1}{2} \left[C_w^2 \right]_1^R + \int_1^R \frac{C_w^2}{R} \, dR$$

At exit from the rotor, where $C_w = aR + (b/R)$, the right-hand side of the equation becomes

$$-\frac{1}{2} \left[a^2 R^2 + 2ab + \frac{b^2}{R} \right]_1^R + \int_1^R \left[a^2 R + \frac{2ab}{R} + \frac{b^2}{R^3} \right] dR$$

and, finally, the radial distribution of C_a is given by

$$C_{a2}^2 - (C_{a2}^2)_m = -2[a^2 R^2 + 2ab \ln R - a^2] \tag{5.22}$$

Similarly, at inlet to the rotor,

$$C_{a1}^2 - (C_{a1}^2)_m = -2[a^2 R^2 - 2ab \ln R - a^2] \tag{5.23}$$

Note that C_{a2} cannot equal C_{a1} except at the mean radius ($R = 1$). It is not possible, therefore, to use the simple equation (5.18) for the degree of reaction when finding the variation of Λ with radius. To do this it is necessary to revert to the original definition of Λ $[= \Delta T_A/(\Delta T_A + \Delta T_B)]$ and work from first principles as follows. We shall retain the assumption that the stage is designed to give $C_3 = C_1$. Then $\Delta T_S = \Delta T_{0S}$ so that

$$c_p(\Delta T_A + \Delta T_B) = W = U(C_{w2} - C_{w1})$$
$$c_p\Delta T_A = \tfrac{1}{2}(C_1^2 - C_2^2) + U(C_{w2} - C_{w1})$$
$$= \tfrac{1}{2}[(C_{a1}^2 + C_{w1}^2) - (C_{a2}^2 + C_{w2}^2)] + U(C_{w2} - C_{w1})$$
$$\Lambda = 1 + \frac{C_{a1}^2 - C_{a2}^2}{2U(C_{w2} - C_{w1})} - \frac{C_{w2} + C_{w1}}{2U}$$

From equations (5.22) and (5.23), assuming we choose to design with $(C_{a2})_m = (C_{a1})_m$,

$$C_{a1}^2 - C_{a2}^2 = 8ab \ln R$$

From the whirl distributions,

$$C_{w2} - C_{w1} = \frac{2b}{R} \quad \text{and} \quad C_{w2} + C_{w1} = 2aR$$

Substituting in the equation for Λ, and writing $U = U_m R$, we get finally

$$\Lambda = 1 + \frac{2a \ln R}{U_m} - \frac{a}{U_m} \tag{5.24}$$

Designs with the exponent $n = 1$ have been referred to as *first-power designs*.

At this point it can easily be made clear that the constants a and b in the whirl distributions are not arbitrary. Equation (5.24) shows that a is fixed once the degree of reaction at the mean radius is chosen, i.e. at $R = 1$

$$a = U_m(1 - \Lambda_m)$$

And b is fixed when the stage temperature rise is chosen because

$$c_p\Delta T_{0S} = U(C_{w2} - C_{w1}) = 2bU_m$$

and hence $b = c_p\Delta T_{0S}/2U_m$.

When $n = 0$

Proceeding as in the previous case, but with

$$C_{w1} = a - \frac{b}{R} \quad \text{and} \quad C_{w2} = a + \frac{b}{R}$$

we find that

$$(C_{a2}^2) - (C_{a2}^2)_m = -2\left[a^2 \ln R - \frac{ab}{R} + ab\right] \tag{5.25}$$

$$(C_{a1}^2) - (C_{a1}^2)_m = -2\left[a^2 \ln R + \frac{ab}{R} - ab\right]$$ (5.26)

Blading designed on this basis is usually referred to as *exponential* blading. Again proceeding as before, and assuming $(C_{a2})_m = (C_{a1})_m$, it can be shown that

$$\Lambda = 1 + \frac{a}{U_m} - \frac{2a}{U_m R}$$ (5.27)

It should be noted that for all three cases, at the mean radius ($R = 1$), the expression $a = U_m(1 - \Lambda_m)$ holds. If a value is specified for the reaction at the mean radius, Λ_m, the variation of reaction with radius can readily be obtained by substituting for a in equations (5.21), (5.24) and (5.27). The results can be summarized conveniently in the following table.

n	Λ	Blading
-1	$1 - \dfrac{1}{R^2}(1 - \Lambda_m)$	Free vortex
0	$1 + \left[1 - \dfrac{2}{R}\right](1 - \Lambda_m)$	Exponential
1	$1 + (2 \ln R - 1)(1 - \Lambda_m)$	First power

It is instructive to evaluate the variation of Λ for all three cases assuming that $\Lambda_m = 0.5$, and the results are given in Fig. 5.12. It can be seen that the free vortex design gives the greatest reduction in Λ for low values of R while the first-power design gives the least reduction. Furthermore, for all

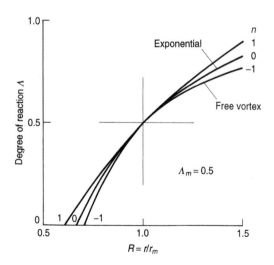

FIG. 5.12 Radial variation of degree of reaction

designs there is a lower limit to R below which the degree of reaction becomes negative. A negative value implies a *reduction* in static pressure in the rotor. For the free vortex design, this limiting value of R is given by $0 = 1 - (1/R^2)(1 - \Lambda_m)$, from which $R = 0.707$ when $\Lambda_m = 0.50$. As mentioned earlier, however, hub–tip ratios as low as 0.4 are common at entry to the compressor of a jet engine. Noting that

$$r/r_t = (r/r_m)(r_m/r_t) = R(r_m/r_t) = R[1 + (r_r/r_t)]/2$$

the variation in reaction can readily be evaluated in terms of r/r_t for a specified value of hub–tip ratio r_r/r_t. Figure 5.13 shows the radial distribution of Λ for a free vortex design of hub–tip ratio 0.4, for a series of values of Λ_m. It is evident that a stage of low hub–tip ratio must have a high value of Λ_m to provide satisfactory conditions at the root section.

So far the discussion has been based upon a choice of C_a, C_w and h_0 (i.e. W) as the primary variables. This choice is not essential, however, and another approach could be to specify a variation of degree of reaction with radius rather than C_w. One such design method was based on (*a*) axial velocity, (*b*) work input and (*c*) degree of reaction, all being independent of radius; this was referred to as a *constant reaction design*. It led to less twisted blading than the free vortex design, but consideration of the analysis leading to equation (5.17) and thence to equation (5.19) shows that radial equilibrium cannot be satisfied: conditions (*a*) and (*b*) involve a varying Λ when equation (5.16) is satisfied. The fact is that the flow will adjust itself to satisfy radial equilibrium between the blade rows. If the design does not allow for this correctly, the actual air angles will not agree with the design values on which the blade angles are based and the efficiency is likely to be reduced.

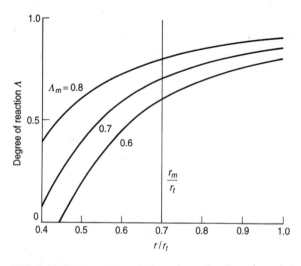

FIG. 5.13 Free vortex variation of reaction for $r_r/r_t = 0.4$

While it is normally desirable to retain the constant specific work-input condition, a fan for a turbofan engine of high bypass ratio might provide an exception to this rule. In this case, the inner stream, feeding the high-pressure core, may be designed for a lower pressure ratio than the outer stream which is supplying air to the bypass duct and where the blade speed is high. The work then varies with radius, dh_0/dr is not zero, and equation (5.15) must be used instead of equation (5.16).

Further discussion of three-dimensional design procedures will follow in section 5.7 after an example is given to show how the material presented in the preceding sections can be used in the design of an axial flow compressor.

5.7 Design process

The theory presented in the previous sections will now be applied to the design of an axial flow compressor, and it will be seen that the process requires continuous judgement by the designer. The design of a compressor suitable for a simple low-cost turbojet will be considered.

A typical gas turbine design procedure was outlined in Fig. 1.23. Assuming that market research has shown there is a need for a low-cost turbojet with a take-off thrust of about 12 000 N, preliminary studies will show that a single-spool all-axial flow arrangement is satisfactory, using a low pressure ratio and a modest turbine inlet temperature to keep the cost down, as discussed in section 3.3. From cycle calculations, a suitable design point under sea-level static conditions (with $p_a = 1 \cdot 01$ bar and $T_a = 288$ K) may emerge as follows:

Compressor pressure ratio   4·15
Air mass flow   20 kg/s
Turbine inlet temperature   1100 K

With these data specified, it is now necessary to investigate the aerodynamic design of the compressor, turbine and other components of the engine. It will be assumed that the compressor has no inlet guide vanes, to keep both weight and noise down. The design of the turbine for this engine will be considered in Chapter 7.

The complete design process for the compressor will encompass the following steps:

(i)   choice of rotational speed and annulus dimensions;
(ii)   determination of number of stages, using an assumed efficiency;
(iii)   calculation of the air angles for each stage at the mean radius;
(iv)   determination of the variation of the air angles from root to tip;
(v)   investigation of compressibility effects;
(vi)   selection of compressor blading, using experimentally obtained cascade data;

(vii)   check on efficiency previously assumed, using the cascade data;
(viii)   estimation of off-design performance;
(ix)   rig testing.

Steps (i) to (v) will be outlined in this section and the remaining steps will be covered in later sections. In practice, the process will be one of continued refinement, coupled with feedback from other groups such as the designers of the combustion system and turbine, metallurgists and stress analysts, and those concerned with mechanical problems associated with whirling speeds, bearings, stiffness of structural members and so on.

Determination of rotational speed and annulus dimensions

Reviewing the theory presented earlier, it is seen that there is no equation which enables the designer to select a suitable value of rotational speed. This can be found, however, by assuming values for the blade tip speed, and the axial velocity and hub–tip ratio at inlet to the first stage. The required annulus area at entry is obtained from the specified mass flow, assumed axial velocity, and ambient conditions, using the continuity equation.

Previous experience will suggest that a tip speed, U_t, of around 350 m/s will lead to acceptable stresses and that the axial velocity, C_a, could range from 150 to 200 m/s. Without IGVs there will be no whirl component of velocity at inlet, and this will increase the Mach number relative to the blade (see Fig. 5.5) so it may be advisable to use a modest value of 150 m/s for C_a. The hub–tip ratio at entry may vary between 0.4 and 0.6, and for a specified annulus area the tip radius will be a function of the hub–tip ratio. For a fixed blade speed, then, the rotational speed will also be a function of hub–tip ratio. Thus the designer will, in very short order, be presented with a wide range of solutions and must use engineering judgement to select the most promising. At the same time the turbine designer will be examining a suitable turbine and the compressor and turbine designers must keep in close contact while establishing preliminary designs.

To satisfy continuity

$$m = \rho_1 A C_{a1} = \rho_1 \pi r_t^2 \left[1 - \left(\frac{r_r}{r_t} \right)^2 \right] C_a$$

$$r_t^2 = \frac{m}{\pi \rho_1 C_{a1} [1 - (r_r/r_t)^2]}$$

At sea-level static conditions, $T_{01} = T_a = 288$ K. Assuming no loss in the intake, $p_{01} = p_a = 1.01$ bar. With

$$C_1 = C_{a1} = 150 \text{ m/s} \ (C_{w1} = 0)$$

$$T_1 = 288 - \frac{150^2}{2 \times 1.005 \times 10^3} = 276.8 \text{ K}$$

$$p_1 = p_{01} \left[\frac{T_1}{T_{01}}\right]^{\gamma/(\gamma-1)} = 1\cdot01 \left[\frac{276\cdot8}{288}\right]^{3\cdot5} = 0\cdot879 \text{ bar}$$

$$\rho_1 = \frac{100 \times 0\cdot879}{0\cdot287 \times 276\cdot8} = 1\cdot106 \text{ kg/m}^3$$

$$r_t^2 = \frac{20}{\pi \times 1\cdot106 \times 150[1 - (r_r/r_t)^2]} = \frac{0\cdot038\ 37}{[1 - (r_r/r_t)^2]}$$

The tip speed, U_t, is related to r_t by $U_t = 2\pi r_t N$, and hence if U_t is chosen to be 350 m/s,

$$N = \frac{350}{2\pi r_t}$$

Evaluating r_t and N over a range of hub–tip ratios the following table results:

r_r/r_t	r_t [m]	N [rev/s]
0·40	0·2137	260·6
0·45	0·2194	253·9
0·50	0·2262	246·3
0·55	0·2346	237·5
0·60	0·2449	227·5

At this point it would be pertinent to consider the turbine design. The example in Chapter 7 shows that a speed of 250 rev/s results in quite an adequate single-stage unit, and the outer radius at the turbine inlet is found to be 0·239 m. Referring to the table above, a hub–tip ratio of 0·50 would give a compatible compressor tip radius of 0·2262 m although the rotational speed is 246·3 rev/s. There was nothing sacrosanct about the choice of 350 m/s for the tip speed, and the design could be adjusted for a rotational speed of 250 rev/s. With the speed slightly altered, then

$$U_t = 2\pi \times 0\cdot2262 \times 250 = 355\cdot3 \text{ m/s}$$

For a simple engine of the type under consideration, there would be no merit in using a low hub–tip ratio; this would merely increase the mismatch between the compressor and turbine diameters, and also complicate both the mechanical and aerodynamic design of the first stage. On the other hand, using a high hub–tip ratio would unnecessarily increase the compressor diameter and weight. But it should be realized that the choice of 0·50 for hub–tip ratio is arbitrary, and merely provides a sensible starting point; later considerations following detailed analysis could cause an adjustment, and a considerable amount of design optimization is called for.

At this stage it is appropriate to check the Mach number relative to the rotor tip at inlet to the compressor. Assuming the axial velocity to be constant across the annulus, which will be the case where there are no inlet guide vanes,

$$V_{1t}^2 = U_{1t}^2 + C_{a1}^2 = 355\cdot3^2 + 150^2, \quad \text{and} \quad V_{1t} = 385\cdot7 \text{ m/s}$$

$$a = \sqrt{(\gamma R T_1)} = \sqrt{1\cdot4 \times 0\cdot287 \times 1000 \times 276\cdot8} = 331\cdot0 \text{ m/s}$$

$$M_{1t} = \frac{V_{1t}}{a} = \frac{385\cdot7}{331\cdot0} = 1\cdot165$$

Thus the Mach number relative to the rotor tip is 1·165 and the first stage is transonic; this level of Mach number should not present any problem, and methods of dealing with shock losses will be covered in a later section.

With the geometry selected, i.e. a hub–tip ratio of 0·50 and a tip radius of 0·2262 m, it follows that the root radius is 0·1131 m and the mean radius is 0·1697 m. It is instructive now to estimate the annulus dimensions at exit from the compressor, and for these preliminary calculations it will be assumed that the mean radius is kept constant for all stages. The compressor delivery pressure, $p_{02} = 4\cdot15 \times 1\cdot01 = 4\cdot19$ bar. To estimate the compressor delivery temperature it will be assumed that the polytropic efficiency of the compressor is 0·90. Thus

$$T_{02} = T_{01}\left[\frac{p_{02}}{p_{01}}\right]^{(n-1)/n}, \quad \text{where} \quad \frac{(n-1)}{n} = \frac{1}{0\cdot90} \times \frac{0\cdot4}{1\cdot4} = 0\cdot3175$$

so that

$$T_{02} = 288\cdot0(4\cdot15)^{0\cdot3175} = 452\cdot5 \text{ K}$$

Assuming that the air leaving the stator of the last stage has an axial velocity of 150 m/s and no swirl, the static temperature, pressure and density at exit can readily be calculated as follows:

$$T_2 = 452\cdot5 - \frac{150^2}{2 \times 1\cdot005 \times 10^3} = 441\cdot3 \text{ K}$$

$$p_2 = p_{02}\left[\frac{T_2}{T_{02}}\right]^{\gamma/(\gamma-1)} = 4\cdot19\left[\frac{441\cdot3}{452\cdot5}\right]^{3\cdot5} = 3\cdot838 \text{ bar}$$

$$\rho_2 = \frac{100 \times 3\cdot838}{0\cdot287 \times 441\cdot3} = 3\cdot03 \text{ kg/m}^3$$

The exit annulus area is thus given by

$$A_2 = \frac{20}{3\cdot031 \times 150} = 0\cdot0440 \text{ m}^2$$

With $r_m = 0\cdot1697$ m, the blade height at exit, h, is then given by

$$h = \frac{0\cdot044}{2\pi r_m} = \frac{0\cdot044}{2\pi \times 0\cdot1697} = 0\cdot0413 \text{ m}$$

The radii at exit from the last stator are then

$$r_t = 0.1697 + (0.0413/2) = 0.1903 \text{ m}$$
$$r_r = 0.1697 - (0.0413/2) = 0.1491 \text{ m}$$

At this point we have established the rotational speed and the annulus dimensions at inlet and outlet, on the basis of a constant mean diameter. To summarize:

$$N = 250 \text{ rev/s} \qquad \left.\begin{array}{l} r_t = 0.2262 \text{ m} \\ r_r = 0.1131 \text{ m} \end{array}\right\} \text{inlet}$$

$$U_t = 355.3 \text{ m/s}$$

$$C_a = 150 \text{ m/s} \qquad \left.\begin{array}{l} r_t = 0.1903 \text{ m} \\ r_r = 0.1491 \text{ m} \end{array}\right\} \text{outlet}$$

$$r_m = 0.1697 \text{ m (constant)}$$

Estimation of number of stages

With the assumed polytropic efficiency of 0.90, the overall stagnation temperature rise through the compressor is $452.5 - 288 = 164.5$ K. The stage temperature rise ΔT_{0S} can vary widely in different compressor designs, depending on the application and the importance or otherwise of low weight: values may vary from 10 to 30 K for subsonic stages and may be 45 K or higher in high-performance transonic stages. Rather than choosing a value at random, it is instructive to estimate a suitable ΔT_{0S} based on the mean blade speed

$$U = 2 \times \pi \times 0.1697 \times 250 = 266.6 \text{ m/s}$$

We will adopt the simple design condition $C_{a1} = C_{a2} = C_a$ throughout the compressor, so the temperature rise from equation (5.9) is given by

$$\Delta T_{0S} = \frac{\lambda U C_a(\tan \beta_1 - \tan \beta_2)}{c_p} = \frac{\lambda U(C_{w2} - C_{w1})}{c_p}$$

With a purely axial velocity at entry to the first stage, in the absence of IGVs,

$$\tan \beta_1 = \frac{U}{C_a} = \frac{266.6}{150}$$

$$\beta_1 = 60.64°$$

$$V_1 = \frac{C_a}{\cos \beta_1} = \frac{150}{\cos 60.64} = 305.9 \text{ m/s}$$

In order to estimate the maximum possible deflection in the rotor, we will apply the de Haller criterion $V_2/V_1 \not< 0.72$. On this basis the minimum allowable value of $V_2 = 305.9 \times 0.72 = 220$ m/s, and the corresponding rotor blade outlet angle is given by

$$\cos \beta_2 = \frac{C_a}{V_2} = \frac{150}{220}, \qquad \beta_2 = 47.01°$$

Using this deflection and neglecting the work-done factor for this crude estimate

$$\Delta T_{0S} = \frac{266 \cdot 6 \times 150(\tan 60 \cdot 64 - \tan 47 \cdot 01)}{1 \cdot 005 \times 10^3} \approx 28 \text{ K}$$

A temperature rise of 28 K per stage implies $164 \cdot 5/28 = 5 \cdot 9$ stages. It is likely, then, that the compressor will require six or seven stages and, in view of the influence of the work-done factor, seven is more likely. An attempt will therefore be made to design a seven-stage compressor.

With seven stages and an overall temperature rise of 164·5 K the average temperature rise is 23·5 K per stage. It is normal to design for a somewhat lower temperature rise in the first and last stages, for reasons which will be discussed at the end of this section. A good starting point would be to assume $\Delta T_0 \approx 20$ K for the first and last stages, leaving a requirement for $\Delta T_0 \approx 25$ K in the remaining stages.

Stage-by-stage design

Having determined the rotational speed and annulus dimensions, and estimated the number of stages required, the next step is to evaluate the air angles for each stage *at the mean radius.* It will then be possible to check that the estimated number of stages is likely to result in an acceptable design.

From the velocity diagram, Fig. 5.6, it is seen that $C_{w1} = C_a \tan \alpha_1$ and $C_{w2} = C_{w1} + \Delta C_w$. For the first stage $\alpha_1 = 0$, because these are no inlet guide vanes. The stator outlet angle for each stage, α_3, will be the inlet angle α_1 for the following rotor. Calculations of stage temperature rise are based on rotor considerations only, but care must be taken to ensure that the diffusion in the stator is kept to a reasonable level. The work-done factors will vary through the compressor and reasonable values for the seven stages would be 0·98 for the first stage, 0·93 for the second, 0·88 for the third and 0·83 for the remaining four stages.

Stages 1 and 2

Recalling the equation for the stage temperature rise in terms of change in whirl velocity $\Delta C_w = C_{w2} - C_{w1}$, we have

$$\Delta C_w = \frac{c_p \Delta T_0}{\lambda U} = \frac{1 \cdot 005 \times 10^3 \times 20}{0 \cdot 98 \times 266 \cdot 6} = 76 \cdot 9 \text{ m/s}$$

Since $C_{w1} = 0$, $C_{w2} = 76 \cdot 9$ m/s and hence

$$\tan \beta_1 = \frac{U}{C_a} = \frac{266 \cdot 6}{150} = 1 \cdot 7773, \qquad \beta_1 = 60 \cdot 64°$$

$$\tan \beta_2 = \frac{U - C_{w2}}{C_a} = \frac{266 \cdot 6 - 76 \cdot 9}{150} = 1 \cdot 264, \quad \beta_2 = 51 \cdot 67°$$

$$\tan \alpha_2 = \frac{C_{w2}}{C_a} = \frac{76 \cdot 9}{150} = 0 \cdot 513, \qquad\qquad \alpha_2 = 27 \cdot 14°$$

The velocity diagram for the first stage therefore appears as in Fig. 5.14(a).

The deflection in the rotor blades is $\beta_1 - \beta_2 = 8 \cdot 98°$, which is modest. The diffusion can readily be checked using the de Haller number as follows:

$$\frac{V_2}{V_1} = \frac{C_a/\cos \beta_2}{C_a/\cos \beta_1} = \frac{\cos \beta_1}{\cos \beta_2} = \frac{0 \cdot 490}{0 \cdot 260} = 0 \cdot 790$$

This value of de Haller number indicates a relatively light aerodynamic loading, i.e. a low rate of diffusion. It is not necessary to calculate the diffusion factor at this stage, because the de Haller number gives an adequate preliminary check. After the pitch chord ratio (s/c) is determined from cascade data, as explained in a later section, the diffusion factor can be calculated readily from the known velocities.

At this point it is convenient to calculate the pressure ratio of the stage $(p_{03}/p_{01})_1$, the suffix outside the parentheses denoting the number of the stage, and then the pressure and temperature at exit which will also be the values at inlet to the second stage. The isentropic efficiency of the stage is approximately equal to the polytropic efficiency of the compressor, which has been assumed to be 0·90, so we have

$$\left(\frac{p_{03}}{p_{01}}\right)_1 = \left(1 + \frac{0 \cdot 90 \times 20}{288}\right)^{3 \cdot 5} = 1 \cdot 236$$

$$(p_{03})_1 = 1 \cdot 01 \times 1 \cdot 236 = 1 \cdot 249 \text{ bar}$$

$$(T_{03})_1 = 288 + 20 = 308 \text{ K}$$

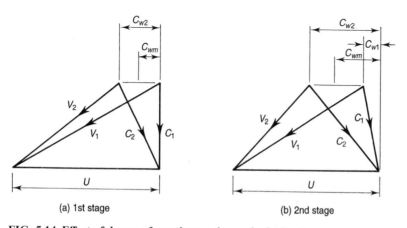

(a) 1st stage   (b) 2nd stage

FIG. 5.14 Effect of degree of reaction on shape of velocity diagram

We have finally to choose a value for the air angle at outlet from the stator row, α_3, which will also be the direction of flow, α_1, into the second stage. Here it is useful to consider the degree of reaction. For this first stage, with the prescribed axial inlet velocity, C_3 will not equal C_1 (unless α_3 is made zero) whereas our equations for Λ were derived on the assumption of this equality of inlet and outlet velocities. Nevertheless, C_3 will not differ markedly from C_1, and we can arrive at an *approximate* value of Λ by using equation (5.18).

$$\Lambda \approx 1 - \frac{C_{w2} + C_{w1}}{2U} = 1 - \frac{76 \cdot 9}{2 \times 266 \cdot 6} = 0 \cdot 856$$

The degree of reaction is high, but we have seen from Fig. 5.13 that this is necessary with low hub–tip ratios to avoid a negative value at the root radius. We shall hope to be able to use 50 per cent reaction stages from the third or fourth stage onwards, and an appropriate value of Λ for the second stage may be about 0·70.

For the second stage $\Delta T_{0S} = 25$ K and $\lambda = 0 \cdot 93$, and we can determine β_1 and β_2 using equations (5.9) and (5.11). From (5.9)

$$25 = \frac{0 \cdot 93 \times 266 \cdot 2 \times 150}{1 \cdot 005 \times 10^3} (\tan \beta_1 - \tan \beta_2)$$

$$\tan \beta_1 - \tan \beta_2 = 0 \cdot 6756$$

And from (5.11)

$$0 \cdot 70 \approx \frac{150}{2 \times 266 \cdot 6} (\tan \beta_1 + \tan \beta_2)$$

$$\tan \beta_1 + \tan \beta_2 \approx 2 \cdot 4883$$

Solving these simultaneous equations we get

$$\beta_1 = 57 \cdot 70° \quad \text{and} \quad \beta_2 = 42 \cdot 19°$$

Finally, using (5.2) and (5.3)

$$\alpha_1 = 11 \cdot 06° \quad \text{and} \quad \alpha_2 = 41 \cdot 05°$$

The whirl velocities at inlet and outlet are readily found from the velocity diagram,

$$C_{w1} = C_a \tan \alpha_1 = 150 \tan 11 \cdot 06° = 29 \cdot 3 \text{ m/s}$$

$$C_{w2} = C_a \tan \alpha_2 = 150 \tan 41 \cdot 05° = 130 \cdot 6 \text{ m/s}$$

The required change in whirl velocity is 101·3 m/s, compared with 76·9 m/s for the first stage; this is due to the higher stage temperature rise and the lower work-done factor. The fluid deflection in the rotor blades has increased to 15·51°. It appears that α_3 for the first stage should be 11·06°. This design gives a de Haller number for the second-stage rotor blades of $\cos 57 \cdot 70° / \cos 42 \cdot 19° = 0 \cdot 721$, which is satisfactory. With the stator outlet angle for the first-stage stator now known, the de Haller number for the first-stage stator would be

$$\frac{C_3}{C_2} = \frac{\cos \alpha_2}{\cos \alpha_3} = \frac{\cos 27\cdot15}{\cos 11\cdot06} = 0\cdot907$$

implying a small amount of diffusion. This is a consequence of the high degree of reaction in the first stage.

The velocity diagram for the second stage appears as in Fig. 5.14(b) and the outlet pressure and temperature become

$$\left(\frac{p_{03}}{p_{01}}\right)_2 = \left(1 + \frac{0\cdot90 \times 25}{308}\right)^{3\cdot5} = 1\cdot280$$
$$(p_{03})_2 = 1\cdot249 \times 1\cdot280 = 1\cdot599 \text{ bar}$$
$$(T_{03})_2 = 308 + 25 = 333 \text{ K}$$

At this point we do not know α_3 for the second stage, but it will be determined from the fact that it is equal to α_1 for the third stage, which we will now proceed to consider.

Before doing so, it is useful to point out that the degree of reaction is directly related to the shape of the velocity diagram. It was previously shown that for 50 per cent reaction the velocity diagram is symmetrical. Writing $C_{wm} = (C_{w1} + C_{w2})/2$, equation (5.18) can be rewritten in the form $\Lambda = 1 - (C_{wm}/U)$. Referring to Figs 5.14(a) and (b) it can be seen that when C_{wm}/U is small, and the corresponding reaction is high, the velocity diagram is highly skewed; the high degree of reaction in the first stage is a direct consequence of the decision to dispense with inlet guide vanes and use a purely axial inlet velocity. The degree of reaction is reduced in the second stage, and we would eventually like to achieve 50 per cent reaction in the later stages where the hub–tip ratios are higher.

Stage 3

Using a stage temperature rise of 25 K and a work-done factor of 0·88, an attempt will be made to use a 50 per cent reaction design for the third stage.

Proceeding as before,

$$\tan \beta_1 - \tan \beta_2 = \frac{\Delta T_{0s} c_p}{\lambda U C_a} = \frac{25 \times 1\cdot005 \times 10^3}{0\cdot88 \times 266\cdot6 \times 150} = 0\cdot7140$$

$$\tan \beta_1 + \tan \beta_2 = \Lambda \frac{2U}{C_a} = \frac{0\cdot5 \times 2 \times 266\cdot6}{150} = 1\cdot7773$$

yielding $\beta_1 = 51\cdot24°$ and $\beta_2 = 28\cdot00°$. The corresponding value of the de Haller number is given by $\cos 51\cdot24/\cos 28\cdot00 = 0\cdot709$. This is rather low, but could be deemed satisfactory for a preliminary design. It is instructive, however, to investigate the possibilities available to the designer for reducing the diffusion. One possibility is to consider changing the degree of reaction, but it is found that the de Haller number is not strongly influenced by the degree of reaction chosen; as Λ had a value of ≈ 0.70 for the second stage it might appear that a suitable value for the third stage might be between 0.70 and 0.50. Repeating the above calculations for a range of Λ, however, shows that $\Lambda = 0.55$ results in a further *decrease* of the de Haller number to 0.706; referring again to Fig. 5.14 it can be observed that for a specified axial velocity, the required diffusion increases with reaction. A de Haller number of 0.725 can be achieved for $\Lambda = 0.40$, but it is undesirable to use such a low degree of reaction. A more useful approach might be to accept a slightly lower temperature rise in the stage, and reducing ΔT_{0S} from 25 K to 24 K while keeping $\Lambda = 0.50$ gives

$$\tan \beta_1 - \tan \beta_2 = 0.6854$$

yielding $\beta_1 = 50.92°$, $\beta_2 = 28.63°$ and a de Haller number of 0.718, which is satisfactory for this preliminary design. Other methods of reducing the aerodynamic loading include increases in blade speed or axial velocity, which could readily be accommodated.

With a stage temperature rise of 24 K, the performance of the third stage is then given by

$$\left(\frac{p_{03}}{p_{01}}\right)_3 = \left(1 + \frac{0.90 \times 24}{333}\right)^{3.5} = 1.246$$
$$(p_{03})_3 = 1.599 \times 1.246 = 1.992 \text{ bar}$$
$$(T_{03})_3 = 333 + 24 = 357 \text{ K}$$

From the symmetry of the velocity diagram $\alpha_1 = \beta_2 = 28.63°$ and $\alpha_2 = \beta_1 = 50.92°$. The whirl velocities are given by

$$C_{w1} = 150 \tan 28.63 = 81.9 \text{ m/s}$$
$$C_{w2} = 150 \tan 50.92 = 184.7 \text{ m/s}$$

Stages 4, 5 and 6

A work-done factor of 0.83 is appropriate for all stages from the fourth onwards, and 50 per cent reaction can be used. The design can be simplified by using the same mean diameter velocity diagrams for stages 4 to 6, although each blade will have a different length due to the continuous increase in density. The seventh and final stage can then be designed to give the required overall pressure ratio. It is not necessary to repeat all the calculations for stages 4–6, but it should be noted that the reduction in work-done factor to 0.83, combined with the desired stage temperature rise of 25 K, results in an unacceptably low de Haller number of 0.695. Reducing the stage temperature rise to 24 K increases the de Haller number to 0.705, which will be considered to be just acceptable for the preliminary design.

Proceeding as before,

$$\tan \beta_1 - \tan \beta_2 = \frac{24 \times 1 \cdot 005 \times 10^3}{0 \cdot 83 \times 266 \cdot 6 \times 150} = 0 \cdot 7267$$

$$\tan \beta_1 + \tan \beta_2 = 0 \cdot 5 \times 2 \times \frac{266 \cdot 6}{150} = 1 \cdot 7773$$

yielding $\beta_1 = 51 \cdot 38°$ ($=\alpha_2$) and $\beta_2 = 27 \cdot 71°$ ($=\alpha_1$). The performance of the three stages can be summarized below:

Stage	4	5	6
p_{01} (bar)	1·992	2·447	2·968
T_{01} (K)	357	381	405
(p_{03}/p_{01})	1·228	1·213	1·199
p_{03} (bar)	2·447	2·968	3·560
T_{03} (K)	381	405	429
$p_{03} - p_{01}$ (bar)	0·455	0·521	0·592

It should be noted that although each stage is designed for the same temperature rise, the pressure ratio decreases with stage number; this is a direct consequence of the increasing inlet temperature as flow progresses through the compressor. The pressure *rise*, however, increases steadily.

Stage 7

At entry to the final stage the pressure and temperature are 3·560 bar and 429 K. The required compressor delivery pressure is $4 \cdot 15 \times 1 \cdot 01 = 4 \cdot 192$ bar. The pressure ratio of the seventh stage is thus given by

$$\left(\frac{p_{03}}{p_{01}} \right)_7 = \frac{4 \cdot 192}{3 \cdot 560} = 1 \cdot 177$$

The temperature rise required to give this pressure ratio can be determined from

$$\left(1 + \frac{0 \cdot 90 \Delta T_{0S}}{429} \right)^{3 \cdot 5} = 1 \cdot 177$$

giving $\Delta T_{0S} = 22 \cdot 8$ K.

The corresponding air angles, assuming 50 per cent reaction, are then $\beta_1 = 50 \cdot 98°$ ($=\alpha_2$), $\beta_2 = 28 \cdot 52°$ ($=\alpha_1$) with a satisfactory de Haller number of 0·717.

With a 50 per cent reaction design used for the final stage, the fluid will leave the last stator with an angle $\alpha_3 = \alpha_1 = 28 \cdot 52°$, whereas ideally the flow should be axial at entry to the combustion chamber. The flow can be straightened by incorporating vanes after the final compressor stage and these can form part of the necessary diffuser at entry to the combustion chamber.

All the preliminary calculations have been carried out on the basis of a constant *mean* diameter. Another problem now arises: a sketch, approximately to scale, of the compressor and turbine annuli (Fig. 5.15) shows that the combustor will have an awkward shape, the required changes in flow direction causing additional pressure losses. A more satisfactory solution might be to design the compressor for a constant *outer* diameter; both solutions are shown in the figure. The use of a constant outer diameter results in the mean blade speed increasing with stage number, and this in turn implies that for a given temperature rise ΔC_w is reduced. The fluid deflection is correspondingly reduced with a beneficial increase in de Haller number. Alternatively, because of the higher blade speed a higher temperature rise could be achieved in the later stages; this might permit the required pressure ratio to be obtained in six stages rather than seven. The reader should be aware that the simple equations derived on the basis of $U = \text{constant}$ are then not valid, and it would be necessary to use the appropriate values of U_1 and U_2; the stage temperature rise would then be given by $\lambda(U_2 C_{w2} - U_1 C_{w1})/c_p$.

Compressors which use constant inner diameter, constant mean diameter or constant outer diameter will all be found in service. The use of a constant *inner* diameter is often found in industrial units, permitting the use of rotor discs of the same diameter, which lowers the cost. It would be important to minimize the number of turbine stages, again for reasons of cost, and with a subsonic compressor it is very probable that the turbine diameter would be noticeably larger than the compressor diameter. The difference in turbine and compressor diameter is not critical, however, because frontal area is unimportant and with reverse flow combustion chambers large differences in diameter can be easily accommodated. Constant *outer* diameter compressors are used where the minimum number of stages is required, and these are commonly found in aircraft engines.

The compressor annulus of the Olympus 593 engine used in Concorde employs a combination of these approaches; the LP compressor annulus

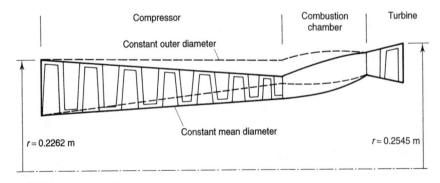

FIG. 5.15 Annulus shape

has a virtually constant inner diameter, while the HP compressor has a constant outer diameter. The accessories are packed around the HP compressor annulus and the engine when fully equipped is almost cylindrical in shape, with the compressor inlet and turbine exit diameters almost equal. In this application, frontal area is of critical importance because of the high supersonic speed.

Variation of air angles from root to tip

In section 5.6 various distributions of whirl velocity with radius were considered, and it was shown that the designer had quite a wide choice. In the case of the first stage, however, the choice is restricted because of the absence of IGVs; this means that there is no whirl component at entry to the compressor and the inlet velocity will be constant across the annulus. For all other stages the whirl velocity at entry to the rotor blades will be determined by the axial velocity and the stator outlet angle from the previous stage, giving more freedom in the aerodynamic design of the stage.

The material developed in section 5.6 will be applied to the first stage, which is a special case because of the axial inlet velocity, and the third stage, which is typical of the later stages. The first stage will be investigated using a free vortex design, noting that the condition $C_w r = $ constant is satisfied for $C_w = 0$. Attention will then be turned to the design of the third stage, recalling that the mean radius design was based on $\Lambda_m = 0.50$. The third stage will be investigated for three different design approaches, viz. (i) free vortex, $\Lambda_m = 0.50$, (ii) constant reaction, $\Lambda_m = 0.50$ with radial equilibrium ignored and (iii) exponential blading, $\Lambda_m = 0.50$.

Considering the first stage, the rotor blade angle at inlet (β_1) is obtained directly from the axial velocity (150 m/s) and the blade speed. The blade speeds at root, mean and tip, corresponding to radii of 0.1131, 0.1697 and 0.2262 m, are 177.7, 266.6 and 355.3 m/s respectively. Thus

$$\tan \beta_{1r} = \frac{177.7}{150}, \qquad \beta_{1r} = 49.83°$$

$$\tan \beta_{1m} = \frac{266.6}{150}, \qquad \beta_{1m} = 60.64°$$

$$\tan \beta_{1t} = \frac{355.3}{150}, \qquad \beta_{1t} = 67.11°$$

Air angles at any radius can be calculated as above. For our purposes the calculations will be restricted to root, mean and tip radii, although for detailed design of the blading it would be necessary to calculate the angles at intermediate radii also.

To calculate the air angles β_2 and α_2 it is necessary to determine the radial variation of C_{w2}. For the free vortex condition $C_{w2}r = $ constant, and the value of C_{w2m} was previously determined to be 76·9 m/s. Because of the reduction of annulus area through the compressor, the blade height at exit from the rotor will be slightly less than at inlet, and it is necessary to calculate the tip and root radii at exit from the rotor blades to find the relevant variation of C_{w2}. The stagnation pressure and temperature at exit from the first stage were found to be 1·249 bar and 308 K. Recalling that a stator exit angle of 11·06° was established,

$$C_3 = \frac{150}{\cos 11\cdot06} = 152\cdot8 \text{ m/s}$$

$$T_3 = 308 - (152\cdot8)^2/(2 \times 1\cdot005 \times 10^3) = 296\cdot4 \text{ K}$$

$$p_3 = 1\cdot249\left(\frac{296\cdot4}{308}\right)^{3\cdot5} = 1\cdot092 \text{ bar}$$

$$\rho_3 = \frac{100 \times 1\cdot092}{296\cdot4 \times 0\cdot287} = 1\cdot283 \text{ kg/m}^3$$

$$A_3 = \frac{m}{\rho_3 C_{a3}} = \frac{20}{1\cdot283 \times 150} = 0\cdot1039 \text{ m}^3$$

$$h = \frac{0\cdot1039}{2\pi \times 0\cdot1697} = 0\cdot0974 \text{ m}$$

$$r_t = 0\cdot1697 + \frac{0\cdot0974}{2} = 0\cdot2184 \text{ m}$$

$$r_r = 0\cdot1697 - \frac{0\cdot0974}{2} = 0\cdot1210 \text{ m}$$

These radii refer to conditions at the *stator* exit. With negligible error it can be assumed that the radii at exit from the rotor blades are the mean of those at rotor inlet and stator exit.† Thus at exit from the rotor,

$$r_t = \frac{0\cdot2262 + 0\cdot2184}{2} = 0\cdot2223 \text{ m}, \qquad U_t = 349\cdot2 \text{ m/s}$$

$$r_r = \frac{0\cdot1131 + 0\cdot1210}{2} = 0\cdot1171 \text{ m}, \qquad U_r = 183\cdot9 \text{ m/s}$$

From the free vortex condition,

$$C_{w2r} = 76\cdot9 \times \frac{0\cdot1697}{0\cdot1171} = 111\cdot4 \text{ m/s}$$

$$C_{w2t} = 76\cdot9 \times \frac{0\cdot1697}{0\cdot2223} = 58\cdot7 \text{ m/s}$$

The stator inlet angle is given by $\tan \alpha_2 = C_{w2}/C_a$, and the rotor exit angle by $\tan \beta_2 = (U - C_{w2})/C_a$. Hence,

† For turbines, where there is considerably greater flare in the annulus, it is necessary to make an allowance for the spacing between the rotor and stator; this is done in the turbine design example presented in Chapter 7.

$$\tan \alpha_{2r} = \frac{111\cdot4}{150}, \qquad \alpha_{2r} = 36\cdot60°$$

$$\tan \alpha_{2m} = \frac{76\cdot9}{150}, \qquad \alpha_{2m} = 27\cdot14°$$

$$\tan \alpha_{2t} = \frac{58\cdot7}{150}, \qquad \alpha_{2t} = 21\cdot37°$$

$$\tan \beta_{2r} = \frac{183\cdot9 - 111\cdot4}{150}, \qquad \beta_{2r} = 25\cdot80°$$

$$\tan \beta_{2m} = \frac{266\cdot6 - 76\cdot9}{150}, \qquad \beta_{2m} = 51\cdot67°$$

$$\tan \beta_{2t} = \frac{349\cdot2 - 58\cdot7}{150}, \qquad \beta_{2t} = 62\cdot69°$$

The radial variation of air angles is shown in Fig. 5.16, which shows both the increased deflection $(\beta_1 - \beta_2)$ at the root and the requirement for considerable blade twist along the height of the blade to ensure that the blade angles are in agreement with the air angles.

The degree of reaction at the root can be approximated (approximate because $C_3 \neq C_1$) by $\Lambda_r = 1 - (C_{w2}/2U)$ giving

$$\Lambda_r \approx 1 - \frac{111\cdot4}{2 \times 183\cdot9} \approx 0\cdot697$$

As expected, with the high value of Λ_m of 0·856 there is no problem with too low a degree of reaction at the root.

When calculating the annulus area at exit from the stage it was assumed that the density at the mean radius could be used in the continuity equation. Although the *stagnation* pressure and temperature are assumed

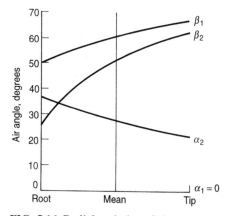

FIG. 5.16 Radial variation of air angles, 1st stage

to be constant across the height of the blade, the *static* pressure and temperature, and hence the density, will vary. This effect is small in compressor stages, with their low pressure ratio, but is more pronounced in turbine stages. The method of dealing with radial variation of density will be described in the turbine example of Chapter 7. The density variation from root to tip is about 4 per cent for the first stage of the compressor and would be even less for later stages of higher hub–tip ratio.

The velocity diagrams for the first stage are shown (to scale) in Fig. 5.17. The increase in fluid deflection and diffusion at the root section are readily visible from the vectors V_1 and V_2. Since V_1 is a maximum at the tip and C_2 a maximum at the root, it is apparent that the maximum relative Mach numbers occur at r_t for the rotor blade and at r_r for the stator blade. Compressibility effects will be discussed in the next sub-section.

Turning our attention to the *third stage*, it is useful to summarize conditions at inlet and outlet before examining different distributions of whirl velocity. From the mean radius design, with $\Lambda_m = 0.50$,

$$p_{01} = 1.599 \text{ bar}, \qquad p_{03}/p_{01} = 1.246, \qquad p_{03} = 1.992 \text{ bar}$$
$$T_{01} = 333 \text{ K}, \qquad T_{03}(=T_{02}) = 357 \text{ K}$$
$$\alpha_1 = 26.63° \ (=\beta_2), \qquad \beta_1 = 50.92° \ (=\alpha_2)$$
$$C_{w1} = 81.9 \text{ m/s}, \qquad C_{w2} = 184.7 \text{ m/s}, \qquad \Delta C_w = 102.8 \text{ m/s}$$

It is not necessary to repeat the calculations for rotor radii and blade speed at root and tip for this stage, and it will be sufficient to summarize the results as follows:

	$\dfrac{r_r}{[\text{m}]}$	$\dfrac{r_t}{[\text{m}]}$	r_r/r_t	$\dfrac{U_r}{[\text{m/s}]}$	$\dfrac{U_t}{[\text{m/s}]}$
Inlet	0.1279	0.2115	0.605	200.9	332.2
Exit	0.1341	0.2053	0.653	205.8	322.5

The resulting air angles for a *free vortex design* are obtained from the calculations tabulated below, and the values already obtained at the mean radius are included for completeness.

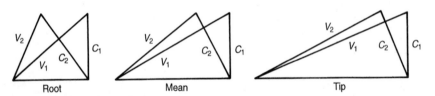

FIG. 5.17 Velocity diagrams, 1st stage

	Root	Mean	Tip
$C_{w1} = C_{w1}r_m/r \, (\text{m/s})$	108·7	81·9	65·7
$\tan \alpha_1 = C_{w1}/C_a$	108·7/150		65·7/150
α_1	35·93	28·63	23·65
$\tan \beta_1 = (U - C_{w1})/C_a$	$\dfrac{(200\cdot9 - 108\cdot7)}{150}$		$\dfrac{(332\cdot2 - 65\cdot7)}{150}$
β_1	31·58	50·92	60·63
$C_{w2} = C_{w2}r_m/r \, (\text{m/s})$	233·7	184·7	152·7
$\tan \alpha_2 = C_{w2}/C_a$	233·7/150		152·7/150
α_2	57·31	50·92	45·51
$\tan \beta_2 = (U - C_{w2})/C_a$	$\dfrac{(205\cdot8 - 233\cdot7)}{150}$		$\dfrac{(322\cdot5 - 152\cdot7)}{150}$
β_2	−10·54	28·63	48·54
$\beta_1 - \beta_2$	42·12	22·29	12·09
$\Lambda = 1 - \dfrac{1}{R^2}(1 - \Lambda_m)$	0·161	0·50	0·668

The radial variation of air angles shown in Fig. 5.18 clearly indicates the very large fluid deflection $(\beta_1 - \beta_2)$ required at the root and the substantial blade twist from root to tip.

It is instructive to consider a *constant reaction design*, with radial equilibrium ignored. With $\Lambda_m = 0.50$, conditions at the mean radius will be the same as those previously considered for the free vortex design. As before, it will be assumed that 50 per cent reaction gives a symmetrical velocity diagram and this will hold across the annulus. From symmetry, $U = \Delta C_w + 2C_{w1}$, and from constant work input at all radii $U\Delta C_w = U_m \Delta C_{wm} = \text{constant}$.

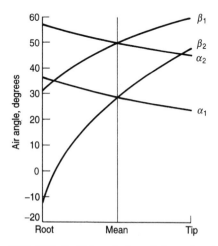

FIG. 5.18 Radial variation of air angles: free vortex, 3rd stage

Considering the root section at inlet to the blade,

$$\Delta C_{wr} = \Delta C_{wm}\frac{r_m}{r_r} = 102{\cdot}8 \times \frac{0{\cdot}1697}{0{\cdot}1279} = 136{\cdot}4 \text{ m/s}$$

$$C_{w1r} = \frac{U_r - \Delta C_{wr}}{2} = \frac{205{\cdot}8 - 136{\cdot}4}{2} = 34{\cdot}7 \text{ m/s}$$

$$\tan\alpha_{1r} = \frac{C_{w1r}}{C_a} = \frac{34{\cdot}70}{150}, \qquad \alpha_{1r} = 13{\cdot}03° = \beta_{2r}$$

$$C_{w2r} = C_{w1r} + \Delta C_{wr} = 34{\cdot}70 + 136{\cdot}4 = 171{\cdot}1 \text{ m/s}$$

$$\tan\alpha_{2r} = \frac{C_{w2r}}{C_a} = \frac{171{\cdot}1}{150}, \qquad \alpha_{2r} = 48{\cdot}76° = \beta_{1r}$$

Repeating these calculations at the tip ($r_t = 0{\cdot}2115\,\text{m}$, $U = 332{\cdot}2\,\text{m/s}$) yields $\alpha_{1t} = 39{\cdot}77° = \beta_{2t}$ and $\alpha_{2t} = 54{\cdot}12° = \beta_{1t}$. The radial variations of the air angles are shown in Fig. 5.19, which shows the greatly reduced twist compared with the free vortex variation shown in Fig. 5.18. It will be recalled, however, that the radial equilibrium condition is not satisfied by the constant reaction design, and with the relatively low hub–tip ratio for this stage the constant reaction design is not likely to be the most efficient. It should be pointed out, also, that a small error is introduced by assuming a symmetrical velocity diagram based on the blade speeds at rotor inlet, because at hub and tip radii the blade speed is not constant throughout the stage due to the taper of the annulus. The reader should recognize that reaction is a useful concept in design but it does not matter if the actual value of reaction is not quite equal to the theoretical value.

To conclude this subsection, a third type of design will be examined, namely exponential blading with $\Lambda_m = 0{\cdot}5$. The value of R at the root section is given by $R = 0{\cdot}1279/0{\cdot}1697 = 0{\cdot}754$. The reaction at the root is then given by

$$\Lambda_r = 1 + \left(1 - \frac{2}{0{\cdot}754}\right)(1 - 0{\cdot}50)$$
$$= 0{\cdot}174$$

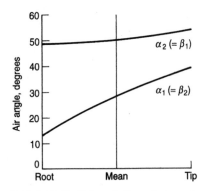

FIG. 5.19 Radial variation of air angles: constant 50 per cent reaction, 3rd stage

This is slightly higher than the value obtained for the free vortex design (0·164), as would be expected from Fig. 5.12. With exponential blading $n = 0$, $C_{w1} = a - (b/R)$ and $C_{w2} = a + (b/R)$. Evaluating the constants a and b,

$$a = U_m(1 - \Lambda_m) = 266 \cdot 6(1 - 0 \cdot 50) = 133 \cdot 3 \, \text{m/s}$$

$$b = \frac{c_p \Delta T}{2 U_m \lambda} = \frac{1 \cdot 005 \times 1000 \times 24}{2 \times 266 \cdot 6 \times 0 \cdot 88} = 51 \cdot 4 \, \text{m/s}$$

The following table gives the resulting whirl velocity distribution.

	Root	Mean	Tip
Inlet R	0·754	1·00	1·246
$C_{w1} = a - \dfrac{b}{R}$ (m/s)	65·1	81·9	92·0
Exit R	0·790	1·00	1·210
$C_{w2} = a + \dfrac{b}{R}$ (m/s)	198·4	184·7	175·8

Setting $C_{a1m} = 150 \, \text{m/s}$ as before, from equation (5.26), at inlet

$$(C_{a1})^2 - (C_{a1m})^2 = -2 \left[a^2 \ln R + \frac{ab}{R} - ab \right]$$

Hence

$$\left(\frac{C_{a1}}{C_{a1m}} \right)^2 = 1 - \frac{2}{C_{a1m}^2} \left[a^2 \ln R + \frac{ab}{R} - ab \right]$$

Thus, at the root

$$\left(\frac{C_{a1}}{C_{a1m}} \right)^2 = 1 \cdot 247, \quad C_{a1r} = 167 \cdot 5 \, \text{m/s}$$

at the tip $(C_{a1}/C_{a1m})^2 = 0 \cdot 773$, $C_{a1t} = 131 \cdot 9 \, \text{m/s}$.

From equation (5.25), at exit from the rotor,

$$\left(\frac{C_{a2}}{C_{a2m}} \right)^2 = 1 - \frac{2}{C_{a2m}^2} \left[a^2 \ln R - \frac{ab}{R} + ab \right]$$

giving

$$C_{a2r} = 185 \cdot 8 \, \text{m/s}$$

$$C_{a2t} = 115 \cdot 5 \, \text{m/s}$$

From the velocities calculated above, the rotor air angles can readily be calculated to give $\beta_{1r} = 39 \cdot 03°$, $\beta_{2r} = 2 \cdot 28°$, $\beta_{1t} = 61 \cdot 23°$ and $\beta_{2t} = 51 \cdot 79°$.

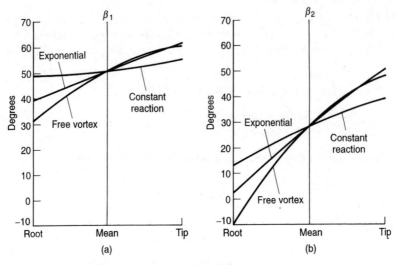

FIG. 5.20 Comparison of rotor air angles, 3rd stage

The rotor air angles for the free vortex, constant reaction and exponential designs are compared in Fig. 5.20, both at inlet and exit from the rotor. It can be seen that the free vortex exhibits the most marked twist over the blade span, with the constant reaction showing the least; the exponential design gives a compromise between the two. The fluid deflection, $\beta_1 - \beta_2$, for the three designs can be summarized as below.

	Root	Mean	Tip
Free vortex	42·12	22·29	12·09
Constant reaction	35·73	22·29	14·35
Exponential	37·75	22·29	9·44

The aerodynamic loading at the root section of the free vortex is substantially higher than that for either of the other two designs.

It was pointed out earlier that the maximum relative inlet Mach numbers occur at the rotor blade tip and the stator blade root. Without detailing the calculations we may summarize the results as follows.

At rotor tip (i.e. at r_t):

	C_{a1} [m/s]	C_{w1} [m/s]	V_1 [m/s]	T_1 [K]	M
Free vortex	150·0	65·7	305·8	319·7	0·853
Constant reaction	150·0	82·5	291·3	314·0	0·820
Exponential	131·9	92·0	274·0	320·1	0·764

At stator tip (i.e. at r_r):

	$\dfrac{C_{a2}}{[m/s]}$	$\dfrac{C_{w2}}{[m/s]}$	$\dfrac{C_2}{[m/s]}$	$\dfrac{T_2}{[K]}$	M
Free vortex	150·0	233·7	277·7	318·6	0·776
Constant reaction	150·0	171·1	227·5	331·2	0·624
Exponential	185·8	198·4	271·8	320·2	0·758

Clearly the rotor tip value is the critical one, and it can be seen that the exponential design yields an appreciably lower Mach number. The importance of this will be discussed further in section 5.10. We have been considering the third stage here as an example, and the Mach numbers in the later stages will be lower because of the increased temperature and hence increased acoustic velocity.

The exponential design results in a substantial variation in axial velocity, both across the annulus and through the stage. We have based the annulus area on the premise of constant axial velocity, and this would have to be recalculated for the case of non-constant axial velocity. At any radius r, the rate of flow δm through an element of width δr is given by $\delta m = \rho 2\pi r \delta r C_a$ and

$$m = 2\pi \int_{r_r}^{r_t} \rho r C_a \, dr$$

The annulus area required to pass the known flow could be determined by numerical integration and would be slightly different to the value assumed for the previous cases of free vortex and constant reaction.

The three design methods considered all have advantages and the final choice of design would depend to a large extent on the design team's previous experience. The constant reaction design looks quite attractive, but it must be borne in mind that radial equilibrium has been ignored; ignoring radial equilibrium will result in flow velocities not being in agreement with the predicted air angles, leading to some loss in efficiency.

It is appropriate at this point to return to the question of the reduced loading in the first and last stages. The first stage is the most critical because of the high Mach number at the tip of the rotor; in addition, at certain flight conditions (e.g. yaw, high angle of attack or rapid turns of the aircraft) the inlet flow may become distorted with significant variations in axial velocity across the compressor annulus. By somewhat reducing the design temperature rise, and hence the aerodynamic loading, these problems can be partially alleviated. In the case of the last stage, the flow is delivered to the diffuser at entry to the combustion chamber; it would be desirable to have a purely axial velocity at exit, and certainly the swirl should be kept as low as possible. Once again, a slight reduction in the required temperature rise eases the aerodynamic design problem.

In this section we have been deciding on the air angles likely to lead to a satisfactory design of compressor. It is now necessary to discuss methods of obtaining the blade shapes which will lead to these air angles being achieved, and this will be done in the next section.

5.8 Blade design

Having determined the air angle distributions that will give the required stage work, it is now necessary to convert these into blade angle distributions from which the correct geometry of the blade forms may be determined.

The most obvious requirements of any particular blade row are: firstly, that it should turn the air through the required angle ($(\beta_1 - \beta_2)$ in the case of the rotor and $(\alpha_2 - \alpha_3)$ in the case of the stator); and secondly that it should carry out its diffusing process with optimum efficiency, i.e. with a minimum loss of stagnation pressure. With regard to the first requirement, we shall see that due allowance must be made for the fact that the air will not leave a blade precisely in the direction indicated by the blade outlet angle. As far as the second requirement is concerned, certainly the air and blade angles at inlet to a blade row must be similar to minimize losses. But when choosing the blade angle, the designer must remember that the compressor has to operate over a wide range of speed and pressure ratio. The air angles have been calculated for the design speed and pressure ratio, and under different operating conditions both the fluid velocities and blade speed may change with resulting changes in the air angles. On the other hand, the blade angles, once chosen, are fixed. It follows that to obtain the best performance over a range of operating conditions it may not be the best policy to make the blade inlet angle equal to the design value of the relative air angle.

The number of variables involved in the geometry of a compressor blade row is so large that the design becomes to a certain extent dependent on the particular preference and previous experience of the designer, who will have at his or her disposal, however, correlated experimental results from wind tunnel tests on single blades or rows of blades. In the former case, the effects of adjacent blades in the row have to be accounted for by the application of empirical factors. The second type of data, from tests on rows or *cascades* of blades, is more widely used. Although it might appear desirable to test a cascade of blades in an annular form of wind tunnel, in an attempt to simulate conditions in an actual compressor, the blades are generally tested in the form of a straight cascade. An annular tunnel would not satisfactorily reproduce conditions in a real compressor and a considerable range of hub–tip ratios would have to be tested. By using a straight cascade, mechanical complication of the test rig is considerably reduced, and the two-dimensional flow conditions obtained in a tunnel of rectangular section greatly simplify the interpretation of the test results.

In both Britain and the United States, much experimental research has been devoted to cascade tests on compressor blading. It is proposed to give a broad outline of this work and to show how the results obtained may be correlated in a fashion suitable for direct use by the compressor designer. Cascade tests result in two main items of information. These are (*a*) the angle through which the air is turned for a minimum loss, and (*b*) the corresponding profile drag coefficient from which the cascade efficiency may be estimated. When high velocities in the region of the sonic velocity are used, the tests also yield valuable information on compressibility effects. A typical cascade tunnel and the type of test result obtainable from it will now be described.

Compressor blade cascade tunnel and typical test results

The tunnel consists essentially of an arrangement whereby a stream of air can be sucked or blown through a number of blades set out in the form of a straight cascade (Fig. 5.21). Means are provided for traversing pressure and flow direction measuring instruments over two planes usually a distance of one blade chord upstream and downstream of the cascade. The height and length of the cascade are made as large as the available air

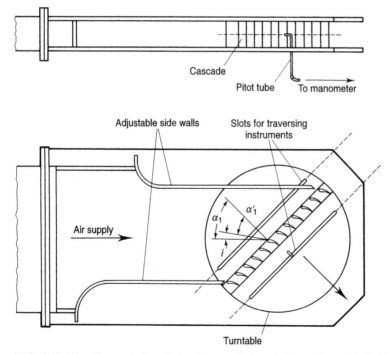

FIG. 5.21 Elevation and plan of simple cascade tunnel (traversing paths indicated by dotted lines)

supply will allow, in an attempt to eliminate interference effects due to the tunnel walls. Boundary layer suction on the walls is frequently applied to prevent contraction of the air stream as it passes through the tunnel.

The cascade is mounted on a turntable so that its angular direction relative to the inflow duct can be set to any desired value. This device enables tests to be made on the cascade over a range of incidence angles of the air entering the cascade. In other more complex tunnels provision is also made for variation of the geometry of the blade row, such as the spacing between the blades and their setting angle, without removing the cascade from the tunnel. Pressure and velocity measurements are made by the usual form of L-shaped pitot and pitot-static tubes. Air directions are found by various types of instruments, the most common being the claw and cylindrical yawmeters illustrated in Fig. 5.22. The principle of operation is the same in both and consists of rotating the instrument about its axis until a balance of pressure from the two holes is obtained. The bisector of the angle between the two holes then gives the air direction.†

A cross-section of three blades forming a part of a typical cascade is shown in Fig. 5.23 which also includes details of the various angles, lengths and velocities associated with the cascade experiments. For any particular test, the blade *camber* angle θ, its *chord c* and the *pitch* (or

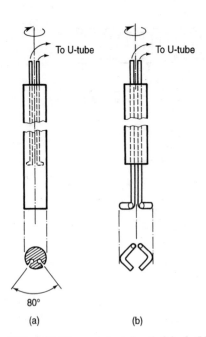

To U-tube   To U-tube

80°

(a)   (b)

FIG. 5.22 Yawmeters: (a) cylindrical, (b) claw

† For a full description of cascade tunnel testing see Todd, K. W. [Ref. (7)] and Gostelow, J. P. [Ref. (8)].

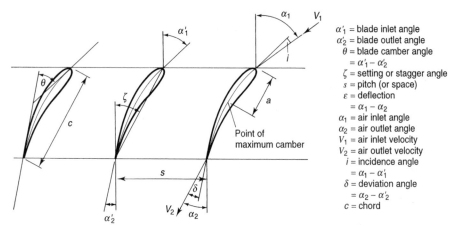

α'_1 = blade inlet angle
α'_2 = blade outlet angle
θ = blade camber angle
$\quad = \alpha'_1 - \alpha'_2$
ζ = setting or stagger angle
s = pitch (or space)
ε = deflection
$\quad = \alpha_1 - \alpha_2$
α_1 = air inlet angle
α_2 = air outlet angle
V_1 = air inlet velocity
V_2 = air outlet velocity
i = incidence angle
$\quad = \alpha_1 - \alpha'_1$
δ = deviation angle
$\quad = \alpha_2 - \alpha'_2$
c = chord

FIG. 5.23 Cascade notation

space) s will be fixed and the blade inlet and outlet angles α'_1 and α'_2 determined by the chosen setting or *stagger* angle ζ. The angle of *incidence*, i, is then fixed by the choice of a suitable air inlet angle α_1 since $i = \alpha_1 - \alpha'_1$. This can be done by an appropriate setting of the turntable on which the cascade is mounted. With the cascade in this position, the pressure and direction-measuring instruments are then traversed along the blade row in the upstream and downstream positions, and the test results plotted as in Fig. 5.24. This shows the variation of loss of stagnation pressure, and *air deflection* $\varepsilon = \alpha_1 - \alpha_2$, covering two blades at the centre of the cascade. As the loss will be dependent on the magnitude of the velocity of the air entering the cascade, it is convenient to express it in a dimensionless form by dividing by the inlet dynamic head, i.e.

$$\text{loss} = \frac{p_{01} - p_{02}}{\frac{1}{2}\rho V_1^2} = \frac{w}{\frac{1}{2}\rho V_1^2} \tag{5.28}$$

This facilitates correlation of test results covering a range of values of V_1.

The curves of Fig. 5.24 can now be repeated for different values of incidence angle, and the whole set of results condensed to the form shown in Fig. 5.25 in which the mean loss $\bar{w}/\frac{1}{2}\rho V_1^2$ and mean deflection ε are plotted against incidence for a cascade of fixed geometrical form. These curves show that the mean loss remains fairly constant over a wide range of incidence, rising rapidly when the incidence has a large positive or negative value. At these extreme incidences the flow of air around the blades breaks down in a manner similar to the stalling of an isolated aerofoil. The mean deflection rises with increasing incidence, reaching a maximum value in the region of the positive stalling incidence.

Test results in the form given in Fig. 5.25 are obtained for a wide range of geometrical forms of the cascade, by varying the camber, pitch/chord

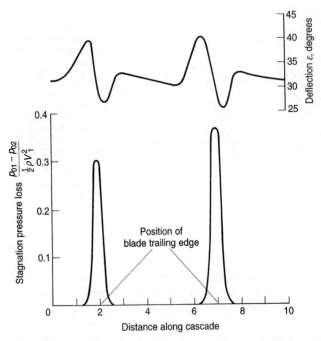

FIG. 5.24 Variations of stagnation pressure loss and deflection for cascade at fixed incidence

ratio, etc. Reduction of the resulting data to a set of design curves is achieved by the following method.

From curves of the form given in Fig. 5.25, a single value of deflection which is most suitable for use with that particular form of cascade is

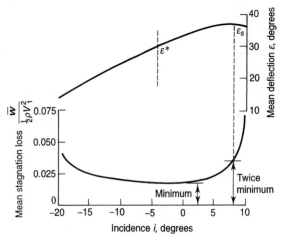

FIG. 5.25 Mean deflection and mean stagnation pressure loss for cascade of fixed geometrical form

selected. As the object of the cascade is to turn the air through as large
an angle as possible with a minimum loss, the use of the maximum
deflection is not possible because of the high loss due to stalling.
Furthermore, remembering that in a compressor the blade will have to
operate over a range of incidence under off-design conditions, the design
incidence should be in the region of the flat portion of the loss curve.
The practice has been to select a deflection which corresponds to a
definite proportion of the stalling deflection. The proportion found to be
most suitable is eight-tenths, so that the selected or *nominal deflection*
$\varepsilon^* = 0 \cdot 8\varepsilon_S$ where ε_S is the stalling deflection. Difficulty is sometimes
experienced in deciding on the exact position of the stall, so its position is
standardized by assuming that the stall occurs when the loss has reached
twice its minimum value.

Analysis of the values of the nominal deflection ε^* determined from a
large number of tests covering different forms of cascade has shown that
its value is mainly dependent on the pitch/chord ratio and air outlet
angle α_2. Its variation with change of other factors determining the
geometrical form of the cascade, such as blade camber angle, is
comparatively small. On this basis, the whole set of test results is
reducible to the form shown in Fig. 5.26 in which the nominal
deflection is plotted against air outlet angle with pitch/chord ratio as
parameter. This set of master curves, as they may well be called, is of
great value to the designer because, having fixed any two of the three
variables involved, an appropriate value for the third may be

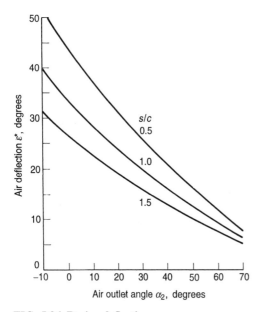

FIG. 5.26 Design deflection curves

determined. For instance, if the air inlet and outlet angles have been fixed by the air angle design, a suitable pitch/chord ratio can be read from the diagram.

As an example, consider the design of the third-stage rotor blade required for the free vortex design carried out in the previous section. At the mean radius of 0·1697 m, where $\beta_1 = 50\cdot92°$ and $\beta_2 = 28\cdot63°$, $\varepsilon^* = \beta_1 - \beta_2 = 22\cdot29°$ and from Fig. 5.26, with an air outlet angle of 28·63°, $s/c = 0\cdot9$.

Determination of the chord length will now depend on the pitch, which itself is clearly dependent on the number of blades in the row. When making a choice for this number, the *aspect ratio* of the blade, i.e. the ratio of length to chord, has to be considered because of its effect on secondary losses. This will be discussed in greater detail in the next section on stage performance. For the purpose of our example it will be assumed that an aspect ratio h/c of about 3 will be suitable. The blade height can be obtained from the previously determined annulus dimensions as 0·0836 m, so that the chord becomes

$$c = \frac{0\cdot0836}{3} = 0\cdot0279 \text{ m}$$

and the pitch is

$$s = 0\cdot9 \times 0\cdot0279 = 0\cdot0251 \text{ m}$$

The number of blades n is then given by

$$n = \frac{2\pi \times 0\cdot1697}{0\cdot0251} = 42\cdot5$$

It is desirable to avoid numbers with common multiples for the blades in successive rows to reduce the likelihood of introducing resonant forcing frequencies. One method of doing this is to choose an even number for the stator blades and a prime number for the rotor blades. An appropriate number for the rotor blades in this stage would therefore be 43, and recalculation in the reverse order gives

$$s = 0\cdot0248 \text{ m}, \qquad c = 0\cdot0276 \text{ m}, \qquad \text{and} \qquad h/c = 3\cdot03$$

A similar procedure could also be carried out for the stator blades.

The use of prime numbers for rotor blades is less common than used to be the case. This is a result of developments in mechanical design which have resulted in the ability to replace damaged rotor blades in the field without the need for rebalancing the rotor system. In the case of a fan for a high bypass turbofan engine, for example, an even number of rotor blades is frequently used and airlines keep a stock of replacement blades in balanced pairs. In the event of a blade failure, the defective blade and the one diametrically opposite are replaced.

In the above example, the chord was determined simply from aerodynamic considerations. The chord chosen for the first stage, however, may well be determined by stringent requirements for resistance to Foreign Object Damage (FOD). This is often the case for aircraft engines, and it may result in a somewhat lower than optimum aspect ratio.

One further item of information is necessary before the design of the blade forms at this radius can be completed. Whereas the blade inlet angle α'_1 will be known from the air inlet angle and chosen incidence (usually taken as zero so that $\alpha'_1 = \alpha_1$), the blade outlet angle α'_2 cannot be determined from the air outlet angle α_2 until the deviation angle $\delta = \alpha_2 - \alpha'_2$ has been determined. Ideally, the mean direction of the air leaving the cascade would be that of the outlet angle of the blades, but in practice it is found that there is a deviation which is due to the reluctance of the air to turn through the full angle required by the shape of the blades. This will be seen from Fig. 5.23. An analysis of the relation between the air and blade outlet angles from cascade tests shows that their difference is mainly dependent on the blade camber and the pitch/chord ratio. It is also dependent on the shape of the camber-line of the blade section and on the air outlet angle itself. The whole of this may be summed up in the following empirical rule for deviation:

$$\delta = m\theta\sqrt{(s/c)} \tag{5.29}$$

where

$$m = 0.23\left(\frac{2a}{c}\right)^2 + 0.1\left(\frac{\alpha_2}{50}\right)$$

a is the distance of the point of maximum camber from the leading edge of the blade, which feature is illustrated in Fig. 5.23, and α_2 is in degrees. Frequently a circular arc camber-line is chosen so that $2a/c = 1$, thereby simplifying the formula for m, but its general form as given above embraces all shapes including a parabolic arc which is sometimes used. (For inlet guide vanes, which are essentially nozzle vanes giving accelerating flow, the power of s/c in equation (5.29) is taken as unity instead of 0.5 and m is given a constant value of 0.19.)

Construction of blade shape

On the assumption of a *circular arc* camber-line, the deviation in the current example will be

$$\delta = \left[0.23 + 0.1 \times \frac{28.63}{50}\right]\sqrt{(0.9)}\theta = 0.273\theta$$

With this information it is now possible to fix the main geometrical parameters of the rotor blade row of our example. The procedure is as follows.

Since $\theta = \alpha_1' - \alpha_2'$ and $\alpha_1' = \alpha_1 - \delta$

$$\theta = \alpha_1' - \alpha_2 + \delta$$
$$= \alpha_1' - \alpha_2 + 0{\cdot}273\theta$$
$$0{\cdot}727\theta = \alpha_1' - \alpha_2$$
$$= 50{\cdot}92 - 28{\cdot}63$$

(assuming zero incidence, $\alpha_1' = \alpha_1$).

It follows that $\theta = 30{\cdot}64°$ and $\alpha_2' = \alpha_1' - \theta = 20{\cdot}28°$. The deviation angle is $8{\cdot}63°$.

The position of the blade chord can be fixed relative to the axial direction by the stagger angle ζ given by

$$\zeta = \alpha_1' - \frac{\theta}{2}$$
$$= 50{\cdot}92 - (30{\cdot}64/2)$$
$$= 35{\cdot}60°$$

Referring to Fig. 5.27, the chord line AB is drawn $0{\cdot}0276$ m long at $35{\cdot}60°$ to the axial direction. The lines AC and BD at angles α_1' and α_2' are then added and a circular arc constructed tangential to these lines and having AB as chord. This arc will now be the camber-line of the blade around which an aerofoil section can be built up.

The manner of specifying the base profile is shown in Fig. 5.27, ordinates being given at definite positions along the camber-line. The RAF 27 profile and the so-called 'C series' of profiles have been widely used in British practice, and the NACA series in the USA. For moderately loaded stages, in which the velocities are well removed from sonic values, small variations in form are found to have little effect on the final performance of the compressor. Further details of base profiles, together with the geometrical relations necessary when parabolic camber-lines are employed, can be found in Ref. (9). For stages operating with relative Mach numbers in the transonic range, it has been found that double circular arc blade sections (see Fig. 5.35) offer the best performance.

The method outlined could now be applied to a selected number of points along the blade length, bearing in mind that having once fixed the pitch at the mean diameter by the choice of a particular number of blades, the pitch at all other radii is determined. As the s/c ratio is derived from the air angles, the length of the chord of the blade at any particular radius will be determined from the pitch. By this means a complete picture of the blade can be built up. The blade stresses can then be accurately assessed.

Now that the reader understands the basis for selecting the pitch/chord ratio, it is appropriate to check the diffusion factor for a couple of stages. Recalling for convenience equation (5.7),

$$D = 1 - \frac{V_2}{V_1} + \frac{\Delta C_w}{2V_1}\frac{s}{c}$$

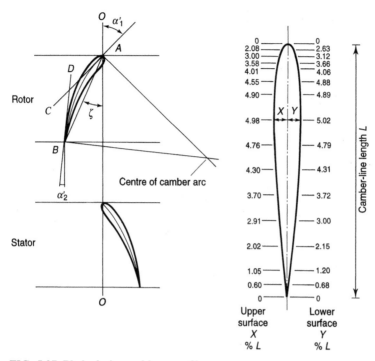

Upper surface X % L	Lower surface Y % L
0	0
2.08	2.63
3.00	3.12
3.58	3.66
4.01	4.06
4.55	4.88
4.90	4.89
4.98	5.02
4.76	4.79
4.30	4.31
3.70	3.72
2.91	3.00
2.02	2.15
1.05	1.20
0.60	0.68
0	0

FIG. 5.27 Blade design and base profile

the velocity terms are all available from the calculated velocity triangles. Considering the free vortex design of the third stage, at the tip section $C_a = 150 \, \text{m/s}$, $\beta_1 = 60\cdot63°$, $\beta_2 = 48\cdot54°$, $\beta_1 - \beta_2 = 12\cdot09°$ and $\Delta C_w = 87 \, \text{m/s}$. From Fig. 5.26 a suitable value of s/c would be about $1\cdot1$ and the values of V_2 and V_1 can be readily calculated from $C_a / \cos\beta$ to be $226\cdot5$ and $305\cdot8 \, \text{m/s}$ respectively. Hence

$$D = 1 - \frac{226\cdot5}{305\cdot8} + \frac{87 \times 1\cdot1}{2 \times 305\cdot8} = 0\cdot42$$

Referring back to Fig. 5.8, it can be seen that this value is quite satisfactory. Repeating the calculations at the root results in a slightly higher value of $0\cdot45$, primarily due to the increase in ΔC_w. It is worth noting that the de Haller numbers for the third stage were close to the limit of $0\cdot72$, and the diffusion factors are also near the point at which losses increase rapidly. At the tip of the first-stage rotor, the diffusion factor can be calculated to be about $0\cdot23$, and this again correlates well with the light aerodynamic loading of this stage; the transonic relative Mach number, however, leads to further losses which will be discussed in section 5.10. The experienced compressor designer will have a good idea of the expected pitch/chord ratios at the outset of the design process, and the diffusion factor is an excellent preliminary check on aerodynamic loading.

5.9   Calculation of stage performance

After completion of the stage design, it will now be necessary to check over the performance, particularly in regard to the efficiency which for a given work input will completely govern the final pressure ratio. This efficiency is dependent on the total drag coefficient for each of the blade rows comprising the stage, and in order to evaluate these quantities it will be necessary to revert to the loss measurements in cascade tests. From the measured values of mean loss $\bar{w}$, two coefficients can be obtained. These are the lift and profile drag coefficients C_L and C_{Dp}, formulae for which may be obtained as follows.

Referring to the diagram of forces acting on the cascade as shown in Fig. 5.28, the static pressure rise across the blades is given by

$$\Delta p = p_2 - p_1$$
$$= \left(p_{02} - \tfrac{1}{2}\rho V_2^2\right) - \left(p_{01} - \tfrac{1}{2}\rho V_1^2\right)$$

The incompressible flow formula is used because the change of density is negligible. Hence, still using cascade notation for velocities and angles,

$$\Delta p = \tfrac{1}{2}(V_1^2 - V_2^2) - \bar{w}$$
$$= \tfrac{1}{2}\rho V_a^2(\tan^2 \alpha_1 - \tan^2 \alpha_2) - \bar{w} \tag{5.30}$$

the axial velocity V_a being assumed the same at inlet and outlet.

The axial force *per unit length* of each blade is $s\Delta p$ and, from consideration of momentum changes, the force acting *along* the cascade per unit length is given by

$$F = s\rho V_a \times \text{change in velocity component along cascade}$$
$$= s\rho V_a^2(\tan \alpha_1 - \tan \alpha_2) \tag{5.31}$$

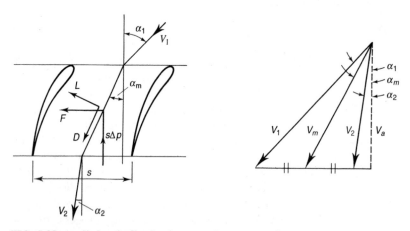

FIG. 5.28 Applied and effective forces acting on cascade

The coefficients C_L and C_{Dp} are based on a vector mean velocity V_m defined by the velocity triangles in Fig. 5.28. Thus

$$V_m = V_a \sec \alpha_m$$

where α_m is given by

$$\tan \alpha_m = \left[\tfrac{1}{2}(V_a \tan \alpha_1 - V_a \tan \alpha_2) + V_a \tan \alpha_2\right]/V_a = \tfrac{1}{2}(\tan \alpha_1 + \tan \alpha_2)$$

If D and L are the drag and lift forces along and perpendicular to the direction of the vector mean velocity, then resolving along the vector mean gives

$$D = \tfrac{1}{2}\rho V_m^2 cC_{Dp} \text{ (by definition of } C_{Dp})$$
$$= F \sin \alpha_m - s\Delta p \cos \alpha_m$$

Hence from equations (5.30) and (5.31),

$$\tfrac{1}{2}\rho V_m^2 cC_{Dp} = s\rho V_a^2(\tan \alpha_1 - \tan \alpha_2) \sin \alpha_m$$
$$- \tfrac{1}{2}\rho V_a^2 s(\tan^2 \alpha_1 - \tan^2 \alpha_2) \cos \alpha_m + \bar{w}s \cos \alpha_m$$

Since

$$\tan^2 \alpha_1 - \tan^2 \alpha_2 = (\tan \alpha_1 - \tan \alpha_2)(\tan \alpha_1 + \tan \alpha_2)$$
$$= 2(\tan \alpha_1 - \tan \alpha_2) \tan \alpha_m$$

the first two terms in the expressions for C_{Dp} are equal and the equation reduces to

$$C_{Dp} = \left(\frac{s}{c}\right)\left(\frac{\bar{w}}{\tfrac{1}{2}\rho}\right)\left(\frac{\cos \alpha_m}{V_m^2}\right)$$
$$= \left(\frac{s}{c}\right)\left(\frac{\bar{w}}{\tfrac{1}{2}\rho}\right)\left(\frac{\cos^3 \alpha_m}{V_a^2}\right)$$

i.e.

$$C_{Dp} = \left(\frac{s}{c}\right)\left(\frac{\bar{w}}{\tfrac{1}{2}\rho V_1^2}\right)\left(\frac{\cos^3 \alpha_m}{\cos^2 \alpha_1}\right) \qquad (5.32)$$

Also, resolving perpendicularly to the vector mean,

$$L = \tfrac{1}{2}\rho V_m^2 cC_L \text{ (by definition of } C_L)$$
$$= F \cos \alpha_m + s\Delta p \sin \alpha_m$$

Therefore

$$\tfrac{1}{2}\rho V_m^2 cC_L = s\rho V_a^2(\tan \alpha_1 - \tan \alpha_2) \cos \alpha_m$$
$$+ \tfrac{1}{2}\rho V_a^2 s(\tan^2 \alpha_1 - \tan^2 \alpha_2) \sin \alpha_m - \bar{w}s \sin \alpha_m$$

which on reduction finally gives

$$C_L = 2(s/c)(\tan \alpha_1 - \tan \alpha_2) \cos \alpha_m - C_{Dp} \tan \alpha_m \qquad (5.33)$$

Using these formulae the values of C_L and C_{Dp} may be calculated from the data given by the curves of Fig. 5.25. Since α'_1 is known from the geometry of the blade row, the following data may be found for any incidence angle i:

$$\alpha_1 = \alpha'_1 + i$$

$$\alpha_2 = \alpha_1 - \varepsilon^*$$

$$\alpha_m = \tan^{-1}[\tfrac{1}{2}(\tan \alpha_1 + \tan \alpha_2)]$$

Then by using values of $\bar{w}/\tfrac{1}{2}\rho V_1^2$ read from the curve and the known value of s/c for the cascade, C_{Dp} and C_L may be calculated from equations (5.32) and (5.33) and plotted against incidence as in Fig. 5.29.

Because the value of the term $C_{Dp} \tan \alpha_m$ in equation (5.33) is negligibly small, it is usual to use the more convenient 'theoretical' value of C_L given by

$$C_L = 2(s/c)(\tan \alpha_1 - \tan \alpha_2) \cos \alpha_m \qquad (5.34)$$

in which the effect of profile drag is ignored. Using this formula, curves of C_L can be plotted for nominal (or design) conditions to correspond with the curves of deflection given in Fig. 5.26. These curves, which are again plotted against air outlet angle α_2 for fixed values of pitch/chord ratio s/c, are given in Fig. 5.30.

Before these coefficients can be applied to the blade rows of the compressor stage, two additional factors must be taken into account. These are the additional drag effects due to the walls of the compressor annulus, and the secondary loss due to trailing vortices and tip clearance. The flow effects which give rise to these losses are illustrated in Fig. 5.31. Analysis of compressor performance figures have shown that the

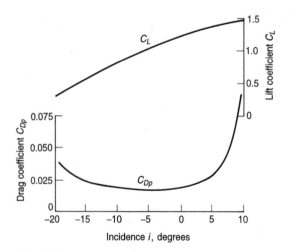

FIG. 5.29 Lift and drag coefficient for cascade of fixed geometrical form

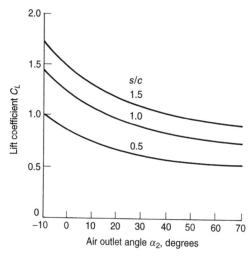

FIG. 5.30 Design lift coefficients

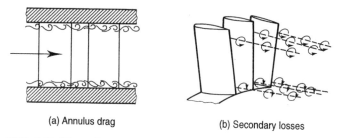

(a) Annulus drag   (b) Secondary losses

FIG. 5.31 Three-dimensional flow effects in compressor annulus

secondary loss is of major importance and that its magnitude is of the same order as that incurred by the profile drag of the blades. It is greatly influenced by tip clearance which should consequently be kept as small as possible—in the region of 1–2 per cent of the blade height. For typical axial compressor designs, the following empirical formula for the additional drag coefficient arising from secondary losses has been derived:

$$C_{DS} = 0\cdot018C_L^2 \tag{5.35}$$

where C_L is the lift coefficient as given by equation (5.34) and the curves of Fig. 5.30.

The loss due to annulus drag is naturally dependent on the relative proportions of the blade row, its influence increasing as the blades become shorter relative to their chord length. It has been found convenient to relate the drag coefficient, resulting from this loss, to the blade row dimensions by the empirical formula

$$C_{DA} = 0\cdot020(s/h) \tag{5.36}$$

where s and h are the pitch and height of the blades respectively. Therefore, on inclusion of these factors, an overall drag coefficient can be determined which is given by

$$C_D = C_{Dp} + C_{DA} + C_{DS} \qquad (5.37)$$

The argument used in deriving equation (5.32) for the profile drag coefficient in the case of a straight cascade will apply equally well to the annular case if C_D is substituted for C_{Dp}. Hence for the annular case

$$C_D = \left(\frac{s}{c}\right)\left(\frac{\bar{w}}{\frac{1}{2}\rho V_1^2}\right)\left(\frac{\cos^3 \alpha_m}{\cos^2 \alpha_1}\right) \qquad (5.38)$$

This enables the loss coefficient $\bar{w}/\frac{1}{2}\rho V_1^2$ for the blade row to be determined. The *theoretical* static pressure rise through the blade row is found by putting the loss $\bar{w}$ equal to zero in equation (5.30), which gives

$$\Delta p_{th} = \frac{1}{2}\rho V_a^2(\tan^2 \alpha_1 - \tan^2 \alpha_2)$$
$$= \frac{1}{2}\rho V_a^2(\sec^2 \alpha_1 - \sec^2 \alpha_2)$$

Therefore

$$\frac{\Delta p_{th}}{\frac{1}{2}\rho V_a^2 \sec^2 \alpha_1} = 1 - \frac{\sec^2 \alpha_2}{\sec^2 \alpha_1}$$

so that the theoretical pressure rise in terms of the inlet dynamic head and cascade air angles becomes

$$\frac{\Delta p_{th}}{\frac{1}{2}\rho V_1^2} = 1 - \frac{\cos^2 \alpha_1}{\cos^2 \alpha_2}$$

The efficiency of the blade row, η_b, which is defined as the ratio of the actual pressure rise to the theoretical pressure rise, can then be found from $\eta_b = (\Delta p_{th} - \bar{w})/\Delta p_{th}$. Or in non-dimensional terms,

$$\eta_b = 1 - \frac{\bar{w}/\frac{1}{2}\rho V_1^2}{\Delta p_{th}/\frac{1}{2}\rho V_1^2} \qquad (5.39)$$

When dealing with cascade data, efficiency is evaluated from pressure rises, whereas compressor stage efficiency is defined in terms of temperature rises. Furthermore, the stage efficiency encompasses both a rotor and a stator row. It is therefore not obvious how the efficiency of a blade row obtained from cascade tests can be related to the stage efficiency. Initially we will consider this problem in the context of a stage designed for 50 per cent reaction at the mean diameter. In such a case, because of the symmetrical blading, η_b (evaluated from equation (5.39) for conditions obtaining at the mean diameter) will be virtually the same for both the rotor and stator blade row.† What we shall now proceed to show is that this value of η_b can be regarded as being equal to the isentropic efficiency of the whole stage, η_S.

If p_1 and p_2 are the static pressures at inlet to and outlet from the rotor,

$$\eta_b = \frac{p_2 - p_1}{p_2' - p_1}$$

where p_2' denotes the ideal pressure at outlet corresponding to no losses. If the stage efficiency η_S is defined as the ratio of the isentropic static temperature rise to the actual static temperature rise,

$$\frac{p_2}{p_1} = \left[1 + \frac{\eta_S \Delta T_S}{2T_1}\right]^{\gamma/(\gamma-1)}$$

because $\Delta T_S/2$ will be the temperature rise in the rotor for 50 per cent reaction. Also

$$\frac{p_2'}{p_1} = \left[1 + \frac{\Delta T_S}{2T_1}\right]^{\gamma/(\gamma-1)}$$

Therefore

$$\eta_b = \left[\frac{p_2}{p_1} - 1\right] \Big/ \left[\frac{p_2'}{p_1} - 1\right]$$

$$= \left\{\left[1 + \frac{\eta_S \Delta T_S}{2T_1}\right]^{\gamma/(\gamma-1)} - 1\right\} \Big/ \left\{\left[1 + \frac{\Delta T_S}{2T_1}\right]^{\gamma/(\gamma-1)} - 1\right\}$$

After expanding and neglecting second-order terms, this reduces to

$$\eta_b = \eta_S\left[1 - \frac{1}{\gamma - 1} \times \frac{\Delta T_S}{4T_1}(1 - \eta_S)\right]$$

But ΔT_S is of the order 20 K and T_1 about 300 K, so that the second term in the bracket is negligible and

$$\eta_b = \eta_S$$

For cases other than 50 per cent reaction at the mean diameter, an approximate stage efficiency can be deduced by taking the arithmetic mean of the efficiencies of the two blade rows, i.e.

$$\eta_S = \tfrac{1}{2}(\eta_{b \text{ rotor}} + \eta_{b \text{ stator}})$$

---

† Owing to slight differences in blade pitch and height between the rotor and stator rows of a stage, C_{DA} from equation (5.36) may differ marginally. But as the following numerical example will show, C_{DA} is only a small proportion of C_D and this difference will have little effect on η_b.

If the degree of reaction is far removed from the 50 per cent condition, then a more accurate expression for the stage efficiency may be taken as

$$\eta_S = \Lambda \eta_{b\ rotor} + (1 - \Lambda)\eta_{b\ stator}$$

where Λ is the degree of reaction as given in section 5.5.

Using this theory, we will estimate the performance of the *third stage* of the compressor used as an example of compressor design, for which the degree of reaction was 50 per cent at the mean diameter. From the blade design given in the previous section, and remembering that β_1 and β_2 are the air inlet and outlet angles α_1 and α_2 respectively in cascade terminology,

$$\tan \alpha_m = \tfrac{1}{2}(\tan \alpha_1 + \tan \alpha_2) = \tfrac{1}{2}(\tan 50 \cdot 92 + \tan 28 \cdot 63) = 0 \cdot 889$$

whence

$$\alpha_m = 41 \cdot 63°$$

From Fig. 5.30, C_L at $s/c = 0 \cdot 9$ and $\alpha_2 = 28 \cdot 63°$ is equal to $0 \cdot 875$. Hence from equation (5.33)

$$C_{DS} = 0 \cdot 018(0 \cdot 875)^2 = 0 \cdot 0138$$

Recalling that $s = 0 \cdot 0248$ m and $h = 0 \cdot 0836$ m we have from equation (5.36)

$$C_{DA} = \frac{0 \cdot 020 \times 0 \cdot 0248}{0 \cdot 0836} = 0 \cdot 0059$$

From Fig. 5.29, at zero incidence $C_{Dp} = 0 \cdot 018$; hence the total drag coefficient is

$$C_D = C_{DP} + C_{DA} + C_{DS} = 0 \cdot 018 + 0 \cdot 0059 + 0 \cdot 0138 = 0 \cdot 0377$$

Therefore, from equation (5.38)

$$\frac{\bar{w}}{\tfrac{1}{2}\rho V_1^2} = C_D \bigg/ \left[\left(\frac{s}{c}\right) \frac{\cos^3 \alpha_m}{\cos^2 \alpha_1} \right] = \frac{0 \cdot 0377 \cos^2 50 \cdot 92°}{0 \cdot 9 \cos^3 41 \cdot 63°} = 0 \cdot 0399$$

We also have

$$\frac{\Delta p_{th}}{\tfrac{1}{2}\rho V_1^2} = 1 - \frac{\cos^2 \alpha_1}{\cos^2 \alpha_2} = 1 - \frac{\cos^2 50 \cdot 92°}{\cos^2 28 \cdot 63°} = 0 \cdot 4842$$

Therefore, from equation (5.39)

$$\eta_b = 1 \cdot 0 - \frac{0 \cdot 0399}{0 \cdot 4842} = 0 \cdot 918$$

The efficiency of the stator row will be similar, and hence we may take

$$\eta_S = 0 \cdot 92$$

The static temperature at entry to the stage can be found to be 318·5 K, and with a stage temperature rise of 24 K the *static* pressure ratio of the stage is found to be

$$R_S = \left[1 + \frac{\eta_S \Delta T_S}{T_1}\right]^{\gamma/(\gamma-1)} = \left[1 + \frac{0\cdot92 \times 24}{318\cdot5}\right]^{3\cdot5} = 1\cdot264$$

We have now seen how the design point performance of a stage may be estimated. It will have been noticed that the results are in terms of an isentropic stage efficiency based on *static* temperatures and a stage pressure ratio based on *static* pressures. To the compressor designer, however, it is the efficiency based on *stagnation* temperature rise and the *stagnation* pressure ratio which are of interest. Making use of the relations $p_0/p = (T_0/T)^{\gamma/(\gamma-1)}$ and $T_0 = T + C^2/2c_p$ it is of course possible to transform a static pressure ratio into a stagnation pressure ratio when the inlet and outlet velocities are known. But for the common case where these velocities are equal (i.e. $C_3 = C_1$) there is no need for this elaboration. $\Delta T_{0S} = \Delta T_S$, and it is easy to show as follows that η_S is virtually the same whether based on static or stagnation temperatures. Then merely by substituting T_{01} for T_1 in the foregoing equation for R_S the stagnation pressure ratio is obtained directly.

Referring to Fig. 5.32, where the stage inlet and outlet states are denoted by 1 and 3 (static) and 01 and 03 (stagnation), we have

$$\eta_S = \frac{T'_3 - T_1}{T_3 - T_1} = \frac{T_3 - T_1 - x}{T_3 - T_1} = 1 - \frac{x}{\Delta T_S}$$

And when based on stagnation temperatures,

$$\eta_S = \frac{T'_{03} - T_{01}}{T_{03} - T_{01}} = 1 - \frac{y}{\Delta T_{0S}}$$

With $C_3 = C_1$, $\Delta T_{0S} = \Delta T_S$. Furthermore, since the pressure ratio per stage is small the constant p_{03} and p_3 lines are virtually parallel between 3'

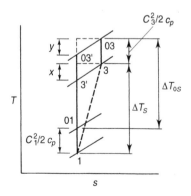

FIG. 5.32

and 3 so that $y \approx x$. It follows that η_S has the same value on either basis. For the stage of our example, T_{01} was 333 K, and hence the stagnation pressure ratio is

$$R_S = \left[1 + \frac{0.92 \times 20}{333}\right]^{3.5} = 1.252$$

It will be remembered that it was necessary to make use of two assumed efficiencies at the start of the design process: a polytropic efficiency for the compressor as a whole, and a stage efficiency. As a first approximation these were taken to be equal and a value of 0.90 was assumed. The estimated value of η_S for the third stage, i.e. 0.92, is in sufficient agreement, bearing in mind the uncertainties in predicting the secondary and annulus losses. If similar agreement were to be obtained for all the stages, we might conclude that the design had been conservative and that the compressor should have no difficulty in achieving the specified performance. To obtain an estimate of the overall efficiency it would be necessary to repeat the foregoing for all the stages. The product of the stage pressure ratios would then yield the overall pressure ratio from which the overall isentropic temperature rise, and hence the overall efficiency, could be calculated.

Before continuing, it may be helpful to summarize the main steps in the design procedure described in the previous sections. Having made appropriate assumptions about the efficiency, tip speed, axial velocity, and so on, it was possible to size the annulus at inlet and outlet of the compressor and calculate the air angles required for each stage at the mean diameter. A choice was then made of a suitable vortex theory to enable the air angles to be calculated at various radii from root to tip. Throughout this work it was necessary to ensure that limitations on blade stresses, rates of diffusion and Mach number were not exceeded. Cascade test data were used to determine a blade geometry which would give these air angles, and also the lift and drag coefficients for a two-dimensional row of blades of this form. Finally, empirical correction factors were evaluated to enable these coefficients to be applied to the annular row at the mean diameter so that the stage efficiency and pressure ratio could be estimated.

5.10   Compressibility effects

Over a long period of time, the development of high-performance gas turbines has led to the use of much higher flow rates per unit frontal area and blade tip speeds, resulting in higher velocities and Mach numbers in compressors. While early units had to be designed with subsonic velocities throughout, Mach numbers exceeding unity are now found in the compressors of industrial gas turbines and Mach numbers as high as 1.5 are used in the design of fans for turbofans of high bypass ratio. It is not within the scope of this book to cover transonic design in detail, and the treatment will be confined to a brief introduction and provision of some key references; again, it should be realized that much of the relevant information is of a proprietary nature and is not available in the open literature.

High-speed cascade testing is required to provide experimental data on compressibility effects, and in particular to determine the values of the Mach numbers, corresponding to entry velocities relative to the blades, which bring about poor cascade performance. The first high velocity of interest is that corresponding to what is called the 'critical' Mach number M_c; at entry velocities lower than this, the performance of the cascade differs very little from that at low speeds. Above this velocity, the losses begin to show a marked increase until a point is reached where the losses completely cancel the pressure rise and the blade ceases to be any use as a diffuser. The corresponding Mach number is then referred to as the 'maximum' value M_m. For a typical subsonic compressor cascade at zero incidence, the values of these Mach numbers are in the region of 0·7 and 0·85 respectively. A typical variation of fluid deflection and pressure loss for subsonic blading was shown in Fig. 5.25, representing flow at low Mach numbers. As the Mach number increases, two important effects take place: first, the overall level of losses increases substantially, and second, the range of incidence for which losses are acceptable is drastically reduced. This means that off-design performance of the compressor may be seriously affected. Figure 5.33 shows test results for a subsonic blade section over a range of Mach number from 0·5 to 0·8; the usable range of incidence at $M\,0·8$ can be seen to be very narrow and clearly compressor

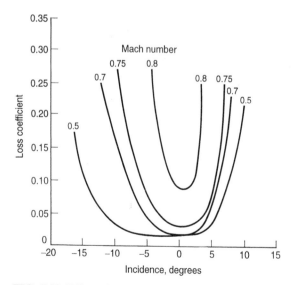

FIG. 5.33 Effect of Mach number on losses

blading of this type could not be used at Mach numbers approaching or exceeding unity. It should be pointed out that for compressible flow the denominator in the loss coefficient is $(p_{01} - p_1)$ rather than $\rho V_1^2/2$.

Compressibility effects will be most important at the front of the compressor where the inlet temperature, and hence the acoustic velocity, are lowest. The Mach number corresponding to the velocity relative to the tip of the rotor is the highest encountered and is important both from the viewpoint of shock losses and noise. The stator Mach number is generally highest at the hub radius because of the increased whirl velocity normally required to give constant work input at all radii in the rotor. The rotor and stator Mach numbers corresponding to the first stage of the compressor designed in section 5.7 are shown in Fig. 5.34, which shows that the flow relative to the rotor is supersonic over a considerable length of the blade. The stator Mach numbers can be seen to be significantly lower, this being due to the fact that they are unaffected by the blade speed.

Analysis of a large amount of compressor tests by NACA [Refs (10, 11)] showed that losses for subsonic conditions correlated well on the basis of diffusion factor, as shown in Fig. 5.8, but were significantly higher for transonic conditions. It was deduced that this increase in loss must be due to shock losses, but it was also found that the spacing of the blades had a considerable effect; a reduction in solidity (i.e. an increase in pitch/chord ratio) caused a rapid increase in loss. A simple method for predicting the loss was developed and can be explained with reference to Fig. 5.35.

A pair of double circular arc (DCA) blades are shown, with a supersonic velocity entering in a direction aligned with the leading edge. The supersonic expansion along the uncovered portion of the suction side can be analysed by means of the Prandtl–Meyer relations discussed in Appendix A.8, and the Mach number will increase as the flow progresses along the suction surface.

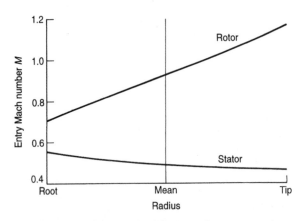

FIG. 5.34 Variation of entry Mach numbers

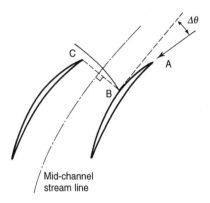

FIG. 5.35 Shock loss model

A shock structure is assumed in which the shock stands near the entrance to the blade passage, striking the suction surface at B, extending in front of the blade at C and then bending back similar to a bow wave. It is then assumed that the loss across the shock can be approximated by the normal shock loss taken for the average of the Mach numbers at A and B. At A the Mach number is taken to be the inlet relative Mach number. The value at B can be calculated from the inlet Mach number and the angle of supersonic turning to the point B, where the flow is tangential to the surface, using the Prandtl–Meyer relations. The location of the point B is taken to be at the intersection of the suction surface and a line drawn normal to the mean passage camber-line and through the leading edge of the next blade. The requirement, then, is to establish the supersonic turning from inlet to point B and this clearly depends on the blade spacing; moving the blades further apart (reducing solidity) results in point B moving back, increasing both the turning and Mach number, resulting in an increased shock loss. Figure 5.36 shows the relationship between inlet Mach number, supersonic turning and shock loss. If we consider the tip of our first-stage rotor, the relative Mach number was 1.165 and the rotor air angles were $67.11°$ at inlet and $62.69°$ at exit. The total turning was therefore $4.42°$, and the supersonic turning to point B would depend on the blade spacing; making a reasonable assumption of $2°$, from Fig. 5.36 it is found that $M_B = 1.22$ and $w = 0.015$. The additional loss due to shocks can be significant and would increase rapidly with increased supersonic turning; this is another reason for the designer to select a slightly reduced temperature rise in the first stage.

The fans of high bypass turbofans require a large diameter, and even though they run at a lower rotational speed than a conventional jet engine, the tip speeds are significantly higher; typical Mach numbers at the tip would be of the order of 1.4–1.6 and DCA blading is not satisfactory. Blade sections have been specially developed to meet this high Mach number requirement and are not based on aerofoil sections. To keep shock

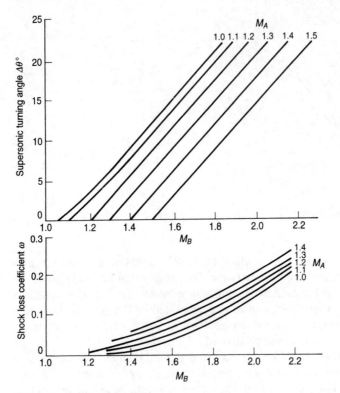

FIG. 5.36 Shock loss coefficient

losses to an acceptable level, it is necessary to provide supersonic diffusion to a Mach number of about 1·2 prior to the normal shock at entry to the blade passage. It is shown in Appendix A.2 that supersonic diffusion requires a *decrease* in flow area. This may be accomplished in two ways, either by decreasing the annulus area or by making the suction side slightly concave. Both methods may be used together. A typical blade profile is shown in Fig. 5.37, where it can be seen that the amount of turning is very small. Until recently it was thought that shock losses

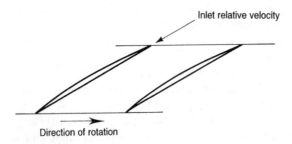

FIG. 5.37 Blade section for supersonic flow

explained the decrease in efficiency with Mach number which has been observed, but it is now realized that the shocks only account for part of the additional losses. There are also significant losses resulting from shock–boundary layer interaction and the net effect appears to be a magnification of the viscous losses. Kerrebrock [Ref. (12)] gives a comprehensive review of the aerodynamic problems of transonic compressors and fans.

High bypass ratio turbofans were introduced in the early 1970s. The fan blades were rather long and flexible, requiring the use of part-span dampers to prevent excessive vibration and torsional displacement. These were located in a region of high Mach number, resulting in significant local decreases in efficiency. Figure 5.38(a) shows the effect of a seagull

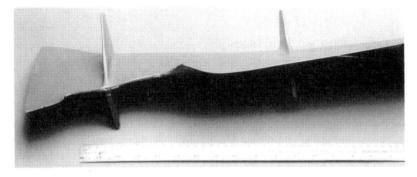

FIG. 5.38(a)

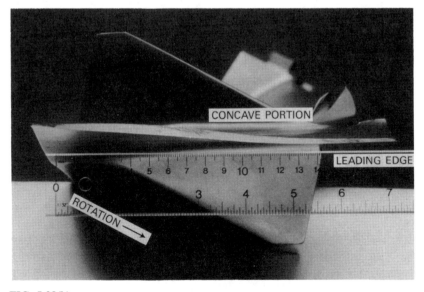

FIG. 5.38(b)

strike on a large turbofan during take-off and gives some idea of the impact forces involved. The shape of the tip section is shown in Fig. 5.38(b) and the use of a concave portion on the suction side can be seen. Improvements in manufacturing methods and stress analysis have resulted in the development of wide-chord fan blades that do not need dampers, giving both an improvement in aerodynamic performance and resistance to bird strikes, a key certification requirement. The fan of the PW 530 (Fig. 1.12(a)) is built as a wide-chord integrally bladed rotor (IBR), with the fan machined from a solid forging. The very large fan blades on the Trent (Fig. 1.16) are of hollow construction, with internal stiffening, to keep the blade weight to an acceptable level. This fan has an even number of blades, so that damaged blades can be replaced by a balanced pair and the engine does not need to be rebalanced in the field.

5.11   Off-design performance

Attention so far has focused on the aerothermodynamic design of axial flow compressors to meet a particular design point of mass flow, pressure ratio and efficiency. It must be realized at the outset, however, that any compressor will be required to operate at conditions far removed from the design point including engine starting, idling, reduced power, maximum power, acceleration and deceleration. Thus it is clear that the compressor must be capable of satisfactory operation over a wide range of rotational speeds and inlet conditions. With the compressor blading and annulus dimensions chosen to satisfy the design point condition, it is obvious that these will not be correct for conditions far removed from design. Previous sections showed that compressor blading had a limited range of incidence before losses became unacceptably high, resulting in low compressor efficiency. What is even more important, however, is that the problems of blade stalling may lead to surging of the whole compressor, preventing operation of the engine at particular conditions; this may lead to severe engine damage or, in the case of aircraft, a critical safety hazard.

An overall picture of compressor off-design performance can be built up from consideration of the behaviour of individual stages and the interaction between a series of stages. Recalling equation (5.5), the temperature rise in a stage is given by

$$\Delta T_{0S} = \frac{UC_a}{c_p}(\tan \beta_1 - \tan \beta_2)$$

which is expressed in terms of the rotor air angles. This can be recast to give

$$\Delta T_{0S} = \frac{U}{c_p}[U - C_a(\tan \alpha_1 + \tan \beta_2)]$$

The angle α_1 is the outlet air angle from the preceding stator, β_2 is the rotor outlet air angle, and these can be considered essentially constant, being determined by the blading geometry; β_1, on the other hand, will vary widely as C_a and U change. Dividing the previous equation throughout by U^2 and rearranging we get

$$\frac{c_p \Delta T_{0S}}{U^2} = 1 - \frac{C_a}{U}(\tan \alpha_1 + \tan \beta_2)$$

The term C_a/U is known as the *flow coefficient* (ϕ) and ($c_p \Delta T_{0S}/U^2$) as the *temperature coefficient* (ψ). With the stage operating at the design value of ϕ the incidence will be at its design value and a high efficiency will be achieved. With the assumption that α_1 and β_2 are constant,

$$\psi = 1 - \phi k$$

where $k = \tan \alpha_1 + \tan \beta_2$. This relationship is shown by the dotted straight line in Fig. 5.39. It is obvious that the temperature coefficient increases as the flow coefficient decreases and reaches a value of 1 when $\phi = 0$. The temperature coefficient can be interpreted from the shape of the velocity diagram. Recalling that

$$\Delta T_{0S} = \frac{U \Delta C_w}{c_p}$$

it immediately follows that

$$\frac{c_p \Delta T_{0S}}{U^2} = \frac{\Delta C_w}{U}$$

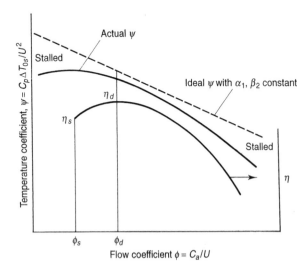

FIG. 5.39 Stage characteristic

Thus, if $\psi = 1$, $\Delta C_w = U$ and from earlier examination of velocity diagrams it will be recognized that this results in excessive diffusion in the blade passage and efficiency will decrease. For satisfactory operation, $\Delta C_w/U$, and hence ψ, should be around 0·3–0·4.

The pressure rise across the stage, Δp_{0S}, is determined by the temperature rise and the isentropic efficiency of the stage.

$$\frac{p_{01} + \Delta p_{0S}}{p_{01}} = 1 + \frac{\Delta p_{0S}}{p_{01}} = \left(1 + \frac{\eta_S \Delta T_{0S}}{T_{01}}\right)^{\gamma/(\gamma-1)}$$

Expanding by means of the binomial theorem, assuming that $\Delta T_{0S} \ll T_{01}$, we have

$$\frac{\Delta p_{0S}}{p_{01}} = \frac{\gamma}{\gamma - 1} \eta_S \frac{\Delta T_{0S}}{T_{01}}$$

Making use of the equation of state and the relationship $(\gamma - 1)/\gamma = R/c_p$, it follows that

$$\frac{\Delta p_{0S}}{\rho_{01}} = \eta_S c_p \Delta T_{0S}$$

and hence

$$\frac{\Delta p_{0S}}{\rho_{01} U^2} = \eta_S \frac{c_p \Delta T_{0S}}{U^2}$$

The term $\Delta p_{0S}/\rho_{01} U^2$ is referred to as the *pressure coefficient*, which can be seen to be the product of the stage efficiency and temperature coefficient. Thus the overall performance of a stage can be expressed in terms of flow coefficient, temperature coefficient and either a pressure coefficient or the stage efficiency. The resulting form of the stage characteristic is shown in Fig. 5.39; in practice, α_1 and β_2 will not remain constant due to increased deviation as conditions change from the design point. In regions of blade stalling, both at positive and negative incidence, there will be a considerable departure from the linear relationship giving the shape shown. Choking will occur at a high value of flow coefficient, leading to a very rapid drop in efficiency and placing an upper limit on the flow which can be passed at a given blade speed. Stage characteristics may be obtained from single-stage tests, by analysis of interstage data on a complete compressor or by prediction using cascade data. In practice, not all the constant speed lines would collapse into a single curve as shown, but for the following brief discussion a single line characteristic will be assumed.

The ψ–ϕ curve in Fig. 5.39 is drawn for the case where the efficiency is a maximum at the design flow coefficient, ϕ_d. Moving away from ϕ_d results in a change in incidence and increased losses. Reducing ϕ results in increased positive incidence and stall at ϕ_s; increasing ϕ eventually results in choking of the stage and a severe drop in efficiency. It is essential that all individual stages of a compressor operate in the region of high efficiency without

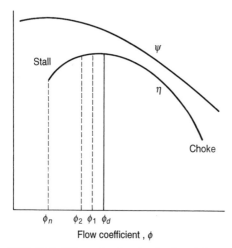

FIG. 5.40 Effect of reduction in ϕ at inlet

encountering either stall or choke at normal operating conditions; at conditions far removed from design it may not be possible to achieve this without remedial action involving changes in compressor geometry. The difficulties involved in achieving correct matching of the stages can be understood by considering the operation of several identical stages in series. This procedure is known as 'stage stacking' and is an invaluable tool to the aerodynamicist who is concerned with the overall performance of a compressor and determining the reasons for sub-standard performance; stage stacking can be used to produce overall compressor characteristics, showing the regions in which individual stages stall. In this elementary treatment we will only touch on the use of this procedure. An idealized stage characteristic is shown in Fig. 5.40 and we will consider that at the design point all stages operate at ϕ_d. If the compressor should be subjected to a decrease in mass flow, this will result in a reduction in the flow coefficient entering the first stage to ϕ_1, which will result in an increase in pressure ratio causing the density at entry to the second stage to be increased. The axial velocity at entry to the second stage is determined by the continuity equation and both effects combine to give a further decrease in flow coefficient for that stage to ϕ_2. This effect is propagated through the compressor and eventually some stage will stall at ϕ_n. Increasing the flow coefficient has the opposite effect and will drive some stage into choke.

5.12 Axial compressor characteristics

The characteristic curves of an axial compressor take a form similar to those of the centrifugal type, being plotted on the same non-dimensional basis, i.e. pressure ratio p_{02}/p_{01} and isentropic efficiency η_c against the non-dimensional mass flow $m\sqrt{T_{01}}/p_{01}$ for fixed values of the non-dimensional speed $N/\sqrt{T_{01}}$.

A typical set of such curves is shown in Fig. 5.41 and in comparison with Fig. 4.10 it will be observed that the characteristics for fixed values of $N/\sqrt{T_{01}}$ cover a much narrower range of mass flow than in the case of the centrifugal compressor. At high rotational speeds the constant speed lines

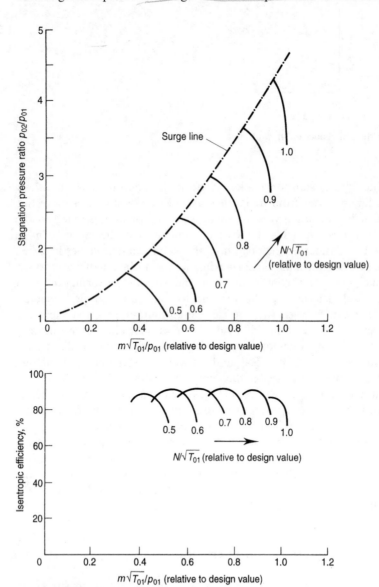

FIG. 5.41 **Axial compressor characteristics**

become very steep and ultimately may be vertical. The same limitations occur at either end of the $N/\sqrt{T_{01}}$ lines due to surging and choking. The surge points, however, are normally reached before the curves reach a maximum value and, because of this, the design operating point, which is always near the peak of the characteristic, is also very near the surge line. Consequently the range of stable operation of axial compressors is narrow and gas turbine plant incorporating this type of compressor calls for great care in matching the individual components if instability is to be avoided at operating conditions removed from the design point. Chapter 8 deals with this aspect in detail. The mechanism of surging in axial compressors is complex and is still not yet fully understood. It is often difficult to distinguish between surging and stalling and one phenomenon may easily lead to the other. The phenomenon of rotating stall referred to in section 4.6 may also be encountered: it can lead to loss of performance and severe blade vibration without actually causing surge. A comprehensive review of compressor stall phenomena is given by Greitzer in Ref. (13).

The complete compressor characteristic can be obtained only if the compressor is driven by an external power source; it will be shown in Chapter 8 that the running range of the compressor is severely restricted when it is combined with a combustion chamber and turbine. The driving unit must be capable of variable speed operation, with the speed continuously variable and closely controlled. Compressor test rigs in the past have been driven by electric motors, steam turbines or gas turbines. One of the major problems of testing compressors is the very large power requirements as shown in the following examples. The compressors of the Olympus 593 in Concorde absorb about 75 MW at take-off, with approximately 25 MW going to the LP compressor and 50 MW to the HP compressor. A typical fan stage from a 350 kN turbofan will compress 1100 kg/s through a pressure ratio of 1·7 requiring about 60 MW. The compressor for a 250 MW industrial unit, similar to that in Example 2.3, would require approximately 300 MW. Two methods of reducing the total power requirement are to throttle the intake to the compressor or to test a scale model. Throttling the intake reduces the density and hence the mass flow, reducing the power input; a major problem that occurs, however, is the large drop in Reynolds number with reduced density. The Reynolds number may well be reduced by a factor of 4 or more, which has the effect of increasing the relative magnitude of the viscous losses, giving unrealistically low values of compressor efficiency. Siemens-Westinghouse used throttled intake testing for the development of their Advanced Turbine System compressor, with the power required reduced from about 300 MW to 25 MW. ABB used a 30 per cent scale model of the compressor of their 250 MW reheat gas turbine, which would result in a power requirement of 25–30 MW. Single-stage fans are more likely to be tested as scale models, especially for the evaluation of new aerodynamic design techniques; fans may also be heavily instrumented and tested on development engines. Another problem with compressor testing is that the operating environment of the compressor in the test rig is not the same as that encountered in the engine, due to effects such as change in the tip-clearances as a result of different casing temperatures and shaft axial movements due to thermal expansion. Some manufacturers prefer to carry out compressor tests on a gas generator rig, using the actual combustion system and turbine, but this requires the use of a variable exhaust nozzle to cover a reasonable range of the compressor characteristic; the compressor, however, is exposed to realistic engine operating conditions. It can be seen that compressor testing is both complex and expensive, but essential for the development of a high-performance engine. A typical compressor development programme for an industrial gas turbine is described in Ref. (15).

Some deductions from the compressor characteristics

The axial flow compressor consists of a series of stages, each of which has its own characteristic: stage characteristics are similar to the overall characteristic but have much lower pressure ratios. The mass flow through the compressor is limited by choking in the various stages and under some conditions this will occur in the early stages and under others in the rear stages.

We have noted that if an axial flow compressor is designed for a constant axial velocity through all stages, the annulus area must progressively decrease as the flow proceeds through the compressor, because of the increasing density. The annulus area required for each stage will be determined for the design condition and clearly at any other operating condition the fixed area will result in a variation of axial velocity through the compressor. When the compressor is run at a speed lower than design, the temperature rise and pressure ratio will be lower than the design value. The effect of the reduction in density will be to increase the axial velocity in the rear stages, where choking will eventually occur and limit the mass flow. Thus at low speeds the mass flow will be determined by choking of the rear stages, as indicated in Fig. 5.42. As the speed is increased, the density in the stages is increased to the design value and the rear stages of the compressor can pass all the flow provided by the early stages. Eventually, however, choking will occur at the inlet; the vertical constant speed line in Fig. 5.42 is due to choking at the inlet of the compressor.

When the compressor is operating at its design condition, all stages are operating at the correct value of C_a/U and hence at the correct incidence. If we consider moving from design point A to point B on the surge line at the design speed it can be seen that the density at the compressor exit will be increased due to the increase in delivery pressure, but the mass flow is slightly reduced. Both of these effects will reduce the axial velocity in the last stage and hence increase the incidence as shown by the velocity

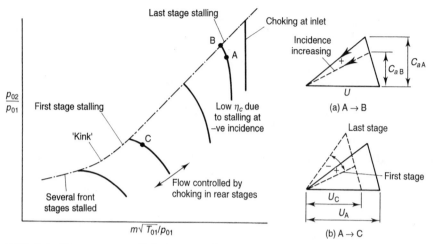

FIG. 5.42 Phenomena at off-design operation

triangle (a) in Fig. 5.42. A relatively small increase in incidence will cause the rotor blade to stall, and it is thought that surge at high speeds is caused by stalling of the last stage.

When speed is reduced from A to C, the mass flow generally falls off more rapidly than the speed, and the effect is to decrease the axial velocity at inlet and cause the incidence on the first-stage blade to increase as shown in Fig. 5.42(b). The axial velocity in the latter stages, however, is increased because of the lower pressure and density, so causing the incidence to decrease as shown. Thus at low speeds, surging is probably due to stalling of the first stage. It is possible for axial compressors to operate with several of the early stages stalled, and this is thought to account for the 'kink' in the surge line which is often encountered with high-performance compressors (see Fig. 5.42). A detailed discussion of the relation between stage stalling and the surge line using stage stacking techniques is given in Ref. (14).

At conditions far removed from surge the density will be much lower than required and the resulting high axial velocities will give a large decrease in incidence, which will eventually result in stalling at negative incidences. The efficiency will be very low in these regions.

The twin-spool compressor

It will be evident that as the design pressure ratio is increased, the difference in density between design and off-design conditions will be increased, and the probability of blades stalling due to incorrect axial velocities will be much higher. The effects of increased axial velocity towards the rear of the compressor can be alleviated by means of *blow-off*, where air is discharged from the compressor at some intermediate stage through a valve to reduce the mass flow through the later stages. Blow-off is wasteful, but is sometimes necessary to prevent the engine running line intersecting the surge line; this will be discussed in Chapter 8. A more satisfactory solution is to use a twin-spool compressor.

We have seen that reduction of compressor speed from the design value will cause an increase of incidence in the first stage and a decrease of incidence in the last stage; clearly the effect will increase with pressure ratio. The incidence could be maintained at the design value by increasing the speed of the last stage and decreasing the speed of the first stage, as indicated in Fig. 5.43. These conflicting requirements can be met by splitting the compressor into two (or more) sections, each being driven by a separate turbine, as shown in Fig. 1.7 for example. In the common twin-spool configuration the LP compressor is driven by the LP turbine and the HP compressor by the HP turbine. The speeds of the two spools are mechanically independent but a strong aerodynamic coupling exists which has the desired effect on the relative speeds when the gas turbine is operating off the design point. This will be discussed in Chapter 9.

The variable geometry compressor

An alternative approach to satisfying the off-design performance of high pressure ratio compressors is to use several rows of *variable stators* at the front of the compressor, permitting pressure ratios of over 16:1 to be achieved in a single spool. If the stators are rotated away from the axial direction, increasing α_1 in Fig. 5.4, the effect is to decrease the axial velocity and mass flow for a given rotational speed. This delays stalling of the first few stages and choking in the last stages, at low rotational speeds. The earlier discussion on stage characteristics showed that ideally

$$\psi = 1 - k\phi$$

where $k = \tan\alpha_1 + \tan\beta_2$. Using variable stators, it is possible to increase α_1 with β_2 remaining constant. The effect is to decrease the temperature coefficient for a given flow coefficient; the pressure coefficient will also be reduced because of the reduction in temperature coefficient. The stage characteristic will be shifted to the left as shown in Fig. 5.44 because of the

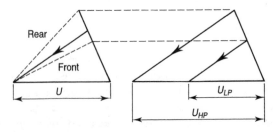

FIG. 5.43

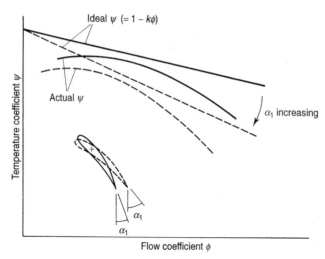

FIG. 5.44 Effect of variable stators

reduction in mass flow. The stage stacking techniques mentioned in section 5.11 can be used to predict the effects of incorporating variable geometry on the overall compressor characteristics. One of the major benefits is an improvement in the surge margin at low engine speeds, which is particularly important during starting and operation at idle conditions: these topics will be discussed further in Chapter 8.

The compressor of the Ruston Tornado, described in Ref. (15), uses variable IGVs followed by four rows of variable stators to achieve a pressure ratio of 12:1. The design speed of the compressor is 11 085 rev/min and at speeds below 10 000 rev/min the stators start to close progressively until they are in the fully closed position for all speeds below 8000 rev/min: the IGVs are rotated through 35° and the stators through 32°, 25°, 25° and 10° from the first to the fourth stage. While the effect of variable stators can be predicted with reasonable accuracy, it is necessary to optimize the settings of each row by rig-testing the compressor.

The compressor shown in Fig. 5.1 is from the GE LM 2500 aero-derivative gas turbine, based on the HP compressor of the TF-39 military turbofan. The pressure ratio of 16 was achieved in 16 stages, using variable IGVs and six stages of variable stators; the mass flow was about 70 kg/s, resulting in a power output of 23 MW. In the late 1990s GE introduced an uprated version, the LM 2500+, with a rating of 29 MW and a thermal efficiency of around 38 per cent. This was achieved by adding a stage in front of the existing first stage, supercharging the existing compressor; the process is referred to as *zero staging*. The mass flow was increased to 85 kg/s and the pressure ratio to 23. The design and development of the LM 2500+ compressor is described in Ref. (16). The zero stage is quoted as having a pressure ratio of 1·438, the tip relative Mach number is 1·19, and the compressor has a polytropic efficiency of 91 per cent. Preliminary mechanical design studies showed that blade-fixing stresses in a conventional design with individual blades would require a hub–tip ratio of 0.45; by using an integrally bladed disc (known as a 'blisk' in GE terminology) the hub–tip ratio could be reduced to 0.368, permitting a slightly lower axial velocity and reduced tip Mach number. The blisk design results in wider chord blades with increased Foreign Object Damage (FOD) tolerance.

5.13   Closure

It must be emphasized that the foregoing treatment of axial compressor theory and design has been kept at an elementary level to serve as an introduction to what has now become an extremely complex field of study. Since the early pioneering work of NGTE and NACA in establishing rational design techniques based on empirical data, much effort has gone into the development of sophisticated methods making use of digital computers. While digital computers have been used in compressor development for many years, the enormous improvements in computing speed and storage capacity in recent years have had a major impact on turbomachinery design; this effort is truly international in scope. Two of these methods are referred to as 'streamline curvature' and 'matrix through flow' techniques. These attempt to determine the flow pattern in the so-called 'meridional plane' which is the plane containing the axial and radial directions. This is in contrast to earlier methods which were essentially concerned with flow patterns in the peripheral–axial planes at various positions along the blade span. For a summary of these techniques the reader is referred to Ref. (17) which compares them from the user's point of view. A third approach involving advanced numerical analysis is the so-called 'time marching' method, in which an initial estimate of the flow pattern is made and unsteady aerodynamic theory is used to predict succeeding solutions until equilibrium is finally achieved, Ref. (18). Reviews of computational methods and their application are given in Refs (19) and (20). Modern methods of numerical analysis have resulted in developments such as 'controlled diffusion' and 'end bend' blading; the first of these makes use of supercritical airfoil sections developed from high subsonic wing sections and the second alters the shape of the blades at both inner and outer annulus walls to allow for the decreased axial velocity in the boundary layer. Computational methods will play an ever-increasing role in compressor design in the future, but it is unlikely that empiricism or experimental tests will ever be completely displaced.

Finally, it should be recognized that the very limited number of references supplied is intended to be central to the foundation of compressor design. The literature is expanding very rapidly and those actively involved in compressor design have to follow this closely, making it all too easy to forget the source of this technology. The books by Horlock [Ref. (5)], updated in 1982, and Gostelow [Ref. (8)] both contain extensive bibliographies. Dunham [Ref. (21)] has recently reviewed the pioneering work of Howell at NGTE and highlights the fact that in the days before large-scale computers, designers were forced to analyse a great deal of experimental results and interpret the underlying physics. Modern CFD techniques now permit in-depth predictions of flows that could not even have been thought of at the start of the gas turbine era.

NOMENCLATURE

For velocity triangle notation (U, C, V, α, β) see Fig. 5.4.
For cascade notation (α', δ, ε, ζ, θ, i, s, c) see Fig. 5.23.

C_D	overall drag coefficient
C_{DA}	annulus drag coefficient
C_{Dp}	profile drag coefficient
C_{DS}	secondary loss coefficient
C_L	lift coefficient
D	diffusion factor
h	blade height, specific enthalpy
n	number of blades
N	number of stages
r	radius
R	r/r_m as well as pressure ratio and gas constant
w	stagnation pressure loss
h/c	aspect ratio
s/c	pitch/chord ratio
λ	work-done factor
Λ	degree of reaction ($\Delta T_{\text{rotor}}/\Delta T_{\text{stage}}$)
ϕ	flow coefficient (C_a/U)
ψ	temperature coefficient ($c_p \Delta T_{0S}/U^2$)

Suffixes

a, w	axial, whirl, component
b	blade row
m	mean, vector mean
s	stage
r	root radius, radial component
t	tip radius

6

Combustion systems

The design of a gas turbine combustion system is a complex process involving fluid dynamics, combustion and mechanical design. For many years the combustion system was much less amenable to theoretical treatment than other components of the gas turbine, and any development programme required a considerable amount of trial and error. With the very high cycle temperatures of modern gas turbines, mechanical design remains difficult and a mechanical development programme is inevitable. The rapidly increasing use of Computational Fluid Dynamics (CFD) in recent years has had a major impact on the design process, greatly increasing the understanding of the complex flow and so reducing the amount of trial and error required. CFD methods are beyond the scope of this introductory text, but their importance should be recognized.

The main purpose of this chapter is to show how the problem of design is basically one of reaching the best compromise between a number of conflicting requirements, which will vary widely with different applications. Aircraft and ground-based gas turbine combustion systems differ in some respects and are similar in others. The most common fuels for gas turbines are liquid petroleum distillates and natural gas, and attention will be focused on combustion systems suitable for these fuels. In the mid 1990s there was considerable interest in the development of systems to burn gas produced from coal. Gasification requires large amounts of steam so that it is a long-term option for modifying combined cycle plant currently burning natural gas.

In the early days of gas turbine design, major goals included high combustion efficiency and the reduction of visible smoke, both of which were largely solved by the early 1970s. A much more demanding problem has been the reduction of oxides of nitrogen, and on-going research programmes are essential to meet the ever more stringent pollution limits while maintaining existing levels of reliability and keeping costs affordable. The chapter will close with a brief treatment of methods currently under development for low-emission combustion systems.

6.1   Operational requirements

Chapter 2 discussed the thermodynamic design of gas turbines, emphasizing the importance of high cycle temperature and high component efficiencies. For combustion systems the latter implies the need for high combustion efficiency and low pressure loss, typical values assumed for cycle calculations being 99 per cent and 2–8 per cent of the compressor delivery pressure. Although the effect of these losses on cycle efficiency and specific output is not so pronounced as that of inefficiencies in the turbomachinery, the combustor is a critical component because it must operate reliably at extreme temperatures, provide a suitable temperature distribution at entry to the turbine (to be discussed in section 7.3), and create the minimum amount of pollutants over a long operating life.

Aircraft (and ships) must carry the fuel required for their missions, and this has resulted in the universal use of liquid fuels. Proposals have been made for hydrogen-powered aircraft, and a few experimental flights have been made using hydrogen-fuelled engines, but it is very unlikely that hydrogen will be widely used. Aircraft gas turbines face the problems associated with operating over a wide range of inlet pressure and temperature within the flight envelope of Mach number and altitude. A typical subsonic airliner will operate at a cruise altitude of 11 000 m, where the ambient pressure and temperature for ISA conditions are 0·2270 bar and 216·8 K, compared with the sea-level values of 1·013 bar and 288·15 K. Thus the combustor has to operate with a greatly reduced air density and mass flow at altitude, while using approximately the same fuel/air ratio as at sea level to maintain an appropriate value of turbine inlet temperature. Atmospheric conditions will change quite rapidly during climb and descent, and the combustor must deal with a continuously varying fuel flow without allowing the engine to flame-out or exceed temperature limits; high-performance fighters, for example, may climb from sea level to 11 000 m in less than 2 minutes. In the event of a flame-out in flight, the combustor must be capable of relighting over a wide range of flight conditions. Supersonic aircraft operate under very different conditions at high Mach number and altitude; because of the large ram pressure rise the compressor inlet pressure may be almost equivalent to the Sea Level Static value and the inlet temperature is significantly higher. It is clear that a supersonic aircraft will operate over a much wider range of conditions than its subsonic counterpart, and emissions of oxides of nitrogen at high altitudes proposed for future supersonic transports could be a serious enough problem to terminate proposed ventures.

Stationary land-based gas turbines have a wider choice of fuel. It should be recognized, however, that altitude effects can still be significant: some engines are operated at altitudes of nearly 4000 m in regions of South America, and in Western Canada many engines operate at altitudes of 1000 m. Natural gas is the preferred fuel for applications such as pipeline compression, utility power and cogeneration. If natural gas is not available, the most widely used fuel is a liquid distillate. A relatively small number of gas turbines burn residual fuels, but such fuels generally require pre-treatment which is costly. Units for continuous operation would normally use natural gas, but peak-load applications may use liquid fuels requiring the storage of substantial quantities. Gas turbines may be designed for *dual-fuel* operation, with normal operation on natural gas and an option to switch to liquid fuels for short periods. The combustion system may have to be designed to burn the two fuels, sometimes simultaneously, and also to accommodate water or steam injection for emission control.

There are few commercial applications of gas turbines in ships, and nearly all marine gas turbines are aircraft derivatives operating in warships. They universally use marine diesel as fuel. Propulsion plants for warships are also becoming liable for emissions produced in harbour, and are subject to similar requirements as land-based units.

6.2   Types of combustion system

Combustion in the normal, open-cycle, gas turbine is a continuous process in which fuel is burned in the air supplied by the compressor; an electric spark is required only for initiating the combustion process, and thereafter the flame must be self-sustaining. The designer has considerable latitude in choosing a combustor configuration and the different requirements of aircraft and ground-based units with respect to weight, volume and frontal area can result in widely different solutions. In recent years the effect of stringent restrictions on emissions of oxides of nitrogen (NO_x) has had a major impact on combustion design, for both industrial and aircraft applications.

The earliest aircraft engines made use of *can* (or *tubular*) combustors, as shown in Fig. 6.1, in which the air leaving the compressor is split into a number of separate streams, each supplying a separate chamber. These chambers are spaced around the shaft connecting the compressor and turbine, each chamber having its own fuel jet fed from a common supply line. This arrangement was well suited to engines with centrifugal compressors, where the flow was divided into separate streams in the diffuser. The Rolls-Royce Dart, shown in Fig. 1.10, is an example. A major advantage of can-type combustors was that development could be carried out on a single can using only a fraction of the overall airflow and fuel flow. In the aircraft application, however, the can type of combustor is undesirable in terms of weight, volume and frontal area and is no longer used in current designs. Small gas turbines, such as auxiliary power units (APUs) and those proposed for vehicles, have often been designed with a single combustion can.

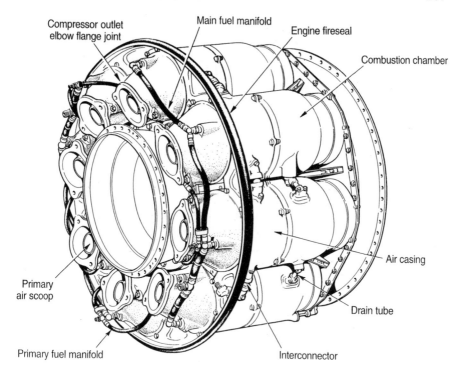

Compressor outlet
elbow flange joint

Main fuel manifold

Engine fireseal

Combustion chamber

Primary
air scoop

Air casing

Drain tube

Primary fuel manifold

Interconnector

FIG. 6.1 Can-type combustor [courtesy Rolls-Royce plc]

Separate combustion cans are still widely used in industrial engines, but recent designs make use of a *cannular* (or *tubo-annular*) system, where individual flame tubes are uniformly spaced around an annular casing. The Alstom Typhoon, shown in Fig. 1.13, uses this system; the General Electric and Westinghouse families of industrial gas turbines also use this arrangement. Figure 1.13 shows the reverse flow nature of the airflow after leaving the diffuser downstream of the axial compressor; the use of a reverse flow arrangement allows a significant reduction in the overall length of the compressor–turbine shaft and also permits easy access to the fuel nozzles and combustion cans for maintenance.

The ideal configuration in terms of compact dimensions is the *annular* combustor, in which maximum use is made of the space available within a specified diameter; this should reduce the pressure loss and result in an engine of minimum diameter. Annular combustors presented some disadvantages, which led to the development of cannular combustors initially. Firstly, although a large number of fuel jets can be employed, it is more difficult to obtain an even fuel/air distribution and an even outlet temperature distribution. Secondly, the annular chamber is inevitably weaker structurally and it is difficult to avoid buckling of the hot flame tube walls. Thirdly, most of the development work must be carried out on the complete chamber, requiring a test facility capable of supplying the full engine air mass flow, compared with the testing of a single can in the multi-chamber layout. These problems were vigorously attacked and annular combustors are universally used in modern aircraft engines. The Olympus 593 (Fig. 1.9), PT-6 (Fig. 1.11), PW 530 (Fig. 1.12(a)) and V 2500 (Fig. 1.12(b)) all use annular combustors. The most recent designs by ABB and Siemens have introduced annular combustors in units of over 150 MW.

Large industrial gas turbines, where the space required by the combustion system is less critical, have used one or two large cylindrical combustion chambers; these were mounted vertically and were often referred to as *silo-type* combustors because of their size and physical resemblance to silos. ABB designs used a single combustor, while Siemens used two; Fig. 1.14 shows a typical Siemens arrangement. These large combustors allowed lower fluid velocities, and hence pressure losses, and were capable of burning lower quality fuels. Some later Siemens engines use two large combustion chambers arranged horizontally rather than vertically, and they no longer resemble silos.

Figure 1.16 shows the configurations of both the aero and industrial versions of the Rolls-Royce Trent, and the major differences in the design of the combustion system are clearly shown. The industrial engine uses separate combustion cans arranged radially, this arrangement being used to provide *dry* low emissions (DLE), i.e. without the added complexity of steam or water injection. The aircraft engine uses a conventional annular combustor.

For the remainder of this chapter we shall concentrate mainly on the way in which combustion is arranged to take place inside a flame tube, and need not be concerned with the overall configuration.

6.3   Some important factors affecting combustor design

Over a period of five decades, the basic factors influencing the design of combustion systems for gas turbines have not changed, although recently some new requirements have evolved. The key issues may be summarized as follows.

(*a*)   The temperature of the gases after combustion must be comparatively low to suit the highly stressed turbine materials. Development of improved materials and methods of blade cooling, however, has enabled permissible combustor outlet temperatures to rise from about 1100 K to as much as 1850 K for aircraft applications.

(*b*)   At the end of the combustion space the temperature distribution must be of known form if the turbine blades are not to suffer from local overheating. In practice, the temperature can increase with radius over the turbine annulus, because of the strong influence of temperature on allowable stress and the decrease of blade centrifugal stress from root to tip.

(c) Combustion must be maintained in a stream of air moving with a high velocity in the region of 30–60 m/s, and stable operation is required over a wide range of air/fuel ratio from full load to idling conditions. The air/fuel ratio might vary from about 60:1 to 120:1 for simple cycle gas turbines and from 100:1 to 200:1 if a heat-exchanger is used. Considering that the stoichiometric ratio is approximately 15:1 it is clear that a high dilution is required to maintain the temperature level dictated by turbine stresses.

(d) The formation of carbon deposits ('coking') must be avoided, particularly the hard brittle variety. Small particles carried into the turbine in the high-velocity gas stream can erode the blades and block cooling air passages; furthermore, aerodynamically excited vibration in the combustion chamber might cause sizeable pieces of carbon to break free, resulting in even worse damage to the turbine.

(e) In aircraft gas turbines, combustion must also be stable over a wide range of chamber pressure because of the substantial change in this parameter with altitude and forward speed. Another important requirement is the capability of relighting at high altitude in the event of an engine flame-out.

(f) Avoidance of smoke in the exhaust is of major importance for all types of gas turbine; early jet engines had very smoky exhausts, and this became a serious problem around airports when jet transport aircraft started to operate in large numbers. Smoke trails in flight were a problem for military aircraft, permitting them to be seen from a great distance. Stationary gas turbines are now found in urban locations, sometimes close to residential areas.

(g) Although gas turbine combustion systems operate at extremely high efficiencies, they produce pollutants such as oxides of nitrogen (NO_x), carbon monoxide (CO) and unburned hydrocarbons (UHC) and these must be controlled to very low levels. Over the years, the performance of the gas turbine has been improved mainly by increasing the compressor pressure ratio and turbine inlet temperature (TIT). Unfortunately this results in increased production of NO_x. Ever more stringent emissions legislation has led to significant changes in combustor design to cope with the problem.

Probably the only feature of the gas turbine that eases the combustion designer's problem is the peculiar interdependence of compressor delivery air density and mass flow which leads to the velocity of the air at entry to the combustion system being reasonably constant over the operating range.

For aircraft applications there are the additional limitations of small space and low weight, which are, however, slightly offset by somewhat shorter endurance requirements. Aircraft engine combustion chambers are normally constructed of light-gauge, heat-resisting alloy sheet (approx. 0·8 mm thick), but are only expected to have a life of some 10 000 hours. Combustion chambers for industrial gas turbine plant may be constructed on much sturdier lines but, on the other hand, a life of about 100 000 hours is required. Refractory linings are sometimes used in heavy chambers, although the remarks made in (d) about the effects of hard carbon deposits breaking free apply with even greater force to refractory material.

We have seen that the gas turbine cycle is very sensitive to component inefficiencies, and it is important that the aforementioned requirements should be met without sacrificing combustion efficiency. That is, it is essential that over most of the operating range all the fuel injected should be completely burnt and the full calorific value realized. Any pressure drop between inlet and outlet of the combustor leads to both an increase in *SFC* and a reduction in specific power output, so it is essential to keep the pressure loss to a minimum. It will be appreciated from the following discussion that the smaller the space available for combustion, and hence the shorter the time available for the necessary chemical reactions, the more difficult it is to meet all the requirements and still obtain a high combustion efficiency with low pressure loss. Clearly in this respect designers of combustion systems for industrial gas turbines have an easier task than their counterparts in the aircraft field.

6.4   The combustion process

Combustion of a liquid fuel involves the mixing of a fine spray of droplets with air, vaporization of the droplets, the breaking down of heavy hydrocarbons into lighter fractions, the intimate mixing of molecules of these hydrocarbons with oxygen molecules, and finally the chemical reactions themselves. A high temperature, such as is provided by the combustion of an approximately stoichiometric mixture, is necessary if all these processes are to occur sufficiently rapidly for combustion in a moving air stream to be completed in a small space. Combustion of a gaseous fuel involves fewer processes, but much of what follows is still applicable.

Since the overall air/fuel ratio is in the region of 100:1, while the stoichiometric ratio is approximately 15:1, the first essential feature is that the air should be introduced in stages. Three such stages can be distinguished. About 15–20 per cent of the air is introduced around the jet of fuel in the *primary zone* to provide the necessary high temperature for rapid combustion. Some 30 per cent of the total air is then introduced through holes in the flame-tube in the *secondary zone* to complete the combustion. For high combustion efficiency, this air must be injected carefully at the right points in the process, to avoid chilling the flame locally and drastically reducing the reaction rate in that neighbourhood. Finally, in the *tertiary* or *dilution zone* the remaining air is mixed with the products of combustion to cool them down to the temperature required at inlet to the turbine. Sufficient turbulence must be promoted so that the hot and cold streams are thoroughly mixed to give the desired outlet temperature distribution, with no hot streaks which would damage the turbine blades.

The zonal method of introducing the air cannot by itself give a self-piloting flame in an air stream which is moving an order of magnitude faster than the flame speed in a burning mixture. The second essential feature is therefore a recirculating flow pattern which directs some of the burning mixture in the primary zone back on to the incoming fuel and air. One way of achieving this is shown in Fig. 6.2, which is typical of British practice. The fuel is injected in the same direction as the air stream, and the primary air is introduced through twisted radial vanes, known as *swirl vanes*, so that the resulting vortex motion will induce a region of low pressure along the axis of the chamber. This vortex motion is sometimes enhanced by injecting the secondary air through short tangential chutes in the flame-tube, instead of through plain holes as in the figure. The net result is that the burning gases tend to flow towards the region of low pressure, and some portion of them is swept round towards the jet of fuel as indicated by the arrows.

Many other solutions to the problem of obtaining a stable flame are possible. One American practice is to dispense with the swirl vanes and achieve the recirculation by a careful positioning of holes in the flame-tube downstream of a hemispherical baffle as shown in Fig. 6.3(a). Figure 6.3(b) shows a possible solution using upstream injection which gives good mixing of the fuel and primary air. It is difficult to avoid overheating the fuel injector, however, and upstream injection is employed more for afterburners (or 'reheat') in the jet-pipe of aircraft engines than in main combustion systems. Afterburners operate only for short periods of thrust-

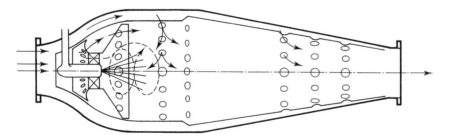

FIG. 6.2 **Combustion chamber with swirl vanes**

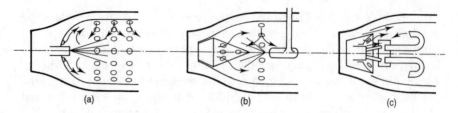

FIG. 6.3 Methods of flame stabilization

boosting. Finally, Fig. 6.3(c) illustrates a vaporizer system wherein the fuel is injected at low pressure into walking-stick shaped tubes placed in the primary zone. A rich mixture of fuel vapour and air issues from the vaporizer tubes in the upstream direction to mix with the remaining primary air passing through holes in a baffle around the fuel supply pipes. The fuel system is much simpler, and the difficulty of arranging for an adequate distribution of fine droplets over the whole operating range of fuel flow is overcome (see 'Fuel injection' in section 6.6). The problem in this case is to avoid local 'cracking' of the fuel in the vaporizer tubes with the formation of deposits of low thermal conductivity leading to overheating and burn-out. Vaporizer schemes are particularly well suited for annular combustors where it is inherently more difficult to obtain a satisfactory fuel–air distribution with sprays of droplets from high-pressure injectors, and they have been used in several successful aircraft engines. The original walking-stick shaped tubes have been replaced in modern engines by more compact and mechanically rugged T-shape vaporizers as shown in Fig. 6.4. Sotheran [Ref. (1)] describes the history of vaporizer development at Rolls-Royce.

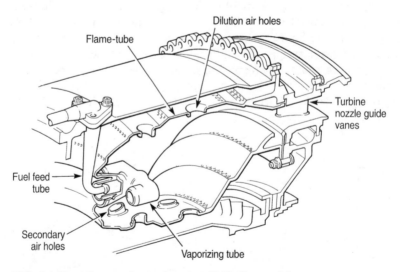

FIG. 6.4 Vaporizer combustor [courtesy Rolls-Royce plc]

Having described the way in which the combustion process is accomplished, it is now possible to see how incomplete combustion and pressure losses arise. When not due simply to poor fuel injector design leading to fuel droplets being carried along the flame-tube wall, incomplete combustion may be caused by local chilling of the flame at points of secondary air entry. This can easily reduce the reaction rate to the point where some of the products into which the fuel has decomposed are left in their partially burnt state, and the temperature at the downstream end of the chamber is normally below that at which the burning of these products can be expected to take place. Since the lighter hydrocarbons into which the fuel has decomposed have a higher ignition temperature than the original fuel, it is clearly difficult to prevent some chilling from taking place, particularly if space is limited and the secondary air cannot be introduced gradually enough. If devices are used to increase large-scale turbulence and so distribute the secondary air more uniformly throughout the burning gases, the combustion efficiency will be improved but at the expense of increased pressure loss. A satisfactory compromise must somehow be reached.

Combustion chamber pressure loss is due to two distinct causes: (i) skin friction and turbulence and (ii) the rise in temperature due to combustion. The stagnation pressure drop associated with the latter, often called the *fundamental loss*, arises because an increase in temperature implies a decrease in density and consequently an increase in velocity and momentum of the stream. A pressure force $(\Delta p \times A)$ must be present to impart the increase in momentum. One of the standard idealized cases considered in gas dynamics is that of a heated gas stream flowing without friction in a duct of constant cross-sectional area. The stagnation pressure drop in this situation, for any given temperature rise, can be predicted with the aid of the Rayleigh-line functions (see Appendix A.4). When the velocity is low and the fluid flow can be treated as incompressible (in the sense that although ρ is a function of T it is independent of p), a simple equation for the pressure drop can be found as follows.

The momentum equation for one-dimensional frictionless flow in a duct of constant cross-sectional area A is

$$A(p_2 - p_1) + m(C_2 - C_1) = 0$$

For incompressible flow the stagnation pressure p_0 is simply $(p + \rho C^2/2)$, and

$$p_{02} - p_{01} = (p_2 - p_1) + \tfrac{1}{2}(\rho_2 C_2^2 - \rho_1 C_1^2)$$

Combining these equations, remembering that $m = \rho_1 A C_1 = \rho_2 A C_2$,

$$p_{02} - p_{01} = -(\rho_2 C_2^2 - \rho_1 C_1^2) + \tfrac{1}{2}(\rho_2 C_2^2 - \rho_1 C_1^2)$$
$$= -\tfrac{1}{2}(\rho_2 C_2^2 - \rho_1 C_1^2)$$

The stagnation pressure loss as a fraction of the inlet dynamic head then becomes

$$\frac{p_{01} - p_{02}}{\rho_1 C_1^2/2} = \left(\frac{\rho_2 C_2^2}{\rho_1 C_1^2} - 1\right) = \left(\frac{\rho_1}{\rho_2} - 1\right)$$

Finally, since $\rho \propto 1/T$ for incompressible flow,

$$\frac{p_{01} - p_{02}}{\rho_1 C_1^2/2} = \left(\frac{T_2}{T_1} - 1\right)$$

This will be seen from Appendix A.4 to be the same as the compressible flow value of $(p_{01} - p_{02})/(p_{01} - p_1)$ in the limiting case of zero inlet Mach number. At this condition $T_2/T_1 = T_{02}/T_{01}$.

Although the assumptions of incompressible flow and constant cross-sectional area are not quite true for a combustion chamber, the result is sufficiently accurate to provide us with the order of magnitude of the fundamental loss. Thus, since the outlet/inlet temperature ratio is in the region of 2–3, it is clear that the fundamental loss is only about 1–2 inlet dynamic heads. The pressure loss due to friction is found to be very much higher—of the order of 20 inlet dynamic heads. When measured by pitot traverses at inlet and outlet with no combustion taking place, it is known as the *cold loss*. That the friction loss is so high is due to the need for large-scale turbulence. Turbulence of this kind is created by the devices used to stabilize the flame, e.g. the swirl vanes in Fig. 6.2. In addition, there is the turbulence induced by the jets of secondary and dilution air. The need for good mixing of the secondary air with the burning gases to avoid chilling has been emphasized. Similarly, good mixing of the dilution air to avoid hot streaks in the turbine is essential. In general, the more effective the mixing the higher the pressure loss. Here again a compromise must be reached: this time between uniformity of outlet temperature distribution and low pressure loss.

Usually it is found that adequate mixing is obtained merely by injecting air through circular or elongated holes in the flame-tube. Sufficient penetration of the cold air jets into the hot stream is achieved as a result of the cold air having the greater density. The pressure loss produced by such a mixing process is associated with the change in momentum of the streams before and after mixing. In aircraft gas turbines the duct between combustion chamber outlet and turbine inlet is very short, and the compromise reached between good temperature distribution and low pressure loss is normally such that the temperature non-uniformity is up to ± 10 per cent of the mean value. The length of duct is often greater in an industrial gas turbine and the temperature distribution at the turbine inlet may be more uniform, although at the expense of increased pressure drop due to skin friction in the ducting. The paper by Lefebvre and Norster in Ref. (2) outlines a method of proportioning a tubular combustion chamber to give the most effective mixing for a given pressure loss. Making use of empirical data from mixing experiments, such as dilution hole discharge coefficients, the authors show how to estimate the optimum ratio of flame-tube to casing diameter, and the optimum pitch/diameter ratio and number of dilution holes.

6.5 Combustion chamber performance

The main factors of importance in assessing combustion chamber performance are (a) pressure loss, (b) combustion efficiency, (c) outlet temperature distribution, (d) stability limits and (e) combustion intensity. We need say no more of (c), but (a) and (b) require further comment, and (d) and (e) have not yet received attention.

Pressure loss

We have seen in section 6.4 that the overall stagnation pressure loss can be regarded as the sum of the fundamental loss (a small component which is a function of T_{02}/T_{01}) and the frictional loss. Our knowledge of friction in ordinary turbulent pipe flow at high Reynolds number would suggest that when the pressure loss is expressed non-dimensionally in terms of the dynamic head it will not vary much over the range of Reynolds number under which combustion systems operate. Experiments have shown, in fact, that the overall pressure loss can often be expressed adequately by an equation of the form

$$\text{pressure loss factor, } PLF = \frac{\Delta p_0}{m^2/2\rho_1 A_m^2} = K_1 + K_2 \left(\frac{T_{02}}{T_{01}} - 1 \right) \quad (6.1)$$

Note that rather than $\rho_1 C_1^2/2$, a conventional dynamic head is used based on a velocity calculated from the inlet density, air mass flow m, and maximum cross-sectional area A_m of the chamber. This velocity—sometimes known as the reference velocity—is more representative of conditions in the chamber, and the convention is useful when comparing results from chambers of different shape. Equation (6.1) is illustrated in Fig. 6.5. If K_1 and K_2 are determined from a combustion chamber on a test rig from a cold run and a hot run, then equation (6.1) enables the pressure loss to be estimated when the chamber is operating as part of a gas turbine over a wide range of conditions of mass flow, pressure ratio and fuel input.

To give an idea of relative orders of magnitude, typical values of PLF at design operating conditions for tubular, tubo-annular and annular combustion chambers are 35, 25 and 18 respectively. There are two points which must be remembered when considering pressure loss data. Firstly, the velocity of the air leaving the last stage of an axial compressor is quite

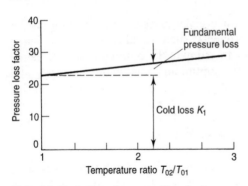

FIG. 6.5 Variation of pressure loss factor

high—say 150 m/s—and some form of diffusing section is introduced between the compressor and combustion chamber to reduce the velocity to about 60 m/s. It is a matter of convention, depending upon the layout of the gas turbine, as to how much of the stagnation pressure loss in this diffuser is included in the *PLF* of the combustion system. In other words, it depends on where the compressor is deemed to end and the combustion chamber begin.

Secondly, it should be appreciated from Chapters 2 and 3 that from the point of view of cycle performance calculations it is Δp_0 as a fraction of the compressor delivery pressure (p_{01} in the notation of this chapter) which is the useful parameter. This is related to the *PLF* as follows:

$$\frac{\Delta p_0}{p_{01}} = \frac{\Delta p_0}{m^2/2\rho_1 A_m^2} \times \frac{m^2/2\rho_1 A_m^2}{p_{01}} = PLF \times \frac{R}{2} \left(\frac{m\sqrt{T_{01}}}{A_m p_{01}} \right)^2 \qquad (6.2)$$

where the difference between ρ_1 and ρ_{01} has been ignored because the velocity is low. By combining equations (6.1) and (6.2) it can be seen that $\Delta p_0/p_{01}$ can be expressed as a function of non-dimensional mass flow at entry to the combustion chamber and combustion temperature ratio: such a relation is useful when predicting pressure losses at conditions other than design, as discussed in Chapter 8. Consider now the two extreme cases of tubular and annular designs. If the values of $\Delta p_0/p_{01}$ are to be similar, it follows from equation (6.2) and the values of *PLF* given above that the chamber cross-sectional area per unit mass flow (A_m/m) can be smaller for the annular design. For aircraft engines, where space and weight are vital, the value of A_m/m is normally chosen to yield a value of $\Delta p_0/p_{01}$ between 4 and 7 per cent. For industrial gas turbine chambers, A_m/m is usually such that $\Delta p_0/p_{01}$ is little more than 2 per cent.

Combustion efficiency

The efficiency of a combustion process may be found from a chemical analysis of the combustion products. Knowing the air/fuel ratio used and the proportion of incompletely burnt constituents, it is possible to calculate the ratio of the actual energy released to the theoretical quantity available. This approach via chemical analysis is not easy, because not only is it difficult to obtain truly representative samples from the high-velocity stream, but also, owing to the high air/fuel ratios employed in gas turbines, the unburnt constituents to be measured are a very small proportion of the whole sample. Ordinary gas analysis apparatus, such as the Orsat, is not adequate and much more elaborate techniques have had to be developed.

If an overall combustion efficiency is all that is required, however, and not an investigation of the state of the combustion process at different stages, it is easier to conduct development work on a test rig on the basis of the combustion efficiency which was defined in Chapter 2, namely

$$\eta_b = \frac{\text{theoretical } f \text{ for actual } \Delta T}{\text{actual } f \text{ for actual } \Delta T}$$

For this purpose, the only measurements required are those necessary for determining the fuel/air ratio and the mean stagnation temperatures at inlet and outlet of the chamber. The theoretical f can be obtained from curves such as those in Fig. 2.17.

It is worth describing how the mean stagnation temperature may be measured: there are two aspects, associated with the adjectives 'mean' and 'stagnation'. Firstly, it should be realized from the discussion in section 2.2 under heading 'Fuel/air ratio, combustion efficiency and cycle efficiency' that the expression for η_b arises from the energy equation which consists of such terms as mc_pT_0. Since in practice there is always a variation in velocity as well as temperature over the cross-section, it is necessary to determine not the ordinary arithmetic mean of a number of temperature readings, but what is known as the 'weighted mean'. If the cross-section is divided into a number of elemental areas $A_1, A_2, \ldots, A_i, \ldots, A_n$, at which the stagnation temperatures are $T_{01}, T_{02}, \ldots, T_{0i}, \ldots, T_{0n}$, and the mass flows are $m_1, m_2, \ldots, m_i, \ldots, m_n$, then the weighted mean temperature T_{0w} is defined by

$$T_{0w} = \frac{\sum m_i T_{0i}}{\sum m_i} = \frac{\sum m_i T_{0i}}{m}$$

where the summations are from 1 to n. We may assume that c_p is effectively constant over the cross-section. It follows that the product mc_pT_{0w} will be a true measure of the energy passing the section per unit time.

A simple expression for T_{0w} in terms of measured quantities can be derived as follows. The velocity at the centre of each elemental area may be found using a pitot-static tube. Denoting the dynamic head $\rho C^2/2$ by p_d, the mass flow for area A_i is then

$$m_i = \rho_i A_i (2p_{di}/\rho_i)^{\frac{1}{2}}$$

If the static pressure is constant over the cross-section, as it will be when there is simple axial flow with no swirl,

$$\rho_i \propto 1/T_i \quad \text{and} \quad m_i \propto A_i(p_{di}/T_i)^{\frac{1}{2}}$$

and thus

$$T_{0w} = \frac{\sum A_i T_{0i}(p_{di}/T_i)^{\frac{1}{2}}}{\sum A_i(p_{di}/T_i)^{\frac{1}{2}}}$$

Since T_i has only a second-order effect on T_{0w}, we can write $T_i = T_{0i}$. Furthermore, it is usual to divide the cross-section into equal areas. The expression then finally reduces to

$$T_{0w} = \frac{\sum (p_{di} T_{0i})^{\frac{1}{2}}}{\sum (p_{di}/T_{0i})^{\frac{1}{2}}}$$

Thus the weighted mean temperature may be determined directly from measurements of dynamic head and stagnation temperature at the centre of each elemental area. It remains to describe how stagnation temperatures may be measured.

In gas turbine work, temperatures are usually measured by thermocouples. The high accuracy of a pitot-static tube is well known, but considerable difficulty has been experienced in designing thermocouples to operate in a high-temperature fast-moving gas stream with a similar order of accuracy. Since the combustion efficiency rarely falls below 98 per cent over much of the operating range, accurate measurements are essential. Chromel–alumel thermocouples have been found to withstand the arduous requirements of combustion chamber testing satisfactorily, and give accurate results up to about 1300 K if special precautions are taken. It must be remembered that the temperature recorded is that of the hot junction of the thermocouple which for various reasons may not be at the temperature of the gas stream in which it is situated, particularly if the velocity of the stream is high.

If it be imagined that the thermocouple junction is moving with the stream of gas, then the temperature of the junction may differ from the static temperature of the gas by an amount depending upon the conduction of heat along the thermocouple wires, the convection between the junction and the gas stream, the radiation from the hot flame to the junction, and the radiation from the junction to the walls of the containing boundary if these are cooler than the junction. There is an additional possible error, because in practice the thermocouple is stationary, and the gas velocity will be reduced by friction in the boundary layer around the thermocouple junction. Kinetic energy is transformed into internal energy, some of which raises the temperature of the junction, while some is carried away by convection. In a high-speed gas stream it is obviously important to know how much of the velocity energy is being measured as temperature. The temperature corresponding to the velocity energy, i.e. the dynamic temperature, is about 40 K for a velocity of 300 m/s.

Since it is the stagnation temperature which is of interest, it is usual to place the thermocouple wires and junction in a metal tube in which the gas stream can be brought to rest adiabatically so that almost the whole of the dynamic temperature is measured, on the same principle as a pitot tube measuring stagnation pressure. Figure 6.6(a) shows one form of stagnation thermocouple which will measure about 98 per cent of the dynamic temperature as against the 60–70 per cent measured by a simple junction placed directly in the gas stream. A large hole facing upstream allows the gas to enter the tube, while a small hole, not more than 5 per cent of the area of the inlet orifice, provides sufficient ventilation without spoiling the pitot effect. This form of thermocouple is excellent for all work where radiation effects are small, such as the measurement of compressor delivery temperature. Where radiation effects are appreciable, as at the outlet of a combustion chamber, it is preferable to use a thermocouple of the type shown in Fig. 6.6(b). A radiation error of the order of 60 K in 1300 K is quite possible with a completely unshielded thermocouple. A short length of polished stainless steel sheet, twisted into a helix and placed in front of the junction, provides an effective radiation shield without impeding the flow of gas into the thermocouple tube. One or more concentric cylindrical shields may also be included. The bending of the wires so that about 2 or 3 centimetres run parallel with the direction of the stream, i.e. parallel with an isothermal, reduces the error due to conduction of heat away from the junction along the wires. This is in most cases quite a small error if the wires are of small diameter. If all these precautions are taken, stagnation temperature measurement up to 1300 K is possible to within ±5 K. While these few remarks do not by any means exhaust all the possibilities of thermocouple design, they at least indicate the extreme care necessary when choosing thermocouples for gas turbine temperature measurement.

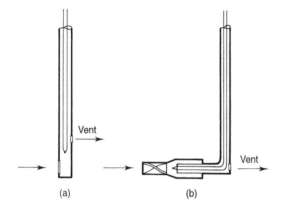

(a)   (b)

FIG. 6.6 Stagnation thermocouples

The temperature at outlet from the combustion chamber is not often measured on an engine, where there is the added problem that mechanical failure of the thermocouple support system could lead to major damage in the turbine. For this reason, on an engine it is normal to measure the temperature downstream of the turbine. For a simple jet engine the temperature at exit from the turbine is known as the *jet pipe temperature* (*JPT*) but the nomenclature *exhaust gas temperature* (*EGT*) is more commonly used for industrial gas turbines. When a twin-spool or free turbine configuration is used it is more common to measure the temperature downstream of the high-pressure turbine, this being referred to as the *inter-turbine temperature* (*ITT*). The combustion outlet temperature (i.e. turbine inlet temperature) can be calculated from the indirect measurements of *ITT* or *EGT* using the methods described in Chapter 8. Either *ITT* or *EGT* is used as a control system limit to prevent overheating of the turbine and these are the only temperature measurements that will normally be available to the user of gas turbines.

Stability limits

For any particular combustion chamber there is both a rich and a weak limit to the air/fuel ratio beyond which the flame is unstable. Usually the limit is taken as the air/fuel ratio at which the flame blows out, although instability often occurs before this limit is reached. Such instability takes the form of rough running, which not only indicates poor combustion, but sets up aerodynamic vibration which reduces the life of the chamber and causes blade vibration problems. The range of air/fuel ratio between the rich and weak limits is reduced with increase of air velocity, and if the air mass flow is increased beyond a certain value it is impossible to initiate combustion at all. A typical stability loop is shown in Fig. 6.7, where the limiting air/fuel ratio is plotted against air mass flow. If a combustion chamber is to be suitable, its operating range defined by the stability loop must obviously cover the required range of air/fuel ratio and mass flow of the gas turbine for which it is intended. Furthermore, allowance must be made for conditions which prevail when the engine is accelerated or decelerated. For example, on acceleration there will be a rapid increase in fuel flow as the 'throttle' is opened while the airflow will not reach its new equilibrium value until the engine has reached its new speed. Momentarily the combustion system will be operating with a very low air/fuel ratio. Most control systems have a built-in device which places an upper limit on the rate of change of fuel flow: not only to avoid blow-out, but also to avoid transient high temperatures in the turbine.

The stability loop is a function of the pressure in the chamber: a decrease in pressure reduces the rate at which the chemical reactions proceed, and consequently it narrows the stability limits. For aircraft engines it is important to check that the limits are sufficiently wide with a chamber

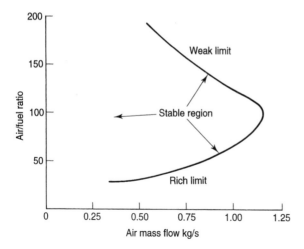

FIG. 6.7 Stability loop

pressure equal to the compressor delivery pressure which exists at the highest operating altitude. Engines of high pressure ratio present less of a problem to the combustion chamber designer than those of low pressure ratio. If the stability limits are too narrow, changes must be made to improve the recirculation pattern in the primary zone.

Combustion intensity

The size of combustion chamber is determined primarily by the rate of heat release required. The nominal heat release rate can be found from $mfQ_{net,p}$ where m is the air mass flow, f the fuel/air ratio and $Q_{net,p}$ the net calorific value of the fuel. Enough has been said for the reader to appreciate that the larger the volume which can be provided the easier it will be to achieve a low pressure drop, high efficiency, good outlet temperature distribution and satisfactory stability characteristics.

The design problem is also eased by an increase in the pressure and temperature of the air entering the chamber, for two reasons. Firstly, an increase will reduce the time necessary for the 'preparation' of the fuel and air mixture (evaporation of droplets, etc.), making more time available for the combustion process itself. Note that since the compressor delivery temperature is a function of the compressor delivery pressure, the pressure (usually expressed in atmospheres) is an adequate measure of both.

Secondly, we have already observed under the heading 'Stability limits' above, that the combustion chamber pressure is important because of its effect on the rate at which the chemical reactions proceed. An indication of the nature of this dependence can be obtained from chemical kinetics, i.e. kinetic theory applied to reacting gases. By calculating the number of molecular collisions per unit time and unit volume which have an energy exceeding a certain activation value E, it is possible to obtain the following expression for the rate r at which a simple bimolecular gas reaction proceeds [Ref. (3)].

$$r \propto m_j m_k \rho^2 \sigma^2 T^{1/2} M^{-3/2} e^{E/\tilde{R}T}$$

ρ and T have their usual meaning and the other symbols denote:

m_j, m_k	local concentrations of molecules j and k
σ	mean molecular diameter
M	mean molecular weight
E	activation energy
$\tilde{R}$	molar (universal) gas constant

Substituting for ρ in terms of p and T we can for our purpose simplify the expression to

$$r \propto p^2 f(T)$$

Now T is maintained at a high value by having an approximately stoichiometric mixture in the primary zone: we are concerned here with the independent variable p. It is not to be expected that the theoretical exponent 2 will apply to the complex set of reactions occurring when a hydrocarbon fuel is burnt in air, and experiments with homogeneous mixtures in stoichiometric proportions suggest that it should be 1·8. At first sight it appears therefore that the design problem should be eased as the pressure is increased according to the law $p^{1.8}$. In fact there is reason to believe that under design operating conditions the chemical reaction rate is not a limiting factor in an actual combustion chamber where physical mixing processes play such an important role, and that an exponent of unity is more realistic. This is not to say that under extreme conditions— say at high altitude—the performance will not fall off more in accordance with the $p^{1.8}$ law.

A quantity known as the *combustion intensity* has been introduced to take account of the foregoing effects. One definition used is

$$\text{combustion intensity} = \frac{\text{heat release rate}}{\text{comb. vol.} \times \text{pressure}} \text{ kW/m}^3 \text{ atm}$$

Another definition employs $p^{1.8}$, with the units kW/m^3 atm$^{1.8}$. However it is defined, certainly the lower the value of the combustion intensity the easier it is to design a combustion system which will meet all the desired requirements. It is quite inappropriate to compare the performance of different systems on the basis of efficiency, pressure loss, etc., if they are operating with widely differing orders of combustion intensity. In aircraft systems the combustion intensity is in the region of 2–5×10^4 kW/m^3 atm,† while in industrial gas turbines the figure can be much lower because of

---

† Note that 1 kW/m^3 atm $= 96 \cdot 62$ Btu/h ft^3 atm.

the larger volume of combustion space available; a further reduction would result if a heat-exchanger were used, requiring a significantly smaller heat release in the combustor.

6.6 Some practical problems

We will briefly describe some of the problems which have not so far been mentioned but which are none the less important. These are concerned with (i) flame-tube cooling, (ii) fuel injection, (iii) starting and ignition and (iv) the use of cheaper fuels.

Flame-tube cooling

One problem which has assumed greater importance as permissible turbine inlet temperatures have increased is that of cooling the flame-tube. The tube receives energy by convection from the hot gases and by radiation from the flame. It loses energy by convection to the cooler air flowing along the outside surface and by radiation to the outer casing, but this loss is not sufficient to maintain the tube wall at a safe temperature. A common practice is to leave narrow annular gaps between overlapping sections of the flame-tube so that a film of cooling air is swept along the inner surface; corrugated 'wigglestrip', spot welded to successive lengths of flame-tube, provides adequate stiffness with annular gaps which do not vary too much with thermal expansion, as shown in Fig. 6.8(a). Another method is to use a ring of small holes with an internal splash ring to deflect the jets along the inner surface, as shown in Fig. 6.8(b). A more recent development is the use of *transpiration cooling*, allowing cooling air to enter a network of passages within the flame-tube wall before exiting to form an insulating film of air; this method may permit a reduction in cooling flow of up to 50 per cent.

Although empirical relations are available from which it is possible to predict convective heat transfer rates when film cooling a plate of known temperature, the emissivities of the flame and flame-tube can vary so widely that prediction of the flame-tube temperature from an energy balance is not possible with any accuracy. Even in this limited aspect of combustion chamber design, final development is still a matter of trial and error on the test rig. The emissivity of the flame varies with the type of fuel, tending to

(a)   (b)

FIG. 6.8 Film cooling of flame-tube

increase with the specific gravity. Carbon dioxide and water vapour are the principal radiating components in non-luminous flames, and soot particles in luminous flames. It is worth noting that vaporizer systems ease the problem, because flames from pre-mixed fuel vapour–air mixtures have a lower luminosity than those from sprays of droplets.

Higher turbine inlet temperatures imply the use of lower air/fuel ratios, with consequently less air available for film cooling. Furthermore, the use of a higher cycle temperature is usually accompanied by the use of a higher cycle pressure ratio to obtain the full benefit in terms of cycle efficiency. Thus the temperature of the air leaving the compressor is increased and its cooling potential is reduced. At projected levels of turbine inlet temperature, up to 1850 K, the use of transpiration cooling to reduce the required cooling flow may become essential.

Fuel injection

Most combustion chambers employ high-pressure fuel systems in which the liquid fuel is forced through a small orifice to form a conical spray of fine droplets in the primary zone. The fuel is said to be 'atomized' and the burner is often referred to as an 'atomizer'. An alternative is the vaporizer system, but it should be realized that even this requires an auxiliary starting burner of the atomizing type.

In the simplest form of atomizing burner, fuel is fed to a conical vortex chamber via tangential ports which impart a swirling action to the flow. The vortex chamber does not run full but has a vapour/air core. The combination of axial and tangential components of velocity causes a hollow conical sheet of fuel to issue from the orifice, the ratio of the components determining the cone angle. This conical sheet then breaks up in the air stream into a spray of droplets, and the higher the fuel pressure the closer to the orifice does this break-up occur. There will be a certain minimum fuel pressure at which a fully developed spray will issue from the orifice, although for the following reason the effective minimum pressure may well be higher than this.

The spray will consist of droplets having a wide range of diameter, and the degree of atomization is usually expressed in terms of a mean droplet diameter. In common use is the *Sauter mean diameter*, which is the diameter of a drop having the same surface/volume ratio as the mean value for the spray: 50–100 microns is the order of magnitude used in practice. The higher the supply pressure, the smaller the mean diameter. If the droplets are too small they will not penetrate far enough into the air stream, and if too large the evaporation time may be too long. The effective minimum supply pressure is that which will provide the required degree of atomization.

The object is to produce an approximately stoichiometric mixture of air and fuel uniformly distributed across the primary zone, and to achieve this over the whole range of fuel flow from idling to full load conditions.

Herein lies the main problem of burner design. If the fuel is metered by varying the pressure in the fuel supply line, the simple type of burner just described (sometimes referred to as the *simplex*) will have widely differing atomizing properties over the range of fuel flow. It must be remembered that the flow through an orifice is proportional to the square root of the pressure difference across it, and thus a fuel flow range of 10:1 implies a supply pressure range of 100:1. If the burner is designed to give adequate atomization at full load, the atomization will be inadequate at low load. This problem has been overcome in a variety of ways.

Perhaps the most commonly used solution to the problem is that employed in the *duplex* burner, an example of which is shown in Fig. 6.9(a). Two fuel manifolds are required, each supplying independent orifices. The small central orifice is used alone for the lower flows, while the larger annular orifice surrounding it is additionally brought into operation for the higher flows. The sketch also shows a third annulus formed by a shroud through which air passes to prevent carbon deposits building up on the face of the burner. This feature is incorporated in most burners. An alternative form of duplex burner employs a single vortex chamber and final orifice, with the two fuel supply lines feeding two sets of tangential ports in the vortex chamber.

Figure 6.9(b) illustrates a second practical method of obtaining good atomization over a wide range of fuel flow: the *spill* burner. It is virtually a simplex burner with a passage from the vortex chamber through which excess fuel can be spilled off. The supply pressure can remain at the high value necessary for good atomization while the flow through the orifice is reduced by reducing the pressure in the spill line. One disadvantage of this system is that when a large quantity of fuel is being recirculated to the pump inlet there may be undesirable heating and consequent deterioration of the fuel.

Dual-fuel burners are used in industrial gas turbines where gas is the normal fuel, but oil is required for short periods when the gas supply may be interrupted. The gas and liquid fuels would be supplied through separate concentric annuli; an additional annulus may also be provided for water or steam injection for emission control. Such engines can operate on either fuel separately or on both simultaneously. A typical dual-fuel nozzle for an aero-derivative gas turbine is shown in Fig. 6.10.

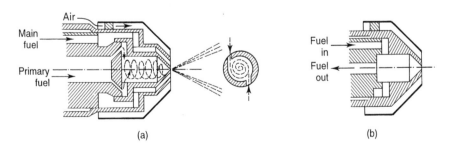

(a)   (b)

FIG. 6.9 Duplex and spill burners

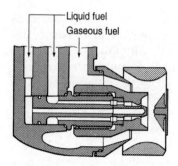
Liquid fuel
Gaseous fuel

FIG. 6.10 Dual-fuel burner [courtesy Rolls-Royce plc]

Burners of the type described here by no means exhaust all the possibilities. For example, for small gas turbines having a single chamber, it has been found possible to use rotary atomizers. Here fuel is fed to a spinning disc or cup and flung into the air stream from the rim. High tip speeds are required for good atomization, but only a low-pressure fuel supply is required.

Starting and ignition

Under normal operating conditions, gas turbine combustion is continuous and self-sustaining. An ignition system, however, is required for starting, and the ignition and starting systems must be closely integrated. The first step in starting a gas turbine is to accelerate the compressor to a speed that gives an airflow capable of sustaining combustion; during the period of acceleration the ignition system is switched on and fuel is fed to the burners when the rotational speed reaches about 15–20 per cent of normal. An igniter plug is situated near the primary zone in one or two of the flame-tubes or cans. Once the flame is established, suitably placed interconnecting tubes between the cans permit 'light round', i.e. flame propagation from one flame-tube to the other. Light round presents few problems in annular combustors. When the engine has achieved self-sustaining operation, the ignition system is turned off. Aircraft gas turbines have two additional requirements to meet: (i) re-ignition must be possible under wind-milling conditions if for any reason the flame is extinguished at altitude, and (ii) operation at idle power must be demonstrated while ingesting large amounts of water. This latter requirement is to prevent flame-out during final approach to an airport in very heavy rain, and it is normal operating procedure to turn on the ignition system in adverse weather at low altitude during both climb and descent. Engine shutdown normally requires the engine to be brought back to idle followed by shutting off the fuel; shutdowns from full power should be avoided because of the possibility of differential expansion/contractions leading to seal rubs or seizure of the rotor.

The starting systems for aircraft and industrial gas turbines are quite different, with compact size and low weight being critical for the former and very large powers sometimes required for the latter. Starting devices include electric motors, compressed air or hydraulic starters, diesel engines, steam turbines or gas expansion turbines. Early civil aircraft were dependent on ground power supplies for starting, and some engines used direct air impingement on the turbine blades; modern aircraft normally use an air turbine starter, which is connected to the main rotor by a reduction gearbox and clutch. The supply of compressed air may be from a ground cart or auxiliary power unit (APU) or may be bled from the compressor of an engine already running. Military aircraft use similar systems, but early aircraft used cartridge-type starters which provided a flow of hot, high-pressure gas for up to 30 seconds; this was expanded through a small turbine geared to the main rotor via a clutch.

The type of starting system required for an industrial gas turbine depends on the configuration; for units with a free power turbine it is only necessary to accelerate the gas generator. A single-shaft unit for electrical power generation, however, requires that the gas turbine and electric generator be accelerated as a single train. Power requirements for a 150 MW unit can be as high as 5 MW; for the largest units it is now common for the generator to be wound so that it can also be used as a motor, which is then used as the starting device. Diesel or steam turbine starting units with a power of 400–500 kW may be used for 60–80 MW units. The starter requirements for a free turbine engine are much less, and this is even more pronounced for a twin-spool gas generator where only the high-pressure rotor has to be turned over; a 30 MW unit may then require as little as 20 kW for starting.

The ignition performance can be expressed by an ignition loop which is similar to the stability loop of Fig. 6.6 but lying inside it. That is, at any given air mass flow the range of air/fuel ratio within which the mixture can be ignited is smaller than that for which stable combustion is possible once ignition has occurred. The ignition loop is very dependent on combustion chamber pressure, and the lower the pressure the more difficult the problem of ignition. Relighting of an aircraft engine at altitude is thus the most stringent requirement. Although high-tension sparking plugs similar to those used in piston engines are adequate for ground starting, a spark of much greater energy is necessary to ensure ignition under adverse conditions. A *surface-discharge igniter*, yielding a spark having an energy of about 3 joules at the rate of one per second, is probably the most widely used type for aircraft gas turbines in which fuel is injected as a spray of droplets.

One example of a surface-discharge igniter is shown in Fig. 6.11. It consists of a central and an outer electrode separated by a ceramic insulator except near the tip where the separation is by a layer of semiconductor material. When a condenser voltage is applied, current

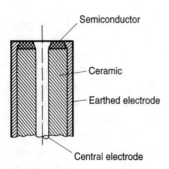

FIG. 6.11 Surface-discharge igniter

flows through the semiconductor which becomes incandescent and provides an ionized path of low resistance for the energy stored in the capacitor. Once ionization has occurred, the main discharge takes place as an intense flashover. To obtain good performance and long life the location of the igniter is critical: it must protrude through the layer of cooling air on the inside of the flame-tube wall to the outer edge of the fuel spray, but not so far as to be seriously wetted by the fuel.

For vaporizing combustors, some form of *torch igniter* is necessary. This comprises a spark plug and auxiliary spray burner in a common housing, resulting in a bulkier and heavier system than the surface-discharge type.

The normal spark rate of a typical ignition system is between 60 and 100 sparks per minute. Each discharge causes progressive erosion of the igniter electrodes, making periodic replacement of the igniter plug necessary; it is for this reason that the ignition system is switched off in normal operation. The pilot, however, must be provided with the capability of re-engaging the ignition system in the event of flame-out or extremely heavy rain.

Use of cheap fuels

Gas turbines rapidly supplanted piston engines in aircraft because of their major advantages in power and weight, permitting much higher flight speeds. Penetration of industrial markets was much slower, initially because of uncompetitive thermal efficiency and later, as performance improved, because of the need for expensive fuel. Gas turbines became established in applications such as peak-load or emergency electricity generation, where the running hours were short; the oil crises of the 1970s, resulting in soaring oil costs, saw a large number of units mothballed because of high fuel costs.

Natural gas, although a relatively expensive fuel, is ideal for use in stationary gas turbines, containing very few impurities such as sulphur and not requiring atomization or vaporization as do liquid fuels. As a result, gas turbines rapidly became the prime mover of choice for gas compression duties on pipelines, making use of the high-pressure gas flowing through the pipeline as fuel. Natural gas is now used widely for base-load electrical power generation, using combined cycle plant with capacities in excess of 2000 MW and thermal efficiencies of around 55 per cent. The long-term availability of natural gas is somewhat controversial, but these stations could eventually be converted to burning gas obtained from coal.

Market penetration would be greatly enhanced if gas turbines could burn *residual oil*. This cheap fuel is the residue from crude oil following the extraction of profitable light fractions. Some of its undesirable characteristics are:

(*a*)   high viscosity requiring heating before delivery to the atomizers;

(*b*)   tendency to polymerize to form tar or sludge when overheated;

(*c*)   incompatibility with other oils with which it might come into contact, leading to jelly-like substances which can clog the fuel system;

(*d*)   high carbon content leading to excessive carbon deposits in the combustion chamber;

(*e*)   presence of vanadium, the vanadium compounds formed during combustion causing corrosion in the turbine;

(*f*)   presence of alkali metals, such as sodium, which combine with sulphur in the fuel to form corrosive sulphates;

(*g*)   relatively large amount of ash, causing build-up of deposits on the nozzle blades with consequent reduction in air mass flow and power output.

The problems arising from characteristics (*a*), (*b*), (*c*) and (*d*) can be overcome without excessive difficulty. A typical residual fuel may require heating to about 140 °C, and for a large station this would require steam heating both for the storage tanks and prior to delivery to the atomizers. It is the major problems arising from (*e*), (*f*) and (*g*) that have greatly restricted the use of residual oil. The rate of corrosion from (*e*) and (*f*) increases with turbine inlet temperature, and early industrial gas turbines designed for residual oil operated with temperatures around 900 K to avoid the problem. Such a low cycle temperature inevitably meant a low cycle efficiency. It has now been found that the alkali metals can be removed, and that fuel additives such as magnesium compounds can neutralize the vanadium; sulphur cannot be removed at this stage, and would need to be done as part of the refinery process. There are two basic methods of treating the alkali metals. The first is to wash the fuel oil with water, followed by centrifuging of the mixture to separate the heavier liquids containing the alkali metals. The second is a process in which the fuel is mixed with water, pumped to a high pressure and then subjected to a high-voltage static discharge resulting in the required separation. Both methods require a sophisticated fuel treatment plant which means a considerable capital expense plus additional operating costs, so the actual cost of the apparently cheap fuel is significantly increased. Residual oil is not used very often, but may be used as a back-up fuel with natural gas as the prime fuel. In a typical modern large plant using residual oil as back-up, the site rating of the gas turbine was reduced from 139 MW to 116 MW owing to the need to operate at reduced turbine inlet temperature. It is important to realize that this dual-fuel type of operation is often dictated by the supply of natural gas at reduced cost on an *interruptible* basis; this means that the gas supplier can cut off gas in peak periods, so the customer must have a capability of switching fuels. It is common to find gas turbines designed to switch *automatically* from gas to oil while continuing to operate at full power.

In the early days of gas turbines, much effort went into experiments with burning coal; these used coal in the form of a pulverized solid, and were unsuccessful owing to the severe erosion caused by hard particles. For many years coal was not considered as a fuel, but recently a major effort has been expended on developing schemes for producing gas from coal, i.e. *coal gasification*. The establishment of the combined cycle has made this feasible, as many gasification schemes require large quantities of steam. This will be discussed further in the last section of this chapter.

6.7 Gas turbine emissions

Gas turbine combustion is a steady flow process in which a hydrocarbon fuel is burned with a large amount of excess air, to keep the turbine inlet temperature at an appropriate value. This is essentially a clean and efficient process and for many years there was no concern about emissions, with the exception of the need to eliminate smoke from the exhaust. Recently, however, control of emissions has become probably the most important factor in the design of industrial gas turbines, as the causes and effects of industrial pollution become better understood and the population of gas turbines increases. Emissions from aero-engines are also important, but the problems and solutions are quite different from those for land-based gas turbines.

Combustion equations express conservation of mass in molecular terms following the rearrangement of molecules during the combustion process. The basic principles of combustion are described in standard texts on thermodynamics, e.g. Ref. (4). The oxygen required for stoichiometric combustion can be found from the general equation

$$C_xH_y + nO_2 \rightarrow aCO_2 + bH_2O$$

where

$$a = x, \ b = (y/2) \text{ and } n = x + (y/4)$$

Each kilogram of oxygen will be accompanied by $(76 \cdot 7/23 \cdot 3)$ kg of nitrogen, which is normally considered to be inert and to appear unchanged in the exhaust; at the temperatures in the primary zone, however, small amounts of oxides of nitrogen are formed. The combustion equation assumes *complete* combustion of the carbon to CO_2, but *incomplete* combustion can result in small amounts of carbon monoxide (CO) and unburned hydrocarbons (UHC) being present in the exhaust. The gas turbine uses a large quantity of excess air, resulting in considerable oxygen in the exhaust; the amount can be deduced from the total oxygen in the incoming air less that required for combustion. Thus the exhaust of any gas turbine consists primarily of CO_2, H_2O, O_2 and N_2 and the composition can be expressed in terms of either gravimetric (by mass) or molar (by volume) composition. Concern about the possible harmful effects of CO_2 as a 'greenhouse' gas leading to global warming has been mentioned in section 1.7.

The pollutants appearing in the exhaust will include oxides of nitrogen (NO_x), carbon monoxide (CO) and unburned hydrocarbons (UHC); any sulphur in the fuel will result in oxides of sulphur (SO_x), the most common of which is SO_2. Although these all represent a very small proportion of the exhaust, the large flow of exhaust gases produces significant quantities of pollutants which can become concentrated in the area close to the plant. The oxides of nitrogen, in particular, can react in the presence of sunlight to produce 'smog' which can be seen as a brownish cloud; this problem was originally identified in the Los Angeles area, where the combination of vehicle exhausts, strong sunlight, local geography and temperature inversions resulted in severe smog. This led to major efforts to clean up the emissions from vehicle exhausts and stringent restrictions on emissions from *all* types of power plant. Oxides of nitrogen also cause acid rain, in combination with moisture in the atmosphere, and ozone depletion at high altitudes, which may result in a reduction in the protection from ultraviolet rays provided by the ozone layer, leading to increases in the incidence of skin cancer. UHC may also contain carcinogens, and CO is fatal if inhaled in significant amounts. With an ever-increasing worldwide demand for power and transportation, control of emissions is becoming essential.

Operational considerations

The problem of controlling emissions is complicated by the fact that gas turbines may be operated over a wide range of power and ambient conditions. The off-design performance of gas turbines will be discussed in Chapters 8 and 9, where it will be shown that different engine configurations will have widely varying operating characteristics. As a simple example, power changes on a single-shaft gas turbine driving a generator will occur with the compressor operating at constant speed and approximately constant airflow, while an engine with a free power turbine must operate at different compressor speeds, and hence airflows, as the power setting is changed. Multi-spool and variable geometry compressor systems introduce further problems. In the past, the function of the engine control system was to provide the correct amount of fuel for all operating conditions, both steady state and transient; emission control was a function of the combustor design. With modern engines, however, the control system plays a major role in adjusting fuel/air ratios to minimize emissions over the complete operating range; this may involve supplying fuel to different zones of the combustor at different operating conditions, and is only possible with the advent of sophisticated digital fuel control systems.

With gas turbines being increasingly used for large base-load electricity generation, they consume very large quantities of fuel; they are normally operated at a fixed rating for long periods, but may be operated at lower loads for shorter periods. Cogeneration installations are generally designed to operate at full load on a continuous basis, but are more likely to be located close to residential areas than is the case for power stations. Pipeline gas turbines are often located in remote regions, and will mostly operate at steady power settings; at the present time they have not been subjected to the same emissions standards as power stations, but it is likely that this will change in the near future.

Aircraft engines have two quite different requirements. The first is for very high combustion efficiency at low power, because of the large amounts of fuel burned during taxiing and ground manoeuvring. The primary problem here is the reduction of UHC. At take-off power, climb and cruise the main concern is NO_x. The International Civil Aviation Organization (ICAO) sets standards on a worldwide basis, for both the take-off and landing cycles and also for cruise at high altitude; the first is concerned with air quality in the vicinity of airports and the second with ozone depletion in the upper atmosphere.

Pollutant formation

For many years the attention of combustion engineers was focused on the design and development of high-efficiency combustors that were rugged and durable, followed by relatively simple solutions to the problem of smoke. When the requirements for emission control emerged, much basic research was necessary to establish the fundamentals of pollutant formation.

The single most important factor affecting the formation of NO_x is the *flame temperature*; this is theoretically a maximum at stoichiometric conditions and will fall off at both rich and lean mixtures. Unfortunately, while NO_x could be reduced by operating well away from stoichiometric, this results in increasing formation of both CO and UHC, as shown in Fig. 6.12. The rate of formation of NO_x varies *exponentially* with the flame

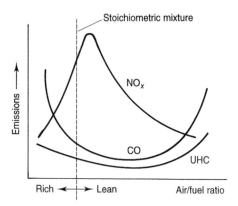

FIG. 6.12 Dependence of emissions on fuel/air ratio

temperature, so the key to reducing NO_x is reduction of the flame temperature. The formation of NO_x is slightly dependent on the *residence time* of the fluid in the combustor, decreasing in a linear fashion as residence time is reduced; an increase in residence time, however, has a favourable effect on reducing both CO and UHC emissions. Increasing the residence time implies an increase in combustor cross-sectional area or volume.

It is important to understand the relationship between emissions and the key cycle parameters of pressure ratio and turbine inlet temperature. Initial attempts to correlate emissions were based on the pollutants produced for a given power level, but it was soon realized that the rate of formation of pollutants depends on the internal conditions in the combustor, which are functions of the basic cycle parameters. One of the most widely used correlations was due to Lipfert [Ref. (5)], who found that NO_x emissions increased with combustor inlet temperature (i.e. compressor delivery temperature). His correlation was based on results from a range of engines from small APUs to the high bypass turbofans of the period such as the JT-9D. This work led to the unwelcome conclusion that the use of a high pressure ratio to obtain high efficiency will have a deleterious effect on emissions. Fortunately this is no longer true. Once designers of combustion systems began to understand the problem and incorporate the necessary measures to minimize NO_x, it was found that cycle pressure ratio did not have a major effect: it is the *flame* temperature which is important. Figure 6.13, a correlation from Leonard and Stegmaier [Ref. (6)], shows that NO_x emissions can be more than halved by reducing the flame temperature from 1900 K to 1800 K. In this figure NO_x is measured in units of 'corrected ppmvd', which means 'parts per million by volume of dry exhaust gas corrected to standard pressure and temperature'. In the next sub-section we will consider some of the methods used to minimize emissions.

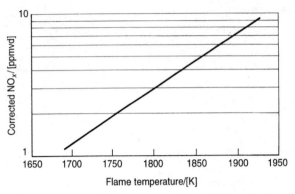

FIG. 6.13 **Effect of flame temperature on NO$_x$ emissions**

Methods for reducing emissions

Emissions control requirements for NO$_x$ were first applied to stationary gas turbines in the Los Angeles area in the early 1970s, where it was found that an emissions level of about 75 parts per million by volume of dry exhaust (75 ppmvd) when burning oil could be achieved by injecting water into the combustor to lower the flame temperature. Because of this the Environmental Protection Agency (EPA) set this level as a standard for new installations; with the increasing use of gas turbines and the longer running hours associated with base-load or cogeneration plant these limits have become ever more restrictive. Areas which are particularly environmentally sensitive, such as Southern California and Japan, have promulgated even lower levels. European countries also began to introduce their own standards, in this case specifying pollutants in terms of mg/m^3 of exhaust flow. Thus there are many varied requirements to be met in different parts of the world and even in different areas of the same country.

The picture is rather more coherent with respect to emissions from civil aircraft, where standards are set on a worldwide basis by the International Civil Aviation Organization (ICAO) with limits being set following extensive deliberations by multi-national committees. Limits are specified in terms of the amount of pollutants produced per unit thrust or unit mass of fuel burned, with emphasis on both the take-off/landing cycle and the high-altitude cruise condition.

The emission of pollutants into the atmosphere may be tackled either during the combustion process, or post combustion by exhaust clean-up; the latter method is widely used in coal-burning steam plant, e.g. as flue gas desulphurization. Gas turbine designers have chosen to attack the problem by focusing on new combustor designs, although sometimes exhaust clean-up is also used to obtain very low emissions. There are basically three major methods of minimizing emissions: (i) water or steam injection into the combustor, (ii) selective catalytic reduction (SCR) and (iii) dry low NO$_x$ (so called because no water is involved).

(i)   Water or steam injection

As stated earlier, the purpose of water injection is to provide a substantial decrease in flame temperature. In the first installations where this was used it was found that to obtain 75 ppmvd it was necessary to use half as much water as fuel, resulting in approximately 40 per cent reduction in NO_x. The amounts of water required are substantial, and demineralized water must be used to prevent corrosive deposits in the turbine. For lower levels of NO_x the water/fuel ratio may be 1·0 or even higher. Unfortunately, the small increase in power due to the higher mass flow through the turbine is offset by a decrease in thermal efficiency. It is also found that increasing the water/fuel ratio, while continuing to decrease NO_x, increases both CO and UHC emissions. In many locations water is scarce or expensive, and is obviously difficult to use on a year-round basis in countries where the ambient temperature is well below freezing for months on end. To give some indication of the amounts of water required, a 4 MW gas turbine needs about 4 million litres of water annually. It is clear that water injection introduces many new problems for the operator, but it has been successful in allowing operation at low levels of NO_x while better methods are developed.

Steam injection operates on the same principle, and is often available at high pressure from the heat recovery steam generator (HRSG) in combined cycle or cogeneration installations. With high-efficiency engines operating at compressor delivery pressures in excess of 30 bar, the availability of steam at even higher pressures is necessary. As an example, a 40 MW aero-derivative gas turbine may use about 25 per cent of the steam produced in the HRSG for NO_x control.

(ii)   Selective Catalytic Reduction

SCR has been used in situations where extremely low (<10 ppmvd) limits of NO_x have been specified. This is a system for exhaust clean-up, where a catalyst is used together with injection of controlled amounts of ammonia (NH_3) resulting in the conversion of NO_x to N_2 and H_2O. The catalytic reaction only occurs in a limited temperature range (285–400 °C), and the system is installed midway through the HRSG; because of the limited temperature range, SCR can be used *only* with waste heat recovery applications. The use of SCR introduces a whole range of new problems including increased capital cost, handling and storage of a noxious fluid, control of NH_3 flow and difficulty in dealing with variable loads. SCR systems have been used in gas turbines burning natural gas, and although proposed for units burning oil it appears that none have actually been built at the time of writing. The need arose because of the installation of dual-fuel engines which may be required to operate for relatively short periods on oil when the gas supply is interrupted. It appears likely that SCR will be a method which is superseded as dry low NO_x systems enter the market.

(iii)   Dry low NO$_x$

Currently all designers of gas turbines are heavily involved in research and development into combustor designs capable of operation at low levels of NO$_x$ without any requirement for water, i.e. *dry* systems. Remembering the exponential variation of NO$_x$ emissions with flame temperature, it is possible to consider either lean burning or rich burning in the primary zone to achieve the necessary reduction in flame temperature. Both approaches have been investigated; in the case of rich burning, there is the probability of smoke being produced in the primary zone. This led to investigation of the rich burn/quick quench concept in which the combustor is designed with two axial 'stages', with rich combustion in the first followed by large amounts of dilution air in the second. Most manufacturers, however, have decided to use the lean burn approach, with the major modification of pre-mixing the air and fuel prior to combustion. The use of lean burn, however, leads to problems of maintaining stable combustion as power is reduced; this leads to significant increase in the complexity of the engine control system. Many different design approaches have been used by the principal engine manufacturers, and some of these will be considered in the next sub-section.

Design of dry low NO$_x$ systems

There are three approaches to the design of dry low NO$_x$ systems, depending on the type of gas turbine. With the industrial gas turbine, considerable space is available and the fuel is usually natural gas. For aircraft engines volume and frontal area must be kept to a minimum, and liquid fuel is always used. The third type requiring special consideration is the aero-engine which has been modified for use as a land-based unit: it is then necessary for the new combustor to be capable of being retrofitted in place of the original. There are a large number of these aero-derivative gas turbines in use which may be subject to more severe emissions limits than when they entered service. This is also true for industrial engines, where the problem is eased by the less stringent demands on space.

It is only possible to give a very brief overview of some of the approaches taken, but the wide range of solutions arrived at gives some indication of the amount of on-going research and development. An important concept for emissions reduction is *fuel staging*, where the total engine fuel flow is divided into two parts which are supplied separately to two distinct combustion zones. One of these is fuelled continuously, providing fuel for starting and idling, acting as a pilot stage. The bulk of the fuel is burned in the second zone, which serves as the main stage of combustion. Fuel staging is widely used in modern combustors, and many different arrangements are found in practice.

(i)   Industrial gas turbines

General Electric have always used multiple combustors in a cannular configuration in their heavy industrial gas turbines, so their approach is based on modifications to the individual cans. The original cans had a single burner, but this has been replaced by a ring of six primary dual-fuel burners surrounding a single secondary dual-fuel burner. The combustor is shown schematically in Fig. 6.14. A convergent–divergent section at the end of the primary zone serves to accelerate the flow to prevent upstream propagation of the flame from the second stage to the first, that could happen under some modes of operation. The venturi also produces a recirculation zone on its downstream face to stabilize the flame. The split of fuel between the primary and secondary burners changes with load. At start-up, and up to 20 per cent load, all the fuel is supplied to the primary burners and combustion takes place in the first stage. From about 20–40 per cent load, about 30 per cent of the fuel is supplied to the secondary burner and lean combustion takes place in both stages. At about 40 per cent load, all fuel is supplied to the secondary burner with no combustion in the first stage; this is a transient situation leading to fully pre-mixed operation. From 40–100 per cent load, fuel is supplied to all burners (approximately 83 per cent to the primary set), the fuel and air are pre-mixed in the primary stage and then combustion takes place only in the second stage. The development of this combustor is described by Davis and Washam [Ref. (7)].

Siemens and ABB have adopted the approach of developing single burners, which can then be used in the numbers required for the specified power output. The Siemens development is referred to as a hybrid burner, operating in different modes at low and high power. At low power only a pilot burner is lit, providing a diffusion flame. At about 40 per cent load, the mode is changed to pre-mixed lean burning, giving very low emissions of both NO_x and CO. The burners are designed to burn gas, liquids, or gas and liquids simultaneously, and also to use water or steam for NO_x reduction. The variation of emission levels with load is shown in Fig. 6.15. A rotating shutter ring is used to control fuel/air ratio at low powers to

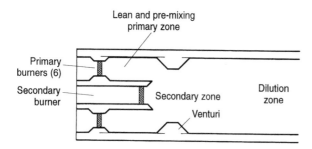

FIG. 6.14 Can for General Electric low NO_x combustor

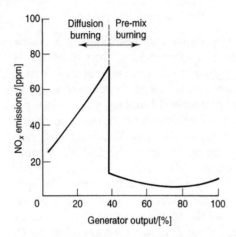

FIG. 6.15 Variation of emissions with load (hybrid burner)

keep emissions at an acceptable level, and the engine airflow is modified by use of variable inlet guide vanes (VIGVs) in the compressor. Maghon *et al.* [Ref. (8)] describe the development of this burner. The ABB approach is novel, using a double-cone burner where flame stabilization is achieved by vortex breakdown at exit from the burner as shown in Fig. 6.16. This is another example of a pre-mixed lean burn system, and the burner is used in both silo-type and fully annular combustors. The burners are arranged in concentric circles, and various segments are lit in sequence as power is increased. Development of this concept is described by Sattelmeyer *et al.* [Ref. (9)].

The methods described above are all based on the requirements of single-shaft units which operate at constant speed. Solar Gas Turbines build smaller units, up to 14 MW, most of which use free turbines requiring variation in compressor speed, and hence airflow, as load is changed. Their solution also uses the pre-mixed lean burn concept, but requires the use of

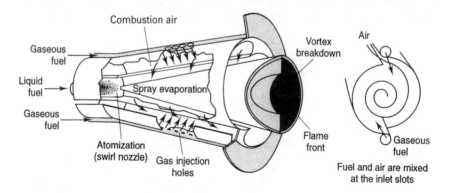

FIG. 6.16 ABB dual-fuel double-cone burner [courtesy ABB]

overboard bleed from compressor discharge at low power to provide the correct fuel/air ratio to maintain the specified emission levels. This is a simple and reliable method, but a penalty in thermal efficiency is incurred when using bleed. Etheridge [Ref. (10)] describes field experience with this system, including the effects of ambient temperature on emissions; NO_x increases at low temperatures, whereas CO increases at high temperatures. This once again illustrates the many problems facing the combustion designer.

(ii) Aircraft gas turbines

ICAO sets standards for emissions for both the take-off/landing cycle and the cruise condition. The take-off/landing cycle prescribes standard times at take-off and approach power, and the limits are specified in terms of g/kN of take-off thrust. At typical subsonic cruise conditions a modern turbofan produces about 12 g of NO_x per kg of fuel burned, compared to about 34 g/kg at take-off; this is primarily due to the reduction in turbine inlet temperature at cruise. Emissions of UHC and CO are negligible in comparison, as low as 0·1 and 0·6 g/kg respectively, at both take-off and cruise. These data are from Bahr [Ref. (11)], who also shows that for a modern twin-engine transport operating over an 800 km range approximately 25 per cent of the emissions are produced during the take-off/landing cycle, with the remainder during climb/cruise/descent; approximately 86 per cent of the total emissions are NO_x.

The aircraft problem can be solved by using lean primary zone mixtures, but great care must be taken that this does not result in unacceptable losses in ignition and blow-out performance or unacceptable increases in UHC and CO emissions at idle. These conflicting requirements led to the development of combustors using two or more combustion zones which may be staged, either in parallel or in series.

General Electric developed a double annular combustor for civil aircraft engines, resulting in *parallel* staging. The arrangement is shown schematically in Fig. 6.17(a); during starting, idle and relighting at altitude only the outer ring of burners is fuelled. At normal operating conditions

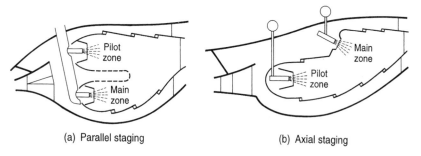

(a) Parallel staging   (b) Axial staging

FIG. 6.17 Low-emission aero-engine combustors

both annuli are fuelled and the fuel flow split can be adjusted to provide lean fuel/air ratios in both zones at high powers. Future requirements for lower emissions may require pre-mixing as well, but this adds considerably to the complexity of the combustor design; Bahr, [Ref. (11)] discusses some of the problems which must be overcome.

The alternative approach of *axially* staged combustion was adopted by International Aero Engines for the design of a low NO_x combustor for the V 2500, as described by Segalman *et al.* [Ref. (12)]. The combustor layout is shown in Fig. 6.17(b), and results in a longer combustor than could be achieved with a double annular arrangement. This layout, however, was chosen because it was considered to have greater potential for emissions reduction, and the length penalty could be reduced by using the staggered inline arrangement shown.

(iii)   Aero-derivative engines

Aero-derivative units in service in the mid 1990s were based on aero-engines developed in the early 1970s which incorporated very compact annular combustors. These engines were not designed with low emissions in mind, and having high design pressure ratios the level of emissions was high. This was overcome by using water or steam injection, especially in cogeneration applications. As pointed out earlier, water or steam injection introduces many operational problems and is particularly inappropriate for pipeline applications. Manufacturers of aero-derivative engines have therefore been forced to develop dry low NO_x systems, which have to be retrofittable to engines in the field.

The well-proven concept of lean pre-mixed combustion has been adapted. It is interesting to contrast the solutions arrived at by General Electric and Rolls-Royce. The General Electric system increases the combustor volume to get longer residence times for low UHC and CO, and this is done by increasing the depth of the annulus; three concentric circles of burners are used, with a centre-body between each circle. The outer two circles each have 30 burners, while the inner has 15; the burners are mounted on a stalk which may have either two or three burners, to allow for the reduced number in the inner circle. As in the case of the ABB design, they can be lit in sequence as power is increased. Leonard and Stegmaier [Ref. (6)] discuss the development of the low emissions combustor for the LM 6000, and the same technology will be applied to the other aero-derivatives in the General Electric stable (LM 2500 and LM 1600).

Rolls-Royce have moved away from the fully annular combustor to separate combustion cans arranged radially inwards; this gives the increased volume required for increased residence time, without the need for any increase in length. This system was originally developed for the RB-211 [Ref. (11)], but a similar system is used on the industrial version of the Trent. Figure 1.16 shows the aero and industrial versions on the same centreline and clearly shows the differences between the two combustors.

Corbett and Lines [Ref. (14)] show that the need for introducing the fuel in stages results in a very sophisticated control system; the problems are more complex than on the single-shaft machines described earlier, because the compressor speeds and airflow are changing continually on the multi-spool engine as load is increased.

Combustion noise

It was mentioned earlier that the design of low-emission combustion systems presents a number of conflicting requirements to be met so that both low emissions and high turbine inlet temperatures can be achieved. Dry low emission (DLE) systems are preferable to steam injection or SCR from the operational point of view, but their introduction into service resulted in a new phenomenon of combustion instability with significant pressure pulsations and combustion noise; this is colloquially referred to as *humming*. The combustor may exhibit a resonant acoustic mode at different operating conditions, causing mechanical damage to the combustor and possible secondary damage to the downstream components. The problem of combustion noise arose in many different designs and results from the use of lean pre-mixed combustion systems. Scarinci and Halpin [Ref. (15)] state that a key difference between lean pre-mixed and conventional combustors is the distribution of heat release within the combustor volume. In a lean pre-mixed combustor heat release occurs abruptly across a flame front whereas in conventional combustors the heat release is smeared across a much wider region.

Reference (15) shows that the frequency and amplitude of the resonant modes are strongly affected by the axial distribution of heat release and temperature inside the combustor. Various manufacturers have developed different solutions to the problem of humming but little has been published at the time of writing.

6.8   Coal gasification

The prospect of future shortages of natural gas has led to a resurgence of interest in the use of coal. Several closed-cycle gas turbines, with external combustion, were built in Germany and ran successfully for over 120 000 hours. All but one have been decommissioned. They were small units of 2–17 MW and were not economic with the current relative price of coal and gas. This German experience is described in Ref. (16).

The approach receiving most attention at the present time is integration of a coal gasification process with a combined cycle plant, described by the acronym IGCC. The principle was discussed in section 1.8, and Fig. 1.21 showed a diagrammatic sketch of a possible scheme. The actual gasification takes place in a pressure vessel where coal is reacted with an oxidant (steam, oxygen or air) yielding a 'dirty' gas and slag or ash. The dirty gas may contain particulates which must be removed by cyclone separators to avoid damage to the turbine and also undesirable chemicals which cause corrosion and pollution. In particular, sulphur is likely to be present in the form of either H_2S or SO_2; coal with up to 7 per cent sulphur may have to be used. In many systems the sulphur is reduced to its elemental form and is then sold as a valuable by-product. The cleaned fuel gas is then delivered to the combustor. It should be noted that the purification (clean-up) process often requires cooling of the dirty gas, and the heat removed is transferred to the steam-raising process.

Three types of gasifier are under development: (i) moving bed, (ii) fluidized bed and (iii) entrained bed in which the coal and oxidant are introduced together. The oxidant preferred seems to be either air or oxygen, rather than steam. Some relevant characteristics of four gasifiers are given in the following table:

	Bed type	Feed	Oxidant	O_2 flow (kg/kg coal)	Gas temp. (K)
Texaco	Entrained	Slurry	O_2	0·9	1480
Shell	Entrained	Dry	O_2	0·85	1750
Combustion Engineering	Entrained	Dry	Air	0·7	1280
British Gas/Lurgi	Moving	Dry	O_2	0·5	1000

The gas temperature is that following the gasification process, and a high value requires more complex integration into the steam-raising process. The quantity of oxygen required per unit mass of coal is important, as an oxygen-blown gasifier requires the incorporation of an air separation unit (ASU), which separates the oxygen from the nitrogen in some of the compressor delivery air. The ASU is a complex and expensive component with a considerable power demand; the chemical process required for separation is beyond the scope of this book. With air-blown gasifiers the resulting gas has a low calorific value (4500–5500 kJ/m^3) because of the large nitrogen content; oxygen-blown units give gas with a significantly higher calorific value (9000–13 000 kJ/m^3).

Two major demonstration projects have been the Cool Water plant in California and the Buggenum plant in Holland. Cool Water used a Texaco gasifier mated to a General Electric gas turbine, which demonstrated that a standard 80 MW gas turbine could operate in IGCC configuration on an electric utility system. The plant was operated from 1984 to 1989, and completed 27 000 hours using four different coals. A dual-fuel system was used, permitting operation on distillate fuel during periods when gasifier maintenance was required. The emissions achieved were excellent, with NO_x values of 20 ppm; the ash produced was non-hazardous and saleable. The Cool Water plant was purchased by Texaco and will continue to be used to provide electricity while gaining further operational experience on the gasifier. The coal consumption was about 1100 tonnes/day and the maximum output was 118 MW; this is too small for a commercial plant, but successfully demonstrated the concept at large scale. It is anticipated that a modern gas turbine and improved gasifier could give an IGCC efficiency of 40–42 per cent, compared with conventional coal-fired steam plant efficiencies of 36–38 per cent.

A larger demonstration unit, with an output of 250 MW, started operations at Buggenum in Holland at the end of 1993. The IGCC uses a Siemens gas turbine combined with a Shell gasifier; the gas turbine has to be capable of operating on both natural gas and coal gas, with natural gas used for start-up and operation when the gasifier requires maintenance. The steam turbine is also supplied by Siemens, and the heat to the steam cycle is transferred from both the gas turbine exhaust and the heat rejection from the gas clean-up process. The gasification unit has a capacity of 2000 tonnes/day and oxygen for the gasifier is supplied by an ASU with a capacity of 1700 tonnes/day. The air supply for the ASU is bled from the compressor discharge, saving the need for a separate compressor. The coal gas is desulphurized to produce sulphur as a saleable commodity. Figure 6.18 shows the plant layout, where it can be seen that the gas turbine/steam turbine/generator is dwarfed by the gasification plant, the ASU and the sulphur processing plant.

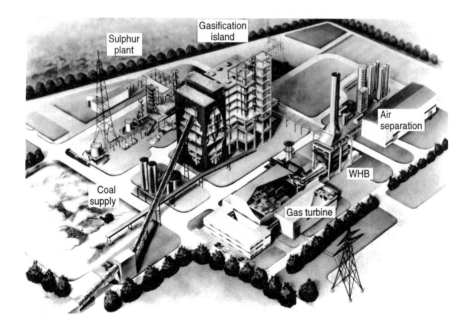

FIG. 6.18 IGCC plant, Buggenum [courtesy Demkolec]

The clean coal gas, after sulphur removal, has a composition of approximately 65% CO, 30% H_2, 1% CO_2, 1% H_2O and 3% N_2 + argon. Both CO and H_2 have adiabatic flame temperatures that are higher than for natural gas, which would result in a substantial increase in NO_x. The technology of pre-mixing the fuel and air to minimize NO_x cannot be used because of the high concentration of H_2, which would form an explosive mixture. The adiabatic flame temperature is lowered by another method, dilution with inert constituents; 16 per cent of the compressor delivery air is used for the ASU and the nitrogen resulting from the separation process is used to dilute the coal gas. Further dilution is provided by water vapour resulting from saturating the gas with warm water, and the low flame temperature results in very favourable levels of NO_x.

Although designed as a demonstration plant, to prove the technology at large scale, a very creditable thermal efficiency of 43 per cent is obtained. It is predicted that with developments in both the gas turbine and the gas purification process the efficiency could be raised to 48 per cent. It also appears that the technology used would be capable of being scaled up to 400 MW, and possibly even to 600 MW.

While coal gasification has still to be fully proved in commercial service, progress to date is extremely encouraging and suggests that it may be economically viable for large power stations. Many of the large natural gas-fired combined cycles commissioned in the early 1990s could eventually be converted to coal gasification if the supply of natural gas becomes inadequate to meet growing demand. When planning expansion of an electric power system, by adding power stations, utilities may do this in phases. In the first phase, simple cycle units may be installed quickly to provide the initial increase in capacity; they do not, of course, have a high enough thermal efficiency for long-term use as base-load plant. In the second phase, these could be converted to combined cycle operation; steam turbines take longer to manufacture and install, largely because of the extensive civil engineering work required. The second phase may be completed about one or two years after the first phase. Finally, if the supply of natural gas becomes a problem the combined cycle plants could be converted to an IGCC system.

7

Axial and radial flow turbines

As with the compressor, there are two basic types of turbine—radial flow and axial flow. The vast majority of gas turbines employ the axial flow turbine, so most of the chapter will be devoted to the theory of this type. The simple mean-diameter treatment is described first, and the essential differences between steam and gas turbine design are pointed out. The application of vortex theory is then discussed, followed by a description of the method of applying cascade test data to complete the design procedure and provide an estimate of the isentropic efficiency. Recent developments in the calculation of blade profiles to give specified velocity distributions are also mentioned. Blade stresses are considered briefly because they have a direct impact upon the aerodynamic design. The chapter closes with a section on the cooled turbine, followed by some material on the radial flow turbine.

The radial turbine can handle low mass flows more efficiently than the axial flow machine and has been widely used in the cryogenic industry as a turboexpander, and in turbochargers for reciprocating engines. Although for all but the lowest powers the axial flow turbine is normally the more efficient, when mounted back-to-back with a centrifugal compressor the radial turbine offers the benefit of a very short and rigid rotor. This configuration is eminently suitable for gas turbines where compactness is more important than low fuel consumption. Auxiliary power units (APUs), for aircraft generating sets of up to 3 MW, and mobile power plants are typical applications. Future microturbines will be another application for the radial flow turbine.

7.1 Elementary theory of axial flow turbine

Figure 7.1 shows the velocity triangles for one axial flow turbine stage and the nomenclature employed. The gas enters the row of nozzle blades† with

---

† These are also known as 'stator blades' and 'nozzle guide vanes'. We shall use suffix N when necessary to denote quantities associated with the nozzle row, but will often use the term 'stator' in the text. Suffix s will be used to denote 'stage'.

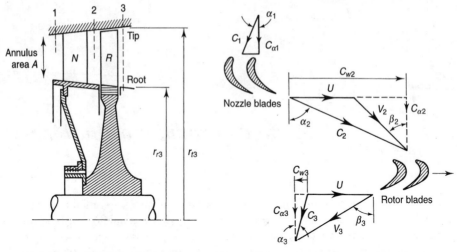

FIG. 7.1 Axial flow turbine stage

a static pressure and temperature p_1, T_1 and a velocity C_1, is expanded to p_2, T_2 and leaves with an increased velocity C_2 at an angle α_2.† The rotor blade inlet angle will be chosen to suit the direction β_2 of the gas velocity V_2 relative to the blade at inlet. β_2 and V_2 are found by vectorial subtraction of the blade speed U from the absolute velocity C_2. After being deflected, and usually further expanded, in the rotor blade passages, the gas leaves at p_3, T_3 with relative velocity V_3 at angle β_3. Vectorial addition of U yields the magnitude and direction of the gas velocity at exit from the stage, C_3 and α_3. α_3 is known as the *swirl angle*.

In a single-stage turbine C_1 will be axial, i.e. $\alpha_1 = 0$ and $C_1 = C_{a1}$. If on the other hand the stage is typical of many in a multi-stage turbine, C_1 and α_1 will probably be equal to C_3 and α_3 so that the same blade shapes can be used in successive stages: it is then sometimes called a *repeating stage*. Because the blade speed U increases with increasing radius, the shape of the velocity triangles varies from root to tip of the blade. We shall assume in this section that we are talking about conditions at the mean diameter of the annulus, and that this represents an average picture of what happens to the total mass flow m as it passes through the stage. This approach is valid when the ratio of the tip radius to the root radius is low, i.e. for short blades, but for long blades it is essential to account for three-dimensional effects as shown in subsequent sections.

$(C_{w2} + C_{w3})$ represents the change in whirl (or tangential) component of momentum per unit mass flow which produces the useful torque. The change in axial component $(C_{a2} - C_{a3})$ produces an axial thrust on the

† In the early days of gas turbines the blade angles were measured from the tangential direction following steam turbine practice. It is now usual to measure angles from the axial direction as for axial compressor blading.

rotor which may supplement or offset the pressure thrust arising from the pressure drop $(p_2 - p_3)$. In a gas turbine the net thrust on the turbine rotor will be partially balanced by the thrust on the compressor rotor, so easing the design of the thrust bearing. In what follows we shall largely restrict our attention to designs in which the axial flow velocity C_a is constant through the rotor. This will imply an annulus flared as in Fig. 7.1 to accommodate the decrease in density as the gas expands through the stage. With this restriction, when the velocity triangles are superimposed in the usual way we have the velocity diagram for the stage shown in Fig. 7.2.

The geometry of the diagram gives immediately the relations

$$\frac{U}{C_a} = \tan \alpha_2 - \tan \beta_2 = \tan \beta_3 - \tan \alpha_3 \tag{7.1}$$

Applying the principle of angular momentum to the rotor, the stage work output per unit mass flow is

$$W_s = U(C_{w2} + C_{w3}) = UC_a(\tan \alpha_2 + \tan \alpha_3)$$

Combining with (7.1) we have W_s in terms of the gas angles associated with the rotor blade, namely

$$W_s = UC_a(\tan \beta_2 + \tan \beta_3) \tag{7.2}$$

Note that the 'work-done factor' required in the case of the axial compressor is unnecessary here. This is because in an accelerating flow the effect of the growth of the boundary layer along the annulus walls is much less than when there is a decelerating flow with an adverse pressure gradient.

From the steady flow energy equation we have $W_s = c_p \Delta T_{0s}$, where ΔT_{0s} is the stagnation temperature drop in the stage, and hence

$$c_p \Delta T_{0s} = UC_a(\tan \beta_2 + \tan \beta_3) \tag{7.3}$$

When the stage inlet and outlet velocities are equal, i.e. $C_1 = C_3$, (7.3) also gives the static temperature drop in the stage ΔT_s. In Chapter 2 we used as typical of combustion gases the average values

$$c_p = 1 \cdot 148 \text{ kJ/kg K}; \quad \gamma = 1 \cdot 333 \text{ or } \gamma/(\gamma - 1) = 4$$

and noted that they were consistent with a gas constant R of $0 \cdot 287$ kJ/kg K. For preliminary design calculations these values are quite adequate

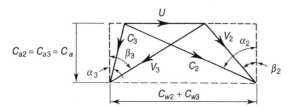

FIG. 7.2 Velocity diagram

and we shall use them throughout this chapter. With velocities in m/s, ΔT_{0s} in Kelvin units is conveniently given by

$$\Delta T_{0s} = 8 \cdot 71 \left(\frac{U}{100} \right) \left(\frac{C_a}{100} \right) (\tan \beta_2 + \tan \beta_3) \qquad (7.4)$$

The stagnation pressure ratio of the stage p_{01}/p_{03} can be found from

$$\Delta T_{0s} = \eta_s T_{01} \left[1 - \left(\frac{1}{p_{01}/p_{03}} \right)^{(\gamma-1)/\gamma} \right] \qquad (7.5)$$

where η_s is the isentropic stage efficiency based on stagnation (or 'total') temperature. Equation (7.5) is simply equation (2.12) applied to a stage, and η_s is often called the *total-to-total stage efficiency*. It is the appropriate efficiency if the stage is followed by others in a multi-stage turbine because the leaving kinetic energy $(C_3^2/2)$ is utilized in the next stage. It is certainly also relevant if the stage is part of a turbojet engine, because the leaving kinetic energy is used in the propelling nozzle. Even if it is the last stage of an industrial plant exhausting to atmosphere, the leaving kinetic energy is substantially recovered in a diffuser or volute and, as explained under the heading 'Compressor and turbine efficiencies', section 2.2, we can put $p_{03} = p_a$ and regard η_s as the combined efficiency of the last stage and diffuser. (Although we shall not use it, it should be noted that a *total-to-static* isentropic efficiency is sometimes quoted for a turbine as a whole and for a stage, and it would be used where it is desirable to separate the turbine and exhaust diffuser losses. Applied to the stage, we would have

$$\text{total-to-static efficiency} = \frac{T_{01} - T_{03}}{T_{01} - T_3'}$$

where T_3' is the static temperature reached after an isentropic expansion from p_{01} to p_3. It assumes that as far as the turbine is concerned all the leaving kinetic energy is wasted, and its value is somewhat less than the total-to-total efficiency which we shall use here.)

There are three dimensionless parameters found to be useful in turbine design. One, which expresses the work capacity of a stage, is called the *blade loading coefficient* or *temperature drop coefficient* ψ. We shall adopt NGTE practice and define it as $c_p \Delta T_{0s} / \frac{1}{2} U^2$, although $c_p \Delta T_{0s} / U^2$ is also used. Thus from equation (7.3),

$$\psi = \frac{2 c_p \Delta T_{0s}}{U^2} = \frac{2 C_a}{U} (\tan \beta_2 + \tan \beta_3) \qquad (7.6)$$

Another useful parameter is the *degree of reaction* or simply the *reaction* Λ. This expresses the fraction of the stage expansion which occurs in the rotor, and it is usual to define it in terms of static temperature (or enthalpy) drops rather than pressure drops, namely

$$\Lambda = \frac{T_2 - T_3}{T_1 - T_3}$$

For the type of stage we are considering, where $C_{a2} = C_{a3} = C_a$ and $C_3 = C_1$, a simple expression for Λ can be derived as follows. From (7.4),

$$c_p(T_1 - T_3) = c_p(T_{01} - T_{03}) = UC_a(\tan \beta_2 + \tan \beta_3)$$

Relative to the rotor blades the flow does no work and the steady flow energy equation yields

$$c_p(T_2 - T_3) = \tfrac{1}{2}(V_3^2 - V_2^2)$$
$$= \tfrac{1}{2}C_a^2(\sec^2 \beta_3 - \sec^2 \beta_2)$$
$$= \tfrac{1}{2}C_a^2(\tan^2 \beta_3 - \tan^2 \beta_2)$$

and thus

$$\Lambda = \frac{C_a}{2U}(\tan \beta_3 - \tan \beta_2) \tag{7.7}$$

The third dimensionless parameter often referred to in gas turbine design appears in both equations (7.6) and (7.7): it is the ratio C_a/U, called the *flow coefficient* ϕ. (It plays the same part as the blade speed ratio U/C_1 used by steam turbine designers.) Thus (7.6) and (7.7) can be written as

$$\psi = 2\phi(\tan \beta_2 + \tan \beta_3) \tag{7.8}$$

$$\Lambda = \frac{\phi}{2}(\tan \beta_3 - \tan \beta_2) \tag{7.9}$$

The gas angles can now be expressed in terms of ψ, Λ and ϕ as follows. Adding and subtracting (7.8) and (7.9) in turn we get

$$\tan \beta_3 = \frac{1}{2\phi}(\tfrac{1}{2}\psi + 2\Lambda) \tag{7.10}$$

$$\tan \beta_2 = \frac{1}{2\phi}(\tfrac{1}{2}\psi - 2\Lambda) \tag{7.11}$$

Then using relations (7.1),

$$\tan \alpha_3 = \tan \beta_3 - \frac{1}{\phi} \tag{7.12}$$

$$\tan \alpha_2 = \tan \beta_2 + \frac{1}{\phi} \tag{7.13}$$

Even with the restrictions we have already introduced (i.e. $C_{a3} = C_{a2}$ and $C_3 = C_1$), and remembering that stressing considerations will place a limit on the blade speed U, there is still an infinite choice facing the designer. For example, although the overall turbine temperature drop will be fixed by cycle calculations, it is open to the designer to choose one or two stages of large ψ or a larger number of smaller ψ. To limit still further our discussion at this point, we may observe that any turbine for a gas turbine power plant is essentially a low pressure ratio machine by steam turbine standards (e.g. in the region of 10:1 compared with over 1000:1

even for cycles operating with subcritical steam pressures). Thus there is little case for adopting impulse stages ($\Lambda = 0$) which find a place at the high-pressure end of steam turbines. Impulse designs are the most efficient type for that duty, because under such conditions the 'leakage losses' associated with rotor blade tip clearances are excessive in reaction stages. It must be remembered that at the high-pressure end of an expansion of large pressure ratio the stage pressure *differences* are considerable even though the stage pressure ratios are modest. Let us then rule out values of Λ near zero, and for the moment consider 50 per cent reaction designs. Our general knowledge of the way nature behaves would suggest that the most efficient design is likely to be achieved when the expansion is reasonably evenly divided between the stator and rotor rows. We shall see later that the reaction will vary from root to tip of the blade, but here we are thinking of 50 per cent reaction at the mean diameter.

Putting $\Lambda = 0.5$ in equation (7.9) we have

$$\frac{1}{\phi} = \tan \beta_3 - \tan \beta_2 \tag{7.14}$$

Direct comparison with relations (7.1) then shows that

$$\beta_3 = \alpha_2 \text{ and } \beta_3 = \alpha_3 \tag{7.15}$$

and the velocity diagram becomes symmetrical. Further, if we are considering a repeating stage with $C_3 = C_1$ in direction as well as magnitude, we have $\alpha_1 = \alpha_3 = \beta_2$ also, and the stator and rotor blades then have the same inlet and outlet angles. Finally, from (7.10) and (7.15) we have for $\Lambda = 0.5$,

$$\psi = 4\phi \tan \beta_3 - 2 = 4\phi \tan \alpha_2 - 2 \tag{7.16}$$

and from (7.11) and (7.15) we get

$$\psi = 4\phi \tan \beta_2 + 2 = 4\phi \tan \alpha_3 + 2 \tag{7.17}$$

Equations (7.15), (7.16) and (7.17) give all the gas angles in terms of ψ and ϕ. Figure 7.3 shows the result of plotting nozzle outlet angle α_2 and stage outlet swirl angle α_3 on a ψ–ϕ basis.

Because the blade shapes are determined within close limits by the gas angles, it is possible from results of cascade tests on families of blades to predict the losses in the blade rows and estimate the stage efficiency of the range of 50 per cent reaction designs covered by Fig. 7.3. One such estimate is shown by the efficiency contours superimposed on the ψ–ϕ plot. The values of η_s on the contours represent an average of detailed estimates quoted in Refs (1) and (2). Many assumptions have to be made about blade profile, blade aspect ratio (height/chord), tip clearance and so on, and no reliance should be placed upon the absolute values of efficiency shown. Nevertheless a knowledge of the general trend is valuable and even essential to the designer. Similar curves for other values of reaction are given in Ref. (2).

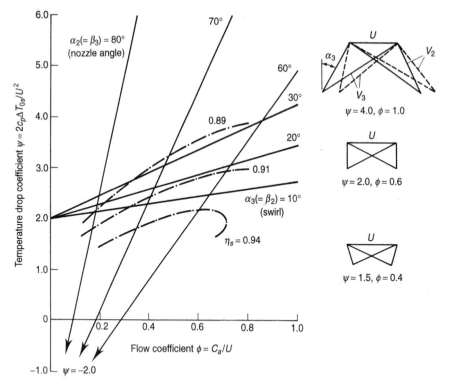

FIG. 7.3 50 per cent reaction designs

We may note that designs having a low ψ and low ϕ yield the best stage efficiencies. Referring to the comparative velocity diagrams also given in Fig. 7.3 (drawn for a constant blade speed U), we can see that low values of ψ and ϕ imply low gas velocities and hence reduced friction losses. But a low ψ means more stages for a given overall turbine output, and a low ϕ means a larger turbine annulus area for a given mass flow. For an industrial gas turbine when size and weight are of little consequence and a low *SFC* is vital, it would be sensible to design with a low ψ and low ϕ. Certainly in the last stage a low axial velocity and a small swirl angle α_3 are desirable to keep down the losses in the exhaust diffuser. For an aircraft propulsion unit, however, it is important to keep the weight and frontal area to a minimum, and this means using higher values of ψ and ϕ. The most efficient stage design is one which leads to the most efficient power plant for its particular purpose, and strictly speaking the optimum ψ and ϕ cannot be determined without detailed calculations of the performance of the aircraft as a whole. It would appear from current aircraft practice that the optimum values for ψ range from 3 to 5, with ϕ ranging from 0·8 to 1·0. A low swirl angle ($\alpha_3 < 20°$) is desirable because swirl increases the losses in the jet pipe and propelling nozzle; to maintain the required high value of ψ and low value of α_3 it might be necessary to use a degree of reaction somewhat less than 50 per cent. The dotted lines in the velocity diagram for $\psi = 4$ indicate what happens when the proportion of the expansion carried out in the rotor is reduced and V_3 becomes more equal to V_2, while maintaining U, ψ and ϕ constant.

We will close this section with a worked example showing how a first tentative 'mean-diameter' design may be arrived at. To do this we need some method of accounting for the losses in the blade rows. Two principal parameters are used, based upon temperature drops and pressure drops respectively. These parameters can best be described by sketching the processes in the nozzle and rotor blade passages on the T–s diagram as in Fig. 7.4. The full and dashed lines connect stagnation and static states respectively. $T_{02} = T_{01}$ because no work is done in the nozzles; and the short horizontal portion of the full line represents the stagnation pressure drop $(p_{01} - p_{02})$ due to friction in the nozzles. The losses are of course exaggerated in the figure. When obtaining the temperature equivalent of the velocity of the gas leaving the nozzle row, we may say that ideally the gas would be expanded from T_{01} to T_2' but that due to friction the temperature at the nozzle exit is T_2, somewhat higher than T_2'. The *loss coefficient for the nozzle blades* may be defined by either

$$\lambda_N = \frac{T_2 - T_2'}{C_2^2/2c_p} \text{ or } Y_N = \frac{p_{01} - p_{02}}{p_{02} - p_2} \tag{7.18}$$

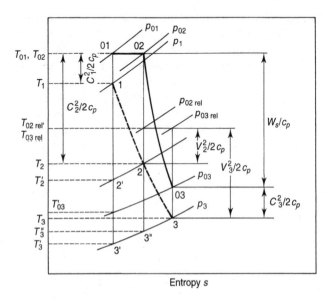

FIG. 7.4 *T–s diagram for a reaction stage*

Both λ and Y express the proportion of the *leaving* energy which is degraded by friction. Y_N can be measured relatively easily in cascade tests (the results being modified to allow for three-dimensional effects on the same lines as described for axial compressors), whereas λ_N is the more easily used in design. It will be shown that λ_N and Y_N are not very different numerically.

Returning to Fig. 7.4, we see that further expansion in the moving blade passages reduces the pressure to p_3. Isentropic expansion in the whole stage would result in a final temperature T_3', and in the rotor blade passages alone T_3''. Expansion with friction leads to a final temperature T_3. The *rotor blade loss* can be expressed by

$$\lambda_R = \frac{T_3 - T_3''}{V_3^2/2c_p}$$

Note that it is defined as a proportion of the leaving kinetic energy *relative* to the row so that it can be related to cascade test results. As no work is done by the gas relative to the blades, $T_{03\mathrm{rel}} = T_{02\mathrm{rel}}$. The *loss coefficient in terms of pressure drop for the rotor blades* is defined by

$$Y_R = \frac{p_{02\mathrm{rel}} - p_{03\mathrm{rel}}}{p_{03\mathrm{rel}} - p_3}$$

We may show that λ and Y are not very different numerically by the following argument (which applies equally to the stator and rotor rows of blades although given only for the former).

$$Y_N = \frac{p_{01} - p_{02}}{p_{02} - p_2} = \frac{(p_{01}/p_{02}) - 1}{1 - (p_2/p_{02})}$$

Now

$$\frac{p_{01}}{p_{02}} = \frac{p_{01}}{p_2}\frac{p_2}{p_{02}} = \left(\frac{T_{01}}{T_2'}\right)^{\gamma/(\gamma-1)}\left(\frac{T_2}{T_{02}}\right)^{\gamma/(\gamma-1)}$$

$$= \left(\frac{T_2}{T_2'}\right)^{\gamma/(\gamma-1)} \quad \text{because } T_{01} = T_{02}$$

Hence

$$Y_N = \frac{(T_2/T_2')^{\gamma/(\gamma-1)} - 1}{1 - (T_2/T_{02})^{\gamma/(\gamma-1)}}$$

$$= \frac{\left[1 + \dfrac{T_2 - T_2'}{T_2'}\right]^{\gamma/(\gamma-1)} - 1}{1 - \left[\dfrac{T_2 - T_{02}}{T_{02}} + 1\right]^{\gamma/(\gamma-1)}}$$

Expanding the bracketed expressions binomially and using the first terms only (although not very accurate for the denominator) we have

$$Y_N = \frac{T_2 - T_2'}{T_{02} - T_2} \times \frac{T_{02}}{T_2'} = \lambda_N \left(\frac{T_{02}}{T_2'} \right) \simeq \lambda_N \left(\frac{T_{02}}{T_2} \right) \tag{7.19}$$

From Appendix A, equation (8), we have the general relation

$$\frac{T_{02}}{T_2} = \left(1 + \frac{\gamma - 1}{2} M_2^2 \right)$$

Even if the Mach number at the blade exit is unity, as it might be for the nozzle blades of a highly loaded stage, $\lambda = 0.86 Y$ and thus λ is only 14 per cent less than Y.

The type of information which is available for predicting values of λ or Y is described briefly in section 7.4. λ_N and λ_R *can be related to the stage isentropic efficiency* η_s as follows:

$$\eta_s = \frac{T_{01} - T_{03}}{T_{01} - T_{03}'} = \frac{1}{1 + (T_{03} - T_{03}')/(T_{01} - T_{03})}$$

Now a glance at Fig. 7.4 shows that

$$T_{03} - T_{03}' \simeq (T_3 - T_3') = (T_3 - T_3'') + (T_3'' - T_3')$$

But $(T_2'/T_3') = (T_2/T_3'')$ because both equal $(p_2/p_3)^{(\gamma-1)/\gamma}$. Rearranging and subtracting 1 from both sides we get

$$\frac{T_3'' - T_3'}{T_3'} = \frac{(T_2 - T_2')}{T_2'} \quad \text{or} \quad (T_3'' - T_3') \simeq (T_2 - T_2') \frac{T_3}{T_2}$$

Hence

$$\eta_s \simeq \frac{1}{1 + [(T_3 - T_3'') + (T_3/T_2)(T_2 - T_2')]/(T_{01} - T_{03})}$$

$$\simeq \frac{1}{1 + [\lambda_R(V_3^2/2c_p) + (T_3/T_2)\lambda_N(C_2^2/2c_p)]/(T_{01} - T_{03})} \tag{7.20}$$

Alternatively, substituting $V_3 = C_a \sec \beta_3$, $C_2 = C_a \sec \alpha_2$, and

$$c_p(T_{01} - T_{03}) = UC_a(\tan \beta_3 + \tan \beta_2)$$
$$= UC_a[\tan \beta_3 + \tan \alpha_2 - (U/C_a)]$$

equation (7.20) can be written in the form

$$\eta_s \simeq \frac{1}{1 + \frac{1}{2} \frac{C_a}{U} \left[\frac{\lambda_R \sec^2 \beta_3 + (T_3/T_2)\lambda_N \sec^2 \alpha_2}{\tan \beta_3 + \tan \alpha_2 - (U/C_a)} \right]} \tag{7.21}$$

Because $Y \simeq \lambda$, loss coefficients Y_R and Y_N may replace λ_R and λ_N in equations (7.20) and (7.21) if desired.

For the purpose of the following example we shall assume that $\lambda_N = 0.05$ and $\eta_s = 0.9$. In suggesting $\lambda_N = 0.05$ we are assuming that convergent nozzles are employed and that they are operating with a pressure ratio (p_{01}/p_2) less than the critical pressure ratio $[(\gamma + 1)/2]^{\gamma/(\gamma-1)}$. Convergent–divergent nozzles as in Fig. 7.5(a) are not used, partly because they tend to be inefficient at pressure ratios other than the design value (i.e. at part load), and partly because high values of C_2 usually imply high values of V_2. If the Mach number relative to the moving blades at inlet, M_{V2}, exceeds about 0.75, additional losses may be incurred by the formation of shock waves in the rotor blade passages (see section A.1 of Appendix A). If V_2 is in fact not too high, perhaps because a low value of the flow coefficient ϕ is being used as in an industrial gas turbine, there is no reason why convergent nozzles should not be operated at pressure ratios giving an efflux velocity which is slightly supersonic (namely $1 < M_2 < 1.2$): very little additional loss seems to be incurred. The pressure at the throats of the nozzles is then the critical pressure, and there is further semi-controlled expansion to p_2 after the throat. As depicted in Fig. 7.5(b), the flow is controlled by the trailing edge on the convex side. On the other side the supersonic stream expands as though turning a corner: it is possible to obtain some idea of the deflection of the stream by treating it as a Prandtl–Meyer expansion and using the method of characteristics (see section A.8 of Appendix A, and Ref. (2) for further details).

As an example of turbine design procedure, we will now consider a possible 'mean-diameter' design for the turbine of a small, cheap, turbojet unit, which should be a single-stage turbine if possible. From cycle calculations the following design-point specification is proposed for the turbine.

Mass flow m	20 kg/s
Isentropic efficiency η_t	0.9
Inlet temperature T_{01}	1100 K
Temperature drop $T_{01}-T_{03}$	145 K
Pressure ratio p_{01}/p_{03}	1.873
Inlet pressure p_{01}	4 bar

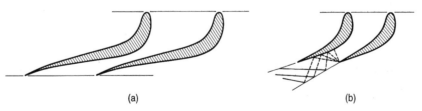

(a)   (b)

FIG. 7.5 Convergent–divergent nozzle, and convergent nozzle operating at a pressure ratio greater than the critical value

In addition to this information, we are likely to have the rotational speed fixed by the compressor, the design of which is always more critical than the turbine because of the decelerating flow.† Also, experience will suggest an upper limit to the blade speed above which stressing difficulties will be severe. Accordingly, we will assume

| Rotational speed N | 250 rev/s |
| Mean blade speed U | 340 m/s |

Finally, we shall assume a nozzle loss coefficient λ_N of 0·05 as a reasonable first guess.

We will start by assuming (a) $C_{a2} = C_{a3}$ and (b) $C_1 = C_3$. As it is to be a single-stage turbine the inlet velocity will be axial, i.e. $\alpha_1 = 0$. From the data, the temperature drop coefficient is

$$\psi = \frac{2c_p \Delta T_{0s}}{U^2} = \frac{2 \times 1 \cdot 148 \times 145 \times 10^3}{340^2} = 2 \cdot 88$$

This is a modest value and there is no difficulty about obtaining the required output from a single stage in a turbojet unit wherein high values of C_a can be used. We will try a flow coefficient ϕ of 0·8 and, because swirl increases the losses in the jet pipe, an α_3 of zero. The degree of reaction that these conditions imply can be found as follows, remembering that because of assumptions (a) and (b) equations (7.8) to (7.13) are applicable. From equation (7.12)

$$\tan \alpha_3 = 0 = \tan \beta_3 - \frac{1}{\phi}$$

$$\tan \beta_3 = 1 \cdot 25$$

From equation (7.10)

$$\tan \beta_3 = \frac{1}{2\phi}(\tfrac{1}{2}\psi + 2\Lambda)$$

$$1 \cdot 25 = \frac{1}{1 \cdot 6}(1 \cdot 44 + 2\Lambda)$$

$$\Lambda = 0 \cdot 28$$

We shall see from section 7.2 that when three-dimensional effects are included the reaction will increase from root to tip of the blades, and a degree of reaction of only 0·28 at the mean diameter might mean too low a value at the root. Negative values must certainly be avoided because this would imply expansion in the nozzles followed by recompression in the rotor and the losses would be large. Perhaps a modest amount of swirl will bring the reaction to a more reasonable value: we will try $\alpha_3 = 10°$.

---

† In practice the compressor and turbine designs interact at this point: a small increase in rotational speed to avoid the use of more than one stage in the turbine would normally be worthwhile even if it meant modifying the compressor design.

$$\tan \alpha_3 = 0\cdot1763; \tan \beta_3 = 0\cdot1763 + 1\cdot25 = 1\cdot426$$

$$1\cdot426 = \frac{1}{1\cdot6}(1\cdot44 + 2\Lambda)$$

$$\Lambda = 0\cdot421$$

This is acceptable: the reaction at the root will be checked when the example is continued in section 7.2. The gas angles can now be established. So far we have

$$\alpha_3 = 10°; \beta_3 = \tan^{-1} 1\cdot426 = 54\cdot96°$$

From equations (7.11) and (7.13)

$$\tan \beta_2 = \frac{1}{1\cdot6}(1\cdot44 - 0\cdot842) = 0\cdot3737$$

$$\tan \alpha_2 = 0\cdot3737 + \frac{1}{0\cdot8} = 1\cdot624$$

$$\beta_2 = 20\cdot49°; \alpha_2 = 58\cdot38°$$

The velocity diagram can now be sketched as in Fig. 7.6, and the next task is to calculate the density at stations 1, 2 and 3 so that the blade height h and tip/root radius ratio (r_t/r_r) can be estimated. We shall commence with station 2 because some modifications will be required if the pressure ratio p_{01}/p_2 across the convergent nozzles is much above the critical value, or if the Mach number relative to the rotor blades at inlet (M_{V2}) exceeds about 0·75.

From the geometry of the velocity diagram,

$$C_{a2} = U\phi = 340 \times 0\cdot8 = 272 \text{ m/s}$$

$$C_2 = \frac{C_{a2}}{\cos \alpha_2} = \frac{272}{0\cdot5242} = 519 \text{ m/s}$$

The temperature equivalent of the outlet velocity is

$$T_{02} - T_2 = \frac{C_2^2}{2c_p} = \frac{519^2}{2296} = 117\cdot3 \text{ K}$$

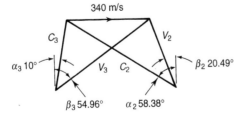

FIG. 7.6

Since $T_{02} = T_{01} = 1100\,\text{K}$, $T_2 = 982\cdot7\,\text{K}$ and

$$T_2 - T_2' = \lambda_N \frac{C_2^2}{2c_p} = 0\cdot05 \times 117\cdot3 = 5\cdot9\,\text{K}$$

$$T_2' = 982\cdot7 - 5\cdot9 = 976\cdot8\,\text{K}$$

p_2 can be found from the isentropic relation

$$\frac{p_{01}}{p_2} = \left(\frac{T_{01}}{T_2'}\right)^{\gamma/(\gamma-1)} = \left(\frac{1100}{976\cdot8}\right)^4 = 1\cdot607$$

$$p_2 = \frac{4\cdot0}{1\cdot607} = 2\cdot49\,\text{bar}$$

Ignoring the effect of friction on the critical pressure ratio, and putting $\gamma = 1\cdot333$ in equation (A.12) we have

$$\frac{p_{01}}{p_c} = \left(\frac{\gamma+1}{2}\right)^{\gamma/(\gamma-1)} = 1\cdot853$$

The actual pressure ratio is $1\cdot607$, well below the critical value. The nozzles are not choking and the pressure in the plane of the throat is equal to p_2.

$$\rho_2 = \frac{p_2}{RT_2} = \frac{100 \times 2\cdot49}{0\cdot287 \times 982\cdot7} = 0\cdot883\,\text{kg/m}^3$$

Annulus area at plane 2 is

$$A_2 = \frac{m}{\rho_2 C_{a2}} = \frac{20}{0\cdot883 \times 272} = 0\cdot0833\,\text{m}^2$$

Throat area of nozzles required is

$$A_{2N} = \frac{m}{\rho_2 C_2} \quad \text{or} \quad A_2 \cos\alpha_2 = 0\cdot0883 \times 0\cdot524 = 0\cdot0437\,\text{m}^2$$

Note that if the pressure ratio had been slightly above the critical value it would be acceptable if a check (given later) on M_{V2} proved satisfactory. ρ_2 and A_2 would be unchanged, but the throat area would then be given by $m/\rho_c C_c$, where ρ_c is obtained from p_c and T_c, and C_c corresponds to a Mach number of unity so that it can be found from $\sqrt{(\gamma R T_c)}$.

We may now calculate the annulus area required in planes 1 and 3 as follows. Because it is not a repeating stage, we are assuming that C_1 is axial and this, together with assumptions (a) and (b) that $C_1 = C_3$ and $C_{a3} = C_{a2}$, yields

$$C_{a1} = C_1 = C_3 = \frac{C_{a3}}{\cos\alpha_3} = \frac{272}{\cos 10°} = 276\cdot4\,\text{m/s}$$

Temperature equivalent of the inlet (and outlet) kinetic energy is

$$\frac{C_1^2}{2c_p} = \frac{276 \cdot 4^2}{2296} = 33 \cdot 3 \, \text{K}$$

$$T_1 = T_{01} - \frac{C_1^2}{2c_p} = 1100 - 33 \cdot 3 = 1067 \, \text{K}$$

$$\frac{p_1}{p_{01}} = \left(\frac{T_1}{T_{01}}\right)^{\gamma/(\gamma-1)} \quad \text{or } p_1 = \frac{4 \cdot 0}{(1100/1067)^4} = 3 \cdot 54 \, \text{bar}$$

$$\rho_1 = \frac{100 \times 3 \cdot 54}{0 \cdot 287 \times 1067} = 1 \cdot 155 \, \text{kg/m}^3$$

$$A_1 = \frac{m}{\rho_1 C_{a1}} = \frac{20}{1 \cdot 155 \times 276 \cdot 4} = 0 \cdot 0626 \, \text{m}^2$$

Similarly, at outlet from the stage we have

$$T_{03} = T_{01} - \Delta T_{0s} = 1100 - 145 = 955 \, \text{K}$$

$$T_3 = T_{03} - \frac{C_3^2}{2c_p} = 955 - 33 \cdot 3 = 922 \, \text{K}$$

p_{03} is given in the data by $p_{01}(p_{03}/p_{01})$ and hence

$$p_3 = p_{03}\left(\frac{T_3}{T_{03}}\right)^{\gamma/(\gamma-1)} = \left(\frac{4}{1 \cdot 873}\right)\left(\frac{922}{955}\right)^4 = 1 \cdot 856 \, \text{bar}$$

$$\rho_3 = \frac{100 \times 1 \cdot 856}{0 \cdot 287 \times 922} = 0 \cdot 702 \, \text{kg/m}^3$$

$$A_3 = \frac{m}{\rho_3 C_{a3}} = \frac{20}{0 \cdot 702 \times 272} = 0 \cdot 1047 \, \text{m}^2$$

The blade height and annulus radius ratio at stations 1, 2 and 3 can now be established. At the mean diameter, which we shall now begin to emphasize by the use of suffix m,

$$U_m = 2\pi N r_m, \text{ so that } r_m = \frac{340}{2\pi 250} = 0 \cdot 216 \, \text{m}$$

Since the annulus area is given by

$$A = 2\pi r_m h = \frac{U_m h}{N}$$

the height and radius ratio of the annulus can be found from

$$h = \frac{AN}{U_m} = \left(\frac{250}{340}\right)A, \text{ and } \frac{r_t}{r_r} = \frac{r_m + (h/2)}{r_m - (h/2)}$$

and the results are as given in the following table.

Station	1	2	3
$A/[\text{m}^2]$	0·0626	0·0833	0·1047
$h/[\text{m}]$	0·046	0·0612	0·077
r_t/r_r	1·24	1·33	1·43

Although we shall leave the discussion of the effects of high and low annulus radius ratio to a later section, we may note here that values in the region of 1·2–1·4 would be regarded as satisfactory. If the rotational speed, which we have assumed to be fixed by the compressor, had led to an ill-proportioned annulus, it would be necessary to rework this preliminary design. For example, r_t/r_r could be reduced by increasing the axial velocity, i.e. by using a higher value of the flow coefficient ϕ. This would also increase the nozzle efflux velocity, but we noted that it was comfortably subsonic and so could be increased if necessary.

The turbine annulus we have arrived at in this example is flared as shown in Fig. 7.7. In sketching this we have assumed a blade height/width ratio of about 3·0 and a space between the stator and rotor blades of about 0·25 of the blade width. The included angle of divergence of the walls then becomes approximately 29°. This might be regarded as rather high, involving a risk of flow separation from the inner wall where the reaction, and therefore the acceleration, is not large; 25° has been suggested as a safe limit, Ref. (3). We shall not pause for adjustment here, because the blade height/width ratio of 3·0 is merely a rough guess to be justified or altered later when the effect of blade stresses on the design has to be considered. Furthermore, the choice of 0·25 for the space/blade width ratio is rather low: a low value is desirable only to reduce the axial length and weight of the turbine. Vibrational stresses are induced in the rotor blades as they pass through the wakes of the nozzle blades, and these stresses increase sharply with decrease in axial space between the blade rows. 0·2 is the lowest value of space/blade width ratio considered to be safe, but a value nearer 0·5 is often used and this would reduce both the vibrational stresses and the annulus flare.

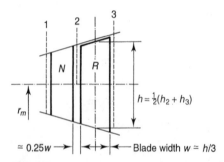

FIG. 7.7

If it were thought desirable to reduce the flare without increasing the axial length of the turbine, then it would be necessary to repeat the calculations allowing the axial velocity to increase through the stage. It would be necessary to check the Mach number at exit from the stage, M_3, because if this is too high the friction losses in the jet pipe become unduly large. For the present design we have

$$M_3 = \frac{C_3}{\sqrt{(\gamma R T_3)}} = \frac{276 \cdot 4}{\sqrt{(1 \cdot 333 \times 0 \cdot 287 \times 922 \times 1000)}} = 0 \cdot 47$$

This could be safely increased to reduce the flare if desired. (It must be remembered that most of the equations we have derived, and in particular (7.8)–(7.13) relating ψ, ϕ and the gas angles, are not valid when $C_{a3} \neq C_{a2}$. It is necessary to revert to first principles or derive new relations. For example, relations (7.1) become

$$\frac{U}{C_{a2}} = \tan \alpha_2 - \tan \beta_2 \quad \text{and} \quad \frac{U}{C_{a3}} = \tan \beta_3 - \tan \alpha_3$$

and the temperature drop coefficient becomes

$$\psi = \frac{2 c_p \Delta T_{0s}}{U^2} = \frac{2}{U}(C_{a2} \tan \alpha_2 + C_{a3} \tan \alpha_3)$$

$$= \frac{2 C_{a2}}{U} \left(\tan \beta_2 + \frac{C_{a3}}{C_{a2}} \tan \beta_3 \right)$$

Additionally, if the flare is not symmetrical U must be replaced by U_{m2} and U_{m3} as appropriate.)

Finally, for this preliminary design we have taken losses into account via λ_N and η_s rather than λ_N and λ_R. The value of λ_R implied by the design can be found by determining $(T_3 - T_3'')$. Thus

$$\frac{T_2}{T_3''} = \left(\frac{p_2}{p_3} \right)^{(\gamma-1)/\gamma} \quad \text{or} \quad T_3'' = \frac{982 \cdot 7}{(2 \cdot 49 / 1 \cdot 856)^{\frac{1}{4}}} = 913 \text{ K}$$

We also require the temperature equivalent of the outlet kinetic energy relative to the blading.

$$V_3 = \frac{C_{a3}}{\cos \beta_3} = \frac{272}{\cos 54 \cdot 96°} = 473 \cdot 5 \text{ m/s}$$

$$\frac{V_3^2}{2 c_p} = \frac{473 \cdot 5^2}{2296} = 97 \cdot 8 \text{ K}$$

Then

$$\lambda_R = \frac{T_3 - T_3''}{V_3^2 / 2 c_p} = \frac{922 - 913}{97 \cdot 8} = 0 \cdot 092$$

Had we used the approximate relation between η_s, λ_N and λ_R, equation (7.20), we would have found λ_R to be 0·108 (which is a useful check on the arithmetic). Note that $\lambda_R > \lambda_N$, which it should be by virtue of the tip leakage loss in the rotor blades.

The next steps in the design are

(a) to consider the three-dimensional nature of the flow in so far as it affects the variation of the gas angles with radius;

(b) to consider the blade shapes necessary to achieve the required gas angles, and the effect of centrifugal and gas bending stresses on the design;

(c) to check the design by estimating λ_N and λ_R from the results of cascade tests suitably modified to take account of three-dimensional flows.

7.2   Vortex theory

Early in the previous section it was pointed out that the shape of the velocity triangles must vary from root to tip of the blade because the blade speed U increases with radius. Another reason is that the whirl component in the flow at outlet from the nozzles causes the static pressure and temperature to vary across the annulus. With a uniform pressure at inlet, or at least with a much smaller variation because the whirl component is smaller, it is clear that the pressure drop across the nozzle will vary, giving rise to a corresponding variation in efflux velocity C_2. Twisted blading designed to take account of the changing gas angles is called *vortex blading*.

It has been common steam turbine practice, except in low-pressure blading where the blades are very long, to design on conditions at the mean diameter, keep the blade angles constant from root to tip, and assume that no additional loss is incurred by the variation in incidence along the blade caused by the changing gas angles. Comparative tests, Ref. (4), have been conducted on a single-stage gas turbine of radius ratio 1·37, using in turn blades of constant angle and vortex blading. The results showed that any improvement in efficiency obtained with vortex blading was within the margin of experimental error. This contrasts with similar tests on a six-stage axial compressor, Ref. (5), which showed a distinct improvement from the use of vortex blading. This was, however, an improvement not so much in efficiency (of about 1·5 per cent) as in the delay of the onset of surging which of course does not arise in accelerating flow. It appears, therefore, that steam turbine designers have been correct in not applying vortex theory except when absolutely necessary at the LP end. They have to consider the additional cost of twisted blades for the very large number of rows of blading required, and they know that the Rankine cycle is relatively insensitive to component losses. Conversely, it is not surprising that the gas turbine designer, struggling to achieve the highest possible component efficiency, has consistently used some form of vortex blading which it is felt intuitively must give a better performance, however small.

Vortex theory has been outlined in section 5.4 where it was shown that if the elements of fluid are to be in radial equilibrium, an increase in static pressure from root to tip is necessary whenever there is a whirl component of velocity. Figure 7.8 shows why the gas turbine designer cannot talk of impulse or 50 per cent reaction stages. The proportion of the stage pressure or temperature drop which occurs in the rotor must increase from root to tip. Although Fig. 7.8 refers to a single-stage turbine with axial inlet velocity and no swirl at outlet, the whirl component at inlet and outlet of a repeating stage will be small compared with C_{w2}: the reaction will therefore still increase from root to tip, if somewhat less markedly.

Free vortex design

Referring to section 5.6, it was shown that if

(a) the stagnation enthalpy h_0 is constant over the annulus (i.e. $dh_0/dr = 0$),
(b) the axial velocity is constant over the annulus,
(c) the whirl velocity is inversely proportional to the radius,

then the condition for radial equilibrium of the fluid elements, namely equation (5.13), is satisfied. A stage designed in accordance with (a), (b) and (c) is called a *free vortex stage*. Applying this to the stage in Fig. 7.8, we can see that with uniform inlet conditions to the nozzles then, since no work is done by the gas in the nozzles, h_0 must also be constant over the annulus at outlet. Thus condition (a) is fulfilled in the space between the nozzles and rotor blades. Furthermore, if the nozzles are designed to give $C_{a2} = \text{constant}$ and $C_{w2}r = \text{constant}$, all three conditions are fulfilled and the condition for radial equilibrium is satisfied in plane 2. Similarly, if the

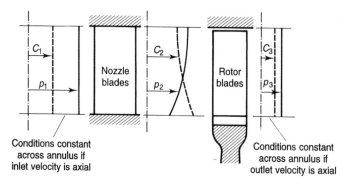

FIG. 7.8 **Changes in pressure and velocity across the annulus**

rotor blades are designed so that C_{a3} = constant and $C_{w3}r$ = constant, it is easy to show as follows that condition (a) will be fulfilled, and thus radial equilibrium will be achieved in plane 3 also. Writing ω for the angular velocity we have

$$W_s = U(C_{w2} + C_{w3}) = \omega(C_{w2}r + C_{w3}r) = \text{constant}$$

But when the work done per unit mass of gas is constant over the annulus, and h_0 is constant at inlet, h_0 must be constant at outlet also: thus condition (a) is met.

It is apparent that a free vortex design is one in which the work done per unit mass of gas is constant over the annulus, and to obtain the total work output this specific value need only be calculated at one convenient radius and multiplied by the mass flow.

In contrast, we may note that because the density varies from root to tip at exit from the nozzles and the axial velocity is constant, an integration over the annulus will be necessary if the continuity equation is to be used in plane 2. Thus, considering a flow δm through an annular element of radius r and width δr,

$$\delta m = \rho_2 2\pi r \delta r C_{a2}$$

$$m = 2\pi C_{a2} \int_{r_r}^{r_t} \rho_2 r \, dr \tag{7.22}$$

With the radial variation of density determined from vortex theory, the integration can be performed although the algebra is lengthy. For detailed calculations it would be normal to use a digital computer, permitting ready calculation of the density at a series of radii and numerical integration of equation (7.22) to obtain the mass flow. For preliminary calculations, however, it is sufficiently accurate to take the intensity of mass flow at the mean diameter as being the mean intensity of mass flow. In other words, the total mass flow is equal to the mass flow per unit area calculated using the density at the mean diameter $(\rho_{2m}C_{a2})$ multiplied by the annulus area (A_2). This is one reason why it is convenient to design the turbine on conditions at mean diameter (as was done in the previous example) and use the relations which will now be derived for obtaining the gas angles at other radii.

Using suffix m to denote quantities at mean diameter, the free vortex variation of nozzle angle α_2 may be found as follows:

$$C_{w2}r = rC_{a2}\tan\alpha_2 = \text{constant}$$

$$C_{a2} = \text{constant}$$

Hence α_2 at any radius r is related to α_{2m} at the mean radius r_m by

$$\tan\alpha_2 = \left(\frac{r_m}{r}\right)_2 \tan\alpha_{2m} \tag{7.23}$$

Similarly, when there is swirl at outlet from the stage,

$$\tan \alpha_3 = \left(\frac{r_m}{r}\right)_3 \tan \alpha_{3m} \tag{7.24}$$

The gas angles at inlet to the rotor blade, β_2, can then be found using equation (7.1), namely

$$\tan \beta_2 = \tan \alpha_2 - \frac{U}{C_{a2}}$$

$$= \left(\frac{r_m}{r}\right)_2 \tan \alpha_{2m} - \left(\frac{r}{r_m}\right)_2 \frac{U_m}{C_{a2}} \tag{7.25}$$

and similarly β_3 is given by

$$\tan \beta_3 = \left(\frac{r_m}{r}\right)_3 \tan \alpha_{3m} + \left(\frac{r}{r_m}\right)_3 \frac{U_m}{C_{a3}} \tag{7.26}$$

To obtain some idea of the free vortex variation of gas angles with radius, equations (7.23)–(7.26) will be applied to the turbine designed in the previous section. We will merely calculate the angles at the root and tip, although in practice they would be determined at several stations up the blade to define the twist more precisely. We will at the same time clear up two loose ends: we have to check that there is some positive reaction at the root radius, and that the Mach number relative to the rotor blade at inlet, M_{V2}, is nowhere higher than say 0·75. From the velocity triangles at root and tip it will be seen that this Mach number is greatest at the root and it is only at this radius that it need be calculated.

From the mean-diameter calculation we found that

$$\alpha_{2m} = 58\cdot38°, \ \beta_{2m} = 20\cdot49°, \ \alpha_{3m} = 10°, \ \beta_{3m} = 54\cdot96°$$

From the calculated values of h and r_m we have $r_r = r_m - (h/2)$ and $r_t = r_m + (h/2)$, and thus

$$\left(\frac{r_m}{r_r}\right)_2 = 1\cdot164, \quad \left(\frac{r_m}{r_t}\right)_2 = 0\cdot877, \quad \left(\frac{r_m}{r_r}\right)_3 = 1\cdot217, \quad \left(\frac{r_m}{r_t}\right)_3 = 0\cdot849$$

Also

$$\frac{U_m}{C_{a2}} = \frac{U_m}{C_{a3}} = \frac{1}{\phi} = 1\cdot25$$

Applying equations (7.23)–(7.26) we get

	α_2	β_2	α_3	β_3
Tip	54·93°	0	8·52°	58·33°
Root	62·15°	39·32°	12·12°	51·13°

The variation of gas angles with radius appears as in Fig. 7.9, which also includes the velocity triangles at root and tip drawn to scale. That

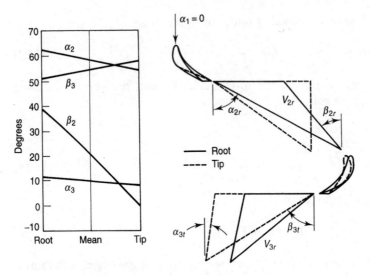

FIG. 7.9 Variation of gas angles with radius

$M_{V2} = V_2/\sqrt{(\gamma RT_2)}$ is greatest at the root is clear from the velocity triangles: V_2 is then a maximum, and $\sqrt{(\gamma RT_2)}$ is a minimum because the temperature drop across the nozzles is greatest at the root. That there is some positive reaction at the root is also clear because $V_{3r} > V_{2r}$. Although there is no need literally to calculate the degree of reaction at the root, we must calculate $(M_{V2})_r$ to ensure that the design implies a safe value. Using data from the example in section 7.1 we have

$$V_{2r} = C_{a2} \sec \beta_{2r} = 272 \sec 39\cdot32° = 352 \text{ m/s}$$

$$C_{2r} = C_{a2} \sec \alpha_{2r} = 272 \sec 62\cdot15° = 583 \text{ m/s}$$

$$T_{2r} = T_{02} - \frac{C_{2r}^2}{2c_p} = 1100 - \frac{583^2}{2294} = 952 \text{ K}$$

$$(M_{V2})_r = \frac{V_{2r}}{\sqrt{(\gamma RT_{2r})}} = \frac{352}{\sqrt{(1\cdot333 \times 0\cdot287 \times 952 \times 1000)}} = 0\cdot58$$

This is a modest value and certainly from this point of view a higher value of the flow coefficient ϕ could safely have been used in the design, perhaps instead of introducing swirl at exit from the stage.

Constant nozzle angle design

As in the case of the axial compressor, it is not essential to design for free vortex flow. Conditions other than constant C_a and $C_w r$ may be used to give some other form of vortex flow, which can still satisfy the requirement for radial equilibrium of the fluid elements. In particular, it may be desirable to make a constant nozzle angle one of the conditions determining the type of vortex, to avoid having to manufacture nozzles of varying outlet angle. This, as will now be shown, requires particular variations of C_a and C_w.

The vortex flow equation (5.15) states that

$$C_a \frac{dC_a}{dr} + C_w \frac{dC_w}{dr} + \frac{C_w^2}{r} = \frac{dh_0}{dr}$$

Consider the flow in the space between the nozzles and blades. As before, we assume that the flow is uniform across the annulus at inlet to the nozzles, and so the stagnation enthalpy at outlet must also be uniform, i.e. $dh_0/dr = 0$ in plane 2. Also, if α_2 is to be constant we have

$$\frac{C_{a2}}{C_{w2}} = \cot \alpha_2 = \text{constant}$$

$$\frac{dC_{a2}}{dr} = \frac{dC_{w2}}{dr} \cot \alpha_2$$

The vortex flow equation therefore becomes

$$C_{w2} \cot^2 \alpha_2 \frac{dC_{w2}}{dr} + C_{w2} \frac{dC_{w2}}{dr} + \frac{C_{w2}^2}{r} = 0 \qquad (7.27)$$

$$(1 + \cot^2 \alpha_2) \frac{dC_{w2}}{dr} + \frac{C_{w2}}{r} = 0$$

$$\frac{dC_{w2}}{C_{w2}} = -\sin^2 \alpha_2 \frac{dr}{r}$$

Integrating this gives

$$C_{w2} r^{\sin^2 \alpha_2} = \text{constant} \qquad (7.28)$$

And with constant α_2, $C_{a2} \propto C_{w2}$ so that the variation of C_{a2} must be the same, namely

$$C_{a2} r^{\sin^2 \alpha_2} = \text{constant} \qquad (7.29)$$

Normally nozzle angles are greater than $60°$, and quite a good approximation to the flow satisfying the equilibrium condition is obtained by designing with a constant nozzle angle and constant angular momentum, i.e. $\alpha_2 = $ constant and $C_{w2} r = $ constant. If this approximation is made and the rotor blades are twisted to give constant angular momentum at outlet also, then, as for free vortex flow, the work output per unit mass flow is the same at all radii. On the other hand, if equation (7.28) were used it would be necessary to integrate from root to tip to obtain the work output. We observed early in section 7.2 that there is little difference in efficiency between turbines of low radius ratio designed with twisted and untwisted blading. It follows that the sort of approximation referred to here is certainly unlikely to result in a significant deterioration of performance.

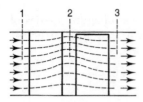

FIG. 7.10

The free vortex and constant nozzle angle types of design do not exhaust the possibilities. For example, one other type of vortex design aims to satisfy the radial equilibrium condition and at the same time meet a condition of *constant mass flow per unit area* at all radii. That is, the axial and whirl velocity distributions are chosen so that the product $\rho_2 C_{a2}$ is constant at all radii. The advocates of this approach correctly point out that the simple vortex theory outlined in section 5.6 assumes no radial component of velocity, and yet even if the turbine is designed with no flare there must be a radial shift of the streamlines as shown in Fig. 7.10. This shift is due to the increase in density from root to tip in plane 2. The assumption that the radial component is zero would undoubtedly be true for a turbine of constant annulus area if the stage were designed for constant mass flow per unit area. It is argued that the flow is then more likely to behave as intended, so that the gas angles will more closely match the blade angles. Further details can be found in Ref. (2). In view of what has just been said about the dubious benefits of vortex blading for turbines of modest radius ratio, it is very doubtful indeed whether such refinements are more than an academic exercise.

7.3 Choice of blade profile, pitch and chord

So far in our worked example we have shown how to establish the gas angles at all radii and blade heights. The next step is to choose stator and rotor blade shapes which will accept the gas incident upon the leading edge, and deflect the gas through the required angle with the minimum loss. An overall blade loss coefficient Y (or λ) must account for the following sources of friction loss.

(a) *Profile loss*—associated with boundary layer growth over the blade profile (including separation loss under adverse conditions of extreme angles of incidence or high inlet Mach number).

(b) *Annulus loss*—associated with boundary layer growth on the inner and outer walls of the annulus.

(c) *Secondary flow loss*—arising from secondary flows which are always present when a wall boundary layer is turned through an angle by an adjacent curved surface.

(d) *Tip clearance loss*—near the rotor blade tip the gas does not follow the intended path, fails to contribute its quota of work output, and interacts with the outer wall boundary layer.

The profile loss coefficient Y_p is measured directly in cascade tests similar to those described for compressor blading in section 5.8. Losses (b) and (c) cannot easily be separated, and they are accounted for by a secondary loss coefficient Y_s. The tip clearance loss coefficient, which normally arises only for rotor blades, will be denoted by Y_k. Thus the total loss coefficient Y comprises the accurately measured two-dimensional loss Y_p, plus the three-dimensional loss $(Y_s + Y_k)$ which must be deduced from turbine stage test results. A description of one important compilation of such data will be given in section 7.4; all that is necessary for our present purpose is a knowledge of the sources of loss.

Conventional blading

Figure 7.11 shows a conventional steam turbine blade profile constructed from circular arcs and straight lines. Gas turbines have until recently used profiles closely resembling this, although specified by aerofoil terminology. One example is shown: the T6 *base profile* which is symmetrical about the centre line. It has a thickness/chord ratio (t/c) of 0·1, a leading edge radius of 12 per cent t and a trailing edge radius of 6 per cent t. When scaled up to a t/c of 0·2 and used in conjunction with a parabolic camber line having

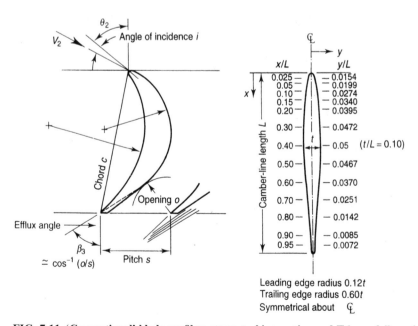

x/L	y/L
0.025	0.0154
0.05	0.0199
0.10	0.0274
0.15	0.0340
0.20	0.0395
0.30	0.0472
0.40	0.05 ($t/L = 0.10$)
0.50	0.0467
0.60	0.0370
0.70	0.0251
0.80	0.0142
0.90	0.0085
0.95	0.0072

Leading edge radius 0.12t
Trailing edge radius 0.60t
Symmetrical about ₵

FIG. 7.11 'Conventional' blade profiles: steam turbine section and T6 aerofoil section

the point of maximum camber a distance of about 40 per cent c from the leading edge, the T6 profile leads to a blade section similar to that shown but with a radiused trailing edge. In particular, the back of the blade after the throat is virtually straight. Other shapes used in British practice have been RAF 27 and C7 base profiles on both circular and parabolic arc camber lines. All such blading may be referred to as *conventional blading*.

It is important to remember that the velocity triangles yield the *gas* angles, not the *blade* angles. Typical cascade results showing the effect of incidence on the profile loss coefficient Y_p for impulse ($\Lambda = 0$ and $\beta_2 \simeq \beta_3$) and reaction type blading are given in Fig. 7.12. Evidently, with reaction blading the angle of incidence can vary from $-15°$ to $+15°$ without increase in Y_p. The picture is not very different even when three-dimensional losses are taken into account. This means that a rotor blade could be designed to have an inlet angle θ_2 equal to say $(\beta_{2r} - 5°)$ at the root and $(\beta_{2t} + 10°)$ at the tip to reduce the twist required by a vortex design. It must be remembered, however, that a substantial margin of safe incidence range must be left to cope with part-load operating conditions of pressure ratio, mass flow and rotational speed.

With regard to the outlet angle, it has been common steam turbine practice to take the gas angle as being equal to the blade angle defined by $\cos^{-1}$ (opening/pitch). This takes account of the bending of the flow as it fills up the narrow space in the wake of the trailing edge; there is no 'deviation' in the sense of that obtained with decelerating flow in a compressor cascade. Tests on gas turbine cascades have shown, however, that the $\cos^{-1}$ (o/s) rule is an over-correction for blades of small outlet angle operating with low gas velocities, i.e for some rotor blades. Figure 7.13 shows the relation between the relative gas outlet angle, β_3 say, and the blade angle defined by $\cos^{-1}$ (o/s). The relation does not seem to be affected by incidence within the working range of $\pm 15°$. This curve is applicable to 'straight-backed' conventional blades operating with a

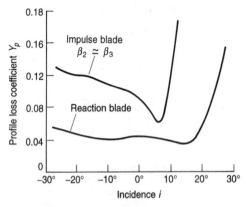

FIG. 7.12 Effect of incidence upon Y_p

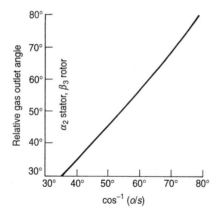

FIG. 7.13 Relation between gas and blade outlet angles

relative outlet Mach number below 0·5. With an exit Mach number of unity the $\cos^{-1}(o/s)$ rule is reasonable for all blade outlet angles, again for straight-backed blades. At Mach numbers intermediate between 0·5 and 1·0, the outlet angle can be assumed to vary linearly between β_3, as given by Fig. 7.13, and $\cos^{-1}(o/s)$.

Modern turbine blades are usually not straight-backed. The suction surface frequently has as much as 12° of 'unguided' or 'uncovered' turning from the throat to the trailing edge and this is reflected in the outlet flow angle. For outlet Mach numbers below 0·5, Ainley and Mathieson in Ref. (3) suggest that β_3 is increased by $4(s/e)$ where e is the radius of curvature of the aft suction-side of the blade. At an exit Mach number of 1·0, the outlet angle is given by $\cos^{-1}(o/s) + f(s/e)\sin^{-1}(o/s)$ where the function f was presented graphically. The function can be approximated by

$$f(s/e) = \frac{0{\cdot}0541(s/e)}{1 - 1{\cdot}49(s/e) + 0{\cdot}742(s/e)^2}$$

At intermediate Mach numbers, linear interpolation is again used.

Recently, Islam and Sjolander [Ref. 6)] compared the Ainley and Mathieson correlations with the outlet flow angles observed for turbine blades of modern design. They concluded that the earlier correlations tend to underestimate the flow-turning ability of modern blades and provide an improved correlation for use at outlet Mach numbers up to about 0·7.

Note that until the *pitch* and *chord* have been established it is not possible to draw a blade section to scale, determine the 'opening', and proceed by trial and error to make adjustments until the required gas outlet angle α_2 or β_3 is obtained. Furthermore, this process must be carried out at a number of radii from root to tip to specify the shape of the blade as a whole. Now the pitch and chord have to be chosen with due regard to (*a*) the effect of the pitch/chord ratio (s/c) on the blade loss coefficient, (*b*) the effect of chord upon the aspect ratio (h/c), remembering that h has already been determined, (c) the effect of rotor blade chord on the blade stresses, and (d) the effect of rotor blade pitch upon the stresses at the point of attachment of the blades to the turbine disc. We will consider each effect in turn.

(a)   'Optimum' pitch/chord ratio

In section 7.4 (Fig. 7.24) are presented cascade data on profile loss coefficients Y_p, and from such data it is possible to obtain the useful design curves in Fig. 7.14. These curves suggest, as might be expected, that the greater the gas deflection required ($(\alpha_1 + \alpha_2)$ for a stator blade and $(\beta_2 + \beta_3)$ for a rotor blade) the smaller must be the 'optimum' s/c ratio to control the gas adequately. The adjective 'optimum' is in inverted commas because it is an optimum with respect to Y_p, not to the overall loss Y. The true optimum value of s/c could be found only by making a detailed estimate of stage performance (e.g. on the lines described in section 7.4) for several stage designs differing in s/c but otherwise similar. In fact the s/c value is not very critical.

For the nozzle and rotor blade of our example turbine we have established that

$$\alpha_{1m} = 0°, \ \alpha_{2m} = 58\cdot38°; \ \beta_{2m} = 20\cdot49°, \ \beta_{3m} = 54\cdot96°$$

From Fig. 7.14 we therefore have at mean diameter

$$(s/c)_N = 0\cdot86 \text{ and } (s/c)_R = 0\cdot83$$

(b)   Aspect ratio (h/c)

The influence of aspect ratio is open to conjecture, but for our purpose it is sufficient to note that, although not critical, too low a value is likely to

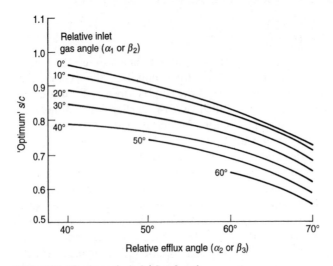

FIG. 7.14 'Optimum' pitch/chord ratio

lead to secondary flow and tip clearance effects occupying an unduly large proportion of the blade height and so increasing Y_s for the nozzle row and $(Y_s + Y_k)$ for the rotor row. On the other hand, too high a value of h/c will increase the likelihood of vibration trouble: vibration characteristics are difficult to predict and depend on the damping provided by the method of attaching the blades to the turbine disc.[†] A value of h/c between 3 and 4 would certainly be very satisfactory, and it would be unwise to use a value below 1.

For our turbine, which is flared, we have the mean heights of the nozzle and rotor blades given by

$$h_N = \tfrac{1}{2}(0.046 + 0.0612) = 0.0536 \, \text{m}$$
$$h_R = \tfrac{1}{2}(0.0612 + 0.077) = 0.0691 \, \text{m}$$

Adopting an aspect ratio (h/c) of 3 we then have

$$c_N = 0.0175 \, \text{m and } c_R = 0.023 \, \text{m}$$

Using these values of chord, in conjunction with the chosen s/c values, gives the blade pitches at the mean radius of $0.216 \, \text{m}$ as

$$s_N = 0.015\,06 \, \text{m and } s_R = 0.0191 \, \text{m}$$

and the numbers of blades, from $2\pi r_m/s$, as

$$n_N = 90 \text{ and } n_R = 71$$

It is usual to avoid numbers with common multiples to reduce the probability of introducing resonant forcing frequencies. A common practice is to use an *even* number for the nozzle blades and a *prime* number for the rotor blades. As it happens the foregoing numbers are satisfactory and there is no need to modify them and re-evaluate the pitch s.

(c) Rotor blade stresses

The next step is to check that the stage design is consistent with a permissible level of stress in the rotor blades. The final design must be checked by laying out the blade cross-sections at several radii between root and tip, and performing an accurate stress analysis on the lines indicated by Sternlicht in Ref. (7). Although we are not concerned with mechanical design problems in this book, simple approximate methods adequate for preliminary design calculations must be mentioned because blade stresses have a direct impact upon the stage design. There are three main sources of stress: (i) centrifugal tensile stress (the largest, but not necessarily the most important because it is a steady stress), (ii) gas bending stress (fluctuating as the rotor blades pass by the trailing edges of the nozzles)

[†] Vibration problems with high aspect ratio blading can be significantly reduced by using tip shrouds which prevent vibrations in the cantilever mode. Shrouds are also sometimes used to reduce tip leakage loss.

and (iii) centrifugal bending stress when the centroids of the blade cross-sections at different radii do not lie on a radial line (any torsional stress arising from this source is small enough to be neglected).

When the rotational speed is specified, the allowable centrifugal tensile stress places a limit on the annulus area but does not affect the choice of blade chord. This somewhat surprising result can easily be shown to be the case as follows. The maximum value of this stress occurs at the root and is readily seen to be given by

$$(\sigma_{ct})_{\text{max}} = \frac{\rho_b \omega^2}{a_r} \int_r^t ar \, dr$$

where ρ_b is the density of blade material, ω is the angular velocity, a is the cross-sectional area of the blade and a_r its value at the root radius. In practice the integration is performed graphically or numerically, but if the blade were of uniform cross-section the equation would reduce directly to

$$(\sigma_{ct})_{\text{max}} = 2\pi N^2 \rho_b A$$

where A is the annulus area and N is the rotational speed in rev/s. A rotor blade is usually tapered in chord and thickness from root to tip such that a_t/a_r is between 1/4 and 1/3. For preliminary design calculations it is sufficiently accurate (and on the safe side) to assume that the taper reduces the stress to 2/3 of the value for an untapered blade. Thus

$$(\sigma_{ct})_{\text{max}} \simeq \tfrac{4}{3}\pi N^2 \rho_b A \qquad (7.30)$$

For the flared turbine of our example we have

$$A = \tfrac{1}{2}(A_2 + A_3) = 0 \cdot 094 \text{ m}^2 \text{ and } N = 250 \text{ rev/s}$$

The density of the Ni–Cr–Co alloys used for gas turbine blading is about 8000 kg/m^3, and so equation (7.30) gives

$$(\sigma_{ct})_{\text{max}} \simeq 200 \text{ MN/m}^2 \text{ (or 2000 bar)}$$

Judgement as to whether or not this stress is satisfactory must await the evaluation of the other stresses.

The force arising from the change in angular momentum of the gas in the tangential direction, which produces the useful torque, also produces a gas bending moment about the axial direction, namely M_w in Fig. 7.15. There may be a change of momentum in the axial direction (i.e. when $C_{a3} \neq C_{a2}$), and with reaction blading there will certainly be a pressure force in the axial direction $((p_2 - p_3)2\pi r/n$ per unit height), so that there will also be a gas bending moment M_a about the tangential direction. Resolving these bending moments into components acting about the principal axes of the blade cross-section, the maximum stresses can be calculated by the method appropriate to asymmetrical sections. A twisted and tapered blade must be divided into strips of height δh and the bending moments calculated from the average force acting on each strip. The gas bending stress σ_{gb} will be

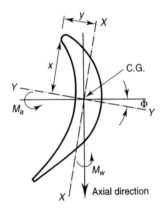

$$\sigma_{gb} = \frac{x}{I_{yy}} (M_a \cos \Phi - M_w \sin \Phi) + \frac{y}{I_{xx}} (M_w \cos \Phi + M_a \sin \Phi)$$

FIG. 7.15 Gas bending stress

tensile in the leading and trailing edges and compressive in the back of the blade, and even with tapered twisted blades the maximum value usually occurs at either the leading or trailing edge of the root section. Because M_w is by far the greater bending moment, and the principal axis XX does not deviate widely from the axial direction (angle Φ is small), a useful approximation for preliminary design purposes is provided by

$$(\sigma_{gb})_{max} \simeq \frac{m(C_{w2m} + C_{w3m})}{n} \times \frac{h}{2} \times \frac{1}{zc^3} \tag{7.31}$$

n is the number of blades, the whirl velocities are evaluated at the mean diameter, and z is the smallest value of the root section modulus (I_{xx}/y) of a blade of unit chord. Clearly σ_{gb} is directly proportional to the stage work output and blade height, and inversely proportional to the number of blades and section modulus. It is convenient to treat the section modulus as the product zc^3 because z is largely a function of blade camber angle ($\simeq$ gas deflection) and thickness/chord ratio.

An unpublished rule for z due to Ainley, useful for approximate calculations, is given in Fig. 7.16. We shall apply this, together with equation (7.31), to our example turbine. Assuming the angle of incidence is zero at the design operating condition, the blade camber angle is virtually equal to the gas deflection, namely at the root

$$\beta_{2r} + \beta_{3r} = 39 \cdot 32° + 51 \cdot 13° \simeq 90°$$

Then from Fig. 7.16, assuming a blade of $t/c = 0 \cdot 2$,

$$(z)_{root} = \frac{(10 \times 0 \cdot 2)^{1 \cdot 27}}{570} = 0 \cdot 004\,23$$

$$m(C_{w2} + C_{w3}) = mC_a(\tan \alpha_2 + \tan \alpha_3)$$

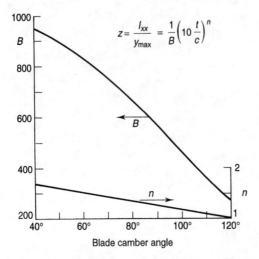

FIG. 7.16 **Approximate rule for section moduli**

which at mean diameter yields

$$20 \times 272(1 \cdot 624 + 0 \cdot 176) = 9800 \, \text{N}$$

For the chosen value of $c_R = 0 \cdot 023$ m, n_R was found to be 71, while $h_R = \frac{1}{2}(h_2 + h_3) = 0 \cdot 0691$ m. Equation (7.31) can now be evaluated to give

$$(\sigma_{gb})_{\text{max}} \simeq \frac{9800}{71} \times \frac{0 \cdot 0691}{2} \times \frac{1}{0 \cdot 004 \, 23 \times 0 \cdot 023^3} \simeq 93 \, \text{MN/m}^2$$

By designing the blade with the centroids of the cross-sections slightly off a radial line, as indicated in Fig. 7.17, it is theoretically possible to design for a centrifugal bending stress which will cancel the gas bending stress. It must be remembered, however, that (a) these two stresses would only cancel each other at the design operating condition, (b) σ_{gb} is only a quasi-steady stress and (c) the centrifugal bending stress is very sensitive to manufacturing errors in the blade and blade root fixing. σ_{gb} is often not regarded as being

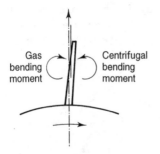

FIG. 7.17

offset by any centrifugal bending stress and usually the latter is merely calculated using the extreme values of manufacturing tolerances to check that it is small and that at least it does not reinforce σ_{gb}.

We have now established a steady centrifugal stress of 200 MN/m² and a gas bending stress of 93 MN/m² which is subject to periodic fluctuation with a frequency dependent on N, n_R and n_N. Creep strength data for possible blade materials will be available: perhaps in the form of Fig. 7.18(a) which shows the time of application of a steady stress at various temperatures required to produce 0·2 per cent creep strain. Fatigue data (e.g. Gerber diagrams) will also be available from which it is possible to assess the relative capacity of the materials to withstand fluctuating stresses. Such data, together with experience from other turbines in service, will indicate how the fluctuating gas bending stress and the steady centrifugal stress can be combined safely. The designer would hope to have a set of curves of the type shown in Fig. 7.18(b) for several safe working lives. The values of temperature on this plot might refer to turbine inlet stagnation temperature T_{01}, allowance having been made for the fact that

(i)   only the leading edge of the rotor blade could theoretically reach stagnation temperature and chordwise conduction in the metal would prevent even the local temperature there from reaching T_{01};

(ii)   even in 'uncooled' turbines (those with no cooling passages in the blades), some cooling air is bled from the compressor and passed over the turbine disc and blade roots: the metal temperature will therefore be appreciably less than 1100 K near the root radius for which the stresses have been estimated.

Furthermore, the values of permissible σ_{gb} and σ_{ct} will be conservative, and include a safety factor to allow for local hot streaks of gas from the combustion system and for the fact that there will be additional thermal

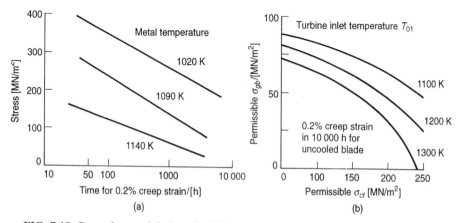

FIG. 7.18 Creep data and designer's aid for assessing preliminary designs

stresses due to chordwise and spanwise temperature gradients in the blade. In our example σ_{ct} and σ_{gb} were found to be 200 and 93 MN/m^2 respectively. If a life of 10 000 h was required, the curve relating to our inlet temperature of 1100 K suggests that the stresses are rather too large. The blade chord could be increased slightly to reduce σ_{gb} if the need for reduced stresses is confirmed by more detailed calculations. As stated earlier, the final design would be subjected to a complete stress analysis, which would include an estimate of the temperature field in the blade and the consequential thermal stresses.

(d)   Effect of pitch on the blade root fixing

The blade pitch s at mean diameter has been chosen primarily to be compatible with required values of s/c and h/c, and (via the chord) of permissible σ_{cb}. A check must be made to see that the pitch is not so small that the blades cannot be attached safely to the turbine disc rim. Only in small turbines is it practicable to machine the blades and disc from a single forging, cast them integrally, or weld the blades to the rim, and Fig. 7.19 shows the commonly used *fir tree* root fixing which permits replacement of blades. The fir trees are made an easy fit in the rim, being prevented from axial movement only (e.g. by a lug on one side and peening on the other). When the turbine is running, the blades are held firmly in the serrations by centripetal force, but the slight freedom to move can provide a useful source of damping for unwanted vibration. The designer must take into account stress concentrations at the individual serrations, and manufacturing tolerances are extremely important; inaccurate matching

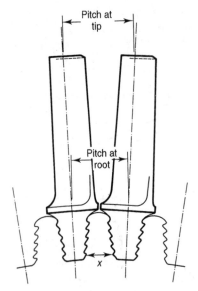

FIG. 7.19 'Fir tree' root

can result in some of the serrations being unloaded at the expense of others. Failure may occur by the disc rim yielding at the base of the stubs left on the disc after broaching (at section x); by shearing or crushing of the serrations; or by tensile stress in the fir tree root itself. The pitch would be regarded as satisfactory when the root stresses can be optimized at a safe level. This need not detain us here because calculation of these centrifugal stresses is straightforward once the size of the blade, and therefore its mass, have been established by the design procedure we have outlined.

Finally, the total centrifugal blade loading on the disc and the disc rim diameter both being known, the disc stresses can be determined to see if the original assumption of a mean blade speed of 340 m/s is satisfactory. Centrifugal hoop and radial stresses in a disc are proportional to the square of the rim speed. Disc design charts (e.g. the Donath chart) are available to permit the nominal stresses to be estimated rapidly for any disc of arbitrary shape; see Ref. (7). They will be 'nominal' because the real stress pattern will be affected substantially by thermal stresses arising from the large temperature gradient between rim and hub or shaft.

Before proceeding to make a critical assessment of the stage design used as an example, which we shall do in section 7.4, it is logical to end this section by outlining briefly recent developments in the prediction and construction of more efficient blade profiles. *This will be a digression from the main theme, however, and the reader may prefer to omit it at the first reading.*

Theoretical approach to the determination of blade profiles and pitch/chord ratio

There is little doubt that the approach to turbine design via cascade test results, which has been outlined here, is satisfactory for moderately loaded turbine stages. Recently, however, advanced aircraft propulsion units have required the use of high blade loadings (i.e. high ψ and ϕ) and cooled blades, which takes the designer into regions of flow involving ever-increasing extrapolation from existing cascade data. Furthermore it is found that under these conditions even minor changes in blade profile, such as the movement of the point of maximum camber towards the leading edge from say 40 to 37 per cent chord, can make a substantial difference to the blade loss coefficient, particularly when the turbine is operating away from the design point, i.e. at part load. Rather than repeat the vast number of cascade tests to cover the more arduous range of conditions, the approach now is to run a few such tests to check the adequacy of theoretical predictions and then apply the theory.

The digital computer has made it possible to develop methods of solving the equations of compressible flow through a blade row, even taking radial components into account to cover the case of a flared annulus. (Earlier incompressible flow solutions, adequate for the small pressure changes in compressor blading, are of little use in turbine design.) At first the approach was towards predicting the potential flow for a blade of given profile. That is, the pressure and velocity distributions outside the boundary layer are calculated in a passage bounded by the prescribed concave surface of one blade and the convex surface of the adjacent blade. Conformal transformation theory, methods of distributed sources and sinks, and stream filament theory have all been applied in one form or another and with varying degrees of success. Once the potential flow pattern has been established, boundary layer theory can be applied to predict the profile loss coefficient. The chief difficulties in this part of the analysis are the determination of the point of transition between laminar and turbulent flow on the convex surface, and the determination of a satisfactory mathematical model for the wake downstream of the trailing edge.

Some idea of the complexity of the calculations involved can be formed from Ref. (2), which also gives detailed references to this work. A not unimportant question is which method takes the least computer time consistent with providing solutions of adequate accuracy. While there is still so much controversy as to the best method, it is not practicable to attempt a simple introductory exposition in a book of this kind. One important outcome of the work, however, can be stated in simple terms with reference to the typical pressure and velocity distributions for conventional blading shown in Fig. 7.20.† The vital feature is the magnitude of the opposing pressure gradient on the convex (suction) surface. If too great it will lead to separation of the boundary layer somewhere on the back of the blade, a

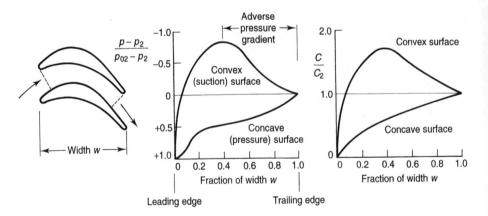

FIG. 7.20 Pressure and velocity distributions on a conventional turbine blade

† It is the flow relative to the blade which is referred to in this section: if a rotor blade is under consideration the parameters would be $(p - p_3)/(p_{03rel} - p_3)$ and V/V_3.

large wake, and a substantial increase in the profile loss coefficient. When trying to design with high aerodynamic blade loadings, which imply low suction surface pressures, it is desirable to know what limiting pressure or velocity distributions on the suction surface will just give separation at the trailing edge of the blade. One guide is provided by Smith, Ref. (8), who also makes a useful comparison of six different criteria for the prediction of separation. (He sidesteps the difficulty of determining the point of transition from a laminar to turbulent boundary layer, however, by assuming a turbulent layer over the whole of the suction surface.) Using the most conservative criterion for separation, namely that of Stratford, he constructed two extreme families of simplified, limiting, velocity distributions as shown by the two inset sketches in Fig. 7.21. Within each family the distribution is defined by a value of (C_{max}/C_2) and the position of the point of maximum velocity, A, along the blade surface. The curves for various Reynolds numbers, only two per family being reproduced here, represent the loci of point A. Any velocity distribution falling in the region below the relevant curve should imply freedom from separation. (Note that, helpfully, separation is delayed by an increase in Reynolds number.) When this type of information is available the value of being able to predict the

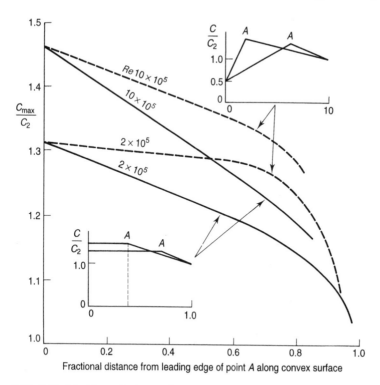

FIG. 7.21 **Limiting velocity profiles on suction surface for boundary layer separation at trailing edge**

velocity distribution is apparent: small changes in a proposed profile or camber-line shape can be made until the velocity distribution falls within the safe region.

What we have been referring to so far are approaches that have been made to *the direct problem*: the prediction of the velocity distribution around a given blade in cascade. Attention is now being focused on *the indirect problem*: the theoretical determination of a blade shape which will give a *prescribed blade surface velocity distribution*. What the ideal velocity distribution should be in various circumstances to give the minimum loss is certainly not yet established, but at least enough is known to avoid such obvious weaknesses as boundary layer separation (e.g. near the blade root where the degree of reaction is low) or the formation of unwanted shock waves. Ultimately it is hoped to be able to build into a computer program such restraints on the possible blade shape as those provided by stressing considerations (blade section area and section moduli) and by a minimum trailing edge thickness (dictated by manufacturing necessity or the need to accommodate a cooling passage). The profile loss coefficient of the restricted range of possible profiles can then be evaluated to enable a final choice to be made. The method of solving the indirect problem which seems to have proved the most capable of useful development is that due to Stanitz, although the approach via stream filament theory initiated by Wu is also receiving attention. A helpful summary of the main steps in these solutions is given by Horlock in Ref. (2).

It is to be hoped that enough has been said here to warn the student who wishes to know more of these topics that a first essential is a thorough grounding in aerodynamics and turbulent boundary layer theory. This should be followed by a study of Sections B and C of Ref. (1), on the 'theory of two-dimensional flow through cascades' and 'three-dimensional flow in turbomachines'. To give the reader a feel for the indirect problem, however, we will end this section with a brief description of an approximate solution due to Stanitz which Horlock has put into terms comprehensible to readers of section 5.9.

The case considered is the relatively simple one of compressible flow in a two-dimensional cascade as depicted in Fig. 7.22. We shall find it convenient to refer to the whirl component of the force acting on unit height of blade which was denoted by F in section 5.9. Unlike the treatment in that section, we shall here be concerned with the way F per unit width (w) changes with x, and moreover we cannot make the incompressible assumption that ρ is constant. For the indirect problem the following data will be specified:

(a) upstream conditions (p_{01}, T_{01}, C_1, α_1) and downstream conditions (C_2, α_2);

(b) surface velocity distributions as a function of x between $x = 0$ and $x = 1$, namely $C_s(x)$ and $C_p(x)$ for the suction and pressure surfaces

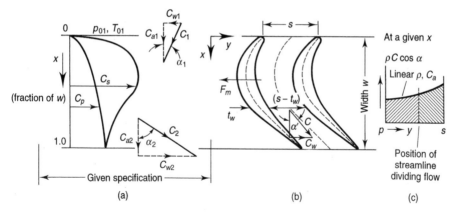

FIG. 7.22 The 'indirect' problem for two-dimensional compressible flow in a turbine cascade

respectively. The directions of C_s and C_p will be unknown because the shape of the surface is unknown, indicating the need for a process of iteration.

Because we are dealing only with the potential flow, the expansion will be isentropic so that p_0 and T_0 will be constant through the passage. For any local value of C, therefore, local values of p and ρ can be found from the isentropic relations

$$T_0 = T + C^2/2c_p; \quad p_0/p = (T_0/T)^{\gamma/(\gamma-1)}; \quad \rho_0/\rho = (p_0/p)^{1/\gamma}$$

We note that the blade profile will be completely determined when (*a*) the pitch/width ratio (s/w) is established, and (*b*) both the camber-line angle and blade thickness/pitch ratio have been calculated for various values of x between 0 and 1. The procedure is as follows.

First approximation

(i)   Determine ρ_1 from the given inlet conditions p_{01}, T_{01} and C_1; and note that for continuity in two-dimensional flow

$$m = \rho_1 s C_1 \cos \alpha_1 = \rho_2 s C_2 \cos \alpha_2$$

where m is the mass flow per unit height of passage.

(ii)   Using the isentropic relations, calculate the pressures p_p and p_s for a series of values of x between 0 and 1 from the given surface velocity distributions of C_p and C_s.

(iii) Integrate numerically $p_p(x)$ and $p_s(x)$ from $x = 0$ to $x = 1$, and so determine the mean force per unit height of blade in the whirl direction from $F_m = \Delta p_m \times w$.

(iv)   Determine s/w by equating F_m to the overall change of momentum in the whirl direction, i.e.

$$\Delta p_m \times w = m(C_{w1} + C_{w2}) = \rho_1 s C_1 \cos \alpha_1 (C_1 \sin \alpha_1 + C_2 \sin \alpha_2)$$

$$\frac{s}{w} = \frac{\Delta p_m}{\rho_1 C_1 \cos \alpha_1 (C_1 \sin \alpha_1 + C_2 \sin \alpha_2)}$$

(v)   For various values of x between 0 and 1, integrate the pressure distributions from the leading edge to x, and so obtain the pressure differences Δp which act over the series of areas $(xw) \times 1$.

(vi)   Determine the mean whirl velocity C_w at the values of x used in (v) from

$$\Delta p \times xw = \rho_1 s C_1 \cos \alpha_1 (C_1 \sin \alpha_1 \pm C_w)$$

(The negative sign refers to values near the leading edge where C_w is in the same direction as C_{w1}.)

(vii)   Finding the corresponding mean values of C from $(C_p + C_s)/2$, the values of α' at the various values of x are given by

$$\sin \alpha' = C_w/C$$

(viii)   For convenience the blade thickness is measured in the whirl direction. Thus t_w at the various values of x can be found as a fraction of the pitch s from the continuity equation:

$$\rho_1 C_1 \cos \alpha_1 = \rho \left[1 - \left(\frac{t_w}{s} \right) \right] C \cos \alpha'$$

where the mean density ρ at any x is found from the corresponding mean values of p and C by using the isentropic relations.

In this first approximation the directions of the surface velocities C_p and C_s have been assumed the same and equal to that of the mean velocity C, namely α'. Furthermore the properties have been assumed constant across the passage at the mean value. Having established an approximate blade profile it is possible to refine these assumptions and obtain a better approximation to the true profile. The value of s/w did not depend on the assumptions and remains unchanged, but the calculation of α' and t_w/s as functions of x must be repeated.

Second approximation
(i)   The flow directions α_p and α_s of the surface velocities C_p and C_s are determined by the geometry of the concave and convex surfaces resulting from the first approximation, and initially the properties are considered to vary linearly across the passage from one surface to the other.

(ii)   The position of the streamline which divides the flow equally between the surfaces is determined by assuming that the axial velocity $C \cos \alpha$ varies linearly across the passage from $C_p \cos \alpha_p$ to $C_s \cos \alpha_s$: see Fig. 7.22(c).

(iii) The velocity on the central streamline, C_m, is determined from the fact that potential flow is irrotational. The criterion of irrotationality is expressed by

$$\frac{\partial}{\partial y}(C\cos\alpha) - \frac{\partial}{\partial x}(C\sin\alpha) = 0$$

where y is the co-ordinate in the whirl direction. The assumption in (ii) implies that the first term is constant. With the additional assumption that $\partial C/\partial x$ varies linearly with y across the passage, the equation can be integrated numerically to give C as a function of y and in particular the value of C on the central streamline (C_m).

(iv) The value of the whirl velocity on the central streamline, C_{wm}, is again determined by equating the change of momentum, from the leading edge to x, to the corresponding pressure force; and the camber-line angle is then obtained from $\tan\alpha' = C_{wm}/C_m$.

(v) The thickness in terms of t_w/s is determined from the continuity equation as before.

When integrating across the passage for steps (iv) and (v), parabolic variations of the relevant properties are used instead of linear variations. The parabolas pass through the values at the two surfaces and at the central streamline.

A third approximation on the same lines as the second can be made if necessary, and the final shape is then determined by rounding off the leading and trailing edges, and subtracting an estimated displacement thickness of the boundary layer, which may be appreciable on the back of the blade where the flow is decelerating.

7.4 Estimation of stage performance

The last step in the process of arriving at the preliminary design of a turbine stage is to check that the design is likely to result in values of nozzle loss coefficient and stage efficiency which were assumed at the outset. If not, the design calculations may be repeated with more probable values of loss coefficient and efficiency. When satisfactory agreement has been reached, the final design may be laid out on the drawing board and accurate stressing calculations can be performed.

Before proceeding to describe a method of estimating the design point performance of a stage, however, the main factors limiting the choice of design, which we have noted during the course of the worked example, will be summarized. The reason we considered a turbine for a turbojet engine was simply that we would thereby be working near those limits to keep size and weight to a minimum. The designer of an industrial gas turbine has a somewhat easier task: he or she will be using lower temperatures and stresses to obtain a longer working life, and this means lower mean blade speeds, more stages, and much less stringent aerodynamic limitations. A power turbine, not mechanically coupled to the gas generator, is another case where much less difficulty will be encountered in arriving at a satisfactory solution. The choice of gear ratio between the power turbine and driven component is normally at the disposal of the turbine designer, and thus the rotational speed can be varied to suit the turbine, instead of the compressor as we have assumed here.

Limiting factors in turbine design

(a) *Centrifugal stresses* in the blades are proportional to the square of the rotational speed N and the annulus area: when N is fixed they place an upper limit on the *annulus area*.

(b) *Gas bending stresses* are (1) inversely proportional to the number of blades and blade section moduli, while being (2) directly proportional to the blade height and specific work output.

   (1) The *number of blades* cannot be increased beyond a point set by blade fixing considerations, but the *section moduli* are roughly proportional to the cube of the *blade chord* which might be increased to reduce σ_{gb}. There is an aerodynamic limit on the *pitch/chord ratio*, however, which if too small will incur a high loss coefficient (friction losses increase because a reduction in s/c increases the blade surface area swept by the gas).

   (2) There remains the *blade height*: but reducing this while maintaining the same annulus area (and therefore the same axial velocity for the given mass flow) implies an increase in the *mean diameter of the annulus*. For a fixed N, the mean diameter cannot be increased without increasing the *centrifugal disc stresses*. There will also be an aerodynamic limit set by the need to keep the *blade aspect ratio* (h/c) and *annulus radius ratio* (r_t/r_r) at values which do not imply disproportionate losses due to *secondary flows, tip clearance* and *friction on the annulus walls* (say not less than 2 and 1·2 respectively). The blade height might be reduced by reducing the annulus area (with the added benefit of reducing the centrifugal blade stresses) but, for a given mass flow, only by increasing the *axial velocity*. An aerodynamic limit on C_a will be set by the need to keep the maximum *relative Mach number at the blade inlet* (namely at the root radius), and the *Mach number at outlet from the stage*, below the levels which mean high friction losses in the blading and jet pipe respectively.

(c) *Optimizing the design*, so that it just falls within the limits set by all these conflicting mechanical and aerodynamic requirements, will lead to an efficient turbine of minimum weight. If it proves to be impossible to meet one or more of the limiting conditions, the required work output must be split between *two stages*. The second design attempt would be commenced on the assumption that the efficiency is likely to be a maximum when the work, and hence the temperature drop, is divided equally between the stages.

(d) The velocity triangles, upon which the rotor blade section depends, are partially determined by the desire to work with an average *degree of reaction of 50 per cent* to obtain low blade loss coefficients and *zero swirl* for minimum loss in the jet pipe. To avoid the need for two stages in a marginal case, particularly if it means adding a bearing on the downstream side, it would certainly be preferable to design with a lower degree of reaction and some swirl. An aerodynamic limit on the minimum value of the reaction at mean diameter is set by the need to ensure some *positive reaction at the blade root radius.*

For what follows in the next section, it will be helpful to have a summary of the results of the design calculations for the turbine of our worked example: such a summary is given in Fig. 7.23 and over the page.

Gas angles	α_1	α_2	α_3	β_2	β_3
root	$0°$	$62.15°$	$12.12°$	$39.32°$	$51.13°$
mean	$0°$	$58.38°$	$10°$	$20.49°$	$54.96°$
tip	$0°$	$54.93°$	$8.52°$	$0°$	$58.33°$

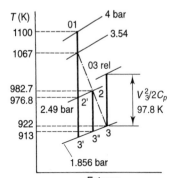

Plane	1	2	3	
p	3.54	2.49	1.856	bar
T	1067	982.7	922	K
ρ	1.155	0.883	0.702	kg/m^3
A	0.0626	0.0833	0.1047	m^2
r_m	$\leftarrow$	0.216	$\rightarrow$	m
r_t/r_r	1.24	1.33	1.43	
h	0.046	0.0612	0.077	m

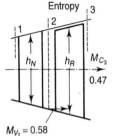

Blade row	Nozzle	Rotor
s/c	0.86	0.83
h (mean)	0.0536	0.0691 m
h/c	3.0	3.0
c	0.0175	0.023 m
s	0.01506	0.0191 m
n	90	71

FIG. 7.23 Summary of data from preliminary design calculations of single-stage turbine

Mean-diameter stage parameters:

$$\psi = 2c_p \Delta T_{0s}/U^2 = 2\cdot88, \quad \phi = C_a/U = 0\cdot8, \quad \Lambda = 0\cdot421$$

$$U = 340\,\text{m/s}$$

$$C_{a1} = C_1 = C_3 = 276\cdot4\,\text{m/s}, \quad C_{a2} = C_{a3} = C_a = 272\,\text{m/s}$$

$$C_2 = 519\,\text{m/s}, \quad V_3 = 473\cdot5\,\text{m/s}$$

The rotor blade is designed for $i = 0°$ with conventional profile having $t/c = 0\cdot2$. At the root section, camber $\simeq \beta_{2r} + \beta_{3r} \simeq 90°$ and hence $z_r \simeq 0\cdot004\,23\,\text{mm}^3/\text{mm}$ chord giving $(\sigma_{gb})_{max} \simeq 93\,\text{MN/m}^2$. $(\sigma_{ct})_{max} \simeq f(N, A) \simeq 200\,\text{MN/m}^2$.

Estimation of design point performance

The method to be outlined here is that due to Ainley and Mathieson, Ref. (9), which estimates the performance on flow conditions at the mean diameter of the annulus. Reference (9) describes how to calculate the performance of a turbine over a range of operating conditions, but we shall be concerned here only to find the efficiency at the design point. A start is made using the two correlations for profile loss coefficient Y_p obtained from cascade data, which are shown in Fig. 7.24. These refer to nozzle-type blades ($\beta_2 = 0$) and impulse-type blades ($\beta_2 = \beta_3$) of conventional profile (e.g. T6) having a thickness/chord ratio (t/c) of 0·2 and a trailing edge thickness/pitch ratio (t_e/s) of 0·02. Rotor blade notation is used in Fig. 7.24, and in what follows, to emphasize that we are thinking of the flow *relative* to any blade row. When the nozzle row is being considered β_2 becomes α_1 and β_3 becomes α_2: there is no need to duplicate equations which apply equally to both rows. The values of Y_p in Fig. 7.24 refer to blades operating at zero incidence, i.e. when the gas inlet angle β_2 is also the blade inlet angle.

Step 1

Estimate $(Y_p)_N$ and $(Y_p)_R$ from the gas angles of the proposed design by using Fig. 7.24 in conjunction with the interpolation formula

$$Y_p = \left\{ Y_{p(\beta_2=0)} + \left(\frac{\beta_2}{\beta_3}\right)^2 \left[Y_{p(\beta_2=\beta_3)} - Y_{p(\beta_2=0)} \right] \right\} \left(\frac{t/c}{0\cdot2}\right)^{\beta_2/\beta_3} \tag{7.32}$$

This equation represents a correction for a change in inlet angle at a constant outlet angle, so that $Y_{p(\beta_2=0)}$ and $Y_{p(\beta_2=\beta_3)}$ are the values for a nozzle- and impulse-type blade having the same *outlet* gas angle β_3 as the actual blade. Equation (7.32) also includes a correction for t/c if it differs from 0·2, a reduction in t/c leading to reduced profile loss for all blades other than nozzle-type blades ($\beta_2 = 0$). The degree of acceleration of the flow in the blading decreases with the degree of reaction as $\beta_2/\beta_3 \rightarrow 1$, and

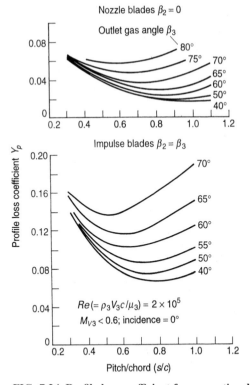

FIG. 7.24 Profile loss coefficient for conventional blading with $t/c = 0.20$

the influence of blade thickness becomes more marked as the acceleration is diminished. The correction is considered reliable only for $0.15 < t/c < 0.25$.

For the nozzle blades of our example, $\alpha_1 = 0$ and so $(Y_p)_N$ can be read straight from Fig. 7.24. $\alpha_2 = 58.38°$ and $(s/c)_N = 0.86$, and hence

$$(Y_p)_N = 0.024$$

For the rotor blades, $\beta_2 = 20.49°$, $\beta_3 = 54.96°$, $(s/c)_R = 0.83$ and $(t/c) = 0.2$, so that

$$(Y_p)_R = \left\{ 0.023 + \left(\frac{20.5}{55}\right)^2 [0.087 - 0.023] \right\} = 0.032$$

Step 2

If it had been decided to design the blades to operate with some incidence at the design point, a correction to Y_p would be required. As this correction is really only important when estimating performance at part load, we shall refer the reader to Ref. (9) for the details. Briefly, it involves using correlations of cascade data to find the stalling incidence i_s for the given blade (i.e. incidence at which Y_p is twice the loss for $i = 0$); and then using a curve of $Y_p/Y_{p(i=0)}$ versus i/i_s to find Y_p for the given i and the value of $Y_{p(i=0)}$ calculated in Step 1.

Step 3

Secondary and tip-clearance loss data for Y_s and Y_k have been correlated using the concepts of lift and drag coefficient which were introduced in section 5.9 for axial compressors. Without repeating the whole argument leading to equation (5.34), it should be possible to see by glancing at section 5.9 that for a turbine cascade (with rotor blade notation)

$$C_L = 2(s/c)(\tan \beta_2 + \tan \beta_3) \cos \beta_m$$

where

$$\beta_m = \tan^{-1}[(\tan \beta_3 - \tan \beta_2)/2]$$

Now, as stated at the beginning of section 7.3, it is convenient to treat Y_s and Y_k simultaneously. The proposed correlation is

$$Y_s + Y_k = \left[\lambda + B\left(\frac{k}{h}\right)\right]\left[\frac{C_L}{s/c}\right]^2 \left[\frac{\cos^2 \beta_3}{\cos^3 \beta_m}\right] \tag{7.33}$$

Reference back to equations (5.35) and (5.38) will make the appearance of the last two bracketed terms seem logical: they occur in the expression for the combined secondary and tip-clearance loss coefficient for a compressor blade row. Considering now the first bracketed term, the tip-clearance component is proportional to k/h where k is the clearance and h the blade height. The constant B is 0·5 for a radial tip clearance, and 0·25 for a shrouded blade with side clearance; see Fig. 7.25. The secondary loss component λ is more complex. We have suggested that secondary flow and annulus wall friction might be affected by the aspect ratio (h/c) and/or annulus radius ratio (r_t/r_r). As we shall see, r_t/r_r is thought by Ainley and Mathieson to be the more relevant parameter. (They argue that h/c is only important in so far as there is a change in h, not in c.) Also, like the profile loss, Y_s is considerably affected by the amount of acceleration of the flow in the blade passage. In general terms, the larger the acceleration the thinner and more stable are the boundary layers, the smaller is the chance of boundary layer separation, and the smaller is the effect of a curved neighbouring surface in setting up the secondary flows. The degree of acceleration is conveniently indicated by the ratio of the area *normal* to the flow at outlet to that at inlet, i.e.

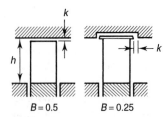

FIG. 7.25

$A_3 \cos \beta_3 / A_2 \cos \beta_2$ where A is the annulus area. It is found that the quantity λ in equation (7.33) is given approximately by

$$\lambda = f\left\{ \left(\frac{A_3 \cos \beta_3}{A_2 \cos \beta_2}\right)^2 \bigg/ \left(1 + \frac{r_r}{r_t}\right) \right\} \tag{7.34}$$

where the function f is given by the curve in Fig. 7.26.

Let us now evaluate $(Y_s + Y_k)$ for the blades in our example.

Nozzle blades: We shall assume that the nozzles are shrouded, with seals supported by a diaphragm at the shaft radius so that the leakage loss is very small. Then B in equation (7.33) can be assumed zero. λ is found as follows.

$$A_2 = 0 \cdot 0833 \text{ m}^2, \ A_1 = 0 \cdot 0626 \text{ m}^2$$

$$\cos \alpha_2 = \cos 58 \cdot 38° = 0 \cdot 524, \ \cos \alpha_1 = \cos 0° = 1 \cdot 0$$

mean r_t/r_r between planes 1 and 2 $= 1 \cdot 29$

$$\left(\frac{A_2 \cos \alpha_2}{A_1 \cos \alpha_1}\right)^2 \bigg/ \left(1 + \frac{r_r}{r_t}\right) = \left(\frac{0 \cdot 0833 \times 0 \cdot 524}{0 \cdot 0626 \times 1 \cdot 0}\right)^2 \bigg/ \left(1 + \frac{1}{1 \cdot 29}\right) = 0 \cdot 274$$

From Fig. 7.26, $\lambda = 0 \cdot 012$

$$\frac{C_L}{s/c} = 2(\tan \alpha_1 + \tan \alpha_2) \cos \alpha_m$$

$$\alpha_m = \tan^{-1} \left[(\tan \alpha_2 - \tan \alpha_1)/2\right]$$

$$= \tan^{-1} \left[(\tan 58 \cdot 38° - \tan 0°)/2\right] = 39 \cdot 08°$$

$$\frac{C_L}{s/c} = 2(\tan 58 \cdot 38° + \tan 0°) \cos 39 \cdot 08° = 2 \cdot 52$$

$$\frac{\cos^2 \alpha_2}{\cos^3 \alpha_m} = \frac{0 \cdot 524^2}{0 \cdot 776^3} = 0 \cdot 589$$

$$[Y_s + Y_k]_N = 0 \cdot 012 \times 2 \cdot 52^2 \times 0 \cdot 589 = 0 \cdot 0448$$

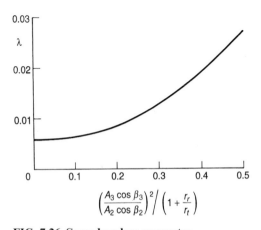

FIG. 7.26 Secondary loss parameter

Rotor blades: We will assume unshrouded rotor blades with radial tip clearance equal to 2 per cent of the mean blade height, so that

$$B(k/h) = 0.5 \times 0.02 = 0.01$$

For these blades, using the data from Fig. 7.23,

$$\left(\frac{A_3 \cos \beta_3}{A_2 \cos \beta_2}\right)^2 \Big/ \left(1 + \frac{r_r}{r_t}\right) = \left(\frac{0.1047 \times \cos 54.96°}{0.0833 \times \cos 20.49°}\right)^2 \Big/ \left(1 + \frac{1}{1.38}\right)$$
$$= 0.334$$

From Fig. 7.26, $\lambda = 0.015$

$$\beta_m = \tan^{-1}[(\tan 54.96° - \tan 20.49°)/2] = 27.74°$$

$$\frac{C_L}{s/c} = 2(\tan 54.96° + \tan 20.49°) \cos 27.74° = 3.18$$

$$\frac{\cos^2 \beta_3}{\cos^3 \beta_m} = \frac{0.574^2}{0.885^3} = 0.475$$

$$[Y_s + Y_k]_R = (0.015 + 0.01)3.18^2 \times 0.475 = 0.120$$

Step 4

The total loss coefficients become

$$Y_N = (Y_p)_N + [Y_s + Y_k]_N = 0.024 + 0.0448 = 0.0688$$
$$Y_R = (Y_p)_R + [Y_s + Y_k]_R = 0.032 + 0.120 = 0.152$$

If the trailing edge thickness/pitch ratio (t_e/s) differs from 0.02, it is at this point that a correction is made for the effect on the losses. 0.02 was the value for the blading to which Fig. 7.24 relates, but trailing edge thickness affects all the losses, not merely the profile loss. The correction curve in Fig. 7.27 has been deduced from turbine test results. There is no reason to suppose that the normal value of 0.02 would be unsuitable for the nozzle and rotor blades of the turbine of our example and no correction is required. The modest turbine inlet temperature indicates a low-cost, long-life, application by aircraft standards and certainly there will be no need to

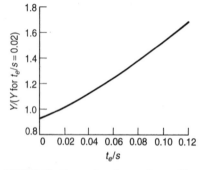

FIG. 7.27 Correction factor for trailing edge thickness

thicken the trailing edge to accommodate cooling passages. Early development tests may indicate vibration troubles which might be overcome by increasing the thickness, and Fig. 7.27 enables the penalty to be paid in loss of performance to be estimated.

Step 5
The stage efficiency can now be calculated using equations (7.19) and (7.20). We first calculate the equivalent loss coefficients defined in terms of temperature. For the nozzles,

$$\lambda_N = \frac{Y_N}{(T_{02}/T_2')} = \frac{0.0688}{(1100/976.8)} = 0.0611$$

For the rotor,

$$\lambda_R = \frac{Y_R}{(T_{03\mathrm{rel}}/T_3'')}$$

We have previously calculated T_3'' to be 913 K, but have not found the value of $T_{03\mathrm{rel}}$. We did, however, find that $(V_3^2/2c_p) = 97.8$ K and $T_3 = 922$ K, so that

$$T_{03\mathrm{rel}} = T_3 + (V_3^2/2c_p) = 1020 \text{ K}$$

$$\lambda_R = \frac{0.152}{1020/913} = 0.136$$

Equation (7.20) now becomes

$$\eta_s = \frac{1}{1 + \left[0.136 \times 97.8 + \left(\dfrac{922}{982.7}\right)0.0611 \times 117.3\right]\bigg/145} = 0.88$$

Thus the design yields $\lambda_N = 0.061$ and $\eta_s = 0.88$, in comparison with the values of 0.05 and 0.9 assumed at the outset. This can be regarded as satisfactory agreement, but minor changes would be looked for to improve the efficiency: perhaps a slight increase in degree of reaction with the reduction in work due to this compensated by designing with some progressive increase in C_a through the stage. The latter would have the added advantage of reducing the flare of the annulus.

Step 6
In conclusion, it must be emphasized that the cascade data and other loss correlations are strictly applicable only to designs where the Mach numbers are such that no shock losses are incurred in the blade passages. It has been suggested, Ref. (10), that the additional loss incurred by designing with a blade outlet relative Mach number greater than unity can be accounted for by adjusting the profile loss coefficient Y_p of the blade row concerned. The correction is given by

$$Y_p = [Y_p \text{ from equation (7.32)}] \times [1 + 60(M-1)^2]$$

where M is M_{V3} for the rotor blades and M_{C2} for the nozzles. There is another restriction on the applicability of the data not yet mentioned: the Reynolds number of the flow should be in the region of 1×10^5 to 3×10^5, with Re defined in terms of blade chord, and density and *relative* velocity at *outlet* of a blade row. If the mean Reynolds number for a turbine, taken as the arithmetic mean of Re for the first nozzle row and the last rotor row (to cover multi-stage turbines), differs much from 2×10^5, an approximate correction can be made to the overall isentropic efficiency by using the expression

$$(1 - \eta_t) = \left(\frac{Re}{2 \times 10^5}\right)^{-0.2} (1 - \eta_t)_{Re = 2 \times 10^5} \tag{7.35}$$

To calculate Re for the nozzle and rotor rows of our example, we need the viscosity of the gas at temperatures $T_2 = 982.7\,\mathrm{K}$ and $T_3 = 922\,\mathrm{K}$. Using data for air, which will be sufficiently accurate for this purpose,

$$\mu_2 = 4.11 \times 10^{-5}\ \mathrm{kg/m\ s} \text{ and } \mu_3 = 3.95 \times 10^{-5}\ \mathrm{kg/m\ s}$$

$$(Re)_N = \frac{\rho_2 C_2 c_N}{\mu_2} = \frac{0.883 \times 519 \times 0.0175}{4.11 \times 10^{-5}} = 1.95 \times 10^5$$

$$(Re)_R = \frac{\rho_3 V_3 c_R}{\mu_3} = \frac{0.702 \times 473.5 \times 0.023}{3.95 \times 10^{-5}} = 1.93 \times 10^5$$

Thus no Reynolds number correction is required.

The Ainley–Mathieson method outlined here has been found to predict efficiencies to within ± 3 per cent of the measured values for aircraft turbines, but to be not so accurate for small turbines which tend to have blades of rather low aspect ratio. Dunham and Came, in Ref. (10), suggest that the method becomes applicable to a wider range of turbines if the secondary and tip-clearance loss correlation, equation (7.33), is modified as follows.

(a) λ, instead of being given by the function expressed in Fig. 7.26, is replaced by

$$0.0334 \left(\frac{c}{h}\right) \left(\frac{\cos \beta_3}{\cos \beta_2}\right)$$

(with $\beta_3 = \alpha_2$ and $\beta_2 = \alpha_1$ for nozzles).

(b) $B(k/h)$ is replaced by

$$B\left(\frac{c}{h}\right)\left(\frac{k}{c}\right)^{0.78}$$

with B equal to 0.47 for radial tip clearances and 0.37 for side clearances on shrouded blades.

When this modification is applied to the turbine of our example there is, as might be expected, very little difference in the predicted efficiency, namely 0.89 as compared with 0.88. Larger differences of up to 5 per cent are obtained with low aspect ratios of about unity.

More recently, Kacker and Okapuu [Ref. (11)] have compared the Ainley–Mathieson and Dunham–Came correlations with the losses observed for turbine blades designed in the early 1980s. They concluded that improvements in turbine design had resulted in, among other things, a roughly one-third reduction in profile losses since Ainley and Mathieson's time. Kacker and Okapuu developed various modifications to the earlier correlations to reflect modern turbine design more accurately. The Kacker–Okapuu modifications are only applicable at the design incidence for the blades, whereas the Ainley–Mathieson correlations represent a complete loss prediction system. As might be expected, the design improvements also affect the off-design behaviour of the blades. Benner, Sjolander and Moustapha [Ref. (12)] have shown that the Ainley–Mathieson correlations generally overpredict the increase in profile losses at off-design incidence for modern blades. Benner and colleagues therefore developed an improved correlation for the off-design profile losses for use with the Kacker–Okapuu design point correlation.

7.5 Overall turbine performance

In the previous section we have described a method of estimating the stage efficiency η_s. If the whole turbine comprises a large number of similar stages, it would be a reasonable approximation to treat the stage efficiency as being equal to the polytropic efficiency $\eta_{\infty t}$ and obtain the overall isentropic efficiency from equation (2.18). This was the approach suggested in section 5.7 for the preliminary design of multi-stage axial compressors. Turbines have few stages, however, and it is preferable to work through the turbine stage by stage, with the outlet conditions from one stage becoming the inlet conditions of the next, until the outlet temperature is established. The overall efficiency η_t is then obtained from the ratio of the actual to isentropic overall temperature drop.

As stated under the heading 'Estimation of design point performance' in section 7.4, it is possible to calculate the performance of a turbine over a range of operating conditions. Whether calculated, or measured on a test rig, the performance is normally expressed by plotting η_t and $M\sqrt{T_{03}}/p_{03}$ against pressure ratio p_{03}/p_{04} for various values of $N/\sqrt{T_{03}}$ as in Fig. 7.28. We are here reverting to cycle notation, using suffixes 3 and 4 to denote turbine inlet and outlet conditions respectively. The efficiency plot shows that η_t is sensibly constant over a wide range of rotational speed and pressure ratio. This is because the accelerating nature of the flow permits turbine blading to operate over a wide range of incidence without much increase in the loss coefficient.

The maximum value of $m\sqrt{T_{03}}/p_{03}$ is reached at a pressure ratio which produces choking conditions at some point in the turbine. Choking may occur in the nozzle throats or, say, in the annulus at outlet from the

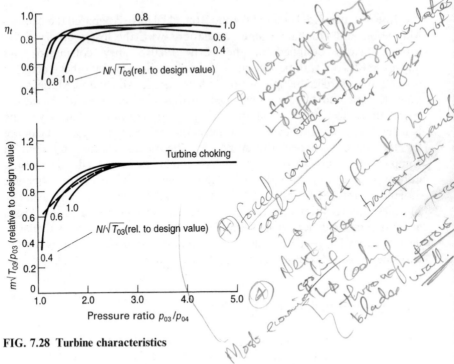

FIG. 7.28 Turbine characteristics

turbine depending on the design. The former is the more normal situation and then the constant speed lines merge into a single horizontal line as indicated on the mass flow plot of Fig. 7.28. (If choking occurs in the rotor blade passages or outlet annulus the maximum mass flow will vary slightly with $N/\sqrt{T_{03}}$.) Even in the unchoked region of operation the separation of the $N/\sqrt{T_{03}}$ lines is not great, and the larger the number of stages the more nearly can the mass flow characteristics be represented by a single curve independent of $N/\sqrt{T_{03}}$. Such an approximation is very convenient when predicting the part-load performance of a complete gas turbine unit as will be apparent from Chapter 8. Furthermore, the approximation then yields little error because when the turbine is linked to the other components the whole operating range shown in Fig. 7.28 is not used. Normally both the pressure ratio and mass flow increase simultaneously as the rotational speed is increased, as indicated by the dotted curve.

7.6   The cooled turbine

It has always been the practice to pass a quantity of cooling air over the turbine disc and blade roots. When speaking of the *cooled turbine*, however, we mean the application of a substantial quantity of coolant to the nozzle and rotor blades themselves. Chapters 2 and 3 should have left the reader in no doubt as to the benefits in reduced *SFC* and increased specific power output (or increased specific thrust in the case of aircraft propulsion units) which follow from an increase in permissible turbine inlet temperature. The benefits are still substantial even when the additional losses introduced by the cooling system are taken into account.

Figure 7.29 illustrates the methods of blade cooling that have received serious attention and research effort. Apart from the use of spray cooling for thrust boosting in turbojet engines, the liquid systems have not proved to be practicable. There are difficulties associated with channelling the liquid to and from the blades—whether as primary coolant for forced convection or free convection open thermosyphon systems, or as secondary coolant for closed thermosyphon systems. It is impossible to eliminate corrosion or the formation of deposits in open systems, and very difficult to provide adequate secondary surface cooling area at the base of the blades for closed systems. The only method used successfully in production engines has been internal, forced convection, air cooling. With 1·5–2 per cent of the air mass flow used for cooling per blade row, the blade temperature can be reduced by between 200 and 300 °C. Using current alloys, this permits turbine inlet temperatures of more than 1650 K to be used. The blades are either cast, using cores to form the cooling passages, or forged with holes of any desired shape produced by electrochemical or laser drilling. Figure 7.30 shows the type of turbine rotor blade introduced in the 1980s. The next step forward is likely to be achieved by transpiration cooling, where the cooling air is forced through a porous blade wall. This method is by far the most economical in cooling air, because not only does it remove heat from the wall more uniformly, but the effusing layer of air insulates the outer surface from the hot gas

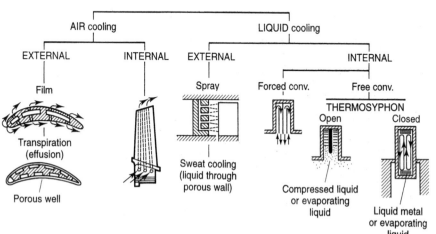

FIG. 7.29 Methods of blade cooling

Pressure side concave
cooling holes

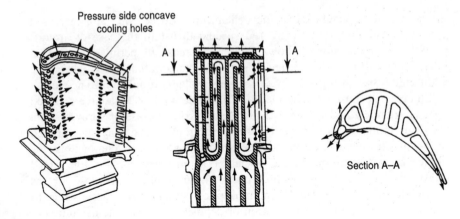

Section A–A

FIG. 7.30 Cooled turbine rotor blade [courtesy General Electric]

stream and so reduces the rate of heat transfer to the blade. Successful application awaits further development of suitable porous materials and techniques of blade manufacture.

We are here speaking mainly of rotor blade cooling because this presents the most difficult problem. Nevertheless it should not be forgotten that, with high gas temperatures, oxidation becomes as significant a limiting factor as creep, and it is therefore equally important to cool even relatively unstressed components such as nozzle blades and annulus walls. A typical distribution of cooling air required for a turbine stage designed to operate at 1500 K might be as follows. The values are expressed as fractions of the entry gas mass flow.

annulus walls	0·016
nozzle blades	0·025
rotor blades	0·019
rotor disc	0·005
	0·065

Figure 7.31(a) illustrates the principal features of nozzle blade cooling. The air is introduced in such a way as to provide jet impingement cooling of the inside surface of the very hot leading edge. The spent air leaves through slots or holes in the blade surface (to provide some film cooling) or in the trailing edge. Figure 7.31(b) depicts a modern cast nozzle blade with intricate inserts forming the cooling passages. It also shows the way the annulus walls are cooled.

It was stated earlier that liquid cooling had not proved to be practical. In 1995, however, General Electric announced the introduction of closed-loop steam cooling in their heavy industrial gas turbines designed for combined cycle operation. The 50 Hz version is scheduled to enter service in the UK in 2001 and the 60 Hz version to follow in the USA in 2002. Both rotor

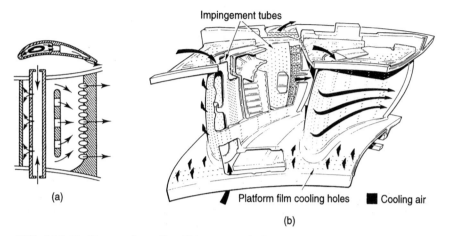

FIG. 7.31 **Turbine nozzle cooling [(b) courtesy Rolls-Royce plc]**

and nozzle blades are steam cooled, using steam extracted from the steam turbine at high pressure; the steam undergoes a pressure loss during the cooling process, but absorbs heat. The heated steam is then returned at a lower pressure to the steam turbine, contributing to the power output and being contained in the steam circuit. This approach is suitable for use with a combined cycle, where steam is readily available, but requires the use of sophisticated sealing technology to prevent loss of steam; the losses due to bleeding high-pressure air from the compressor for use in an air-cooled turbine are eliminated. Steam cooling is at an early stage of development, and attention will now be focused on the simple analysis of the conventional air-cooled turbine.

There are two distinct aspects of cooled turbine design. Firstly, there is the problem of choosing an aerodynamic design which requires the least amount of cooling air for a given cooling performance. One cooling performance parameter in common use is the *blade relative temperature* defined by

$$\text{blade relative temperature} = \frac{T_b - T_{cr}}{T_g - T_{cr}}$$

where T_b = mean blade temperature
T_{cr} = coolant temperature at inlet (i.e. at the root radius r_r)
T_g = mean effective gas temperature relative to the blade ($\simeq$ static temperature $+ 0.85 \times$ dynamic temperature)

The coolant temperature, T_{cr}, will usually be the compressor delivery temperature, and will increase significantly as pressure ratio is raised to reduce specific fuel consumption. Some industrial turbines pass the cooling flow through a water-cooled heat-exchanger to reduce T_{cr} and hence the blade relative temperature. At current levels of turbine inlet temperature three or four stages of a turbine rotor may be cooled, and air would be bled from earlier stages of the compressor to cool the later stages of the turbine, where the cooling air discharges to lower pressures in the gas stream. Bleeding air from earlier stages reduces the work input required to pressurize the cooling air, with beneficial effects on the net output.

Relative to an uncooled turbine, the optimum design might well involve the use of a higher blade loading coefficient ψ (to keep the number of stages to a minimum), a higher pitch/chord ratio (to reduce the number of blades in a row), and a higher flow coefficient ϕ (which implies a blade of smaller camber and hence smaller surface area). The importance of these, and other parameters such as gas flow Reynolds number, are discussed in detail in Ref. (13).

The second aspect is the effect on the cycle efficiency of losses incurred by the cooling process: a pertinent question is whether it is advantageous overall to sacrifice some aerodynamic efficiency to reduce such losses. The sources of loss are as follows.

(a) There is a direct loss of turbine work due to the reduction in turbine mass flow.

(b) The expansion is no longer adiabatic; and furthermore there will be a negative reheat effect in multi-stage turbines.

(c) There is a pressure loss, and a reduction in enthalpy, due to the mixing of spent cooling air with the main gas stream at the blade tips. (This has been found to be partially offset by a reduction in the normal tip leakage loss.)

(d) Some 'pumping' work is done by the blades on the cooling air as it passes radially outwards through the cooling passages.

(e) When considering cooled turbines for cycles with heat-exchange, account must be taken of the reduced temperature of the gas leaving the turbine which makes the heat-exchanger less effective.

Losses (a) and (e) can be incorporated directly into any cycle calculation, while the effect of (b), (c) and (d) can be taken into account by using a reduced value of turbine efficiency. One assessment of the latter, Ref. (14), suggests that the turbine efficiency is likely to be reduced by from 1 to 3 per cent of the uncooled efficiency, the lower value referring to near-impulse designs and the higher to 50 per cent reaction designs. The estimate for reaction designs is substantially confirmed by the tests on an experimental cooled turbine reported in Ref. (15). Cycle calculations have shown that even when all these losses are accounted for, there is a substantial advantage to be gained from using a cooled turbine.†

---

† It is worth remembering that there may be special applications where a cooled turbine might be employed not to raise the cycle temperature but to enable cheaper material to be used at ordinary temperatures: the first researches in blade cooling were carried out in Germany during the Second World War with this aim in mind.

Before either of these two aspects of cooled turbine design can be investigated it is necessary to be able to estimate the cooling airflow required to achieve a specified blade relative temperature for any given aerodynamic stage design. We will end this section with an outline of an approximate one-dimensional treatment, and further refinements can be found in Ref. (13). Figure 7.32 shows the notation employed and the simplifying assumptions made.

Consider the heat flow to and from an elemental length of blade δl a distance l from the root. As the cooling air passes up the blade it increases in temperature and becomes less effective as a coolant, so that the blade temperature increases from root to tip. There must therefore be some conduction of heat along the blade to and from the element δl due to this spanwise temperature gradient. Because turbine blade alloys have a low thermal conductivity, the conduction term will be small and we shall neglect it here. The heat balance for the elemental length δl is then simply

$$h_g S_g (T_g - T_b) = h_c S_c (T_b - T_c) \tag{7.36}$$

where h_g and h_c are the gas-side and coolant-side heat transfer coefficients, and S_g and S_c are the wetted perimeters of the blade profile and combined coolant passages respectively. For the internal airflow m_c we also have

$$m_c c_{pc} \frac{dT_c}{dl} = h_c S_c (T_b - T_c) \tag{7.37}$$

We may first find the variation of T_b with l by eliminating T_c between the two equations. From (7.36) we have

$$T_c = T_b - \frac{h_g S_g}{h_c S_c}(T_g - T_b) \tag{7.38}$$

and hence

$$\frac{dT_c}{dl} = \left(1 + \frac{h_g S_g}{h_c S_c}\right) \frac{dT_b}{dl}$$

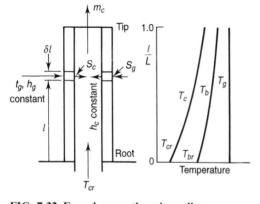

FIG. 7.32 Forced convection air cooling

Substituting in (7.37), remembering that $dT_b/dl = -d(T_g - T_b)/dl$, we get

$$\left(1 + \frac{h_g S_g}{h_c S_c}\right)\frac{d(T_g - T_b)}{dl} + \frac{h_g S_g}{m_c c_{pc}}(T_g - T_b) = 0$$

The solution of this differential equation, with $T_b = T_{br}$ at $l = 0$, is

$$T_g - T_b = (T_g - T_{br})e^{-kl/L} \tag{7.39}$$

where

$$k = \frac{h_g S_g L}{m_c c_{pc}[1 + (h_g S_g/h_c S_c)]}$$

To obtain the variation of T_c with l, we may write (7.38) in the form

$$T_g - T_c = (T_g - T_b)\left[1 + \frac{h_g S_g}{h_c S_c}\right]$$

and substitute (7.39) for $(T_g - T_b)$ to give

$$T_g - T_c = (T_g - T_{br})\left[1 + \frac{h_g S_g}{h_c S_c}\right]e^{-kl/L} \tag{7.40}$$

When $l = 0$, $T_c = T_{cr}$ and hence

$$T_g - T_{cr} = (T_g - T_{br})\left[1 + \frac{h_g S_g}{h_c S_c}\right] \tag{7.41}$$

Combining (7.40) and (7.41) we have the variation of T_c given by

$$T_g - T_c = (T_g - T_{cr})e^{-kl/L} \tag{7.42}$$

Finally, subtracting (7.39) from (7.41),

$$T_b - T_{cr} = (T_g - T_{br})\left[1 + \frac{h_g S_g}{h_c S_c} - e^{-kl/L}\right]$$

and dividing this by (7.41) we have the blade relative temperature given by

$$\frac{T_b - T_{cr}}{T_g - T_{cr}} = 1 - \frac{e^{-kl/L}}{1 + (h_g S_g/h_c S_c)} \tag{7.43}$$

We may note that h_c will be a function of coolant flow Reynolds number and hence of m_c, and that m_c also appears in the parameter k. Thus equation (7.43) is not explicit in m_c, and it is convenient to calculate values of blade relative temperature for various values of m_c rather than vice versa.

The next step is the evaluation of the heat transfer coefficients: we will consider h_c first. For straight cooling passages of uniform cross-section, pipe flow formulae may be used. For this application one recommended correlation is

$$Nu = 0.034(L/D)^{-0.1}(Pr)^{0.4}(Re)^{0.8}(T_c/T_b)^{0.55} \tag{7.44}$$

with fluid properties calculated at the mean bulk temperature. The L/D term accounts for the entrance length effect. The T_c/T_b term is necessary when the difference in temperature between fluid and wall is large, to allow for the effect of variation in fluid properties with temperature. The characteristic dimension D can be taken as the equivalent diameter ($4 \times$ area/periphery) if the cooling passages are of non-circular cross-section. For air ($Pr \simeq 0.71$) and practicable values of L/D between 30 and 100, the equation reduces to the simpler form

$$Nu = 0.020(Re)^{0.8}(T_c/T_b)^{0.55} \tag{7.45}$$

The correlation is for turbulent flow, with the accuracy decreasing at Reynolds numbers below 8000. Note that the mean value of T_c at which the fluid properties should be evaluated, and the mean value of T_b for the T_c/T_b term, are unknown at this stage in the calculations. Guessed values must be used, to be checked later by evaluating T_c and T_b at $l/L = 0.5$ from equations (7.42) and (7.43).

Data for mean blade heat transfer coefficients h_g are available from both cascade tests and turbine tests on a wide variety of blades. The latter yield higher values than the former, presumably because of the greater intensity of the turbulence in a turbine. The full line in Fig. 7.33, taken from Ref. (13), is a useful design curve for the mean value of Nusselt number round the blade surface in terms of the most significant blade shape parameter which is the ratio of inlet/outlet angle β_2/β_3 (or α_1/α_2 for nozzle

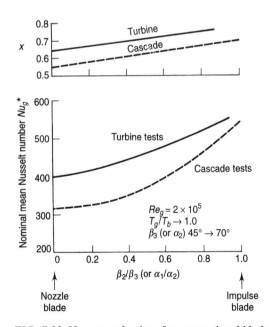

FIG. 7.33 Heat transfer data for conventional blade profiles

blades). Nu_g decreases as the degree of acceleration of the flow increases, because the point of transition from a laminar to a turbulent layer on the convex surface is delayed by accelerating flow. The curve applies to conventional turbine blade profiles, and gives nominal values denoted by Nu_g^* for operating conditions of $Re_g = 2 \times 10^5$ and $T_g/T_b \rightarrow 1$. Nu_g can then be found from

$$Nu_g = Nu_g^* \left(\frac{Re_g}{2 \times 10^5}\right)^x \left(\frac{T_g}{T_b}\right)^y$$

where the exponent x is given by the supplementary curve, and y is given by

$$y = 0.14 \left(\frac{Re_g}{2 \times 10^5}\right)^{-0.4}$$

The characteristic dimension in Nu_g and Re_g is the blade chord, and the fluid properties should be evaluated at the temperature T_g. The velocity in Re_g is the gas velocity relative to the blade at outlet (V_3 or C_2 as the case may be). The quantities required will be known for any given stage design, with the exception of the mean value of T_b for which the guessed value must be used.

All the information has now been obtained to permit the calculation of $(T_b - T_{cr})/(T_g - T_{cr})$, for various values of coolant flow m_c, from equation (7.43). Typical curves of spanwise variation, for values of m_c of 1 and 2 per cent of the gas mass flow per blade row, are shown in Fig. 7.34. Note that the distribution matches requirements because the blade stresses decrease from root to tip. The quantity of heat extracted from the blade row can be found by calculating T_{ct}, i.e. T_c at $l = L$, from equation (7.42), and then evaluating $m_c c_{pc}(T_{ct} - T_{cr})$.

To conclude, we may note that the final design calculation for a cooled blade will involve an estimation of the two-dimensional temperature distribution over the blade cross-section at several values of l/L. Finite difference methods are used to solve the differential equations, and conduction within the blade is taken into account. Figure 7.35 shows a typical temperature distribution at the mid-span of a blade designed to

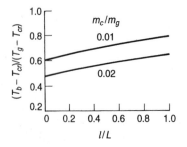

FIG. 7.34 Typical spanwise temperature distributions

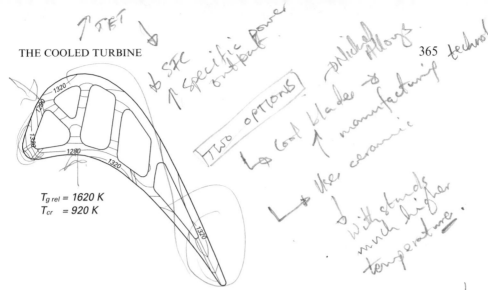

$T_{g\,rel} = 1620\ K$
$T_{cr} = 920\ K$

FIG. 7.35 Typical temperature distribution [courtesy Rolls-Royce plc]

operate with $T_g = 1620\,\text{K}$ and $T_{cr} = 920\,\text{K}$. It emphasizes one of the main problems of blade cooling, i.e. that of obtaining adequate cooling at the trailing edge. Finally, an estimation will be made of the thermal stresses incurred with due allowance for redistribution of stress by creep: with cooled blades the thermal stresses can dominate the gas bending stresses and be comparable with the centrifugal tensile stresses. References to the literature dealing with these more advanced aspects can be found in the paper by Barnes and Dunham in Ref. (16).

It is important to minimize the parasitic losses resulting from air being bled from the compressor to cool the turbine. A great deal of research and development has been carried out to provide improved heat transfer systems and this has inevitably led to blades of increased complexity requiring considerable advances in manufacturing technology. Figure 7.36 shows the continuous development of cooled blades for the RB-211 family of engines; it can be seen that two sources of bleed air are used, an HP feed from compressor delivery and an LP feed from an earlier stage. This approach decreases the useful work required to compress the cooling air.

Finally, mention must be made of the possibility of using ceramic materials for the high-temperature blading; this would permit the use of uncooled blades at higher temperatures than could be achieved using metallic blades. Much effort has been expended on the development of silicon nitride and silicon carbide materials for small turbine rotors (both axial and radial) in which it would be difficult to incorporate cooling passages. Adequate reliability and life are difficult to achieve, but demonstrator engines have been run for short periods. Ceramic rotor blades have been investigated for use in stationary gas turbines for powers up to about 5 MW, and field tests were carried out in the late 1990s. About 1000 hours of endurance running were achieved before the blading was destroyed by impact from a small object which broke loose within the

■ L.P. cooling air   ■ H.P. cooling air

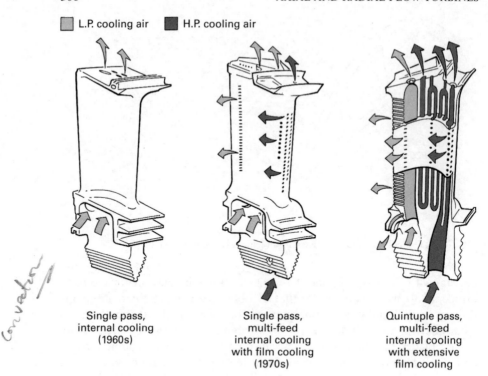

Single pass,
internal cooling
(1960s)

Single pass,
multi-feed
internal cooling
with film cooling
(1970s)

Quintuple pass,
multi-feed
internal cooling
with extensive
film cooling

FIG. 7.36 Development of cooled blades [courtesy Rolls-Royce plc]

combustor. It appears that ceramic blading is still subject to brittle failure; details of the development programme are given in Ref. (17). The use of ceramic turbines in production engines remains an elusive goal more than three decades after optimistic forecasts about their introduction. Research continues, however, and ceramic materials may some day have a significant impact on gas turbine design.

7.7   The radial flow turbine

Figure 7.37 illustrates a rotor having the back-to-back configuration mentioned in the introduction to this chapter, and Ref. (18) discusses the development of a successful family of radial industrial gas turbines. In a radial flow turbine, gas flow with a high tangential velocity is directed inwards and leaves the rotor with as small a whirl velocity as practicable near the axis of rotation. The result is that the turbine looks very similar to the centrifugal compressor, but with a ring of nozzle vanes replacing the diffuser vanes as in Fig. 7.38. Also, as shown there would normally be a diffuser at the outlet to reduce the exhaust velocity to a negligible value.

FIG. 7.37 Back-to-back rotor [courtesy Kongsberg Ltd]

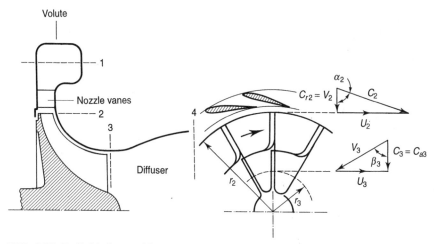

FIG. 7.38 Radial inflow turbine

The velocity triangles are drawn for the normal design condition in which the relative velocity at the rotor tip is radial (i.e. the incidence is zero) and the absolute velocity at exit is axial. Because C_{w3} is zero, the specific work output W becomes simply

$$W = c_p(T_{01} - T_{03}) = C_{w2}U_2 = U_2^2 \qquad (7.46)$$

In the ideal isentropic turbine with perfect diffuser the specific work output would be

$$W' = c_p(T_{01} - T_4') = C_0^2/2$$

where the velocity equivalent of the isentropic enthalpy drop, C_0, is sometimes called the 'spouting velocity' by analogy with hydraulic turbine practice. For this ideal case it follows that $U_2^2 = C_0^2/2$ or $U_2/C_0 = 0.707$. In practice, it is found that a good overall efficiency is obtained if this velocity ratio lies between 0.68 and 0.71, Ref. (19). In terms of the turbine pressure ratio, C_0 is given by

$$\frac{C_0^2}{2} = c_p T_{01} \left[1 - \left(\frac{1}{p_{01}/p_a} \right)^{(\gamma-1)/\gamma} \right]$$

Figure 7.39 depicts the processes in the turbine and exhaust diffuser on the T–s diagram. The overall isentropic efficiency of the turbine and diffuser may be expressed by

$$\eta_0 = \frac{T_{01} - T_{03}}{T_{01} - T_4'} \tag{7.47}$$

because $(T_{01} - T_4')$ is the temperature equivalent of the maximum work that could be produced by an isentropic expansion from the inlet state (p_{01}, T_{01}) to p_a. Considering the turbine alone, however, the efficiency is more suitably expressed by

$$\eta_t = \frac{T_{01} - T_{03}}{T_{01} - T_3'} \tag{7.48}$$

which is the 'total-to-static' isentropic efficiency referred to in section 7.1.

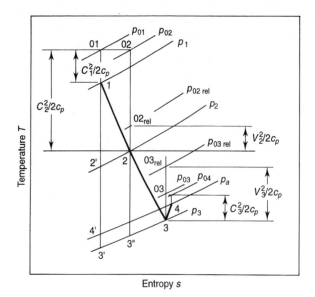

FIG. 7.39 *T–s* **diagram for a radial flow turbine**

Following axial flow turbine practice, the nozzle loss coefficient may be defined in the usual way by

$$\lambda_N = \frac{T_2 - T_2'}{C_2^2/2c_p} \qquad (7.49)$$

Similarly, the rotor loss coefficient is given by

$$\lambda_R = \frac{T_3 - T_3''}{V_3^2/2c_p} \qquad (7.50)$$

Plane 2 is located at the periphery of the rotor, so that the nozzle loss includes not only the loss in the inlet volute but also any friction loss in the vaneless space between the nozzle vane trailing edge and rotor tip. Bearing in mind that for the small constant pressure processes denoted by $2'$–2 and $3'$–$3''$ we can write $c_p \delta T = T \delta s$, λ_N can be alternatively expressed by

$$\lambda_N = \frac{T_3'' - T_3'}{C_2^2/2c_p} \cdot \frac{T_2'}{T_3'} \qquad (7.51)$$

A useful expression for η_t in terms of the nozzle and rotor loss coefficients can be found as follows. The denominator in equation (7.48) may be expanded to yield

$$T_{01} - T_3' = (T_{01} - T_{03}) + (T_{03} - T_3) + (T_3 - T_3'') + (T_3'' - T_3')$$

$$= (T_{01} - T_{03}) + \frac{C_3^2}{2c_p} + \lambda_R \frac{V_3^2}{2c_p} + \lambda_N \frac{C_2^2}{2c_p} \frac{T_3'}{T_2'}$$

Consequently η_t becomes

$$\eta_t = \left[1 + \frac{1}{2c_p(T_{01} - T_{03})} \left\{ C_3^2 + \lambda_R V_3^2 + \lambda_N C_2^2 \frac{T_3'}{T_2'} \right\} \right]^{-1}$$

From the velocity triangles,

$$C_2 = U_2 \operatorname{cosec} \alpha_2, \ V_3 = U_3 \operatorname{cosec} \beta_3, \ C_3 = U_3 \cot \beta_3$$

α_2 is the gas angle at inlet to the impeller and hence the effective outlet angle of the nozzle vanes, while β_3 is the outlet angle of the impeller vanes. Furthermore, $U_3 = U_2 r_3/r_2$ and $c_p(T_{01} - T_{03}) = U_2^2$, so that the expression for efficiency finally becomes

$$\eta_t = \left[1 + \frac{1}{2} \left\{ \left(\frac{r_3}{r_2} \right)^2 (\cot^2 \beta_3 + \lambda_R \operatorname{cosec}^2 \beta_3) + \lambda_N \frac{T_3'}{T_2'} \operatorname{cosec}^2 \alpha_2 \right\} \right]^{-1} \qquad (7.52)$$

Here η_t is expressed in terms of the nozzle and impeller outlet angles, radius ratio of impeller, loss coefficients, and the temperature ratio T_3'/T_2'. The temperature ratio T_3'/T_2' can in turn be expressed in terms of the major design variables. As example 7.1 will show, the term in equation (7.52) containng T_3'/T_2' is so small that it has little effect on η_t. Nevertheless T_3'/T_2' can be determined as follows.

$$\frac{T_3'}{T_2'} \approx \frac{T_3''}{T_2} = 1 - \frac{T_2 - T_3''}{T_2} = 1 - \frac{1}{T_2}[(T_2 - T_3) + (T_3 - T_3'')]$$

Now $(T_2 - T_3)$ may be found by expanding equation (7.46) and making use of the velocity triangles in Fig. 7.38. Thus, since $T_{01} = T_{02}$,

$$U_2^2 = c_p(T_2 - T_3) + \tfrac{1}{2}(C_2^2 - C_3^2)$$
$$= c_p(T_2 - T_3) + \tfrac{1}{2}(V_2^2 + U_2^2) - \tfrac{1}{2}(V_3^2 - U_3^2)$$
$$(T_2 - T_3) = \frac{1}{2c_p}[(V_3^2 - V_2^2) + (U_2^2 - U_3^2)]$$

(Note that $T_{0\mathrm{rel}} = T + V^2/2c_p$ is not the same at inlet and outlet of the rotor as it is in the axial flow machine, because $U_3 \neq U_2$. This is the main difference between Fig. 7.39 and Fig. 7.4.) It follows that

$$\frac{T_3'}{T_2'} \approx \frac{T_3''}{T_2} = 1 - \frac{1}{2c_p T_2}[(V_3^2 - V_2^2) + (U_2^2 - U_3^2) + \lambda_R V_3^2]$$

$$= 1 - \frac{U_2^2}{2c_p T_2}\left[1 + \left(\frac{r_3}{r_2}\right)^2 \{(1 + \lambda_R)\operatorname{cosec}^2 \beta_3 - 1\} - \cot^2 \alpha_2\right]$$

$$(7.53)$$

And, finally, T_2 may be expressed as

$$T_2 = T_{01} - \frac{U_2^2}{2c_p}\operatorname{cosec}^2 \alpha_2$$

λ_N is usually obtained from separate tests on the inlet volute and nozzle vane assembly, enabling λ_R to be deduced from overall efficiency measurements with the aid of equation (7.52) as in the following example.

EXAMPLE 7.1

A radial flow turbine of the following geometry is tested under design conditions with a pressure ratio p_{01}/p_3 of 2·0 and an inlet temperature of 1000 K. The work output, after allowance has been made for mechanical losses, is 45·9 kW when the rotational speed is 1000 rev/s and the gas mass flow is 0·322 kg/s.

Rotor inlet tip diameter	12·7 cm
Rotor exit tip diameter	7·85 cm
Hub/tip ratio at exit	0·30
Nozzle efflux angle α_2	70°
Rotor vane outlet angle β_3	40°

Separate tests on the volute/nozzle combination showed the nozzle loss coefficient λ_N to be 0·070. The turbine isentropic efficiency η_t and the rotor loss coefficient λ_R are to be evaluated from the results.

$$T_{01} - T'_3 = T_0\left[1 - \left(\frac{1}{p_{01}/p_3}\right)^{(\gamma-1)/\gamma}\right] = 1000\left[1 - \left(\frac{1}{2·0}\right)^{\frac{1}{4}}\right] = 159·1\,\text{K}$$

Since $W = mc_p(T_{01} - T_{03})$,

$$T_{01} - T_{03} = \frac{45·9}{0·322 \times 1·148} = 124·2\,\text{K}$$

$$\eta_t = \frac{124·2}{159·1} = 0·781$$

Putting $T'_3/T'_2 = 1·0$ in equation (7.52), with r_3/r_2 given by

$$\frac{r_3}{r_2} = \frac{0·3 \times 7·85 + 7·85}{2 \times 12·7} = 0·402$$

we get

$$0·781 \approx [1 + \tfrac{1}{2}\{0·402^2(\cot^2 40° + \lambda_R \csc^2 40°) + 0·07\csc^2 70°\}]^{-1}$$

$$1·280 = 1 + 0·1148 + 0·1956\lambda_R + 0·0396$$

$$\lambda_R = 0·64$$

It can be seen from the foregoing that the term containing T'_3/T'_2 is small. From the data, $U_2 = \pi \times 1000 \times 12·7 = 390\,\text{m/s}$. Using this in equation (7.53) we get T'_3/T'_2 equal to $(0·921 - 0·028)\lambda_R$. Substituting this result in equation (7.52) and recalculating, we have $\lambda_R = 0·66$. The approximate value of 0·64 is within the margin of experimental uncertainty.

One of the more comprehensive research programmes on radial flow turbines was that carried out by Ricardo and Co. and reported in Ref. (20). Turbine efficiencies of up to 90 per cent were obtained under optimum running conditions with a 12·5 cm diameter rotor having 12 blades. The optimum number of nozzle vanes was found to be 17. Values of λ_N varied from 0·1 to 0·05, decreasing steadily with increase of nozzle angle from 60° to 80°. λ_R varied more widely, from 0·5 to 1·7, increasing rapidly as the nozzle angle was increased above 70°. In spite of this variation in λ_R the overall turbine efficiency was relatively insensitive to nozzle angle: it fell by only 2 per cent as α_2 was increased from 70° to 80°. Also studied in the programme were the effects of varying the axial width of the vanes at the rotor tip, radial width of the vaneless space, and clearance between vanes and casing. The optimum width of vane at the rotor tip was about 10 per cent of the rotor diameter, and the performance seemed insensitive to the radial width of the vaneless space. There was a fall in efficiency of 1 per cent for every increase in clearance of 1 per cent of the rotor vane width (averaged between rotor inlet and outlet). This implies that the radial flow turbine is rather less sensitive to clearance loss than its axial flow counterpart. Because clearances cannot be reduced in proportion to blade height as the size is reduced, as in axial flow turbines, this lower sensitivity is probably the basic reason why radial flow has the advantage for very small turbines.

Finally, various major alterations to the geometry of the rotor were made, including scalloping the disc between the blades to reduce disc stresses, weight and inertia: the reader must turn to Ref. (20) for the consequences of such changes. All we can do here is to give a rough guide to the major dimensions determining the basic shape of the rotor, viz. the hub/tip diameter ratio of the vanes at the rotor exit and the ratio of the vane tip diameter at the exit to the rotor disc diameter. The former should not be much less than 0·3 to avoid excessive blade blockage at the hub, and the latter should be limited to a maximum of about 0·7 to avoid excessive curvature of the rotor passages.

Methods of dealing with the losses other than by the simple use of λ_N and λ_R have been devised, and Ref. (21) provides a useful comparison of the various coefficients that have been used. In particular, if the relative velocity is not radial at inlet to the rotor there is an additional loss, variously described as an 'incidence loss' or 'shock loss', for which a separate loss coefficient is desirable. The use of such a coefficient becomes essential when trying to predict off-design performance because then the flow relative to the rotor vanes may depart substantially from the radial direction. There is no shock in the gas dynamic sense, but there is a shock in the ordinary sense of the word when the fluid impinging on the rotor vanes at some incidence β_2 (Fig. 7.40) is suddenly constrained to move in the radial direction. There is a resulting drop of stagnation pressure and increase in entropy which moves line 2–3 to the right on the T–s diagram. Bridle and Boulter [Ref. (22)] suggest that in the incidence range $\beta_2 = \pm 65°$, this stagnation pressure loss can be accounted for by the expression

$$\frac{\Delta p_0}{(p_{02} - p_2) \cos^2 \beta_2} = (\tan \beta_2 + 0\cdot 1)^2$$

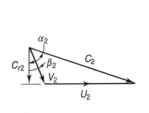

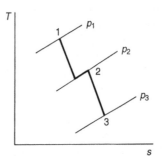

FIG. 7.40 Effect of incidence loss

Making use of the test results in Ref. (20), Bridle and Boulter have also deduced equations for the other components of the rotor loss, i.e. friction loss in the rotor passages and clearance loss, in an attempt to make possible the prediction of off-design performance. Benson in Ref. (23) describes one method of tackling this problem, and includes in an Appendix the Fortran programs necessary for the calculation.

NOMENCLATURE

For velocity triangle notation (U, C, V, α, β) see Fig. 7.1 and Fig. 7.38. For blade geometry notation (s, c, o, t, θ, i) see Fig. 7.11.

C_L	blade lift coefficient
h	blade height, heat transfer coefficient
n	number of blades
S	perimeter
z	section modulus of blade
h/c	aspect ratio
k/h	tip clearance/blade height ratio
s/c	pitch/chord ratio
t/c	thickness/chord ratio
t_e/s	trailing edge thickness/pitch ratio
Y_N	nozzle blade loss coefficient $[(p_{01} - p_{02})/(p_{02} - p_2)]$
Y_R	rotor blade loss coefficient $[(p_{02\mathrm{rel}} - p_{03\mathrm{rel}})/(p_{03\mathrm{rel}} - p_3)]$
Y_k	tip-clearance loss coefficient
Y_p	profile loss coefficient
Y_s	secondary loss coefficient
λ_N	nozzle blade loss coefficient $[(T_2 - T_2')/(C_2^2/2c_p)]$
λ_R	rotor blade loss coefficient $[(T_3 - T_3'')/(V_3^2/2c_p)]$
σ_{ct}	centrifugal tensile stress
σ_{gb}	gas bending stress
Λ	degree of reaction $[(T_2 - T_3)/(T_1 - T_3)]$
ϕ	flow coefficient (C_a/U)
ψ	stage temperature drop coefficient ($2c_p\Delta T_{0s}/U^2$)

Suffixes

a, w	axial, whirl, component
b	blade
c	coolant
m, r, t	at mean, root, tip, radius
N, R	nozzle, rotor, blades
p, s	pressure, suction, surface of blade

8

Prediction of performance of simple gas turbines

From cycle calculations such as those of Chapter 2, it is possible to determine the pressure ratio which for any given maximum cycle temperature will give the greatest overall efficiency, and the mass flow required to give the desired power. When such preliminary calculations have been made, the most suitable design data for any particular application may be chosen. It is then possible to design the individual components of a gas turbine so that the complete unit will give the required performance when running at the *design point*; that is, when it is running at the particular speed, pressure ratio and mass flow for which the components were designed. The problem then remains to find the variation of performance of the gas turbine over the complete operating range of speed and power output, which is normally referred to as *off-design performance*.

The performance characteristics of the individual components may be estimated on the basis of previous experience or obtained from actual tests. When the components are linked together in an engine, the range of possible operating conditions for each component is considerably reduced. The problem is to find corresponding operating points on the characteristics of each component when the engine is running at a steady speed, or in *equilibrium* as it is frequently termed. The equilibrium running points for a series of speeds may be plotted on the compressor characteristic and joined up to form an *equilibrium running line* (or *zone*, depending upon the type of gas turbine and load), the whole forming an *equilibrium running diagram*. When once the operating conditions have been determined, it is a relatively simple matter to obtain performance curves of power output or thrust, and specific fuel consumption.

The equilibrium running diagram also shows the proximity of the operating line or zone to the compressor surge line. If it intersects the surge line the gas turbine will not be capable of being brought up to full speed without some remedial action being taken. It is this phenomenon which was referred to when speaking of 'stability of operation' in the Introduction. Finally, it shows whether the engine is operating in a region of adequate compressor efficiency; ideally the operating line or zone should lie near the locus of points of maximum compressor efficiency shown in Fig. 4.10(a).

The variation of specific fuel consumption with reduction in power, sometimes referred to as *part-load performance*, is of major importance in applications where considerable running at low power settings is required. This would be the case for any vehicular gas turbine, and the poor specific fuel consumption at part load is probably the biggest disadvantage of the gas turbine for vehicular use. The fuel consumption of aircraft gas turbines at reduced power is of critical importance when extensive delays occur either at the point of departure or the destination. In one case the idling fuel consumption when taxiing is important, and in the other the fuel flow at low flight speeds and medium altitudes is critical.

When determining the off-design performance it is important to be able to predict not only the effect on specific fuel consumption of operation at part load, but also the effect of ambient conditions on maximum output. The effects of high and low ambient temperatures and pressures must all be considered. Land-based gas turbines may operate between ambient temperatures of $-60\,^{\circ}\text{C}$ in the Arctic and $50\,^{\circ}\text{C}$ in the tropics, and at altitudes from sea level to about 3000 metres, while aircraft gas turbines have to operate over much wider ranges of inlet temperature and pressure. The variation of maximum power with ambient conditions is clearly of prime importance to the customer, and the manufacturer must be prepared to guarantee the performance available at any specified condition. If, for example, we consider the use of gas turbines for peak-load generation of electricity, the peak loads will occur in the coldest days of winter in Europe, but are much more likely to occur in the summer in the United States because of the heavy demand for air-conditioning and refrigeration systems. The effect of ambient conditions on the performance of aircraft gas turbines has a critical effect on the runway length required and the payload that can be carried, and hence on both safety and economics.

The basic methods for determining equilibrium off-design performance of simple gas turbines will be described here, while more complex arrangements and transient operation are dealt with in Chapter 9. The types of gas turbine discussed in this chapter will be (a) the single-shaft unit delivering shaft power, (b) the free turbine engine, where the gas-generator turbine drives the compressor and the power turbine drives the load, and (c) the simple jet engine where the useful output is a high-velocity jet produced in the propelling nozzle. Schematics of these layouts are shown in Fig. 8.1, and it can readily be observed that the gas generator performs exactly the same function for both the free turbine engine and the jet engine. The flow characteristics of a free turbine and a nozzle are similar, and they impose the same restrictions on the operation of the gas generator, with the result that the free turbine engine and the jet engine are

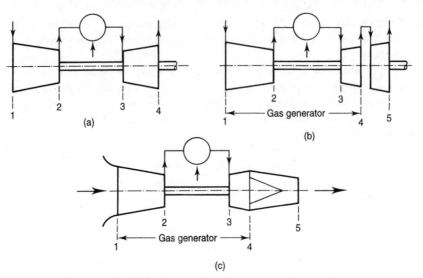

FIG. 8.1 Simple gas turbine units

thermodynamically similar and differ only in the manner in which the output is utilized. Several successful jet engines have been converted to shaft power use by substituting a free power turbine for the propelling nozzle, and this approach has been widely used for pipeline compression, marine propulsion and electricity generation. An important practical result of the flow equivalency of power turbines and propelling nozzles is that overhaul shops can test gas generators on a jet engine test bed, without the need for a complex and expensive dynamometer. In pipeline and marine applications overhaul is normally done by replacing the gas generator while leaving the power turbine *in situ*; it is common to replace a gas generator within a 24 hour period, and some ships actually carry a spare gas generator. Pipelines often own several extra gas generators, which can be installed to meet overhaul requirements at different sites..

All off-design calculations depend on satisfying the essential conditions of compatibility of mass flow, work and rotational speed between the various components. It is logical to deal with the single-shaft engine first, and then proceed to the free turbine engine where there is the added complication of flow compatibility between the gas-generator turbine and the power turbine. Lastly, we will deal with the jet engine where there is the further complication of forward speed and altitude effects.

8.1   Component characteristics

The variation of mass flow, pressure ratio and efficiency with rotational speed of the compressor and turbine is obtained from the compressor and

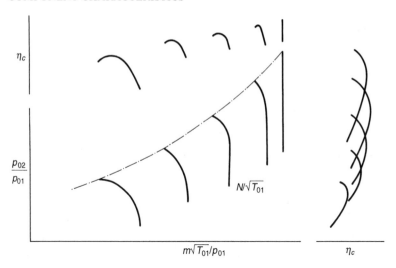

FIG. 8.2 Compressor characteristics

turbine characteristics, examples of which are given in Chapters 4, 5 and 7. It is convenient to represent the compressor characteristic as shown in Fig. 8.2, with the variation of efficiency along each constant speed line plotted against both mass flow and pressure ratio. With high-performance axial compressors the constant speed lines become vertical on a mass flow basis when the inlet is choked, and in this region it is essential to plot the efficiency as a function of pressure ratio. The turbine characteristic can be used in the form given in Fig. 7.28. It is often found in practice, however, that turbines do not exhibit any significant variation in non-dimensional

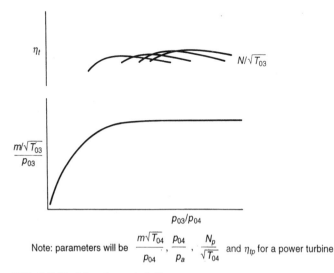

Note: parameters will be $\dfrac{m\sqrt{T_{04}}}{p_{04}}$, $\dfrac{p_{04}}{p_a}$, $\dfrac{N_p}{\sqrt{T_{04}}}$ and η_{tp} for a power turbine

FIG. 8.3 Turbine characteristics

flow with non-dimensional speed, and in most cases the turbine operating region is severely restricted by another component downstream of it. In explaining the method used for off-design performance calculations it will initially be assumed that the mass flow function can be represented by a single curve as in Fig. 8.3. The modification necessary to account for a family of constant speed curves will be discussed in section 8.3.

For accurate calculations it is necessary to consider the variation of pressure losses in the inlet ducting, the combustion chamber and the exhaust ducting. These are essentially secondary effects, however, and the off-design calculations will be introduced on the basis of negligible inlet and exhaust losses and a combustion chamber pressure loss which is a fixed percentage of the compressor delivery pressure. Such approximations are quite adequate for many purposes. For detailed calculations it would be normal to use a digital computer, when methods of allowing for variable pressure losses are easily introduced. Further discussion of this matter will be deferred until section 8.7.

8.2   Off-design operation of the single-shaft gas turbine

Referring to the single-shaft gas turbine shown in Fig. 8.1(a) it can readily be seen that when inlet and exhaust pressure losses are ignored, the pressure ratio across the turbine is determined by the compressor pressure ratio and the pressure loss in the combustion chamber. The mass flow through the turbine will be equal to the mass flow delivered by the compressor, less any bleeds, but supplemented by the fuel flow; it has been pointed out earlier that the bleeds are often approximately equal to the fuel flow. In general terms, the procedure for obtaining an equilibrium running point is as follows.

(a) Select a constant speed line on the compressor characteristic and choose any point on this line; the values of $m \sqrt{T_{01}}/p_{01}$, p_{02}/p_{01}, η_c and $N/\sqrt{T_{01}}$ are then determined.

(b) The corresponding point on the turbine characteristic is obtained from consideration of compatibility of rotational speed and flow.

(c) Having matched the compressor and turbine characteristics, it is necessary to ascertain whether the work output corresponding to the selected operating point is compatible with that required by the driven load; this requires knowledge of the variation of power with speed, which depends on the manner in which the power is absorbed.

The compressor and turbine are directly coupled together, so that compatibility of rotational speed requires

$$\frac{N}{\sqrt{T_{03}}} = \frac{N}{\sqrt{T_{01}}} \times \sqrt{\frac{T_{01}}{T_{03}}} \qquad\qquad (8.1)$$

Compatibility of flow between the compressor and turbine can be expressed in terms of the non-dimensional flows by the identity

$$\frac{m_3\sqrt{T_{03}}}{p_{03}} = \frac{m_1\sqrt{T_{01}}}{p_{01}} \times \frac{p_{01}}{p_{02}} \times \frac{p_{02}}{p_{03}} \times \sqrt{\frac{T_{03}}{T_{01}}} \times \frac{m_3}{m_1}$$

The pressure ratio p_{03}/p_{02} can be obtained directly from the combustion pressure loss, i.e. $p_{03}/p_{02} = 1 - (\Delta p_b/p_{02})$. It will normally be assumed that $m_1 = m_3 = m$, but variation in mass flow at different points in the engine can easily be included if required. Rewriting the previous equation in terms of m, we get

$$\frac{m\sqrt{T_{03}}}{p_{03}} = \frac{m\sqrt{T_{01}}}{p_{01}} \times \frac{p_{01}}{p_{02}} \times \frac{p_{02}}{p_{03}} \times \sqrt{\frac{T_{03}}{T_{01}}} \qquad (8.2)$$

Now $m\sqrt{T_{01}}/p_{01}$ and p_{02}/p_{01} are fixed by the chosen operating point on the compressor characteristic, p_{03}/p_{02} is assumed to be constant and $m\sqrt{T_{03}}/p_{03}$ is a function of the turbine pressure ratio p_{03}/p_{04}. Neglecting inlet and exhaust pressure losses $p_a = p_{01} = p_{04}$, so that the turbine pressure ratio can be calculated from $p_{03}/p_{04} = (p_{03}/p_{02})(p_{02}/p_{01})$. Thus all the terms of equation (8.2) with the exception of $\sqrt{(T_{03}/T_{01})}$ can be obtained from the compressor and turbine characteristics. The turbine inlet temperature T_{03} can therefore be obtained from equation (8.2) when the ambient temperature, which is equal to T_{01}, is specified.

Having determined the turbine inlet temperature, the turbine non-dimensional speed $N/\sqrt{T_{03}}$ is obtained from equation (8.1). The turbine efficiency can then be obtained from the turbine characteristic using the known values of $N/\sqrt{T_{03}}$ and p_{03}/p_{04}, and the turbine temperature drop can be calculated from

$$\Delta T_{034} = \eta_t T_{03}\left[1 - \left(\frac{1}{p_{03}/p_{04}}\right)^{(\gamma-1)/\gamma}\right] \qquad (8.3)$$

The compressor temperature rise for the point selected on the compressor characteristic can be similarly calculated as

$$\Delta T_{012} = \frac{T_{01}}{\eta_c}\left[\left(\frac{p_{02}}{p_{01}}\right)^{(\gamma-1)/\gamma} - 1\right] \qquad (8.4)$$

The net power output corresponding to the selected operating point is then found from

$$\text{net power output} = mc_{pg}\Delta T_{034} - \frac{1}{\eta_m}mc_{pa}\Delta T_{012}\ † \qquad (8.5)$$

---

† Distinguishing suffixes will be added to c_p only in equations where both c_{pa} and c_{pg} appear simultaneously. In other cases it will be clear from the context which mean c_p should be used. γ will be treated similarly.

where η_m is the mechanical efficiency of the compressor–turbine combination, and m is given by $(m\sqrt{T_{01}}/p_{01})(p_a/\sqrt{T_a})$ for prescribed ambient conditions.

Finally, it is necessary to consider the characteristics of the load to determine whether the compressor operating point selected represents a valid solution. If, for example, the engine were run on a test bed coupled to a hydraulic or electrical dynamometer, the load could be set independently of the speed and it would be possible to operate at any point on the compressor characteristic within the temperature limit imposed by safety considerations. With a fixed-pitch propeller load, however, the power absorbed varies as the cube of the rotational speed of the propeller. When the transmission efficiency and gear ratio are known, the load characteristic in terms of the net power actually required from the turbine and the turbine speed can be plotted as in Fig. 8.4. The problem then becomes one of finding the *single* point on each constant speed line of the compressor characteristic which will give the required net power output at that speed; this can only be done by trial and error, taking several operating points on the compressor characteristic and establishing the power output corresponding to each one. If the calculated net power output for any point on the compressor characteristic is not equal to the power required at the selected speed, the engine will not be in equilibrium and will either accelerate or decelerate depending on whether there is a surplus or deficiency of power. Repeating this procedure for a series of constant speed lines, a series of points are obtained, which can be joined up to form the equilibrium running line as shown in Fig. 8.5.

The most common types of load used with single-shaft gas turbines are the electrical generator and variable-pitch aircraft propellers, both of which run at a constant rotational speed with the load varied electrically or by changing pitch at different flight conditions. Thus the equilibrium lines for these loads would correspond to a particular line of constant non-dimensional speed, as shown in Fig. 8.5, and each point on this line would

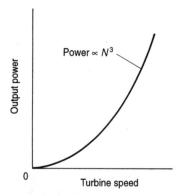

FIG. 8.4 Load characteristics

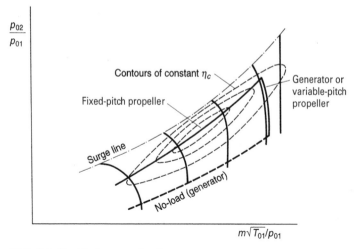

FIG. 8.5 Equilibrium running lines

represent a different value of fuel flow, turbine inlet temperature and power output. At each speed it is possible to find, by trial and error, the compressor operating point corresponding to zero net output, and the no-load running line is also shown in Fig. 8.5.

The equilibrium running lines depicted show that a fixed-pitch propeller implies operation in a zone of high compressor efficiency over a wide range of output, whereas the generator and variable-pitch propeller result in a rapid drop in compressor efficiency as load is reduced. The location of the equilibrium running line relative to the surge line indicates whether the engine can be brought up to full power without any complications. The equilibrium running line for the fixed-pitch propeller lies close to the surge line, and may even intersect it, in which case the engine could not be accelerated to full power. This can be overcome by incorporating a blow-off valve towards the rear of the compressor. Section 8.6 deals with the matter in more detail. The running line for the generator or variable-pitch propeller can be seen to be well away from surge, and the gas turbine could be accelerated to full speed before applying the load without any surge problem being encountered.

The calculations described above determine the values of all the parameters required for a complete performance calculation for any point within the operating range. T_{03} is known, and T_{02} is found from $(\Delta T_{012} + T_{01})$. Thus the combustion temperature rise is known and it is possible to obtain the fuel/air ratio f from the curves of Fig. 2.17 and an assumed value of combustion efficiency. The fuel flow is then given by mf. From the fuel flow and power output at each operating point the variation in specific fuel consumption (or thermal efficiency) with load can be determined. The results refer to operation at the assumed value of T_{01} ($=T_a$) and p_{01} ($=p_a$), but the process could be repeated over the range of values of ambient temperature and pressure likely to be encountered.

The matching calculations for a single-shaft gas turbine are illustrated in the following example.

EXAMPLE 8.1

The following data refer to a single-shaft gas turbine operating at its design speed.

Compressor characteristic			Turbine characteristic	
p_{02}/p_{01}	$m\sqrt{T_{01}}/p_{01}$	η_c	$m\sqrt{T_{03}}/p_{03}$	η_t
5·0	329·0	0·84	139·0	0·87
4·5	339·0	0·79	(both constant over	
4·0	342·0	0·75	range of pressure ratio	
			considered)	

Assuming ambient conditions of 1·013 bar and 288 K, a mechanical efficiency of 98 per cent, and neglecting all pressure losses, calculate the turbine inlet temperature required for a power output of 3800 kW. The 'non-dimensional' flows are expressed in terms of kg/s, K and bar.

The method of solution is to establish, for each point given on the compressor characteristic, the turbine inlet temperature from equation (8.2), the compressor temperature rise from equation (8.4) and the turbine temperature drop from equation (8.3). Once these have been established the power output can be obtained from equation (8.5); it is then necessary to plot turbine inlet temperature against power output to find the required temperature for an output of 3800 kW.

Taking the compressor operating point at a pressure ratio of 5·0,

$$\sqrt{\frac{T_{03}}{T_{01}}} = \frac{(m\sqrt{T_{03}}/p_{03})(p_{03}/p_{01})}{m\sqrt{T_{01}}/p_{01}}$$

With pressure losses neglected $p_{03} = p_{02}$, and hence

$$\sqrt{\frac{T_{03}}{T_{01}}} = \frac{139·0 \times 5·0}{329·0} = 2·11$$

giving $T_{03} = 1285$ K.

The compressor temperature rise is given by

$$\Delta T_{012} = \frac{288}{0·84}[5·0^{1/3·5} - 1] = 200·5 \text{ K}$$

The turbine temperature drop is given by

$$\Delta T_{034} = 0 \cdot 87 \times 1285 \left[1 - \frac{1}{5 \cdot 0^{1/4}} \right] = 370 \cdot 0 \, \text{K}$$

The air mass flow is obtained from the non-dimensional flow entering the compressor,

$$m = 329 \times \frac{1 \cdot 013}{\sqrt{288}} = 19 \cdot 64 \, \text{kg/s}$$

The power output can now be obtained, giving

$$\text{power output} = 19 \cdot 64 \times 1 \cdot 148 \times 370 \cdot 0 - \left(\frac{19 \cdot 64 \times 1 \cdot 005 \times 200 \cdot 5}{0 \cdot 98} \right)$$

$$= 8340 - 4035$$

$$= 4305 \, \text{kW}$$

Thus for a power output of 4305 kW the turbine inlet temperature is 1285 K.

Repeating the calculation for the three points given on the compressor characteristic yields the following results:

p_{02}/p_{01}	T_{03} [K]	ΔT_{012} [K]	ΔT_{034} [K]	m [kg/s]	Power output [kW]
5·0	1285	200·5	370·0	19·64	4305
4·5	982	196·1	267·0	20·25	2130
4·0	761	186·7	194·0	20·4	635

Plotting the value of T_{03} against power output, it is found that for an output of 3800 kW the required turbine inlet temperature is 1215 K.

8.3 Equilibrium running of a gas generator

It was pointed out at the start of this chapter that the gas generator performs the same function for both the free turbine engine and the jet engine, namely the generation of a continuous flow of gas at high pressure and temperature which can be expanded to a lower pressure to produce either shaft work or a high-velocity propulsive jet. Before considering either type of engine, it is appropriate to consider the behaviour of the gas generator alone.

Considerations of compatibility of speed and flow are the same as for the single-shaft engine described in the previous section, and equations (8.1) and (8.2) are applicable. This time, however, the pressure ratio across the turbine is not known, and it must be determined by equating the turbine work to the compressor work. The required turbine temperature drop, in conjunction with the turbine inlet temperature and efficiency, determines the turbine pressure ratio. Thus, instead of equation (8.5) the work requirement is expressed by

$$\eta_m c_{pg} \Delta T_{034} = c_{pa} \Delta T_{012}$$

Rewriting in terms of non-dimensional groups,

$$\frac{\Delta T_{034}}{T_{03}} = \frac{\Delta T_{012}}{T_{01}} \times \frac{T_{01}}{T_{03}} \times \frac{c_{pa}}{c_{pg}\eta_m} \tag{8.6}$$

Equations (8.1), (8.2) and (8.6) are all linked by the temperature ratio T_{03}/T_{01} and it is necessary to determine, by trial and error, the turbine inlet temperature required for operation at any arbitrary point on the compressor characteristic. The procedure is as follows:

(a) Having selected a point on the compressor characteristic, the values of $N/\sqrt{T_{01}}$, p_{02}/p_{01}, $m\sqrt{T_{01}}/p_{01}$, and η_c are determined, and $\Delta T_{012}/T_{01}$ can be calculated from equation (8.4).

(b) If a value of p_{03}/p_{04} is guessed, the value of $m\sqrt{T_{03}}/p_{03}$ can be obtained from the turbine characteristic, enabling the temperature ratio T_{03}/T_{01} to be obtained from the flow compatibility equation (8.2).

(c) This value of T_{03}/T_{01} can now be used to calculate $N/\sqrt{T_{03}}$ from equation (8.1).

(d) With $N/\sqrt{T_{03}}$ and p_{03}/p_{04} known, the turbine efficiency can be obtained from the turbine characteristic.

(e) The non-dimensional temperature drop $\Delta T_{034}/T_{03}$ can be calculated from equation (8.3) and used in conjunction with equation (8.6) to calculate another value of T_{03}/T_{01}.

(f) This second value of T_{03}/T_{01} will not, in general, agree with the first value obtained from equation (8.2), indicating that the guessed value of the pressure ratio p_{03}/p_{04} is not valid for an equilibrium running point.

(g) A new value of p_{03}/p_{04} must now be assumed and the above calculations repeated until the same value of T_{03}/T_{01} is obtained from both equation (8.2) and (8.6).

(h) Agreement signifies that the turbine operating point is compatible with the originally chosen compressor operating point when the fuel flow is such as to produce the iterated value of T_{03}/T_{01}.

The procedure is summarized in the information flow chart of Fig. 8.6(a).

It would be possible to carry out this calculation for a large number of points on the compressor characteristic, and express the results by joining up points of constant T_{03}/T_{01} on the compressor characteristic as indicated by the dotted lines in Fig. 8.7. In practice this is unnecessary. The further requirement for flow compatibility with the component downstream of the gas generator, whether it be a power turbine or

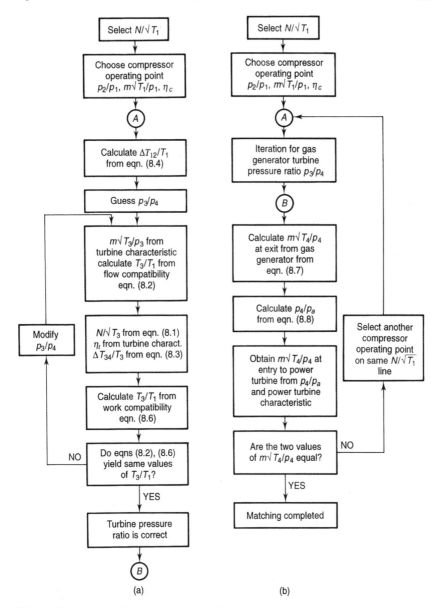

FIG. 8.6(a) Iteration for gas generator (b) Overall iteration procedure for free power turbine unit

propelling nozzle, seriously restricts the operating zone on the compressor characteristic. The foregoing procedure thus comprises only one part of the calculation for the whole unit, and when the whole unit is considered only relatively few points may be needed. The free turbine engine will be considered first.

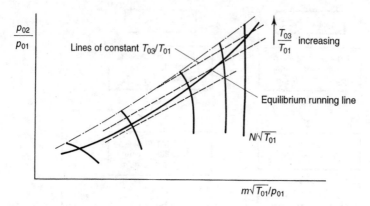

FIG. 8.7 Equilibrium running line for free turbine

Before doing so, it is necessary to emphasize that the matching procedure just outlined has been developed on the assumption that the turbine non-dimensional flow is independent of the non-dimensional speed and is a function only of pressure ratio. If the turbine characteristic does exhibit a variation of $m\sqrt{T_{03}}/p_{03}$ with $N/\sqrt{T_{03}}$, as in Fig. 7.28, the procedure must be modified. There is no change for any turbine operating point falling on the choking point of the mass flow curve, but for others the process would be as follows. Immediately after step (a) it is necessary to guess a value of T_{03}/T_{01}, which permits calculation of $N/\sqrt{T_{03}}$ from equation (8.1) and $m\sqrt{T_{03}}/p_{03}$ from equation (8.2). p_{03}/p_{04} and η_t can then be obtained from the turbine characteristic, enabling $\Delta T_{034}/T_{03}$ to be calculated from equation (8.3). The work compatibility equation (8.6) is then used to provide a value of T_{03}/T_{01} for comparison with the original guess. To avoid obscuring the main principles involved, the use of multi-line turbine flow characteristics will not be discussed further.† Enough has been said to show that they can be accommodated without undue difficulty.

8.4   Off-design operation of free turbine engine

Matching of gas generator with free turbine

The gas generator is matched to the power turbine by the fact that the mass flow leaving the gas generator must equal that at entry to the power turbine, coupled with the fact that the pressure ratio available to the power turbine is fixed by the compressor and gas-generator turbine

---

† The characteristic shown in Fig. 7.28 is typical of that for a turbine in which choking occurs in the stator passages. In some designs choking may occur initially at exit from the rotor passages; this will result in a small variation of the choking value of $m\sqrt{T_{03}}/p_{03}$ with $N/\sqrt{T_{03}}$. The modification to the calculation would then apply also to the choking region.

pressure ratios. The power turbine characteristic will have the same form as Fig. 8.3, but the parameters will be $m\sqrt{T_{04}}/p_{04}$, p_{04}/p_a, $N_p/\sqrt{T_{04}}$ and η_{tp}.

The preceding section described how the gas-generator operating conditions can be determined for any point on the compressor characteristic. The value of $m\sqrt{T_{04}}/p_{04}$ at exit from the gas generator can then be calculated from

$$\frac{m\sqrt{T_{04}}}{p_{04}} = \frac{m\sqrt{T_{03}}}{p_{03}} \times \frac{p_{03}}{p_{04}} \times \sqrt{\frac{T_{04}}{T_{03}}} \qquad (8.7)$$

where

$$\sqrt{\frac{T_{04}}{T_{03}}} = \sqrt{\left(1 - \frac{\Delta T_{034}}{T_{03}}\right)} \quad \text{and} \quad \frac{\Delta T_{034}}{T_{03}} = \eta_t \left[1 - \left(\frac{1}{p_{03}/p_{04}}\right)^{(\gamma-1)/\gamma}\right]$$

The corresponding pressure ratio across the power turbine can also be established from

$$\frac{p_{04}}{p_a} = \frac{p_{02}}{p_{01}} \times \frac{p_{03}}{p_{02}} \times \frac{p_{04}}{p_{03}} \qquad (8.8)$$

(It must be remembered that in the case of a stationary gas turbine with inlet and exit duct losses ignored $p_{01} = p_a$, and the power turbine outlet pressure is also equal to p_a.)

Having found the pressure ratio across the power turbine, the value of $m\sqrt{T_{04}}/p_{04}$ can be found from the power turbine characteristic for comparison with the value obtained from equation (8.7). If agreement is not reached it is necessary to choose another point on the same constant speed line of the compressor characteristic and repeat the procedure until the requirement of flow compatibility between the two turbines is satisfied. The overall procedure for the free turbine engine, including the iteration for the gas generator, is summarized in the information flow chart of Fig. 8.6(b).

For each constant $N/\sqrt{T_{01}}$ line on the compressor characteristic there will be only one point which will satisfy both the work requirement of the gas generator and flow compatibility with the power turbine. If the foregoing calculations are carried out for each constant speed line, the points obtained can be joined up to form the equilibrium running line as shown in Fig. 8.7. *The running line for the free turbine engine is independent of the load and is determined by the swallowing capacity of the power turbine.* This is in contrast to the behaviour of the single-shaft unit, where the running line depends on the characteristic of the load as indicated in Fig. 8.5.

The next step is to calculate the power output and specific fuel consumption for the equilibrium running points. Before discussing this, however, a useful approximation which simplifies the foregoing procedure will be mentioned. It arises from the behaviour of two turbines in series. We introduce it here because it facilitates a better physical understanding of some of the phenomena discussed in subsequent sections.

Matching of two turbines in series

The iterative procedure required for the matching of a gas generator and a free turbine can be considerably simplified if the behaviour of two turbines in series is considered; this approach is also valuable for the analysis of more complex gas turbines considered in Chapter 9. It was shown in the previous sub-section that by using equation (8.7) the value of $m\sqrt{T_{04}}/p_{04}$ at exit from the gas-generator turbine can be obtained for any gas-generator operating point, and in particular that it is a function of $m\sqrt{T_{03}}/p_{03}$, p_{03}/p_{04} and η_t. The value of η_t could be read from the turbine characteristic because $N/\sqrt{T_{03}}$ had been determined for the operating point in the course of the gas-generator calculation. Now in practice the variation of η_t at any given pressure ratio is not large (see Fig. 8.3), particularly over the restricted range of operation of the gas-generator turbine. Furthermore, such a variation has little effect on $m\sqrt{T_{04}}/p_{04}$ because the resulting change in $\sqrt{(T_{03}/T_{04})}$ is very small. It is often sufficiently accurate to take a mean value of η_t at any given pressure ratio, so that $m\sqrt{T_{04}}/p_{04}$ becomes a function only of $m\sqrt{T_{03}}/p_{03}$ and p_{03}/p_{04}. If this is done, a single curve representing the turbine outlet flow characteristic can readily be obtained by applying equation (8.7) to points on the single curve of the inlet flow characteristic. The result is shown by the dotted curve in Fig. 8.8.

The effect of operating two turbines in series is also shown in Fig. 8.8 where it can be seen that the requirement for flow compatibility between the two turbines places a major restriction on the operation of the gas-generator turbine. In particular, as long as the *power turbine is choked* the gas-generator turbine will operate at a *fixed non-dimensional point*, i.e. at the pressure ratio marked (a). With the power turbine unchoked, the gas-generator will be restrained to operate at a fixed pressure ratio for each power turbine pressure ratio, e.g. (b) and (c). Thus the maximum pressure ratio across the gas-generator turbine is controlled by choking of the

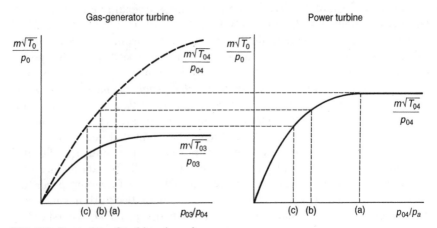

FIG. 8.8 Operation of turbines in series

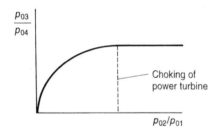

FIG. 8.9

power turbine, and at all times the pressure ratio is controlled by the swallowing capacity of the power turbine.

A further consequence of the fixed relation between the turbine pressure ratios is that it is possible to plot the gas-generator pressure ratio p_{03}/p_{04} against compressor pressure ratio p_{02}/p_{01} by using the identity

$$\frac{p_{03}}{p_{04}} = \frac{p_{03}}{p_{02}} \times \frac{p_{02}}{p_{01}} \times \frac{p_a}{p_{04}} \tag{8.9}$$

p_{03}/p_{02} is determined by the assumed combustion pressure loss, and p_{04}/p_a is obtained from Fig. 8.8 for any value of p_{03}/p_{04}. Such a curve is shown in Fig. 8.9. From it, for any value of compressor pressure ratio, the pressure ratio of the gas-generator turbine can be determined; this in turn fixes the values of $m\sqrt{T_{03}}/p_{03}$ and $\Delta T_{034}/T_{03}$ which are required for use with equations (8.2) and (8.6). Thus it is no longer necessary to carry out the iteration for the pressure ratio of the gas-generator turbine and, for each constant speed line considered, only a *single* iteration is required to find the correct equilibrium running point.

Variation of power output and SFC with output speed

The net power output of the free turbine engine is simply the output from the power turbine, namely

$$\text{power output} = mc_p\Delta T_{045} \tag{8.10}$$

where

$$\Delta T_{045} = \eta_{tp}T_{04}\left[1 - \left(\frac{1}{p_{04}/p_a}\right)^{(\gamma-1)/\gamma}\right]$$

For each equilibrium running point established (one for each compressor speed line) p_{04}/p_a will be known, and T_{04} can be readily calculated from

$$T_{04} = T_{03} - \Delta T_{034} \tag{8.11}$$

The mass flow is obtained from $m\sqrt{T_{01}}/p_{01}$ for assumed values of p_a and T_a. The outstanding unknown is the power turbine efficiency η_{tp}. It can be obtained from the power turbine characteristic, but it depends not only on the pressure ratio p_{04}/p_a but also on $N_p/\sqrt{T_{04}}$, i.e. on the power turbine speed N_p. Free turbine engines are used to drive a variety of loads such as pumps, propellers and electric generators, each with a different power versus speed characteristic. For this reason it is usual to calculate the power output over a range of power turbine speed for each equilibrium running point (i.e. for each compressor speed). The results could be plotted as in Fig. 8.10. Any curve corresponding to a given compressor speed will be fairly flat in the useful upper half of the output speed range where η_{tp} does not vary much with $N_p/\sqrt{T_{04}}$ (see Fig. 8.3).

The fuel consumption can also be calculated for each equilibrium running point, in the same manner as described in section 8.2 for the single-shaft unit. Since it depends only on the gas-generator parameters there will be only one value for each compressor speed. When combined with the power output data to give the *SFC*, however, it is clear that the *SFC*, like the power output, will be a function of both compressor speed and power turbine speed. It is convenient to express the off-design performance by plotting *SFC* against power output for several power turbine speeds as shown in Fig. 8.11. This type of presentation permits the customer to evaluate the performance of the unit when coupled to the specified load by superimposing the load characteristic upon it. The dotted curve in Fig. 8.11 indicates a particular variation of power and speed imposed by a load, and the points of intersection with the N_p curves give the *SFC* versus power output curve for the free turbine engine driving that particular load. Figure 8.11 relates to operation at one ambient condition, and normally the performance calculations would be repeated for extreme values of p_a and T_a.

It should be appreciated that although for convenience of calculation the compressor speed has been chosen as the independent variable, in practice the *fuel flow* is the independent variable. A chosen value of fuel flow (and

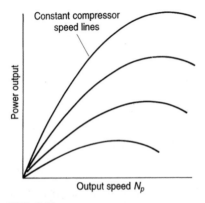

FIG. 8.10

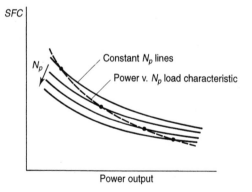

FIG. 8.11

hence T_{03}) determines the compressor speed and ultimately the power output. The *SFC* curves of Fig. 8.11 exhibit an increase in *SFC* as the power is reduced because the reduction in fuel flow leads to a reduction in compressor speed and gas-generator turbine inlet temperature. It will be remembered from Chapter 2 that the efficiency of a real cycle falls as the turbine inlet temperature is reduced. This poor part-load economy is a major disadvantage of the simple gas turbine. Consideration of the substantial improvement arising from the use of more complex cycles will be deferred until Chapter 9. Referring again to Fig. 8.11, the change in *SFC* with N_p at any given power output is not very marked, because the gas-generator parameters change only slightly under this operating condition.

It is useful to consider the variation of the key variables (power, turbine inlet temperature and fuel flow) with gas-generator speed as shown in Fig. 8.12. It can be seen that all increase rapidly as the gas generator approaches its maximum permissible speed. The change in turbine inlet temperature is especially critical because of the effect on creep life of the

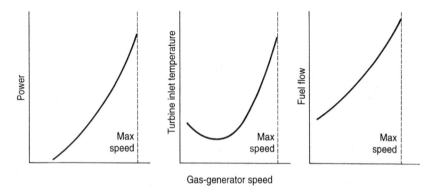

FIG. 8.12 Variation of key parameters with gas-generator speed

first-stage turbine blades and, in general, operation at the maximum speed would only be for limited periods. In electricity generation, for example, this maximum speed would be used for emergency peak duty, and base-load operation would utilize a reduced gas-generator speed with correspondingly increased blade life.

Single-shaft versus twin-shaft engines

The choice of whether to use a single-shaft or twin-shaft (free turbine) power plant is largely determined by the characteristics of the driven load. An electric generator requires a constant rotational speed and an engine designed specifically for this application would make use of a single-shaft configuration (see Fig. 1.24). An alternative, however, is the use of an aircraft derivative with a free power turbine in the place of the propelling nozzle. With this arrangement it is possible to design a power turbine of substantially larger diameter than the gas generator, using an elongated duct between the gas generator and the power turbine; this then permits the power turbine to operate at the required electric generator speed without the need for a reduction gearbox. Turboprops may use either configuration, as shown in Figs 1.10 and 1.11.

The running lines for single-shaft and twin-shaft units were shown in Figs 8.5 and 8.7. It should be noted that in the case of the single-shaft engine driving a generator, reduction in output power results in a slight increase in compressor mass flow; although there is some reduction in compressor pressure ratio, there is little change in compressor temperature rise because the efficiency is also reduced. This means that the compressor power remains essentially fixed. With a twin-shaft engine, however, as Fig. 8.7 shows, reducing net power output involves a reduction in compressor speed and hence in airflow, pressure ratio and temperature rise. The compressor power needed is therefore appreciably lower than for the single-shaft engine. It should also be evident from a comparison of Figs 8.5 and 8.7 that the compressor operates over a smaller range of efficiency in a twin-shaft engine. For these reasons the part-load fuel consumption of a twin-shaft engine is superior when driving a constant speed load.

The two types also have different characteristics regarding the supply of waste heat to a cogeneration plant, primarily due to the differences in exhaust flow as load is reduced; the essentially constant airflow and compressor power in a single-shaft unit results in a larger decrease of exhaust temperature for a given reduction in power, which might necessitate the burning of supplementary fuel in the waste heat boiler under operating conditions where it would be unnecessary with a twin-shaft. In both cases, the exhaust temperature may be increased by the use of variable inlet guide vanes. Cogeneration systems have been successfully built using both single-shaft and twin-shaft units, the latter often being aircraft derivative engines.

Most gas turbines do not spend long periods at low power and the *SFC* differences may not be critical in practice. The *torque characteristics*, however, are very different and the variation of torque with output speed at a given power may well determine the engine's suitability for certain applications, e.g. a high starting torque is particularly important for traction purposes.

The compressor of a *single-shaft engine* is constrained to turn at some multiple of the load speed, fixed by the transmission gear ratio, so that a wide range of load speed implies a reduction in compressor speed. This results in a reduction in mass flow and hence of output torque as shown by curve (a) of Fig. 8.13. This type of turbine is clearly unsuitable for traction purposes. The normal flat torque curve of an internal combustion engine is shown dotted for comparison.

The *free power turbine unit*, however, has a torque characteristic even more favourable than the internal combustion engine. The variation of power output with load speed, at any given compressor speed determined by the fuel flow, is shown in Fig. 8.10. It can be seen that the output power remains relatively constant over a wide load speed range for a fixed compressor speed. This is due to the fact that the compressor can supply an essentially constant flow at a given compressor speed, irrespective of the free turbine speed. Thus at fixed gas-generator operating conditions, reduction in output speed results in an increase in torque as shown by curve (b) in Fig. 8.13. It is quite possible to obtain a stall torque of two to three times the torque delivered at full speed.

The actual range of speed over which the torque conversion is efficient depends on the efficiency characteristic of the power turbine. The typical turbine efficiency characteristic shown in Fig. 8.3 suggests that the fall in

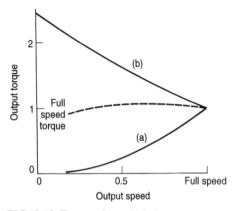

FIG. 8.13 Torque characteristics

efficiency will not be greater than about 5 or 6 per cent over a speed range from half to full speed. Thus quite a large increase in torque can be obtained efficiently when the output speed is reduced to 50 per cent of its maximum value. The efficiency of the torque conversion at low speeds, for example when accelerating a vehicle from rest, will be very low. In some applications a simple two-speed gearbox might be sufficient to overcome this defect, but it is probable that gas turbines for heavy road vehicles will incorporate some form of automatic transmission with five or six speeds.

Further differences between single- and twin-shaft units are brought out in section 9.5 where the transient behaviour of gas turbines is discussed.

EXAMPLE 8.2

A gas turbine with a free power turbine is designed for the following conditions:

air mass flow	30 kg/s
compressor pressure ratio	6·0
compressor isentropic efficiency	0·84
turbine inlet temperature	1200 K
turbine isentropic efficiency (both turbines)	0·87
combustion pressure loss	0·20 bar
mechanical efficiency	0·99
(applies to gas generator and load)	
ambient conditions	1·01 bar, 288 K

Calculate the power developed and the turbine non-dimensional flows $m\sqrt{T_{03}}/p_{03}$ and $m\sqrt{T_{04}}/p_{04}$ at the design point.

If the engine is run at the same mechanical speed at an ambient temperature of 268 K calculate the values of turbine inlet temperature, pressure ratio and power output, assuming the following:

(a) combustion pressure loss remains constant at 0·20 bar;
(b) both turbines choking, with values of $m\sqrt{T_{03}}/p_{03}$ and $m\sqrt{T_{04}}/p_{04}$ as calculated above and no change in turbine efficiency;
(c) at 268 K and the same N, the $N/\sqrt{T_{01}}$ line on the compressor characteristic is a vertical line with a non-dimensional flow 5 per cent greater than the design value;
(d) variation of compressor efficiency with pressure ratio at the relevant value of $N/\sqrt{T_{01}}$ is

p_{02}/p_{01}	6·0	6·2	6·4	6·6
η_c	0·837	0·843	0·845	0·840

The design point calculation is straightforward and only the salient results are presented:

Gas-generator turbine
    pressure ratio   2·373
    inlet pressure   5·86 bar
    temperature drop   203 K
Power turbine
    pressure ratio   2·442
    inlet pressure   2·47 bar
    temperature drop   173·5 K
    inlet temperature   997 K

The power output is then $(30 \times 1·148 \times 173·5 \times 0·99)$ kW or 5910 kW.
    The design point values of $m \sqrt{T_{03}}/p_{03}$ and $m \sqrt{T_{04}}/p_{04}$ are

$$\frac{30 \times \sqrt{1200}}{5·86} = 177·4 \quad \text{and} \quad \frac{30 \times \sqrt{997}}{2·47} = 383·5 \text{ respectively}$$

At 268 K the value of $m \sqrt{T_{01}}/p_{01}$ is

$$1·05 \times \left[\frac{30 \times \sqrt{288}}{1·01} \right] = 529·5$$

If the power turbine remains choked, the gas-generator turbine will be constrained to operate at a fixed non-dimensional point, and thus the value of $\Delta T_{034}/T_{03} = 203/1200 = 0·169$, as for the design condition.
    For work compatibility

$$\frac{\Delta T_{012}}{T_{01}} = \frac{\Delta T_{034}}{T_{03}} \times \frac{T_{03}}{T_{01}} \left[\frac{c_{pg}\eta_m}{c_{pa}} \right] = \frac{0·169 \times 1·148 \times 0·99}{1·005} \times \frac{T_{03}}{T_{01}}$$

and hence

$$\frac{T_{03}}{T_{01}} = 5·23 \frac{\Delta T_{012}}{T_{01}} \tag{A}$$

For flow compatibility

$$\frac{m\sqrt{T_{03}}}{p_{03}} = \frac{m\sqrt{T_{01}}}{p_{01}} \times \frac{p_{01}}{p_{03}} \times \sqrt{\frac{T_{03}}{T_{01}}}$$

$$177·4 = 529·5 \frac{p_{01}}{p_{03}} \times \sqrt{\frac{T_{03}}{T_{01}}}$$

$$\sqrt{\frac{T_{03}}{T_{01}}} = 0·335 \frac{p_{03}}{p_{01}} \tag{B}$$

The problem is thus to find the compressor operating point that will give the same value of T_{03}/T_{01} for both equations (A) and (B). With the variation in efficiency prescribed, the value of $\Delta T_{012}/T_{01}$ can readily be calculated from equation (8.4), and with the constant value of combustion pressure loss the value of p_{03} can also be calculated. The solution is best carried out in tabular form as shown.

$\frac{p_{02}}{p_{01}}$	$\left[\left(\frac{p_{02}}{p_{01}}\right)^{(\gamma-1)/\gamma}-1\right]$	η_c	$\frac{\Delta T_{012}}{T_{01}}$	$\left(\frac{T_{03}}{T_{01}}\right)_A$	p_{02}	p_{03}	$\frac{p_{03}}{p_{01}}$	$\sqrt{\frac{T_{03}}{T_{01}}}$	$\left(\frac{T_{03}}{T_{01}}\right)_B$
6·0	0·669	0·837	0·799	4·18	6·06	5·86	5·80	1·943	3·78
6·2	0·684	0·843	0·812	4·25	6·26	6·06	6·00	2·010	4·05
6·4	0·700	0·845	0·828	4·33	6·46	6·26	6·20	2·078	4·32
6·6	0·715	0·840	0·851	4·45	6·66	6·46	6·40	2·144	4·60

Solving graphically the required pressure ratio is found to be 6·41, with the value of $T_{03}/T_{01} = 4·34$; the corresponding turbine inlet temperature is $4·34 \times 268 = 1163 \, \text{K}$.

Having established the compressor pressure ratio and turbine inlet temperature, it is a straightforward matter to calculate the power developed. Remembering that the gas-generator turbine will still operate at the same non-dimensional point ($\Delta T_{034}/T_{03} = 0·169$, $p_{03}/p_{04} = 2·373$) the power turbine entry conditions and temperature drop can readily be calculated; the resulting temperature drop is 179·6 K. The mass flow is obtained from the non-dimensional flow and the ambient conditions:

$$\frac{m\sqrt{T_{01}}}{p_{01}} = 529·5, \text{ hence } m = \frac{529·5 \times 1·01}{\sqrt{268}} = 32·7 \, \text{kg/s}$$

and power output $= 32·7 \times 1·148 \times 179·6 \times 0·99 = 6680 \, \text{kW}$.

From this example, it can be seen that operation at the design mechanical speed on a cold day results in a decrease of maximum cycle temperature from 1200 to 1163 K, even though the value of T_{03}/T_{01} has increased from 4·17 to 4·34 due to the increase in $N/\sqrt{T_{01}}$. The power has increased from 5910 to 6680 kW and this can be seen to be due to the simultaneous increase in air mass flow and overall pressure ratio. The beneficial effect of low ambient temperature on gas turbine operation is evident; conversely, high ambient temperatures result in significant penalties. The effect of increased ambient temperature on turbojet operation is discussed in the next section under the heading 'Variation of thrust with rotational speed, forward speed and altitude'.

Industrial gas turbines used to generate electricity in regions with high summer temperatures may have to meet very high peak demands due to large air-conditioning loads for about 25 per cent of the year. It is possible to increase power by cooling the air entering the compressor, and electric utilities have found this to be an economical solution compared with the alternative of buying a more powerful gas turbine which would need to run at reduced power and efficiency for 75 per cent of the year. Several different methods have been used. In regions of low relative humidity, cooling can be achieved using an *evaporative cooler*; the incoming air is passed through a wetted filter, and the heat required to evaporate the water leads to a reduction in temperature. This approach, however, is not effective in areas of high relative humidity. An alternative approach, successfully introduced in the mid 1990s, is *ice harvesting*. At off-peak times, primarily overnight, electricity is used to drive refrigeration chillers which produce and store large quantities of ice. When maximum power is needed during the heat of the day, ice is melted and the chilled water is used in heat-exchangers to lower the inlet temperature. This method has proved to be economically attractive, producing substantial increases in power at a low cost per kilowatt [Ref. (1)]. Ice harvesting, however, requires construction of large ice storage containers and a new approach, referred to as *fog cooling*, was introduced in the late 1990s. The 'fog' is produced by injecting finely atomized water droplets, forced through tiny nozzles at pressures up to 200 bar, into the inlet air stream. The use of very small droplets (5–10 microns), in very large quantities, provides a large surface area for evaporation and cooling of the intake air; this approach is clearly superior to evaporative cooling in regions of high relative humidity and any excess water not evaporated will give an increase in mass flow, resulting in more power. It is important that water brought into the airflow is demineralized, to prevent the build-up of deposits on the blading causing performance deterioration due to loss of compressor efficiency. Reference (2) gives a good review of the advantages of inlet cooling.

8.5   Off-design operation of the jet engine

Propelling nozzle characteristics

The propelling nozzle area for a jet engine is determined from the design point calculations as described in Chapter 3, and once the nozzle size has been fixed it has a major influence on the off-design operation. The characteristic for a nozzle, in terms of 'non-dimensional' flow $m\sqrt{T_{04}}/p_{04}$ and pressure ratio p_{04}/p_{05}, can readily be calculated as follows.

The mass flow parameter is given by

$$\frac{m\sqrt{T_{04}}}{p_{04}} = C_5 A_5 \rho_5 \frac{\sqrt{T_{04}}}{p_{04}}$$

$$= \frac{C_5}{\sqrt{T_{04}}} \times \frac{A_5}{R} \times \frac{p_5}{p_{04}} \times \frac{T_{04}}{T_5} \tag{8.12}$$

where A_5 is the effective nozzle area. Making use of equation (3.12), $C_5/\sqrt{T_{04}}$ can be found from

$$\frac{C_5^2}{T_{04}} = 2c_p\eta_j \left[1 - \left(\frac{1}{p_{04}/p_5} \right)^{(\gamma-1)/\gamma} \right] \tag{8.13}$$

and T_5/T_{04} from

$$\frac{T_5}{T_{04}} = 1 - \frac{T_{04} - T_5}{T_{04}} = 1 - \eta_j \left[1 - \left(\frac{1}{p_{04}/p_5} \right)^{(\gamma-1)/\gamma} \right] \qquad (8.14)$$

It follows that for a nozzle of given area and efficiency, $m\sqrt{T_{04}}/p_{04}$ can be calculated as a function of the pressure ratio p_{04}/p_5. But equations (8.13) and (8.14) are valid for pressure ratios only up to the critical value, given by equation (3.14), namely

$$\frac{p_{04}}{p_c} = 1 \Big/ \left[1 - \frac{1}{\eta_j} \left(\frac{\gamma - 1}{\gamma + 1} \right) \right]^{\gamma/(\gamma-1)} \qquad (8.15)$$

and up to this point $p_5 = p_a$. For pressure ratios p_{04}/p_a greater than the critical, $m\sqrt{T_{04}}/p_{04}$ remains constant at the maximum (choking) value, i.e. it is independent of p_{04}/p_a (and incidentally $p_5 = p_c > p_a$). Thus $m\sqrt{T_{04}}/p_{04}$ can be plotted against the *overall* nozzle pressure ratio p_{04}/p_a as in Fig. 8.14, and the similarity between the nozzle flow characteristic and that of a turbine is evident.

Up to the choking condition, T_5/T_{04} is given by equation (8.14), whereas when the nozzle is choking it is given by equation (3.13), namely

$$\frac{T_c}{T_{04}} = \frac{2}{\gamma + 1} \qquad (8.16)$$

Likewise, with the nozzle unchoked $C_5/\sqrt{T_{04}}$ is given by equation (8.13), whereas when it is choked C_5 is the sonic velocity and the Mach number M_5 is unity. Now, recalling that $C = M\sqrt{(\gamma RT)}$ and $T_0 = T[1 + (\gamma - 1)M^2/2]$, we have the general relation

$$\frac{C}{\sqrt{T_0}} = \frac{M\sqrt{(\gamma R)}}{\sqrt{\left(1 + \frac{\gamma - 1}{2} M^2 \right)}} \qquad (8.17)$$

Thus when the nozzle is choked (i.e. $M_5 = 1$) we have

$$\frac{C_5^2}{T_{04}} = \frac{C_5^2}{T_{05}} = \frac{2\gamma R}{\gamma + 1} \qquad (8.18)$$

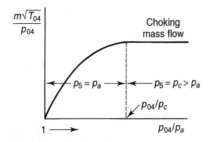

FIG. 8.14 Nozzle characteristics

We shall make use of equations (8.15) to (8.18) later when discussing the determination of the thrust.

Matching of gas generator with nozzle

The similarity between the flow characteristic of a nozzle and a turbine means that the nozzle will exert the same restriction on the operation of the gas generator as a free power turbine. Thus if the operation of a jet engine under *static* conditions is considered, there can be no difference between its behaviour and that of a free power turbine unit. The equilibrium running line can be determined according to the procedure specified in the flow chart of Fig. 8.6(b), with the nozzle characteristic replacing the power turbine characteristic. It follows that Fig. 8.7 can be taken as representative of a typical equilibrium running diagram for a jet engine under static conditions.

The jet engine, of course, is intended for flight at high speeds and it is necessary to consider the effect of forward speed on the equilibrium running line. It is most convenient to express the forward speed in terms of Mach number for the matching calculations, and this can readily be converted to velocity for calculation of the momentum drag and thrust as shown later in the next sub-section.

Forward speed produces a ram pressure ratio which is a function of both flight Mach number and intake efficiency. This ram effect will give rise to an increase in compressor delivery pressure, which will in turn lead to a higher pressure before the propelling nozzle, thereby increasing the nozzle pressure ratio. Once the nozzle is choked, however, the nozzle non-dimensional flow will reach its maximum value and will then be independent of the nozzle pressure ratio and therefore of forward speed. The significance of this is that the turbine operating point will then also be unchanged, because of the requirement for compatibility of flow between the turbine and the nozzle. It follows that as long as the *nozzle is choking* the equilibrium running line will be uniquely determined by the fixed turbine operating point and will be *independent of flight speed.*†

At current levels of cycle pressure ratio, virtually all jet engines operate with the nozzle choked during take-off, climb and cruise, and the nozzle only becomes unchoked when thrust is significantly reduced. Thus the nozzle is liable to be unchoked only when preparing to land or when taxiing. Nevertheless, it is important to consider the effect of forward

---

† It should be noted that the argument based on Fig. 8.8 is equally applicable here. Thus when the nozzle is choked the turbine operating point, and hence $m\sqrt{T_{03}}/p_{03}$, p_{03}/p_{04} and $\Delta T_{034}/T_{03}$ are fixed and any increase in the overall expansion ratio p_{03}/p_a due to increasing ram effect will therefore result in an increase in overall nozzle pressure ratio p_{04}/p_a with the turbine pressure ratio unaffected.

speed on the running line under these conditions because it is at low rotational engine speeds that the running line is in close proximity to the surge line.

The nozzle pressure ratio p_{04}/p_a is linked to the ram pressure ratio by the identity

$$\frac{p_{04}}{p_a} = \frac{p_{04}}{p_{03}} \times \frac{p_{03}}{p_{02}} \times \frac{p_{02}}{p_{01}} \times \frac{p_{01}}{p_a} \tag{8.19}$$

This differs from the corresponding equation for the free turbine unit, equation (8.8), only by the inclusion of the ram pressure ratio p_{01}/p_a. The ram pressure ratio in terms of intake efficiency η_i and flight Mach number M_a is given by equation (3.10b), namely

$$\frac{p_{01}}{p_a} = \left[1 + \eta_i \left(\frac{\gamma - 1}{2}\right) M_a^2\right]^{\gamma/(\gamma-1)} \tag{8.20}$$

It follows from equations (8.19) and (8.20) that for a given intake efficiency, p_{04}/p_a is a function of the gas-generator parameters and flight Mach number. The procedure of Fig. 8.6(b) can be followed, with equation (8.19) substituted for equation (8.8), but for each compressor speed line the calculation is repeated for several values of M_a covering the desired range of flight speed. The result is a fan of equilibrium running lines of constant M_a. As shown in Fig. 8.15, they merge into the single running line obtained with the higher compressor speeds for which the nozzle is choked.

It should be noted that increasing Mach number pushes the equilibrium running line away from the surge line at low compressor speeds. Fundamentally, this is because the ram pressure rise allows the compressor to utilize a lower pressure ratio for pushing the required flow through the nozzle.

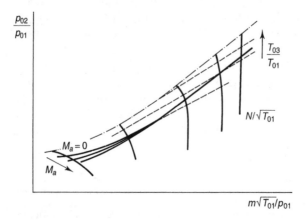

FIG. 8.15 Jet engine running lines

Variation of thrust with rotational speed, forward speed and altitude

We have seen in section 3.1 that when the whole of the expansion from p_{04} to p_a occurs in the propelling nozzle, the net thrust of the jet unit is simply the overall rate of change of momentum of the working fluid, namely $F = m(C_5 - C_a)$ where C_a is the aircraft speed. On the other hand, when part of the expansion occurs outside the propelling nozzle, i.e. when the nozzle is choking and p_5 is greater than p_a, there will be an additional pressure thrust. In this case the net thrust is given by the more general expression

$$F = m(C_5 - C_a) + (p_5 - p_a)A_5 \tag{8.21}$$

To obtain the curves which show the thrust delivered by the jet engine over the complete operating range of inlet pressure and temperature, flight speed and rotational speed, information obtained from the points used to establish the equilibrium running diagram can be used. For each running point on this diagram the values of all the thermodynamic variables such as

$$\frac{p_{01}}{p_a}, \frac{m\sqrt{T_{01}}}{p_{01}}, \frac{p_{02}}{p_{01}}, \frac{T_{03}}{T_{01}}, \frac{T_{04}}{T_{03}}, \frac{p_{03}}{p_{04}} \text{ and } \frac{p_{04}}{p_a}$$

are determined for specified values of $N/\sqrt{T_{01}}$ and M_a (and $N/\sqrt{T_{01}}$ alone in the choked nozzle range). The thrust can ultimately be expressed in terms of these non-dimensional quantities. Thus equation (8.21) can immediately be written in the form

$$\frac{F}{p_a} = \frac{m\sqrt{T_{01}}}{p_{01}} \frac{p_{01}}{p_a} \left[\frac{C_5}{\sqrt{T_{04}}} \sqrt{\left(\frac{T_{04}}{T_{03}} \times \frac{T_{03}}{T_{01}} \right)} - \frac{C_a}{\sqrt{T_{01}}} \right] + \left(\frac{p_5}{p_a} - 1 \right) A_5 \tag{8.22}$$

(A dimensional check will show that the true non-dimensional thrust is $F/(p_a D^2)$, but for an engine with fixed geometry the characteristic dimension can be omitted.) Making use of equation (8.17),

$$\frac{C_a}{\sqrt{T_{01}}} = \frac{C_a}{\sqrt{T_{0a}}} = \frac{M_a\sqrt{(\gamma R)}}{\sqrt{\left(1 + \frac{\gamma - 1}{2} M_a^2 \right)}}$$

When the nozzle is unchoked, $C_5/\sqrt{T_{04}}$ can be obtained from equation (8.13) with p_{04}/p_5 put equal to p_{04}/p_a, and the pressure thrust is zero because $p_5/p_a = 1$.

When the nozzle is choked, $C_5/\sqrt{T_{04}}$ is given by equation (8.18), and p_5/p_a can be calculated from

$$\frac{p_5}{p_a} = \frac{p_5}{p_{04}} \times \frac{p_{04}}{p_a} = \frac{p_c}{p_{04}} \times \frac{p_{04}}{p_a}$$

p_c/p_{04} is the reciprocal of the critical pressure ratio given by equation (8.15), i.e. it is a function only of γ for the gas and the nozzle efficiency η_j.

p_{04}/p_a is one of the known parameters, having been calculated using equation (8.19).

A typical variation of thrust with engine speed and flight speed is illustrated in Fig. 8.16, which shows separate curves of thrust for each Mach number considered. The abscissa can be left as $N/\sqrt{T_{01}}$ as shown, or be put in terms of $N/\sqrt{T_a}$ by making use of the relations

$$\frac{N}{\sqrt{T_a}} = \frac{N}{\sqrt{T_0}} \times \sqrt{\frac{T_{01}}{T_a}} \quad \text{and} \quad \frac{T_{01}}{T_a} = \left(1 + \frac{\gamma - 1}{2} M_a^2\right)$$

It should be noted that although a unique running line is obtained on the compressor characteristic when the propelling nozzle is choked, the thrust for a given value of $N/\sqrt{T_{01}}$ does depend on the flight Mach number. There is a direct dependence due to the increase in momentum drag (mC_a) with increasing flight speed, and an indirect dependence due to the increase in compressor inlet pressure arising from the ram compression. At low rotational speeds the effect of momentum drag predominates and an increase in M_a causes the thrust to decrease, whereas at high values of $N/\sqrt{T_{01}}$ the beneficial effect of ram pressure rise predominates.

Although it is convenient to express the performance in terms of non-dimensional engine speed, it is actual *mechanical* speed upon which a limit is set by the turbine stresses and which must be controlled by a governor. The strong dependence of thrust upon engine speed indicates that accurate control is essential.† If the speed is controlled at a value below the limit, take-off thrust will be substantially reduced. The situation is more serious if the speed exceeds the correct limit: not only do the centrifugal stresses increase with the square of the speed but there is also a rapid increase in turbine inlet temperature. The latter can be seen from the way the running line crosses the T_{03}/T_{01} lines on Fig. 8.15. Typically, an increase in rotor

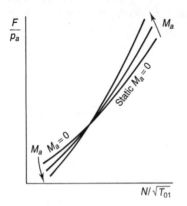

FIG. 8.16 Thrust curves

† It should be realized that the same arguments are applicable in the case of a free turbine engine delivering shaft power.

speed of only 2 per cent above the limit may result in an increase in T_{03} of 50 K. With blade life determined by creep, the time for which high speeds are permitted must be strictly controlled. The maximum permissible speed is normally restricted to periods of less than 5 min, giving the *take-off rating*. The *climb rating* is obtained with a small reduction in fuel flow and hence in rotor speed, and can usually be maintained for a period of 30 min. The *cruise rating* requires a further reduction in fuel flow and rotor speed, resulting in stress and temperature conditions permitting unrestricted operation. Ratings for the Rolls-Royce Viper 20 at Sea Level Static conditions are shown in the following table; this is a simple turbojet with a low pressure ratio. The very rapid drop in thrust with reduction of speed is evident.

	Engine speed (% N_{max})	Thrust [kN]	SFC [kg/kN h]
Take-off (5 min)	100	13·35	100·4
Climb (30 min)	98	12·30	98·2
Cruise	95	10·90	95·1

The effect of ambient temperature on the take-off rating is important to both the manufacturer and the user. With the engine running at its maximum mechanical speed an *increase* in ambient temperature will cause a decrease in $N/\sqrt{T_a}$ and hence $N/\sqrt{T_{01}}$. This will cause the operating point on the compressor characteristic to move along the equilibrium running line to lower values of both $m\sqrt{T_{01}}/p_{01}$ and p_{02}/p_{01}, and it can readily be seen to be equivalent to a decrease in mechanical speed. The actual mass flow entering the engine will be further reduced due to the increase in ambient temperature, because m is given by $(m\sqrt{T_{01}}/p_{01})/(\sqrt{T_{01}}/p_{01})$. The well-known result of all these effects is that an *increase in ambient temperature* results in an appreciable *loss of thrust* as indicated in Fig. 8.16. But the picture is not yet complete. T_{03}/T_{01} also decreases with increase in ambient temperature when N is held constant. The actual temperature T_{03} is given by $(T_{03}/T_{01})T_{01}$, and the decrease in (T_{03}/T_{01}) will be more than offset by the increase in T_{01}. In general, for a fixed mechanical speed T_{03} *will increase with increase in ambient temperature*, and the allowable turbine inlet temperature may be exceeded on a hot day. To keep within the limit it is then necessary to reduce the mechanical speed, giving an even greater reduction in $N/\sqrt{T_{01}}$ and hence thrust.

Turning our attention next to variation in ambient pressure, Fig. 8.16 indicates that the thrust will change in direct proportion to the ambient pressure; no change in the engine operating point is involved, but the mass flow is reduced as the ambient pressure is reduced. Both the pressure and temperature decrease with increasing altitude, the latter levelling off at 11 000 m. Because of the dependence of the thrust on the first power of the pressure, the decrease in thrust due to the decrease in p_a more than outweighs the increase due to the reduction in T_a. Thus the *thrust* of an engine *decreases with increase in altitude*. Airports at high altitudes are often critical with regard to take-off performance and may have substantially longer runways than are required at sea level to allow aircraft to operate with full payloads; well-known examples are Denver, Johannesburg and Mexico City.

Variation of fuel consumption and SFC with rotational speed, forward speed and altitude

The fuel consumption of a jet engine together with the fuel capacity of the aircraft determine the range, and the specific fuel consumption (fuel flow per unit thrust for a jet engine) is a convenient indication of the economy of the unit. From the arguments of the previous section it is apparent that both fuel consumption and specific fuel consumption can be evaluated as functions of $N/\sqrt{T_{01}}$ (or $N/\sqrt{T_a}$) and M_a.

When a value of the combustion efficiency is assumed, the fuel consumption can readily be determined from the airflow, the combustion temperature rise and the fuel/air ratio curves of Fig. 2.17, in the manner described for the other types of gas turbine engine. To do this, values of p_a and T_a have to be assumed to obtain m, T_{03} and T_{02} from the dimensionless parameters. The fuel flow is therefore a function of $N/\sqrt{T_a}$, M_a, p_a and T_a. The dependence of the fuel flow on ambient conditions can in fact be virtually eliminated by plotting the results in terms of the non-dimensional fuel flow $(m_f Q_{net,p}/D^2 p_a \sqrt{T_a})$. For a given fuel and engine the calorific value and linear dimension can be dropped, yielding in practice $m_f/p_a\sqrt{T_a}$. Figure 8.17(a) shows a typical set of fuel consumption curves. Unlike F/p_a, the fuel parameter depends only slightly on M_a because it is due simply to the variation of compressor inlet conditions and not also to

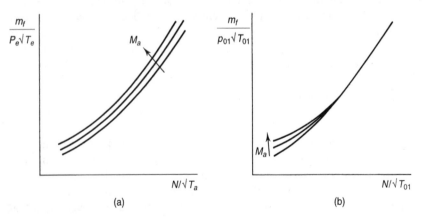

FIG. 8.17 **Fuel consumption curves**

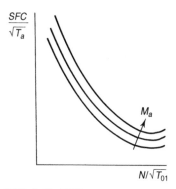

FIG. 8.18 *SFC* curves

momentum drag. Indeed, if $m_f/p_{01}\sqrt{T_{01}}$ is plotted against $N/\sqrt{T_{01}}$ as in Fig. 8.17(b), the curves merge into a single line for the region where the nozzle is choking. It must be mentioned that although the combustion efficiency is high and constant over most of the working range, it can fall drastically at very high altitudes due to low combustion chamber pressures. Such curves as those of Fig. 8.17 may well underestimate the fuel consumption when p_a is low.

Curves of 'non-dimensional' *SFC* may be obtained by combining the data of Figs 8.16 and 8.17(a). Figure 8.18 shows $SFC/\sqrt{T_a}$ plotted against $N/\sqrt{T_{01}}$ for various values of M_a. It is apparent that the *SFC* will improve with increase in altitude due to the decrease in T_a. As the *SFC* is a function only of $\sqrt{T_a}$ and not of p_a, however, the effect of altitude is not so marked as in the case of the thrust. (The actual variation in *SFC* with altitude and Mach number for a simple turbojet, operating at its maximum rotational speed, was shown in Fig. 3.14.) It should be noted that *SFC* increases with Mach number. Thrust initially drops with increasing Mach number and then rises when the ram effect overcomes the increase of momentum drag; the fuel flow, however, will steadily increase with Mach number because of the increased airflow due to the higher stagnation pressure at entry to the compressor, and this effect predominates.

8.6 Methods of displacing the equilibrium running line

It was stated earlier that if the equilibrium running line intersects the surge line it will not be possible to bring the engine up to full power without taking some remedial action. As will be shown in the next chapter, even when clear of the surge line, if the running line approaches it too closely the compressor may surge when the engine is accelerated rapidly.

With most modern compressors, surge is likely to be encountered at low values of $N/\sqrt{T_{01}}$ and is less of a problem at high rotational speeds. Many high-performance axial compressors exhibit a kink in the surge line as shown in Fig. 8.19. A running line, intersecting the surge line at low speeds and at the kink, is also shown. To overcome the problem it is necessary to lower the running line locally in dangerous regions of operation.†

One common method of achieving this is *blow-off*, where air is bled from some intermediate stage of the compressor. Blow-off clearly involves a waste of turbine work, so that the blow-off valve must be designed to operate only over the essential part of the running range. Furthermore, it may be difficult to discharge the air when space around the engine is at a premium as in a nacelle. As an alternative to blow-off a *variable-area propelling nozzle* could be used for a jet engine. It will be shown in the next chapter that variable-area propelling nozzles serve other useful purposes also.

It can be deduced that either of these methods will produce a reduction in pressure ratio at a given compressor speed and hence will lower the running line. This is most easily demonstrated if we consider operation at the high-speed end of the running range, where the constant $N/\sqrt{T_{01}}$ lines on the compressor characteristic are almost vertical (i.e. $m\sqrt{T_{01}}/p_{01} \simeq$ constant) and where both the nozzle and turbine are choked.

If we consider the use of a variable-area nozzle first, Fig. 8.20 shows that increasing the nozzle area will cause an increase in turbine pressure ratio and hence non-dimensional temperature drop $\Delta T_{034}/T_{03}$. From equation (8.2), satisfying compatibility of flow,

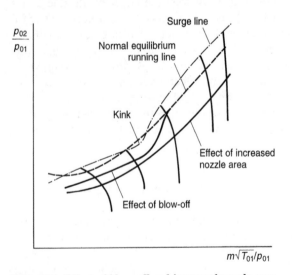

FIG. 8.19 Effect of blow-off and increased nozzle area

† An alternative method is to raise the surge line using variable stators in the compressor.

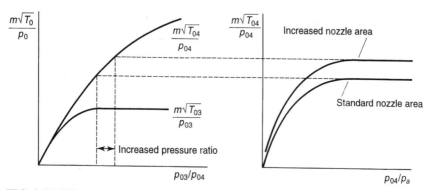

FIG. 8.20 Effect of variable-area propelling nozzle

$$\frac{p_{02}}{p_{01}} = \frac{m\sqrt{T_{01}}}{p_{01}} \times \frac{p_{02}}{p_{03}} \times \sqrt{\frac{T_{03}}{T_{01}}} \times \frac{p_{03}}{m\sqrt{T_{03}}} = K_1 \sqrt{\frac{T_{03}}{T_{01}}} \qquad (8.23)$$

where K_1 is a constant because under the assumed conditions the values of $m\sqrt{T_{01}}/p_{01}$, p_{02}/p_{03} and $m\sqrt{T_{03}}/p_{03}$ are all fixed. From equation (8.6), which satisfies compatibility of work,

$$\frac{T_{03}}{T_{01}} = \frac{\Delta T_{012}}{T_{01}} \times \frac{T_{03}}{\Delta T_{034}} \times \frac{c_{pa}}{c_{pg}\eta_m}$$

Now, at a fixed value of compressor speed N the compressor temperature rise is approximately constant; Fig. 8.21 shows a typical compressor characteristic replotted on a basis of temperature rise v. pressure ratio. The wide range of pressure ratio obtained at almost constant temperature rise is due to the significant variation of efficiency when operating at conditions away from the design point. Thus if we assume $\Delta T_{012}/T_{01}$ is constant,

$$\frac{T_{03}}{T_{01}} = \frac{K_2}{(\Delta T_{034}/T_{03})} \qquad (8.24)$$

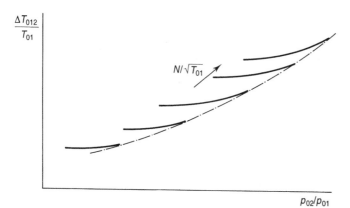

FIG. 8.21 Compressor characteristics

where K_2 is another constant. Combining equations (8.23) and (8.24) gives

$$\frac{p_{02}}{p_{01}} = \frac{K_3}{\sqrt{(\Delta T_{034}/T_{03})}} \qquad (8.25)$$

where K_3 is a constant, obtained from K_1 and K_2.

If the engine speed $N/\sqrt{T_{01}}$ is held constant and the nozzle area is increased, $\Delta T_{034}/T_{03}$ will be increased and equation (8.25) shows that p_{02}/p_{01} will be decreased, hence the running line will be moved away from the surge line. In order to maintain the speed at the required level, it is necessary to reduce the fuel flow; opening the nozzle without reducing the fuel flow would cause the engine to accelerate to a higher speed, as discussed in Chapter 9. It should also be realized that decreasing the nozzle area will move the operating line towards the surge line; reasons for doing this will be discussed in the next chapter.

When considering the effect of bleed, m_1 is no longer equal to m_3 and equation (8.23) is modified to

$$\frac{p_{02}}{p_{01}} = K_1 \sqrt{\frac{T_{03}}{T_{01}}} \times \frac{m_3}{m_1} \qquad (8.26)$$

With the nozzle, or power turbine in the case of a shaft power unit, remaining choked, the turbine operating point remains unchanged and equation (8.26) leads to

$$\frac{T_{03}}{T_{01}} = \frac{m_1}{m_3} \times \frac{\Delta T_{012}}{T_{01}} \times \frac{T_{03}}{\Delta T_{034}} \times \frac{c_{pa}}{c_{pg}\eta_m} = K_4 \frac{m_1}{m_3} \qquad (8.27)$$

where K_4 is a constant, assuming as before that $\Delta T_{012}/T_{01}$ is constant. Combining equations (8.26) and (8.27) gives

$$\frac{p_{02}}{p_{01}} = K_5 \sqrt{\frac{m_3}{m_1}} \qquad (8.28)$$

where K_5 is a constant.

With the use of bleed, m_3 is always less than m_1, and the result will once again be to reduce the pressure ratio and lower the operating line. Equation (8.27) also shows that the use of bleed will increase the turbine inlet temperature. The reduction in turbine mass flow necessitates an increase in temperature drop to provide the compressor work, and with the non-dimensional temperature drop fixed this implies an increase in turbine inlet temperature. The effects of both bleed and increase in nozzle area can be interpreted physically in terms of a decrease in the restriction to the flow, permitting the compressor to operate at a lower pressure ratio for any given rotational speed.

8.7 Incorporation of variable pressure losses

The size of the ducting for components such as the intake and exhaust of a gas turbine will be determined primarily by the requirement for low pressure losses at maximum power, which will normally correspond to maximum mass flow. The designer must compromise between low pressure losses and small duct sizes, and the choice will be largely determined by the application envisaged. It can be shown by cycle calculations that the effect of a given pressure loss is dependent on the pressure level at which it occurs, and it is essential to keep the terms $\Delta p_0/p_0$ at all stations as low as possible. With pressure losses in the intake and exhaust of stationary gas turbines occurring at essentially atmospheric pressure, it will be realized that these losses are much more critical than pressure losses in the combustion chamber or the air side of a heat-exchanger where the air is entering at the compressor delivery pressure. In a typical installation a pressure loss of 2·5 cm H_2O in the intake may cause about 1 per cent reduction in power output.

Having decided on the size of ducting required to give the desired pressure loss at the design point, the variation of loss with changing operating conditions can be predicted for incorporation in the off-design calculation procedures outlined in this chapter. The velocities in such components as the intake and exhaust ducts will be sufficiently low for the flow to be treated as incompressible, so that Δp_0 will be proportional to the inlet dynamic head. Expressed in non-dimensional terms this implies, as shown by equation (6.2), that

$$\frac{\Delta p_0}{p_0} \propto \left(\frac{m\sqrt{T_0}}{p_0}\right)^2$$

Assuming the appropriate values at the design condition to be $(\Delta p_0/p_0)_D$ and $(m\sqrt{T_0}/p_0)_D$, then at any off-design condition

$$\frac{\Delta p_0}{p_0} = \left(\frac{\Delta p_0}{p_0}\right)_D \left[\left(\frac{m\sqrt{T_0}}{p_0}\right) \middle/ \left(\frac{m\sqrt{T_0}}{p_0}\right)_D\right]^2 \tag{8.29}$$

Values of $m\sqrt{T_0}/p_0$ for each component are obtained in the course of the off-design calculations. The pressure loss from equation (8.29) can be incorporated in the same manner as was the combustion pressure loss $\Delta p_b/p_{02}$, namely by making use of such equations as

$$\frac{p_{03}}{p_{02}} = 1 - \frac{\Delta p_b}{p_{02}}$$

Now, however, the pressure ratio across the component is not constant but changes with $m\sqrt{T_0}/p_0$.

In the particular case of the combustion chamber, the more accurate expression of equation (6.1) can be used. This includes the fundamental loss due to heat addition. Combining (6.1) and (6.2) we would have

$$\frac{\Delta p_b}{p_{02}} \propto \left(\frac{m\sqrt{T_{02}}}{p_{02}}\right)^2 \left[K_1 + K_2\left(\frac{T_{03}}{T_{02}} - 1\right)\right]$$

Such refinement would not normally be worthwhile because the fundamental loss is a small proportion of the total.

9 Prediction of performance— further topics

Evaluation of the off-design performance of complex plant incorporating intercooling, heat-exchange and reheat is inevitably more complicated than for the simple gas turbine, although the basic principles described in Chapter 8 are still applicable. The components of such gas turbines can be arranged in a wide variety of ways, about 40 different layouts being possible even with no more than two compressor rotors. A comprehensive survey of the part-load performance and operating stability of many of these possibilities can be found in Ref. (1), wherein use was made of stylized component characteristics. It is as a result of such calculations that it was possible to dismiss certain arrangements as impractical because the running line runs into the surge line under some operating conditions. No attempt will be made to repeat this kind of survey here, and we shall restrict our attention mainly to discussing the prediction of off-design performance in the practical cases of (a) high pressure ratio twin-spool engines and (b) turbofans. Before discussing the matching procedures for twin-spool engines and turbofans, however, methods of improving the part-load performance of gas turbines will be considered briefly.

The chapter ends with an introduction to transient performance, including a brief description of methods for predicting the acceleration or deceleration of rotor systems. Acceleration rates of gas turbine rotors are obviously dependent on mechanical considerations such as the polar moment of inertia and the maximum temperature which the turbine blades can withstand for short periods, but the limiting factor is usually the proximity of the equilibrium running line to the surge line. Thus a thorough understanding of off-design performance is essential before transient behaviour can be investigated and a suitable control system can be designed.

9.1 Methods of improving part-load performance

It was pointed out in the Introduction that the part-load performance of gas turbines intended for vehicular or naval use was of great importance because of the considerable portion of the running time spent at low power. Early studies for both applications resulted in the consideration of complex arrangements incorporating intercooling, heat-exchange and reheat. The sole justification for the marked increase in complexity was the great improvement in part-load specific fuel consumption indicated in Fig. 9.1. Further details regarding the choice of cycle parameters, mechanical layout and development problems can be found in Refs (2) and (3). Such complex arrangements did not prove successful for either of these applications despite their undoubted thermodynamic merit, the main reason being the mechanical complexity involved. It is interesting to note, nevertheless, that the Ford engine described in Ref. (3) was an extremely compact unit of about 225 kW aimed at the heavy truck market. As a result of the problems with the complex cycle the navies of the world focused on simple-cycle gas turbines, using machinery arrangements such as CODOG, COSAG and COGOG to overcome the part-load problems as discussed in the Introduction. In recent years, however, the intercooled regenerative (ICR) cycle has been revived, on the basis that a single engine with very good part-load performance could replace both the cruise and boost gas turbines currently used. The Rolls-Royce WR21 uses a twin-spool gas generator, with intercooling between the two compressors, and a heat-exchanger. The projected performance of the ICR engine was compared with that of a conventional simple-cycle gas turbine used for naval propulsion in Ref. (4). Figure 9.2 shows that the ICR engine demonstrates a superior *SFC* at all power levels, with the most marked improvement at low power. The *SFC* curve is essentially flat between 100 and 40 per cent of design power and the greatly improved low-power *SFC* could be used to increase either range or time at sea before refuelling.

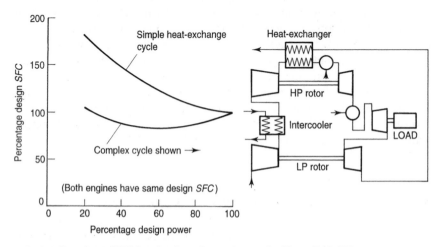

FIG. 9.1 Part-load *SFC* for simple and complex cycles [from Ref. (3)]

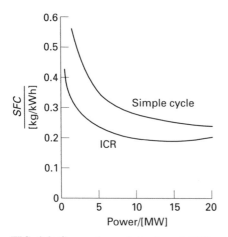

FIG. 9.2 Comparison of simple and ICR cycles

In practice it is found that the vast majority of applications make use of the simple cycle, with pressure ratios steadily increasing over the years. By 2000, industrial engines were operating at pressure ratios of up to 30, and aero engines up to 40. The use of combined cycles for efficiencies approaching 60 per cent, and simple-cycle efficiencies of 40 per cent, have resulted in the virtual elimination of the heat-exchange cycle. The Solar Mercury, a 4 MW heat-exchange cycle with an efficiency in excess of 40 per cent, offers the possibility of the heat-exchanger making a return in the niche market of small but efficient engines; simple-cycle units in this power range would have efficiencies of about 30 per cent. Detailed off-design performance calculations for gas turbines with heat-exchangers would have to account for variation in heat-exchanger effectiveness with engine operating conditions. Heat-exchanger effectiveness depends on the fixed parameters of heat transfer area and configuration (e.g. counter-flow, cross-flow or parallel-flow), and the parameters varying with engine operating conditions such as overall heat transfer coefficient between the two fluids and their thermal capacities (mc_p). Methods for estimating the effectiveness of gas turbine heat-exchangers (and intercoolers) are described in Ref. (5) and will not be dealt with here.

The incorporation of a heat-exchanger will cause an increase in pressure loss between compressor delivery and turbine inlet, and also an increase in turbine outlet pressure. Although the additional pressure losses will result in a reduction in power, they will have little effect on the equilibrium running line, and the part-load behaviour of engines with and without heat-exchanger will be similar. A typical running line for a free turbine engine was shown in Fig. 8.7; the variation of T_{03}/T_{01} with $N/\sqrt{T_{01}}$ can be deduced from this but is more clearly indicated in Fig. 9.3. It was shown in Chapter 2 that the thermal efficiency of a real gas turbine cycle was dependent on the turbine inlet temperature, and the rapid drop in T_{03} with

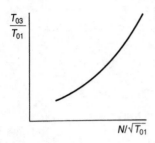

FIG. 9.3

decreasing power is the basic cause of the poor part-load performance of the gas turbine.

To improve the part-load efficiency of gas turbines, therefore, some means must be found of raising the turbine inlet temperature at low powers. In the majority of applications where good part-load economy is required, e.g. vehicular and marine, a free turbine† would be used. If we focus our attention first on the free turbine engine, the turbine inlet temperature at part load can be increased by using variable-area power-turbine stators.

Variable-area power-turbine stators

Variation of area is accomplished by rotating the nozzle blades, and this permits the effective throat area to be reduced or increased as shown in Fig. 9.4. In Chapter 8 it was emphasized that free turbine engines and turbojets were thermodynamically similar because the flow characteristics of a free turbine and a propelling nozzle impose the same operating restrictions on the gas generator; thus variable power-turbine stators will

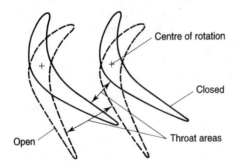

FIG. 9.4 Variable-geometry power-turbine stators

† Single-shaft engines normally operate at constant speed and high power. In practice it may be necessary to operate at fixed exhaust gas temperature for different electrical loads when used in combined cycle or cogeneration applications; this may be achieved by using variable inlet guide vanes at entry to the compressor, permitting some variation in airflow at fixed rotational speed.

have the same effect as a variable-area propelling nozzle. Section 8.6 showed that increasing the propelling nozzle area of a turbojet moved the running line away from the surge line, while decreasing the area displaced the running line towards surge. Reference to Fig. 8.7 shows that the latter will cause an increase in turbine inlet temperature at low powers; it is also likely that the compressor efficiency will be improved as the surge line is approached. Both of these effects will improve the part-load *SFC*.

Ideally, the area variation of the power-turbine stators can be controlled so that the turbine inlet temperature is maintained at its maximum value as power is reduced, as indicated in Fig. 9.5. If the running line at maximum temperature moves to intersect the surge line as shown, it then becomes necessary to reopen the power turbine nozzles for this part of the running range. It should be noted that operation at constant gas-generator turbine inlet temperature with reducing power will cause an increase in temperature at entry to the power turbine, because of the reduced compressor power, and the temperature of the hot gases entering the heat-exchanger will also be raised. Temperature limitations in either of these components may restrict operation in this mode. The use of a variable geometry power turbine is particularly advantageous when combined with a heat-exchanger, because the increased turbine outlet temperature is utilized.

The efficiency of the power turbine will obviously be affected by the position of the variable stators, but with careful design the drop in turbine efficiency can be more than offset by maintaining a higher turbine inlet temperature at part load. Area variations of ±20 per cent can be obtained with acceptable losses in turbine efficiency: see Ref. (6). The variable stators are liable to be exposed to temperatures in the 1000–1100 K region,

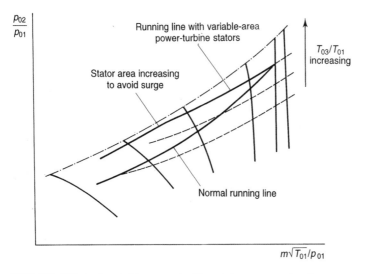

FIG. 9.5 Effect of variable power-turbine stators on running line

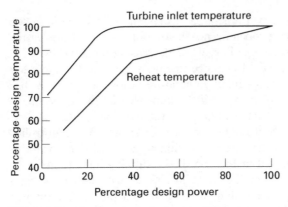

FIG. 9.6 Part-load control of reheat turbine

but the development problems associated with this do not appear to have been too difficult. A facility for increasing the stator area is also advantageous with respect to starting and accelerating the gas generator. If the stators are rotated still further, the gas-generator flow can be directed against the direction of rotation so that the flow impinges on the back of the power turbine blades. This can result in a substantial degree of engine braking which is extremely important for heavy vehicles.

Reheat combustion

The ABB reheat gas turbine is a single-shaft machine designed for electric power generation, primarily used in combined cycle applications. Although normally used at the base-load rating, the use of reheat provides considerable flexibility for operation at reduced power. At the design point approximately 60 per cent of the fuel is burned in the first combustor with 40 per cent in the reheat (or sequential) combustor. Reduction in power can be achieved by maintaining the fuel flow in the first combustor, keeping the maximum cycle temperature constant and modulating flow to the reheat combustor, reducing the temperature at entry to the LP turbine. Figure 9.6, from Ref. (7), shows the variation in both *TIT* and LP *TIT* over the running range; this provides good efficiency down to 40 per cent load. This turbine has VIGVs and three stages of variable stators, and these can be modulated to vary the airflow and maintain constant exhaust temperature to 40 per cent load; this allows the steam turbine to continue to operate at high efficiency as the gas turbine power is reduced.

9.2 Matching procedures for twin-spool engines

The basic methods used for twin-spool matching are similar to those described in Chapter 8. They differ from those for single-spool engines only in that it is also necessary to satisfy compatibility of flow between the spools. This compatibility requirement gives rise to the phenomenon of *aerodynamic coupling* which determines the ratio of the rotor speeds even though the rotors are *mechanically* independent of each other. The station numbering used for twin-spool turbojets is shown in Fig. 9.7 and the rotational speeds will be referred to as N_L and N_H for the LP and HP rotors respectively. The corresponding shaft power unit would have a power turbine between 6 and 7 in place of the propelling nozzle.

In outlining the procedures we will assume single line turbine characteristics, constant turbine efficiencies and constant percentage combustion pressure loss; these are often close approximations in practice and their use makes it easier to understand what is happening in physical terms. Considering work compatibility, the equations for the LP and HP rotors are

$$\eta_{mL} m c_{pg} \Delta T_{056} = m c_{pa} \Delta T_{012}$$

and

$$\eta_{mH} m c_{pg} \Delta T_{045} = m c_{pa} \Delta T_{023}$$

Flow compatibility must be satisfied between the compressors, between the turbines and finally between the LP turbine and nozzle (or power turbine). The problem of flow compatibility between two turbines was dealt with under the heading 'Matching of two turbines in series', section 8.4, and Fig. 9.8 illustrates the further restriction in operating conditions resulting from the introduction of a nozzle (or power turbine).

Calculations for a twin-spool engine can be lengthy, but they may be considerably simplified once it is appreciated that the HP rotor of a twin-spool engine is equivalent to a single-spool turbojet with a fixed nozzle whose area is defined by the throat area of the LP turbine stators. It was shown in Chapter 8 that for a single-spool turbojet a unique running line is defined when the nozzle is choked, and it follows that a similarly unique running line will be defined on the HP compressor characteristic of a twin-spool engine *when the LP turbine stators are choked*. (It should be noted that in practice the LP turbine of a twin-spool unit will be choked over most of the useful running range, but may become unchoked at idle conditions.) The position of the unique running line *on the HP*

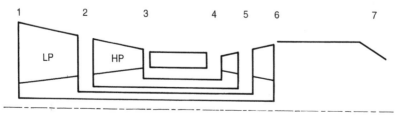

FIG. 9.7 Station numbering for twin-spool turbojet

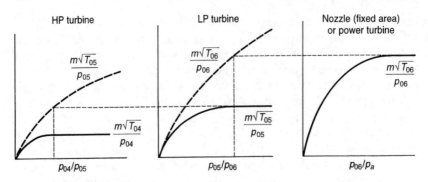

FIG. 9.8 Flow compatibility for twin-spool engine

characteristic can be determined by considering the HP rotor alone, for which compatibility of flow and work yields

$$\frac{m\sqrt{T_{04}}}{p_{04}} = \frac{m\sqrt{T_{02}}}{p_{02}} \times \frac{p_{02}}{p_{03}} \times \frac{p_{03}}{p_{04}} \times \sqrt{\frac{T_{04}}{T_{02}}} \tag{9.1}$$

$$\frac{\Delta T_{045}}{T_{04}} = \frac{\Delta T_{023}}{T_{02}} \times \frac{T_{02}}{T_{04}} \times \frac{c_{pa}}{c_{pg}\eta_{mH}} \tag{9.2}$$

It can be seen that equations (9.1) and (9.2) are the same as equations (8.2) and (8.6) apart from the difference in station numbering, and they can be solved by the trial and error procedure described in section 8.3 to yield a value of T_{04}/T_{02} for each point on a constant $N_H/\sqrt{T_{02}}$ line on the HP compressor characteristic. For the particular case when the LP turbine is choked, the HP turbine will be restricted to operation at a fixed non-dimensional point with the values of p_{04}/p_{05}, $m\sqrt{T_{04}}/p_{04}$ and $\Delta T_{045}/T_{04}$ all fixed. Substituting the relevant values of $m\sqrt{T_{04}}/p_{04}$ and $\Delta T_{045}/T_{04}$ in equations (9.1) and (9.2), the running line defined by LP turbine stator choking can be established on the HP characteristic, and it will be similar to that shown in Fig. 8.7. *The importance of establishing the LP choking line is that it greatly facilitates satisfying the requirement for compatibility of flow between the two compressors* as will be apparent in what follows.

Although much of the procedure to be described in this section is equally applicable to turbojets and shaft power units with a separate power turbine, it is most easily explained for the case of a turbojet with a fully variable propelling nozzle. Any effect resulting from variation of nozzle area will directly influence the LP turbine, and hence the LP compressor; the HP turbine, however, is separated from the nozzle by the LP turbine and, if the LP turbine is choked, the HP rotor will be shielded from disturbances caused by the variable nozzle. The use of a fully variable nozzle permits operation over a wide area of the LP compressor characteristic as in the case of a single-spool engine, even though the operating region on the HP characteristic is a single line determined by choking of the LP turbine. Thus if we assume a fully variable nozzle we can start with any point on the LP characteristic, and the final step in the calculations will be the determination of the required nozzle area. No iteration is required in this procedure, which can be summarized as follows.

(a) Determine the inlet conditions T_{01} and p_{01} from the ambient and flight conditions; for a land-based unit $T_{01} = T_a$ and $p_{01} = p_a$.

(b) Select any point on a line of constant $N_L/\sqrt{T_{01}}$ on the LP compressor characteristic, which will specify values of $m\sqrt{T_{01}}/p_{01}$, p_{02}/p_{01} and η_{cL}. The compressor temperature rise ΔT_{012} and temperature ratio T_{02}/T_{01} can then be calculated using equation (8.4).

(c) The non-dimensional flow at LP compressor exit, $m\sqrt{T_{02}}/p_{02}$, is obtained from the identity

$$\frac{m\sqrt{T_{02}}}{p_{02}} = \frac{m\sqrt{T_{01}}}{p_{01}} \times \frac{p_{01}}{p_{02}} \times \sqrt{\frac{T_{02}}{T_{01}}}$$

and for compatibility of flow between the compressors this is the non-dimensional flow at entry to the HP compressor.

(d) With the operating line on the HP compressor characteristic determined by LP turbine choking, the known value of $m\sqrt{T_{02}}/p_{02}$ defines the operating point on the HP compressor characteristic. This gives the values of p_{03}/p_{02}, $N_H/\sqrt{T_{02}}$ and η_{cH}, and the values of ΔT_{023} and T_{03}/T_{02} can be found.

(e) The overall compressor pressure ratio is now given by

$$\frac{p_{03}}{p_{01}} = \frac{p_{02}}{p_{01}} \times \frac{p_{03}}{p_{02}}$$

(f) The turbine inlet pressure p_{04} can be found from

$$p_{04} = \frac{p_{03}}{p_{01}} \times \frac{p_{04}}{p_{03}} \times p_{01}$$

where p_{04}/p_{03} is obtained from the combustion pressure loss.

(g) The value of $m\sqrt{T_{04}}/p_{04}$ is known, having been determined by flow compatibility between the two turbines, and m can be found from $m\sqrt{T_{01}}/p_{01}$, T_{01} and p_{01}.

(h) The turbine inlet temperature T_{04} can now be found from

$$T_{04} = \left(\frac{m\sqrt{T_{04}}}{p_{04}} \times \frac{p_{04}}{m}\right)^2$$

(i) The HP turbine is operating at a fixed non-dimensional point with p_{04}/p_{05} and $\Delta T_{045}/T_{04}$ determined by LP turbine choking. Thus conditions at entry to the LP turbine, p_{05} and T_{05}, can be obtained from

$$p_{05} = p_{04} \times \frac{p_{05}}{p_{04}}$$

and $T_{05} = T_{04} - \Delta T_{045}$.

(j) We must now satisfy the work requirement for the LP compressor. This is given by

$$\eta_{mL} m c_{pg} \Delta T_{056} = m c_{pa} \Delta T_{012}$$

ΔT_{012} is known, so ΔT_{056} can be calculated.

(k) With ΔT_{056} and T_{05} known, the temperature at exit from the LP turbine and entry to the nozzle is given by

$$T_{06} = T_{05} - \Delta T_{056}$$

(l) The LP turbine pressure ratio p_{05}/p_{06} is found from

$$\Delta T_{056} = \eta_{tL} T_{05}\left[1 - \left(\frac{1}{p_{05}/p_{06}}\right)^{(\gamma-1)/\gamma}\right]$$

Once p_{05}/p_{06} is known p_{06} is given by

$$p_{06} = p_{05} \times \frac{p_{06}}{p_{05}}$$

(m) m, T_{06} and p_{06} are now established. The overall nozzle pressure ratio p_{06}/p_a is then known, and the calculation of the nozzle area given the inlet conditions, pressure ratio and mass flow is quite straightforward (see Chapter 3). Once the required nozzle area has been calculated, the matching procedure is complete and an equilibrium running point has been obtained for that particular nozzle area and initial value of $N_L/\sqrt{T_{01}}$.

(n) The procedure can be repeated for points on other $N_L/\sqrt{T_{01}}$ lines. All the data required for a complete performance calculation, e.g. of the thrust, fuel flow and *SFC* at each running point, are now available.

Note that if a fixed nozzle were used, the area calculated in step (m) would not in general be equal to the specified area; it would then be necessary to return to step (b), try another point on the same $N_L/\sqrt{T_{01}}$ line on the LP characteristic and iterate until the correct nozzle area was obtained.

The effect of the LP turbine operating on the unchoked part of its characteristic is to raise the running line on the HP characteristic (i.e. displace it towards surge), because of the reduced non-dimensional flow at exit from the HP turbine. The calculation described above must be modified somewhat to deal with an unchoked LP turbine, but this condition is only liable to occur at low powers and the modification will not be dealt with here.

Finally, if a shaft power unit were being considered, step (m) yields $m\sqrt{T_{06}}/p_{06}$ at inlet to the power turbine and also the power turbine pressure ratio p_{06}/p_a. If this value of $m\sqrt{T_{06}}/p_{06}$ does not agree with the value from the power turbine characteristic for the known pressure ratio (see Fig. 9.8), again it is necessary to return to step (b) and repeat the calculation.

9.3   Some notes on the behaviour of twin-spool engines

Complete off-design performance calculations for twin-spool engines are obviously time consuming and in practice would be carried out using digital computers. Some aspects of twin-spool behaviour, however, can be deduced from an understanding of the matching procedure and the more important of these will be briefly described.

Aerodynamic coupling of rotor speeds

From step (c) of the matching procedure described in the previous section it can be seen that once an operating point on the LP characteristic has been chosen, the corresponding operating point on the HP characteristic is fixed. Thus for a fixed value of $N_L/\sqrt{T_{01}}$ and nozzle area, the value of $N_H/\sqrt{T_{02}}$ is determined by flow compatibility between the compressors. The HP compressor non-dimensional speed can be expressed in terms of the temperature at engine inlet by

$$\frac{N_H}{\sqrt{T_{01}}} = \frac{N_H}{\sqrt{T_{02}}} \times \sqrt{\frac{T_{02}}{T_{01}}}$$

$N_L/\sqrt{T_{01}}$ and $N_H/\sqrt{T_{01}}$ are then directly proportional† to the actual mechanical speeds of the two rotors.

    A typical variation of $N_L/\sqrt{T_{01}}$ with $N_H/\sqrt{T_{01}}$ for a turbojet is shown in Fig. 9.9; note that the relation between the speeds is dependent on nozzle area, and only for a fixed nozzle will it be unique. At a fixed value of N_H any increase in nozzle area will cause an increase in N_L, the physical reason being that opening the nozzle will increase the pressure ratio across the LP turbine, so causing an increase in LP rotor torque.

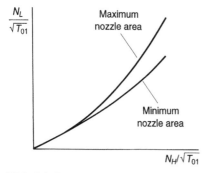

FIG. 9.9 **Speed relationship for twin-spool engine**

---

† If the 'equivalent speed' $N/\sqrt{\theta_0}$ were used instead of $N/\sqrt{T_0}$ (see section 4.6) the numbers would be equal to the mechanical speeds under reference entry conditions.

Effect of variable-area propelling nozzle

We have seen that the use of a fully variable nozzle permits operation over a wide range of the LP compressor characteristic, although the HP running line is not affected by nozzle variation as long as the LP turbine is choked. Typical running lines are shown in Fig. 9.10. It is important to note that increasing the nozzle area moves the LP running line *towards* surge, which is the opposite effect to that obtained on a single-spool engine. The reason for this is the increase in LP turbine power resulting from the redistribution of pressure ratio between the LP turbine and the nozzle. One of the major advantages of the variable nozzle is that it

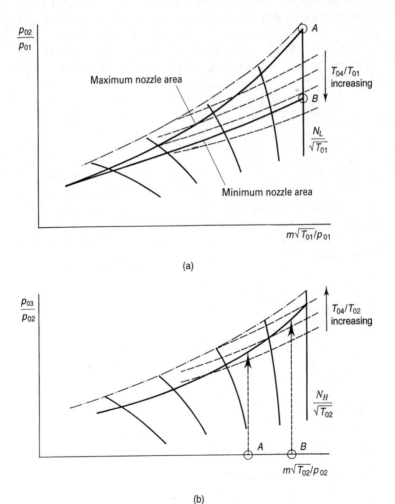

(a)

(b)

FIG. 9.10 Running lines for twin-spool turbojet. (a) LP compressor characteristic, (b) HP compressor characteristic

permits selection of rotor speed and turbine inlet temperature independently; or, in terms more relevant to an aircraft, of airflow and thrust. This is a valuable feature for an engine that has to operate over a wide range of intake temperature which produces significant changes in non-dimensional speed and mass flow. Reference (8) gives an excellent picture of the problems involved in matching the engine and intake for a supersonic transport.

It was shown in Chapter 8 that lines of constant T_{03}/T_{01} can be drawn on the compressor characteristic of a single-spool engine and these represent possible equilibrium running lines when a fully variable nozzle is used. As can be seen from Fig. 8.7, the value of T_{03}/T_{01} increases as lines of constant temperature ratio move towards surge. For a twin-spool engine lines of constant T_{04}/T_{02} can be drawn on the HP compressor characteristic and they behave similarly. If we consider the LP compressor characteristic, however, and plot lines of constant T_{04}/T_{01} as shown in Fig. 9.10(a), it will be seen that they move *away* from surge as the value of T_{04}/T_{01} is increased. Thus again we have behaviour opposite to that found for the single-spool turbojet. This can be explained if we consider operation at a constant value of $N_L/\sqrt{T_{01}}$. With the nozzle fully open the LP compressor operates at point A on Fig. 9.10(a) and closing the nozzle moves the operating point to B. The value of $m\sqrt{T_{01}}/p_{01}$ is constant for the vertical $N_L/\sqrt{T_{01}}$ line shown and the temperature rise across the LP compressor will be approximately fixed by the rotational speed. Thus if the LP compressor pressure ratio is decreased from A to B the value of $m\sqrt{T_{02}}/p_{02}$ will increase because

$$\frac{m\sqrt{T_{02}}}{p_{02}} = \frac{m\sqrt{T_{01}}}{p_{01}} \times \frac{p_{01}}{p_{02}} \times \sqrt{\frac{T_{02}}{T_{01}}}$$

The increase in $m\sqrt{T_{02}}/p_{02}$ results in an increase in HP compressor non-dimensional speed, pressure ratio and T_{04}/T_{02} as shown in Fig. 9.10(b). With the value of T_{02} approximately constant for a fixed rotational speed, it can be seen that T_{04}, and hence T_{04}/T_{01}, must increase as we move the LP operating point away from surge. It is also evident that a decrease in LP pressure ratio due to nozzle variation will be compensated by an increase in HP pressure ratio.

It should be noted that variable-area power turbine stators may also be used with a free turbine following a twin-spool gas generator. This is done in the M1 tank application and also the intercooled regenerative (ICR) marine engine. The area changes would be less than could be achieved with a variable-area propelling nozzle, but the same results apply.

Operation at fixed LP speed

The earlier aero-derivative engines such as the Olympus, FT-4 and RB-211 all combined a twin-spool gas generator with a free power turbine driving the electric generator. The latest large aero-derivatives, the GE LM 6000 and the Rolls-Royce Trent, have the LP shaft directly connected to the generator, equivalent to the configuration shown in Fig. 1.7; the Trent actually has a triple-spool arrangement, with intermediate pressure (IP) and HP compressors. These arrangements require that the LP rotor runs at constant (synchronous) speed throughout the load range, and the other rotor speeds will change with power. At synchronous speed and zero power, the LP compressor provides substantially more flow than the HP (or IP in the Trent) can absorb at low speed, and it is necessary to blow off significant quantities of air until the downstream compressors achieve sufficient speed to accept all the air delivered by the LP compressor. Blow-off is required only during loading and not during normal operation. Environmental considerations require the blow-off system to be adequately silenced.

Some versions of the LM 6000 incorporate spray intercooling (SPRINT) between the compressors. By reducing the HP compressor inlet temperature, and hence the power required, the output power is increased. The SPRINT system is primarily used to increase power on hot days, with as much as 29 per cent increase available.

Presentation of performance

In Chapter 8 it was shown that the performance of a single-spool engine could be presented as a function of $N/\sqrt{T_{01}}$. The performance of a twin-spool fixed geometry turbojet can be presented in terms of either the LP or HP rotor speeds because of the fixed relation between the speeds. For an engine with variable nozzle the performance would have to be presented in terms of any two of N_L, N_H and A_7. An example of a 'carpet' plot showing the variation of thrust with N_L and N_H for a variable nozzle turbojet is shown in Fig. 9.11; other quantities such as fuel flow and SFC could be presented in a similar manner.

For fixed geometry engines the variation of turbine inlet temperature and fuel flow could be plotted versus either N_L or N_H, but because of

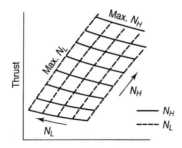

FIG. 9.11

the rapid variation of both with N_H the latter may be more suitable as abscissa. It should be noted that although the HP rotor speed will always be higher than the LP rotor speed and the HP turbine blades are at the higher temperature, they may not be the critical components because the LP turbine blades are usually substantially longer.

The mass flow through a twin-spool engine is primarily determined by the LP compressor speed, and consequently the performance of a twin-spool shaft power unit with a free power turbine is conveniently presented in a manner similar to that of Fig. 8.10 with lines of constant LP compressor speed replacing lines of constant gas-generator speed. Alternatively, the presentation of Fig. 8.12 could be used, with LP speed as the independent variable.

9.4   Matching procedures for turbofan engines

The approach described in section 9.2 is also applicable to turbofans, but in this case we must take into account the division of flow between the bypass duct and the gas generator, which will vary with off-design operating conditions. Only the simplest case, that of the twin-spool turbofan with separate exhausts shown in Fig. 3.15, will be considered here. The additional information required is the bypass or 'cold' nozzle characteristic, as indicated in Fig. 9.12; the turbine characteristics and the hot stream nozzle characteristics will be similar to those of the twin-spool turbojet.

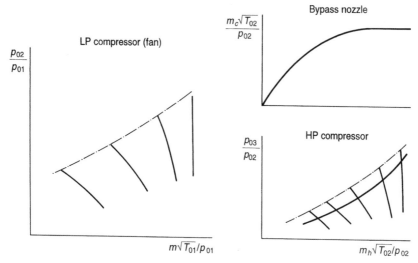

FIG. 9.12 Characteristics required for turbofan matching

Let us consider a turbofan with both nozzles fixed. The running line corresponding to LP turbine choking can be established on the HP characteristic in exactly the same manner as for the twin-spool turbojet. The problem now is to establish the proportion of the total flow (m) passing through the bypass duct (m_c) and the gas generator (m_h). Suffixes h and c refer to 'hot' and 'cold' streams respectively. One possible procedure is as follows.

(a) Select ambient and flight conditions, giving values of p_{01} and T_{01}.
(b) Select a value of $N_L/\sqrt{T_{01}}$ and guess any point on this line. The value of m entering the fan is then known, and the value of $m\sqrt{T_{02}}/p_{02}$ at exit from the fan can be calculated.
(c) From the known value of p_{02}, the pressure ratio across the bypass nozzle, p_{02}/p_a, can be calculated and the value of $m_c\sqrt{T_{02}}/p_{02}$ can be found from the bypass nozzle characteristic. The value of m_c can also be found, and $m_h = m - m_c$.
(d) The gas-generator non-dimensional flow can then be found from

$$\frac{m_h\sqrt{T_{02}}}{p_{02}} = \frac{m\sqrt{T_{02}}}{p_{02}} - \frac{m_c\sqrt{T_{02}}}{p_{02}}$$

(e) Knowing the gas-generator non-dimensional flow we can enter the HP characteristic, and establish the overall pressure ratio and turbine inlet temperature as for the twin-spool turbojet (see steps (e) to (h) of section 9.2). The HP turbine pressure ratio and temperature drop are known as a result of flow compatibility between the two turbines, so that the pressure and temperature at entry to the LP turbine, p_{05} and T_{05}, can be determined as before.
(f) The LP turbine temperature drop is given by

$$\eta_{mL}m_h c_{pg}\Delta T_{056} = mc_{pa}\Delta T_{012}$$

(g) The LP turbine pressure ratio is then found from T_{05}, ΔT_{056} and η_{tL}. We now know p_{06} and T_{06}.
(h) Knowing m_h, p_{06}, T_{06} and p_{06}/p_a the hot stream nozzle area can be calculated; in general this will not agree with the specified value and it is necessary to return to step (b), select a new point on the LP characteristic, and repeat the process until agreement is reached.

The close similarity between the twin-spool and turbofan matching procedures can be seen. The method described may readily be extended to deal with three-spool engines, and the configuration shown in Fig. 3.24(c) is equivalent to a simple turbofan with a twin-spool gas generator. The use of mixed exhausts presents a further complication because it is then necessary to include equations satisfying conservations of energy and momentum for the mixing process. Neither of these topics will be pursued further here.

9.5 Transient behaviour of gas turbines

In certain applications the transient response of gas turbines following a demand for a change in output can be critical; in other applications good response may merely be desirable. Aircraft engines are an obvious example of an application where the transient behaviour is critical; the prime requirement for civil aircraft is for a rapid thrust response to cope with a baulked landing when the aircraft is close to touchdown but is forced to overshoot the runway. Lifting engines for VTOL aircraft provide another example, where an engine failure on a multi-engined installation may result in serious unbalanced forces which have to be adjusted rapidly. Rapid response is also required from gas turbines used for emergency generation of electricity which must often deliver their rated output within 2 minutes of the start signal. The response rate of the gas turbine itself from idle to maximum power must be less than about 10 seconds to allow time for the starting and synchronization sequences. The transient response of a vehicle gas turbine to changes in throttle setting is not critical. Although the relatively slow response of a gas turbine has frequently been regarded as a major disadvantage, for heavy vehicles where it is likely to be competitive the vehicle acceleration will be largely independent of the engine acceleration.

In the early days of gas turbines little attention was paid to the prediction of transient behaviour, and the response rate of an engine was established empirically during development testing; this required extensive test bed running, and frequently engines were damaged. Now, however, the transient behaviour is predicted from a knowledge of the off-design performance, and this can be done during the design phase using estimated component characteristics. An in-depth understanding of the dynamic behaviour at the design stage is essential for the design and development of control systems. Requirements for higher performance have led to multi-spool rotor systems, variable geometry in the compressor and blow-off valves, which complicate prediction of both steady-state and transient performance.

The acceleration of gas turbines is obviously dependent on such factors as the polar moment of inertia of the rotor system and the maximum temperature which the turbine blades can withstand for short periods. Usually the limiting factor on acceleration is the proximity of the surge line to the equilibrium running line, and this is particularly critical at the start of an acceleration from low powers. The configuration of the gas turbine will have a major effect on its transient behaviour; for example, the behaviour of single-shaft and free turbine engines delivering shaft power will be quite different, and a twin-spool engine will respond very differently from a single-spool engine. Only a brief introduction to the problem of engine transients will be given in this book, principally via the single-spool engine with and without free power turbine, and the interested reader must then turn to the specialized literature.

Prediction of transient performance

It will by now be clear that all off-design equilibrium running calculations are based on satisfying the requirements for compatibility of flow and work between the components. During transient operation a gas turbine can be considered to satisfy compatibility of flow† but not of work, and the excess or deficiency of power applied to a rotor can be used to calculate its acceleration or deceleration. The problem then becomes one of calculating increments of net torque associated with increments of fuel flow and integrating to find the change of rotor speed.

The acceleration of the compressor rotor and the excess torque ΔG are related by Newton's Second Law of Motion, namely

$$\Delta G = J\dot{\omega}$$

where J is the polar moment of inertia of the rotor and $\dot{\omega}$ is its angular acceleration. The excess torque ΔG is given by

$$\Delta G = G_t - G_c \text{ (free turbine engine)}$$

or

$$\Delta G = G_t - (G_c + G_l) \text{ (single-shaft engine)}$$

where suffixes t, c and l refer to turbine, compressor and load respectively. *It should be noted that the net torque is the difference between two quantities of similar magnitude, and a small change in either may result in a much larger change in the torque available for acceleration.* The problem now is that of obtaining the torques during transient operation. The turbine torque may be greater or less than the compressor torque, corresponding to acceleration or deceleration of the rotor.

To obtain the torques we may use the methods for off-design performance described in Chapter 8. If we consider *any* point on the compressor characteristic, which will not in general be an equilibrium running point, we can satisfy flow compatibility between compressor and turbine by using the identity

$$\frac{m\sqrt{T_{03}}}{p_{03}} = \frac{m\sqrt{T_{01}}}{p_{01}} \times \frac{p_{01}}{p_{02}} \times \frac{p_{02}}{p_{03}} \times \sqrt{\frac{T_{03}}{T_{01}}}$$

Consider first the turbine non-dimensional flow $m\sqrt{T_{03}}/p_{03}$: it can normally be assumed to be independent of speed and a function of the pressure ratio alone. The value of the gas-generator turbine pressure ratio p_{03}/p_{04} corresponding to the selected value of p_{02}/p_{01} on the compressor

---

† During rapid transients, pressures cannot change instantaneously because of the finite volume between the components, and the assumption of flow compatibility at all times is not exactly true although it is a good approximation. Methods for dealing with non-instantaneous pressure changes are discussed in Ref. (11), and are subject to further research; the effects are confined to a very short period immediately following the start of a transient.

characteristic can be easily obtained for both single-shaft and free turbine engines; for a single-shaft engine it is fixed by the compressor pressure ratio and the combustion pressure loss, and for a free turbine engine it can be found from Fig. 8.9. Thus it can be seen that $m\sqrt{T_{03}}/p_{03}$ is a function of p_{02}/p_{01} for both types of engine. Rearranging the equation for flow compatibility given above,

$$\sqrt{\frac{T_{03}}{T_{01}}} = \frac{m\sqrt{T_{03}}}{p_{03}} \times \frac{p_{02}}{p_{01}} \times \frac{p_{03}}{p_{02}} \bigg/ \frac{m\sqrt{T_{01}}}{p_{01}}$$

With constant percentage combustion pressure loss p_{03}/p_{02} will be constant, and it can be seen that the value of T_{03}/T_{01} satisfying flow compatibility can be obtained for *any* point on the compressor characteristic without reference to work compatibility. The fuel flow required to produce this temperature ratio can be calculated because the airflow, compressor delivery temperature and combustion temperature rise are all known. If the process is repeated for a series of points, lines of constant T_{03}/T_{01} can be drawn on the compressor characteristic. These lines will have a similar form to the constant temperature lines shown for a free turbine engine in Fig. 8.7, but it must be emphasized they are not the *same* lines, because they represent non-equilibrium running conditions. (Constant temperature lines of this kind drawn on the compressor characteristic for a single-shaft engine would, however, represent a series of equilibrium running points for an engine driving a load which can be set independently of speed, e.g. a hydraulic dynamometer.)

With the turbine operating point and turbine inlet temperature known the power developed by the turbine can be calculated. Thus

$$\text{turbine power} = \eta_m mc_{pg}\Delta T_{034}$$

where

$$\Delta T_{034} = \frac{\Delta T_{034}}{T_{03}} \times \frac{T_{03}}{T_{01}} \times T_{01}$$

The compressor power is given by $mc_{pa}\Delta T_{012}$ and for an engine with a free turbine the net torque on the compressor rotor is given by

$$\Delta G = (\eta_m mc_{pg}\Delta T_{034} - mc_{pa}\Delta T_{012})/2\pi N$$

For a single-shaft engine the load torque would also have to be included.

Once the net torque has been obtained, the angular acceleration of the compressor rotor can be calculated; it may be assumed that this will be constant for a small interval of time and the resulting change in speed can be found. The process can then be repeated many times to provide a transient running line starting from some convenient equilibrium running point. The shape of this transient running line will be determined either by the choice of a limiting T_{03} (which might be 50 K higher than the design point value) or by the location of the surge line. The fuel flow required for operation along the desired running line can be calculated and the provision of this fuel during the transient is the responsibility of the designer of the fuel control system. Typical acceleration and deceleration trajectories on the compressor characteristic for a free turbine engine or a single-shaft engine with a propeller load are shown in Fig. 9.13. With an acceleration the initial movement towards surge shown in Fig. 9.13 is due to the rise in temperature following an increase in fuel flow, which takes place before the rotor has had time to increase its speed and provide an increase in mass flow. Too large an initial increase in fuel flow will cause the compressor to surge, resulting in very high temperatures which could destroy the turbine. The acceleration procedure for a single-shaft engine driving an electric generator would be to bring the rotor system up to full speed before applying the load and there should be no problem with regard to surge.

During decelerations the operating point is moving away from surge as shown and the turbine inlet temperature is decreasing; the only problem that may arise is 'flame-out' of the combustion chamber because of very weak mixtures. This can be overcome by scheduling the deceleration fuel flow as a function of rotor speed to prevent too rapid reduction of fuel flow.

We have seen in section 8.6 that if the equilibrium running line intersects the surge line, either blow-off or variable geometry can be used to lower the running line, but it should be noted that neither would have much effect on the *transient* running line. Rates of acceleration will be slowed down by blow-off, because of the reduction in mass flow through the turbine, but can be improved by a facility for increasing the area of the propelling nozzle (or power turbine stators).

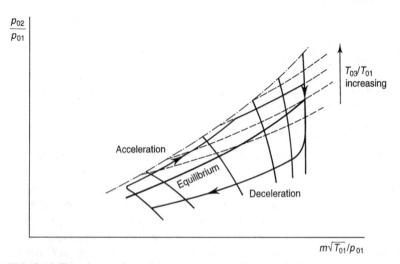

FIG. 9.13 Transient trajectories on compressor characteristic

We pointed out in section 9.1 under the heading 'Variable-area power turbine stators' that gas turbines for vehicular applications may well incorporate a variable-geometry power turbine to improve the part-load economy, achieving this by *decreasing* the stator area as power is reduced. Conversely, on starting, the effect of *increasing* the stator area is to decrease the pressure ratio across the power turbine and increase that across the gas-generator turbine. The resulting transient increase in gas-generator turbine torque provides a substantially greater increase in *net* torque available for acceleration.

Single-shaft versus free turbine engines

The choice of a single-shaft or free turbine configuration will normally be made on the basis of some primary requirement such as low cost or good low-speed torque characteristics, but the transient behaviour might also be a deciding factor.

With a free turbine engine, power reduction must be obtained by reducing the gas-generator speed, because of the fixed running line determined by the flow characteristic of the power turbine. Thus to restore power it is necessary to accelerate the gas generator up to its maximum speed. With a single-shaft engine, however, the load may be varied at constant speed, e.g. as with an electric generator or a variable-pitch propeller. This means that there is no need to accelerate the gas generator and power may be restored very quickly by increasing the fuel flow. Such a feature can be very useful for turboprops where the power can be altered by changing the pitch of the propeller. The single-shaft engine has the obvious additional advantage that in the event of load being shed the compressor acts as a very efficient brake and for this reason regulation of output speed is easier than with a free turbine engine.

The starting power requirements for large gas turbines must be carefully considered; this is particularly important for emergency generating plant where one of the prime requirements is to achieve maximum output quickly. With a single-shaft unit the entire rotating assembly must be brought up to speed, and this may require the use of a steam turbine or diesel engine with a power of the order of 3 MW for a 100 MW unit. The free turbine engine is in a much more advantageous position, because neither the power turbine nor the load are driven by the starter. A twin-spool gas generator is even more favourable: only the HP rotor need be motored and the starter power will be less than 100 kW even for a large unit.

At low powers, with a generator as load, the part-load fuel consumption of a free turbine engine is superior to that of a single-shaft engine. Figures 8.7 and 8.5 show the equilibrium running lines in each case. The running line of the free turbine engine follows the locus of maximum compressor efficiency more closely than that of the single-shaft engine, and in general T_{03} falls off less rapidly.

Single-shaft units are normally used for electric power generation. An exception is where an aero-derivative engine is specified, as on off-shore oil rigs where compactness is essential. The free turbine units can provide power at either 50 or 60 Hz for the European or North American markets, by minor changes in the speed of the power turbine with very small changes in efficiency; their power, however, is restricted to about 50 MW because of the relatively small flow through the gas generator. Single-shaft units can be designed to provide up to 250 MW, and manufacturers offer different machines for the 50 or 60 Hz markets; machines of around 50–75 MW can be designed with output gearboxes suitable for either 50 or 60 Hz.

Twin-spool transient running lines

A full discussion of twin-spool transient behaviour cannot be covered in a book of this nature, but some important differences from single-spool behaviour must be mentioned. The transient behaviour of the HP rotor is similar to that of a simple jet engine and the trajectories on the HP characteristic will be similar to those shown in Fig. 9.13. Consideration of the lines of constant T_{04}/T_{01} on the LP characteristic would suggest that trajectories for accelerations and decelerations would be as shown in Fig. 9.14, with the likelihood of LP surge following a rapid deceleration. The lines of T_{04}/T_{01}, however, apply only to steady-state operation and are irrelevant to transient behaviour. Detailed calculations and experimental investigations, described in Refs (9) and (10), have shown that the LP transient running lines follow the equilibrium running line very closely, both for accelerations and decelerations.

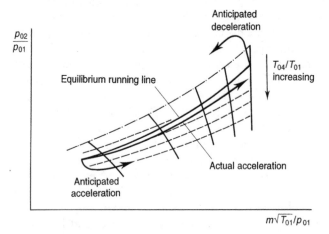

FIG. 9.14 Transient trajectories on LP compressor of twin-spool unit

We have noted that the acceleration of a single-spool turbojet engine can be improved by increasing the nozzle area. Although care must be taken when extending this idea to twin-spool turbojets, considerable advantage can result from judicious manipulation of the nozzle area during transients. Increasing the nozzle area during an acceleration will improve the LP rotor acceleration, because of the increase in LP turbine torque, but it should be realized that the LP running line will be moved towards surge. The mass flow is increased more rapidly because of the improved LP rotor acceleration, and this in turn permits the fuel flow to be increased more rapidly. In practice it is found that the thrust response is primarily determined by the rate of increase of fuel flow, so that the thrust response is improved at the expense of LP surge margin. Detailed calculations have shown that the HP surge margin is improved, however, and it is possible to trade surge margins between the two compressors to get the best response.

When operating at high forward speeds the nozzle will usually be open and the LP operating point will be close to surge. Emergency decelerations from this condition could result in LP surge, which may lead to a flame-out. A study of emergency decelerations in Refs (9) and (10) showed that the best method of avoiding surge was to close the nozzle fully before reducing the fuel flow; this moved the LP operating point down to the running line associated with minimum nozzle area and also reduced the LP rotor speed. Immediate reduction of fuel flow, on the other hand, resulted in a running line very close to surge, as did simultaneous reduction of fuel flow and closing of the nozzle.

Simulation of transient response

The development of a suitable control system requires a deep understanding of the transient behaviour of the gas turbine to be controlled, particularly for new types of engine where no previous experience exists. If a mathematical model or 'simulation' describing the engine dynamics is constructed and stored in a suitable computer, it can provide designers of gas turbines with an extremely versatile tool with which to investigate a wide variety of problems. A prime requirement for the simulation is that it should be capable of covering the entire running range of the engine and also of incorporating such modifications as blow-off and variable geometry as necessary. This can best be achieved by basing the simulation on the methods of off-design performance calculation presented in Chapters 8 and 9; the use of estimated component characteristics makes the simulation flexible in use and easy to understand, and has the further advantage that the component characteristics can be modified in the light of rig testing as the development programme proceeds. Typical problems which can readily be investigated include optimization of acceleration fuel schedules, operation of variable-geometry devices and overspeed protection. Incorrectly chosen operating procedures, particularly under emergency conditions, may be hazardous to both engine and operator.

The information flow chart required for a mathematical model of a simple turbojet engine is shown in Fig. 9.15. It can be seen that the pressure ratios across each component are governed by flow compatibility, the pressure ratios in turn determine the temperature ratios, and the turbine temperature drop (proportional to the power developed by the turbine) is determined by the pressure ratio and turbine inlet temperature (controlled by the fuel flow).

Successful simulations have been carried out using analogue, digital or hybrid computers and appropriate techniques are described in Refs (9) to

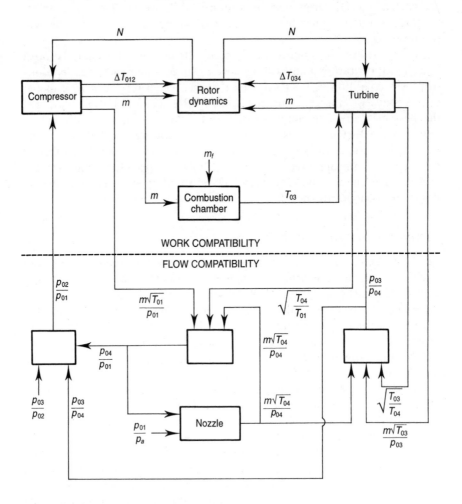

FIG. 9.15 Information flow chart for single-spool turbojet

(12). Early digital computers did not have the computing speed essential for the continuous integration of the net torque, and real-time simulation could only be achieved using analogue computers; the hybrid computer was an intermediate solution which combined the speed of the analogue with the data processing capability of the digital. Now, however, digital computer capabilities are so fast that they are almost universally used for simulation. The judicious use of simulations can give an invaluable insight into transient response and control problems without endangering an engine, resulting in significant savings in test bed development and hence cost.

9.6 Performance deterioration

In Chapter 2 it was emphasized that gas turbine efficiency and output depend strongly on the isentropic efficiencies of the compressors and turbines. If follows that any in-service reduction in component efficiency will have a considerable effect on the overall performance. Degradation of component efficiency can result from build-up of deposits on the compressor blading, increased tip-clearances resulting from rotor-casing rubs, and erosion or corrosion of the blading. These effects will combine to reduce the maximum power available, increase the turbine operating temperature with a consequent reduction in life and increase the fuel consumption. When the deterioration in performance reaches a predetermined level the engine must be overhauled. It is possible to *predict* performance deterioration using the methods for off-design performance which have been developed; the prime requirement is to predict changes in component performance for various defects. The analysis of field data may be extended to the component level if adequate instrumentation is fitted, and it is then possible to pinpoint faulty components. This would permit the exchange of faulty *modules* in a modular engine, without the requirement for a complete overhaul. Transient behaviour will be affected by component deterioration, and analysis of transient performance has proved to be very useful in detecting component faults in fighter engines which spend very little time at steady operating conditions.

The performance of all gas turbines will deteriorate over time, in some applications gradually and in others very rapidly, depending on both the atmospheric conditions and the manner in which the engine is used. Land-based industrial gas turbines may operate in apparently benign locations, such as a rural compressor station on a pipeline, or in hostile locations such as a refinery complex. Marine gas turbines face the problem of salt ingestion, with spray being entrained into the compressor. Aircraft engines normally operate at high altitudes in a clean atmosphere, except during short periods when taking off and landing, but helicopters operate continuously at low level in both salt water and desert conditions where ingestion of salt and sand is a major problem. Even in the benign environment of rural locations it has been found that pollen, tree-sap and insects can form sticky deposits, resulting in compressor *fouling*; in industrial locations this may also be caused by oil vapours or chemical processes. The ingestion of hard particles (e.g. sand) causes *erosion* of the blading, particularly in the early stages of compression; this can result in changes both to the airfoil shape and to the size of the flow passages. Operation in salt water atmospheres can cause severe *corrosion*, especially of the high-temperature blading in the turbine, resulting in a serious loss in blade life or even blade failure. It is obvious that an efficient filtration system is necessary, but it will be remembered from Chapter 2 that any pressure loss in the intake will cause a loss in power. Inlet filter systems are bulky and heavy and cannot be considered on airborne systems; in any case it should be understood that filter systems cannot remove *all* of the contaminants. Helicopters now make use of *inertial separators* upstream of the compressor intake. These are plug-shaped devices that accelerate the inlet air outward in a radial direction and then direct it inward to the eye of the compressor; solid particles are centrifuged to the outer radius where they are collected and discharged overboard. Inertial separators introduce both performance and weight penalties but are essential to provide adequate engine life in the hostile environment of the military helicopter.

Rapid deterioration in performance can occur in applications where the engine is subjected to many starts and short running times per start, such as in emergency electric power generation. This application involves starting from cold metal and achieving maximum power as quickly as possible. Aero-derivative engines were originally used in this market because of their capability of reaching full power from a 'black' start, i.e. with no external power available, in 120 seconds. This, of course, results in high rates of change of temperature in the turbine, causing repeated thermal stresses, leading to thermal fatigue and cracking in the turbine nozzle and rotor blades. An early aero-derivative engine used in this manner required overhaul every 500 hours, while the same engine in pipeline applications achieved an overhaul life of 30 000 hours. The electrical units typically ran for 10–20 minutes per start, while the pipeline units could run for several thousand hours per start. The start cycle for the electrical units was later modified to use a slower increase of load for peaking duty, with the 120 second start used only for emergencies. In aircraft applications the operating cycles of military and civil aircraft are quite different: combat aircraft will be subject to rapid manoeuvring with frequent large changes in power, while civil aircraft will operate for long periods (up to 12 hours) at constant cruise conditions. It should be noted, however, that short-haul aircraft will be exposed to much shorter flight times and will undergo 4–5 times more cycles than their long-haul counterparts, requiring particular attention at the design stage to ensure adequate turbine life. This is discussed in Ref. (6) of Chapter 1. The most spectacular form of rapid performance deterioration has been when civil aircraft unexpectedly flew through a cloud of volcanic dust following a major eruption. There have been several cases of total power loss at high altitude due to the flow passages becoming blocked, causing engine flame-out. Fortunately it was possible to relight one or more engines at low altitude, permitting an emergency landing. In the event of volcanic eruptions it is now common for aircraft to be re-routed to give a wide berth to the plume.

Focusing on industrial applications, performance deterioration falls under two main categories: (a) recoverable and (b) non-recoverable. Recoverable performance loss is primarily the result of compressor fouling, and performance recovery can be obtained by compressor cleaning. This may be done either by abrasive cleaning or by compressor washing. Abrasive cleaning was done by injecting relatively soft materials into the air stream with the engine operating at normal speeds; materials used included walnut shells, apricot stones, rice husks and spent catalyst. The materials broke up on impact and provided a scouring action on the blade surfaces, removing the built-up deposits. Abrasive cleaning eventually damaged the blading, particularly the thin leading edges, and was likely to block internal cooling passages in engines with cooled blades. For these reasons, abrasive cleaning is seldom used today. Compressor washing can be carried out either *on-line* or *off-line*. On-line washing is done by spraying a mixture of clean water and detergent into the compressor inlet with the engine operating at normal load. This is often done in applications where power is needed continuously and there are severe cost penalties associated with shutting the engine down. Off-line (or *soak*) washing requires the engine to be shut down and the water/detergent mixture sprayed in while the rotor is turned by the starting system. This requires several applications and then rinsing with water alone, but appears to give better cleaning. If washing is delayed too long, it is difficult to recover the performance loss, but if done too frequently there are economic penalties. Reference (13) provides a detailed description of compressor washing procedures and benefits.

Non-recoverable performance loss results from mechanical problems such as erosion, corrosion, and tip-clearance increases. If the aerodynamic shape of the blading is altered by erosion or surface finish, the loss in efficiency can only be recovered at overhaul. Increases in tip-clearance result from rotor/casing rubs, which may occur during transients owing to differential growth rates in the rotor and casing, resulting in both a loss in component efficiency and a lowering of the surge line. In-service deterioration will lead to a gradual raising of the steady running line in a jet engine/free turbine application, and this, combined with the lowered surge line, can significantly reduce the amount of excess fuel that can be provided for acceleration.

Reference (14) provides an overview of performance deterioration in industrial gas turbines and suggests methods for detecting performance loss, for example by monitoring compressor delivery pressure or inlet depression, the latter being an indirect measure of airflow. Performance deterioration can be predicted using the normal off-design procedures with modified component characteristics to allow for the effect of fouling and mechanical damage on flow and efficiency. Very little experimental data are available in the open literature, and engines which have suffered performance deterioration are not sufficiently instrumented to measure component data. Reference (15) shows that a 400 kW turboshaft engine experienced a 5 per cent drop in flow and a 1 per cent drop in compressor efficiency over a period of 150 hours; this was in an apparently benign university campus site, and the main cause of fouling was oil leakage from the front bearing. Reference (16) describes experimental work done to predict the effects of mechanical damage in turbine blades, being part of a research effort to develop realistic models for predicting performance degradation. Reference (17) describes computer-based modelling of compressor characteristics, using stage stacking techniques, to predict the effects of fouling.

9.7 Principles of control systems

Figure 9.16 shows the main features of a gas turbine control system. The fundamental requirement is to maintain the safety of the engine, regardless of how the operator moves the throttle lever or how the inlet conditions (e.g. altitude) are changing. The control system must ensure that the critical operating limits of rotational speed and turbine inlet temperature are never exceeded, and that compressor surge is avoided. The sensing of rotational speed presents no problems and a variety of frequency measuring devices are available. Turbine temperature, on the other hand, is very difficult to measure. It is not normally practicable to locate temperature probes at inlet to the turbine, where the temperature is very high, and the temperature is measured at some downstream location, typically at inlet to the power turbine on shaft power units and in the jet

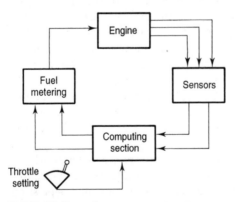

FIG. 9.16 Control system components

pipe on jet engines. Thus the temperatures used to protect the turbines are *indirect* measures of the critical temperatures. The temperatures are usually measured by thermocouples spaced around the annulus: typically six to eight probes are used. In the case of turbines with highly cooled blades, it is the actual *blade* temperature which is important and this may be sensed by radiation pyrometry in advanced engines.

The methods described in Chapters 8 and 9 can be used to provide the control system designer with information about the fuel flow required for steady-state operation over the entire range of operating conditions. Analysis of transient performance can predict the maximum fuel flow which can be used for acceleration without encountering surge or exceeding temperature limits. A typical schedule for fuel flow for a simple jet engine is shown in Fig. 9.17. Mathematical models for simulating the transient behaviour are an essential tool for optimization of fuel schedules, which must be experimentally verified during the engine development programme.

The performance calculations described in Chapters 8 and 9 showed that the variations of all the key parameters are wholly determined by the matching of the compressor, turbine and nozzle characteristics. It is important to realize that if the engine has *fixed geometry*, the steady-state performance cannot be altered in any way by the control system. If, however, devices such as variable IGVs, variable compressor stators or variable nozzles (turbine or propelling) are included, their operation can be integrated with the control system to modify the performance. The level of sophistication required of the control system is strongly dependent on the complexity of the engine.

Figure 9.16 shows that the control system must incorporate both a computing section and a fuel metering section. For many years both of these

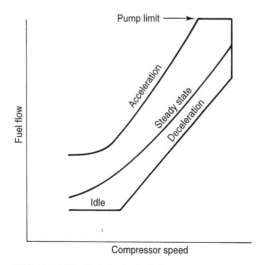

FIG. 9.17 Typical fuel schedule

functions were met by *hydromechanical* systems, in which fuel passing through the unit provided the necessary hydraulic actuation of a variety of pistons, bellows and levers which metered the required fuel to the combustion system. In recent years much development of *digital* control systems has taken place; the increasing computational capacity and rapidly decreasing cost of small digital computers has made them quite feasible for use in control systems. It should be realized, however, that a fuel metering system is still required; the function of the digital computer is limited to computation and the control system must open and close a tap to control fuel flow. Digital control systems with a comprehensive capability for controlling all modes of engine operation, known as Full Authority Digital Engine Controls (FADECs), are becoming widely used on large aero-engines. Although the cost of the actual computer is small, the need for certification of the software leads to very high software development costs. For this reason, FADECs are seldom used on small engines of lower cost.

Digital control systems will increasingly be used for data acquisition, using the data measured by the sensors, and unusual changes may be used in *diagnostic* systems for examining the mechanical condition of the engine. Engine Health Monitoring (EHM) systems use thermodynamic measurements, vibration analysis and chemical analysis of the lubricating oil to assess engine health. Reference (18) discusses the use of thermodynamic models in diagnostic systems. EHM systems have demonstrated significant reductions in maintenance costs and savings due to early detection of engine deterioration before severe damage occurs [Ref. (19)].

To close this section, two very simple examples of the use of the control system for maintaining an engine within operating limits will be considered. Let us consider a turboshaft engine with a free power turbine, used in a helicopter; the gearbox will be very highly stressed to keep weight to a minimum and the maximum power permissible will be strictly limited. The worked example 8.2 showed the significant increase in power on a cold day, and conversely the power will decrease on a hot day. The variation in power would be as shown by the full line in Fig. 9.18. The maximum power limit is shown dotted, and at high ambient temperatures it will also be necessary to limit turbine temperature, causing an even more rapid decline in power, as previously discussed with respect to jet engines. One method of achieving these limits would be by controlling gas-generator speed as shown; the gas generator could only be run at maximum speed where neither limit was exceeded and at both low and high temperatures its value would be reduced.

Aero-engine performance is often quoted on the basis of a thrust (or power) *flat rated* to some temperature, typically 30 °C which is 15 °C above the ISA temperature at sea level. This is done to ensure that the engine is capable of producing adequate thrust on a hot day, while conserving engine life when the ambient temperature is below the flat-rating temperature. The concept is illustrated in Fig. 9.19, where it can be seen

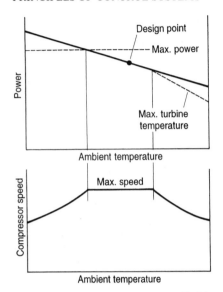

FIG. 9.18 Power and temperature limiting

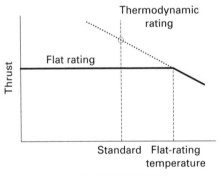

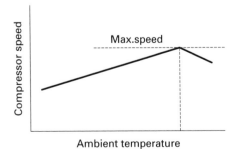

FIG. 9.19 Flat-rating concept

that the thrust is constant (flat) at all temperatures below the rated temperature; it is apparent that at the ISA temperature the engine is thermodynamically capable of considerably higher thrust, referred to as the *thermodynamic rating*. If the engine were controlled on the basis of rotational speed, the maximum speed would only be used at the rated temperature; on cooler days the required thrust could be obtained at a reduced speed and turbine inlet temperature. The thermodynamic rating may be 15–20 per cent higher than the flat rating. One version of the PT-6 turboprop was flat rated at 600 kW to 62·8 °C; this very high temperature was required to provide an engine capable of 600 kW at 12 000 m for a high-speed aircraft. The engine had a thermodynamic rating of 1000 kW but the control system limited the power to 600 kW and the gearbox was designed to meet this lower rating, saving on weight and cost. It can be seen that a high flat-rating temperature implies a significant derating of the engine at ISA conditions.

Control system design is a specialized field which is changing rapidly and the interested reader must turn to the current literature. The control designer, in turn, must have a full understanding of the system to be controlled which necessitates an appreciation of gas turbine performance.

Appendix A
Some notes on gas dynamics

Owing to the increasing tendency towards specialization even at first degree and diploma level, it may be that some readers will not have been exposed to a course in gas dynamics. It is hoped that this Appendix will provide them with an adequate summary of those aspects which are relevant to gas turbine theory, and that it will serve others as useful revision material.

A.1 Compressibility effects (qualitative treatment)

It is well known that when the relative velocity between a gas and a solid body reaches a certain value, the flow behaves in a quite different manner to that expected from a study of hydrodynamics. The effects produced, which manifest themselves as additional loss of stagnation pressure in the stream, do not arise when the fluid is a liquid. This suggests that the phenomena are due to the change in density which accompanies a change in pressure of a gas. The idea is strengthened by the fact that the phenomena only occur at high speeds when the pressure changes set up by the relative motion, and therefore the density changes, become considerable. In consequence, the phenomena here described are known as compressibility effects.

When, in a mass of gas at rest, a small disturbance results in a slight local rise of pressure, it can be shown that a pressure wave is propagated throughout the gas with a velocity which depends upon the pressure and density of the gas. This velocity is the speed of sound in the gas, or *sonic velocity a*, given by

$$a = \sqrt{(\gamma p/\rho)} \text{ or } \sqrt{(\gamma RT)}$$

where p, ρ and T are the local pressure, density and temperature of the gas.

In all processes related to the propagation of pressure waves, the changes take place so rapidly that there is no time for any heat transfer between adjacent layers of fluid; the processes are therefore adiabatic. Also, when the amplitude of the pressure wave is small and there is no material alteration in the pressure and temperature of the gas, as is true of an ordinary sound wave, there is no increase of entropy. The propagation of a sound wave is therefore not only adiabatic but isentropic.

Now consider what happens when a similar disturbance occurs in a gas flowing in one direction with a velocity C. The velocity of propagation of the pressure wave relative to the gas will still be equal to the speed of sound, a. Relative to a fixed point, however, say the walls of the passage confining the gas, the speed of propagation will be $(a + C)$ downstream and $(a - C)$ upstream. It follows that if the velocity of the gas is greater than the sonic velocity, i.e. supersonic, there can be no propagation of the pressure wave upstream at all. This is the usual physical explanation given for the existence of a critical condition in nozzle flow. When once the pressure drop across a nozzle is great enough to cause the gas velocity to reach the local sonic value, no further decrease in outlet pressure will be propagated upstream and no further increase in mass flow will be obtained.

Figure A1 illustrates the effects just described, and a useful picture to have in mind is that of the ever-widening circles of ripples formed by a stone thrown into a pond. When a disturbance, such as an intermittent electric spark, is placed in a gas stream moving with subsonic velocity $(C < a)$, the radius of a spherical pressure wave after time t will be at, while the centre of this wave will have moved downstream a distance Ct. All waves emitted subsequently will lie within the spherical wave front of this wave, as shown in Fig. A1(a). On the other hand, when $C > a$ as in Fig. A1(b), the spherical wave fronts will move downstream at a greater rate than the radii of the waves increase. All the spherical waves will therefore lie within a cone having its apex at the point of the disturbance.

The effect of a small solid particle placed in a stream of gas is that of a disturbance emitting pressure waves continuously, so that the spherical wave fronts of Fig. A1(b) appear as a single conical wave front of semi-angle μ, where μ is given by

$$\mu = \sin^{-1} \frac{at}{Ct} = \sin^{-1} \frac{a}{C}$$

The ratio C/a frequently arises in the mathematical treatment of compressible flow, and it is the well-known *Mach number M*. The angle μ is commonly called the *Mach angle*, and the conical wave front a *Mach wave*. Thus

Mach angle $\mu = \sin^{-1} (1/M)$

So far we have been considering pressure impulses of very small amplitude, such that there is no permanent change in the pressure and temperature of the gas as the wave moves through it, and consequently such that there is no change in entropy. In many practical cases of gas flow relative to a solid body these conditions are not fulfilled; there is a marked pressure and temperature difference across the wave,

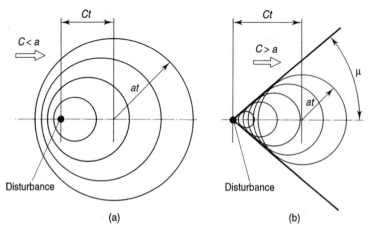

FIG. A.1 **Sound waves in a moving stream**

and there is an increase in entropy indicating an irreversible dissipation of kinetic energy which is manifested by a loss of stagnation pressure. The wave front represents a discontinuity in the flow, and as the change of pressure is to all intents and purposes instantaneous, the wave is termed a *shock wave*. The Mach wave previously discussed can be regarded as the weakest possible form of shock wave. The shock wave formed by a projectile travelling at supersonic speed, for example, is analogous to the bow wave set up by a ship: the water, unable to escape rapidly enough past the sides of the ship, piles up to form a vee-shaped wave which travels along with the ship. In the case of the projectile, the air outside the region enclosed by the conical wave front does not receive a signal warning it of the approach of the solid object creating the disturbance, and hence the formation of the shock wave at the nose of the projectile. It must be stressed that it is the relative motion which is important; it does not matter whether the body or the fluid or both are moving.

We have said that there is a pressure difference across a shock wave. We must now ask whether it is a pressure rise or pressure drop in the direction of gas flow relative to the body; that is, through the shock wave. Both experiment and theory indicate that a shock wave can only be formed when a supersonic flow is decelerated. The velocities in the divergent part of a convergent–divergent nozzle are supersonic, but if the nozzle is operating at the pressure ratio for which it is designed no shock waves will be formed because the flow is accelerating under the influence of the pressure drop. Consider, on the other hand, what happens when the outlet pressure is appreciably above the value which would give just the right amount of expansion to suit the outlet area of the nozzle. Under these conditions the nozzle over-expands the gas so that before the gas can discharge into the surroundings some recompression and deceleration of the gas must occur. This recompression can only be brought about by a shock wave in the divergent part of the nozzle, because a convergent duct is necessary for isentropic diffusion of a supersonic stream. Figure A2 shows typical pressure distributions along a nozzle when the outlet pressure is above the design value. As the outlet pressure is reduced, the plane normal shock wave moves towards the exit, and further reduction towards the design outlet pressure is accompanied by a sudden change to a complex system of oblique shock waves downstream of the exit.

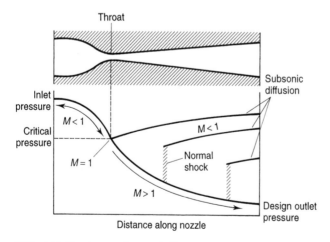

FIG. A.2 Shock wave in nozzle flow

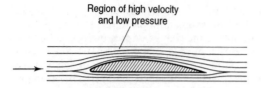

FIG. A.3 Streamline flow over aerofoil

To sum up so far:

(*a*) A shock wave only occurs at supersonic speeds in a decelerating flow.
(*b*) There is a *rise* in *static* pressure and temperature through the wave in the direction of relative motion of the gas, i.e. a shock wave is a compression wave.
(*c*) There is a *drop* in *stagnation* pressure through the wave, some of the kinetic energy being dissipated with a consequent increase in entropy.

Finally, we may now turn to consider the effect of shock waves on the flow of air over an aerofoil. This will be directly applicable to problems associated with the design of axial compressor and turbine blading which are simply rows of aerofoils. First, consider the changes of pressure and velocity which occur around the aerofoil of Fig. A3, assuming that an air stream is flowing past it with subsonic velocity. As indicated by the convergence and divergence of the streamlines, the air moving over the curved top surface must first accelerate and then decelerate, the outer streamlines effectively acting as the wall of a duct. This produces a region of increased velocity and reduced static pressure on the top surface. It is this region of low pressure or suction on the top surface which produces most of the lift of an aerofoil. The amount by which the velocity is increased along the top surface will depend upon the shape and camber of the aerofoil; it is quite possible for the local velocity, somewhere near the point of maximum camber, to be 1·5 times the velocity of the main stream. Furthermore, the acceleration will be accompanied by a fall in static temperature, so that the local sonic velocity $\sqrt{(\gamma R T)}$ will be reduced below that of the free stream. Both effects contribute to an increase in Mach number, and there may well be a region on the top surface where the flow is supersonic when the Mach number of the main stream is only about 0·7. But we have seen that when a deceleration occurs at supersonic speeds in a diverging passage a shock wave will be formed. The effect of this is illustrated in Fig. A4. If under these conditions a pitot tube is traversed across the flow in front of and behind the aerofoil, the loss of stagnation pressure will be found to vary in the manner shown. The loss of head due to the shock wave itself is represented by the part of the curve *AB*; the loss of head represented by the peak of the curve may be very much greater than that across the shock wave, however, and requires some explanation.

When a fluid flows along a surface there is a thin layer of fluid, known as a boundary layer, in which there is a steep velocity gradient due to viscous friction, the velocity dropping to zero at the solid surface. Now the pressure gradient across

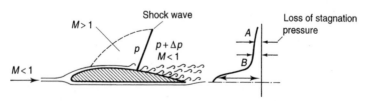

FIG. A.4 Shock wave on aerofoil

the shock wave opposes the direction of flow and consequently, in the boundary layer where the kinetic energy is small, the shock wave may arrest the motion altogether. The boundary layer will thicken just in front of the shock wave, and may break away from the surface at the rear of it. If this breakaway of the boundary layer occurs, it will result in the initiation of a vortex trail involving considerable dissipation of energy. This, then, is the reason for the large loss of stagnation pressure in the wake of the aerofoil, and the reason why the Mach number of the main stream should be kept below the value likely to cause the formation of shock waves with this shape of aerofoil.

We may now turn to the mathematical analysis of compressible flow in a few simple, classical, flow situations. Much of the algebra is too lengthy to be given here, but by its omission we hope to enable the reader to see the wood: for the trees he or she can turn to the many excellent standard texts on gas dynamics, e.g. Refs (3) and (4).

A.2 Basic equations for steady one-dimensional compressible flow of a perfect gas in a duct

A flow can be regarded as *one-dimensional* if

(a) changes in flow area and curvature of the axis are gradual,
(b) all properties are uniform across planes normal to the axis,
(c) any heat transfer per unit mass flow (dQ), across the surface area of the duct (dS), changes the properties uniformly over the cross-section,
(d) the effect of friction can be represented by a shear stress τ at the wall.

The flow is *steady* if there is no change in the mass flowing per unit time at successive planes along the duct, and if the properties of the gas at any plane do not change with time.

Firstly, because we are dealing with a perfect gas we have the *equation of state*

$$\frac{p}{\rho} = RT \text{ or } \frac{dp}{p} = \frac{d\rho}{\rho} + \frac{dT}{T} \tag{1}$$

Secondly, application of the *conservation laws* yields the following equations in integral and differential form (see Fig. A5):

Conservation of mass (continuity equation)

$$m = \rho_1 A_1 C_1 = \rho_2 A_2 C_2 \text{ or } \frac{d\rho}{\rho} + \frac{dC}{C} + \frac{dA}{A} = 0 \tag{2}$$

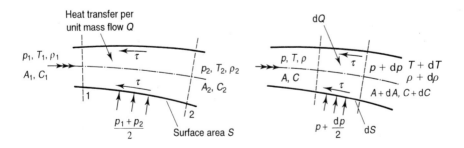

FIG. A.5 One-dimensional flow

Conservation of momentum

$$\left.\begin{array}{l} (\rho_2 C_2^2 A_2 - \rho_1 C_1^2 A_1) + (p_2 A_2 - p_1 A_1) - \frac{1}{2}(p_2 + p_1)(A_2 - A_1) + \tau S = 0 \\ \rho AC\, dC + A\, dp + \tau\, dS = 0 \end{array}\right\} \quad (3)$$

Conservation of energy

$$\left.\begin{array}{l} Q = c_p(T_2 - T_1) + \frac{1}{2}(C_2^2 - C_1^2) = c_p(T_{02} - T_{01}) \\ dQ = c_p\, dT + C\, dC = c_p\, dT_0 \end{array}\right\} \quad (4)$$

Thirdly, the *Second Law of Thermodynamics* must be satisfied, i.e.

$$\text{for adiabatic processes } \Delta s \geqslant 0 \quad (5)$$

For a perfect gas with constant specific heats we have the specific entropy $s = f(p, T)$ given by

$$\Delta s = c_p \ln \frac{T_2}{T_1} - R \ln \frac{p_2}{p_1}$$

and for the special case of a reversible adiabatic (isentropic) process we can put $\Delta s = 0$ to give

$$\frac{p}{T^{\gamma/(\gamma-1)}} = \text{constant or } \frac{p}{\rho^\gamma} = \text{constant} \quad (6)$$

Finally, although no additional physical principle is introduced, the algebra is often simplified if velocities are (a) expressed in terms of Mach number M, or (b) taken into account implicitly by making use of the concepts of stagnation pressure p_0, temperature T_0 and density ρ_0. By definition $M = C/a$. For any fluid, the local sonic velocity a is given by $\sqrt{(dp/d\rho)_s}$, while for a perfect gas it can be expressed variously as

$$a \text{ (or } C/M) = \sqrt{(\gamma p/\rho)} = \sqrt{(\gamma RT)} = \sqrt{[(\gamma - 1)c_p T]} \quad (7)$$

By definition (see under heading 'Stagnation properties' in section 2.2), the stagnation properties in terms of M and the static values become

$$\left.\begin{array}{l} \dfrac{T_0}{T} = 1 + \dfrac{C^2}{2c_p T} = \left[1 + \dfrac{\gamma-1}{2} M^2\right] \\[2ex] \dfrac{p_0}{p} = \left(\dfrac{T_0}{T}\right)^{\gamma/(\gamma-1)} = \left[1 + \dfrac{\gamma-1}{2} M^2\right]^{\gamma/(\gamma-1)} \\[2ex] \dfrac{\rho_0}{\rho} = \left(\dfrac{p_0}{p}\right)^{1/\gamma} = \left[1 + \dfrac{\gamma-1}{2} M^2\right]^{1/(\gamma-1)} \end{array}\right\} \quad (8)$$

We may obtain an important flow equation by combining the differential forms of equations (1) to (4) and relations (7). Thus combining (1) and (2) to eliminate $d\rho/\rho$ we have

$$\frac{dC}{C} + \frac{dA}{A} + \frac{dp}{p} - \frac{dT}{T} = 0$$

From (3) and (7) we get

$$\frac{dp}{p} = -\gamma M^2 \frac{dC}{C} - \frac{\gamma \tau S}{\rho A a^2}$$

From (4) and (7) we get

$$\frac{dT}{T} = \frac{dQ}{c_p T} - (\gamma - 1)M^2 \frac{dC}{C}$$

Combining these three equations we have finally

$$(M^2 - 1)\frac{dC}{C} = \frac{dA}{A} = -\frac{dQ}{c_p T} - \frac{\gamma \tau \, dS}{\rho A a^2} \tag{9}$$

Equation (9) expresses the effect on the velocity of (a) changes in flow area, (b) heat transferred and (c) viscous friction. Considering each of these factors acting in isolation we can draw the following qualitative conclusions.

When $M < 1$ (subsonic flow) the flow will accelerate (+ve dC) if (a) the duct converges (−ve dA) or (b) heat is transferred to the gas (+ve dQ). Conversely, the flow will decelerate if the duct diverges or is cooled. The effect of friction is always to accelerate subsonic flow: the physical reason is that the transformation of directed kinetic energy into internal energy is such that the decrease in density, consequent upon the rise in temperature, predominates and the flow velocity increases to satisfy the continuity equation.

When $M > 1$ (supersonic flow) the flow will accelerate if (a) the duct diverges or (b) the duct is cooled. Conversely, it will decelerate if the duct converges or is heated. The effect of friction is always to decelerate supersonic flow.

When $M = 1$ we have the condition where the flow velocity C equals the local velocity of sound a. Values of all quantities at this condition will be denoted by an asterisk, e.g. T^*, a^*, A^*, p^*. Equation (9) shows that M may be unity when the duct has a throat ($dA = 0$), or when sufficient heat has been added in a duct of constant flow area (supersonic flow decelerates towards $M = 1$ and subsonic flow accelerates towards $M = 1$), or when a constant area adiabatic duct is sufficiently long for friction to decelerate a supersonic flow or accelerate a subsonic flow to the condition $M = 1$. Under all such conditions the duct is said to be *choked* because the mass flow is the maximum which the duct can pass with the given inlet conditions: further heat transfer or friction in a constant area duct merely causes a reduction of mass flow in the duct.

Note that the presence of a throat does not necessarily imply $M = 1$ at the throat, because the duct may be acting as a venturi with subsonic expansion followed by subsonic diffusion. But we are here considering only a *continuous* expansion or compression along the duct (i.e. dC is not changing sign during the process). For this reason also, we cannot expect equation (9) to yield information about discontinuities such as shock waves which will be considered later. Before doing so we will summarize the analysis which gives us quantitative information about the behaviour of a gas undergoing a continuous expansion or compression under the separate actions of changes in flow area (section A.3), heat transfer (A.4), and friction (A.5). This will involve algebraic manipulation of the integrated forms of equations (1)–(4) with appropriate terms omitted, together with relations (5)–(8) as appropriate. In what follows we shall often find it convenient to consider changes which occur when a fluid flowing with an arbitrary Mach number M through area A and with properties denoted by p, p_0, T, T_0, etc., is brought to a reference state where $M = 1$ and the relevant quantities are denoted by A^*, p^*, p_0^*, T^*, etc.

A.3   Isentropic flow in a duct of varying area

For reversible adiabatic flow τ and Q are zero, and the isentropic relations (6) are applicable. The relevant equations become

$$p_1/\rho_1 T_1 = p_2/\rho_2 T_2$$
$$m = \rho_1 A_1 C_1 = \rho_2 A_2 C_2$$

Making use of $p/\rho^\gamma = \text{constant}$, the momentum and energy equations (3) and (4) can be shown to be identical, namely

$$c_p(T_2 - T_1) + \tfrac{1}{2}(C_2^2 - C_1^2) = 0 \text{ or } c_p(T_{02} - T_{01}) = 0$$

We shall consider state 1 to be 'reservoir' or stagnation conditions, and state 2 to be any arbitrary state denoted by suffixless quantities. Then, combining the equations, and making use of relations (6)–(8), we arrive at the well-known equations for flow in nozzles (or diffusers):

$$C = \left[\frac{2\gamma}{\gamma - 1} RT_0 \left\{ 1 - \left(\frac{p}{p_0} \right)^{(\gamma-1)/\gamma} \right\} \right]^{\frac{1}{2}} \tag{10}$$

$$\frac{m\sqrt{T_0}}{Ap_0} = \left[\frac{2\gamma}{\gamma - 1} \frac{1}{R} \left(\frac{p}{p_0} \right)^{2/\gamma} \left\{ 1 - \left(\frac{p}{p_0} \right)^{(\gamma-1)/\gamma} \right\} \right]^{\frac{1}{2}} \tag{11}$$

$m\sqrt{T_0}/Ap_0$ is a maximum at the throat, where $C = a$ (i.e. $M = 1$) and the other quantities are denoted by asterisks. Differentiating (11) with respect to p and equating to zero gives

$$\frac{p^*}{p_0} = \left(\frac{2}{\gamma + 1} \right)^{\gamma/(\gamma-1)} \tag{12}$$

$$\frac{T^*}{T_0} = \left(\frac{2}{\gamma + 1} \right) \tag{13}$$

Substitution of (12) in (11) then yields

$$\frac{m\sqrt{T_0}}{A^* p_0} = \left[\frac{\gamma}{R} \left(\frac{2}{\gamma + 1} \right)^{(\gamma+1)/(\gamma-1)} \right]^{\frac{1}{2}} \tag{14}$$

Dividing (14) by (11), and introducing $p/p_0 = f(M)$ from (8), we obtain the value of A at any M in terms of the throat area A^*, namely

$$\frac{A}{A^*} = \frac{1}{M} \left[\frac{2 + (\gamma - 1)M^2}{(\gamma + 1)} \right]^{(\gamma+1)/2(\gamma-1)} \tag{15}$$

Values of p_0/p, T_0/T, ρ_0/ρ from relations (8) and A/A^* from equation (15) are tabulated in the isentropic flow table of Ref. (1), with M as the argument. The table covers the range $M = 0\cdot01$–4, and is for dry air with $\gamma = 1\cdot403$. Reference (2) is a much fuller set of tables covering a range of values of γ and a wider range of M. Given the passage shape, i.e. values of A/A^* at successive distances along the duct, corresponding values of M and thence p/p_0 can be read from the isentropic flow table. Two values of M and p/p_0 will be obtained from the table for each value of A/A^*, corresponding to the two real roots of equation (15). The pressure distribution can then be plotted as in Fig. A6. Curve BCD refers to the continuous expansion (or compression depending on the direction of flow), while BCE represents the limiting pressure distribution above which the passage is acting as a venturi.

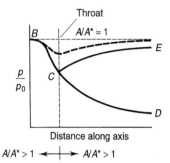

FIG. A.6

A.4   Frictionless flow in a constant area duct with heat transfer

We have seen that when A is constant and τ is zero, heat transfer *to* the gas causes a subsonic flow to accelerate towards $M = 1$ and a supersonic flow to decelerate towards $M = 1$. This idealized flow is referred to as *Rayleigh flow*. One important effect, to which reference was made in Chapter 6, is that heat transfer to a subsonic flow in a duct of constant area must be accompanied by a fall in pressure. The pressure difference is necessary to provide the force required to accelerate the flow, i.e. to satisfy the momentum equation.

The relevant equations for a Rayleigh flow from some arbitrary state M, T, etc., to the state where $M = 1$ and quantities are denoted by asterisks, are as follows:

$$p/\rho T = p^*/\rho^* T^*$$
$$\rho C = \rho^* C^* (= \rho^* a^*)$$
$$(\rho^* C^{*2} - \rho C^2) + (p^* - p) = 0$$
$$Q = c_p(T^* - T) + \tfrac{1}{2}(C^{*2} - C^2) = c_p(T_0^* - T_0)$$

From the momentum equation, in conjunction with relations (7), we have

$$\frac{p}{p^*} = \frac{1+\gamma}{1+\gamma M^2} \tag{16}$$

From this, together with the continuity equation and equation of state, we obtain

$$\frac{T}{T^*} = M^2 \left[\frac{1+\gamma}{1+\gamma M^2}\right]^2 \tag{17}$$

From (16) and (17) and relations (8), by writing

$$\frac{p_0}{p_0^*} = \left(\frac{p_0}{p}\right)\left(\frac{p}{p^*}\right)\left(\frac{p^*}{p_0^*}\right), \quad \text{and likewise for } \frac{T_0}{T_0^*}$$

we obtain

$$\frac{p_0}{p_0^*} = \left[\frac{1+\gamma}{1+\gamma M^2}\right]\left[\frac{2+(\gamma-1)M^2}{\gamma+1}\right]^{\gamma/(\gamma-1)} \tag{18}$$

$$\frac{T_0}{T_0^*} = \frac{(1+\gamma)M^2}{(1+\gamma M^2)^2}[2+(\gamma-1)M^2] \tag{19}$$

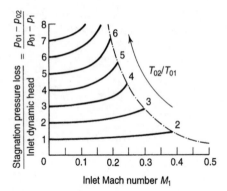

FIG. A.7 Stagnation pressure loss in Rayleigh flow

Finally, from the energy equation we have

$$\frac{Q}{c_p T_0^*} = 1 - \frac{T_0}{T_0^*} \tag{20}$$

Equations (16)–(19) are the Rayleigh functions, and values of p/p^*, T/T^*, p_0/p_0^*, T_0/T_0^* are given in the tables referred to in the previous section. To obtain values of a property at state 2, knowing the value in state 1, we merely apply such equations as

$$\frac{p_2}{p_1} = \frac{p_2/p^*}{p_1/p^*} \quad \text{or} \quad \frac{T_{02}}{T_{01}} = \frac{T_{02}/T_0^*}{T_{01}/T_0^*}$$

For example, given the inlet conditions p_{01}, T_{01} and M_1, we may obtain p_{02} and M_2 for any given final temperature T_{02} (or temperature ratio T_{02}/T_{01}) as follows.

For given M_1 read p_{01}/p_0^* and T_{01}/T_0^*
Evaluate T_{02}/T_0^* from $(T_{02}/T_{01})(T_{01}/T_0^*)$
Read M_2 corresponding to T_{02}/T_0^*
Read p_{02}/p_0^* corresponding to M_2
Evaluate p_{02} from $(p_{02}/p_0^*)(p_0^*/p_{01})p_{01}$

Finally, we could read p_1/p_{01} corresponding to M_1 from the isentropic flow table, and evaluate the 'fundamental pressure loss factor' referred to in Chapter 6, namely $(p_{01} - p_{02})/(p_{01} - p_1)$. If this were done for a series of subsonic values of M_1 and $T_{02}/T_{01} > 1$, the results would appear as in Fig. A7. The chain dotted curve represents the limiting case where the heat transfer is sufficient to accelerate the flow to $M_2 = 1$: this condition is often referred to as *thermal choking*. Note that as the inlet Mach number approaches zero the fundamental pressure loss factor tends to $[(T_{02}/T_{01}) - 1]$: cf. the incompressible flow result obtained in section 6.4.

A.5 Adiabatic flow in a constant area duct with friction

The idealization considered here is referred to as *Fanno flow*. We have seen in section A.2 that friction accelerates subsonic flow towards $M = 1$ and decelerates supersonic flow towards $M = 1$. The length of duct which is sufficient to change the Mach number from any value M to the value unity will be referred to as L^*.

Friction data are normally given in terms of relations between a friction factor f and Reynolds number Re. The definition of f used here (other definitions are also used) is

$$\tau = f \rho C^2 / 2$$

The relevant equations for Fanno flow are

$$p/\rho T = p^*/\rho^* T^*$$
$$\rho C = \rho^* C^* (= \rho^* a^*)$$
$$(\rho^* C^{*2} - \rho C^2) + (p^* - p) + \tau S = 0$$
$$c_p(T^* - T) + \tfrac{1}{2}(C^{*2} - C^2) = c_p(T_0^* - T_0) = 0$$

From the energy equation the stagnation temperature is constant, and this together with the T_0/T relation (8) yields

$$\frac{T}{T^*} = \frac{\gamma + 1}{2 + (\gamma - 1)M^2} \tag{21}$$

Using this in conjunction with the continuity equation and equation of state, we get

$$\frac{p}{p^*} = \frac{1}{M}\left[\frac{\gamma + 1}{2 + (\gamma - 1)M^2}\right]^{\frac{1}{2}} \tag{22}$$

And using the p_0/p relation (8) we obtain

$$\frac{p_0}{p_0^*} = \frac{1}{M}\left[\frac{2 + (\gamma - 1)M^2}{\gamma + 1}\right]^{(\gamma+1)/2(\gamma-1)} \tag{23}$$

The equation expressing L^* for a given M and f is best found from the differential form of the momentum equation (3), with τ replaced by $\tfrac{1}{2} f \rho C^2$. Dividing throughout by A, the momentum equation becomes

$$\rho C\, dC + dp + \frac{f \rho C^2\, dS}{2A} = 0$$

For circular ducts of diameter D,

$$\frac{dS}{A} = \frac{\pi D\, dL}{\pi D^2/4} = \frac{4\, dL}{D}$$

And for non-circular ducts D can be replaced by the *equivalent diameter* ($= 4 \times$ flow area/perimeter). Introducing the equation of state, continuity equation and relations (7) and (8), after much algebra we arrive at a differential equation for $4f\, dL^*/D$ in terms of M. Integrating between M and 1 we have finally

$$\frac{4fL^*}{D} = \frac{1 - M^2}{\gamma M^2} + \frac{\gamma + 1}{2\gamma}\ln\left[\frac{(\gamma + 1)M^2}{2 + (\gamma - 1)M^2}\right] \tag{24}$$

The Fanno functions (21) to (24) are given in the compressible flow tables. To find a length L over which the flow changes from M_1 to M_2 it is only necessary to subtract the two values of $4fL^*/D$, namely

$$\frac{4fL}{D} = \left[\left(\frac{4fL^*}{D}\right)_{M_2} - \left(\frac{4fL^*}{D}\right)_{M_1}\right]$$

It should be appreciated that we have assumed f to be constant, i.e. that it does not change along the duct. Since the Reynolds number is $\rho CD/\mu = GD/\mu$, and the mass velocity G is constant because the area A is constant, Re only changes due to the variation of μ with T. This is small for gases. Furthermore, in turbulent pipe flow f is a weak function of Re. Thus little error is incurred by using an appropriate mean value of f.

In practical situations variation in area, heat addition, and friction may be present simultaneously. Methods for combining the results of the preceding three sections can be found in any text on gas dynamics.

A.6 Plane normal shock waves

When shock waves occur normal to the axis of flow, they are discontinuities which occupy a finite but very short length of duct as depicted in Fig. A8(a). For this reason they can be treated as adiabatic frictionless processes in a duct of constant cross-sectional area. In general, shock waves are formed when the conditions are such that the three conservation laws cannot be satisfied simultaneously with an assumption of reversible flow. What has to be relinquished is the idealization of reversibility, even though the flow is being regarded as frictionless. Then, if the process is adiabatic, all that the Second Law of Thermodynamics requires is that there should be an increase in entropy in the direction of flow.

The relevant equations relating properties on either side of a shock wave are

$$p_1/\rho_1 T_1 = p_2/\rho_2 T_2$$
$$m = \rho_1 C_1 = \rho_2 C_2$$
$$(\rho_2 C_2^2 - \rho_1 C_1^2) + (p_2 - p_1) = 0$$
$$c_p(T_2 - T_1) + \tfrac{1}{2}(C_2^2 - C_1^2) = c_p(T_{02} - T_{01}) = 0$$

From the momentum and continuity equations, together with relations (7), we obtain the pressure ratio across the shock wave as

$$\frac{p_2}{p_1} = \frac{1 + \gamma M_1^2}{1 + \gamma M_2^2} \tag{25}$$

Also, since $T_{02} = T_{01}$ from the energy equation, we have

$$\frac{T_2}{T_1} = \left(\frac{T_2}{T_{02}}\right)\left(\frac{T_{01}}{T_1}\right) = \frac{2 + (\gamma - 1)M_1^2}{2 + (\gamma - 1)M_2^2} \tag{26}$$

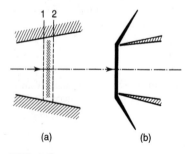

1 2

(a)   (b)

FIG. A.8

At this stage the value of M_2 is unknown but from the continuity equation, equation of state, and relations (7) we can show that M_2 is uniquely related to M_1 by the equation

$$\frac{M_2}{M_1} = \frac{p_1}{p_2} \sqrt{\frac{T_2}{T_1}}$$

On substitution from (25) and (26) we get

$$M_2^2 = \frac{(\gamma - 1)M_1^2 + 2}{2\gamma M_1^2 - (\gamma - 1)} \tag{27}$$

Finally, substituting for M_2 in (25) and (26) we obtain the pressure and temperature ratio in terms of M_1 given by

$$\frac{p_2}{p_1} = \frac{2\gamma M_1^2 - (\gamma - 1)}{\gamma + 1} \tag{28}$$

$$\frac{T_2}{T_1} = \left[2\gamma M_1^2 - (\gamma - 1)\right]\left[\frac{2 + (\gamma - 1)M_1^2}{(\gamma + 1)^2 M_1^2}\right] \tag{29}$$

The pressure ratio p_{02}/p_1 can be found from $(p_{02}/p_2)(p_2/p_1)$, using (28), (27) and the p_0/p relation (8), the result being

$$\frac{p_{02}}{p_1} = \left[\frac{\gamma + 1}{2} M_1^2\right]^{\gamma/(\gamma-1)} \Big/ \left[\frac{2\gamma M_1^2 - (\gamma - 1)}{\gamma + 1}\right]^{1/(\gamma-1)} \tag{30}$$

The normal shock functions (27) to (30) are tabulated in Ref. (1). If required, the stagnation pressure can be found from

$$\frac{p_{02}}{p_{01}} = \left(\frac{p_{02}}{p_1}\right)\left(\frac{p_1}{p_{01}}\right)$$

remembering that p_1/p_{01} can be found from the isentropic table for the given M_1. The fuller tables of Ref. (2) include p_{02}/p_{01} as well as p_{02}/p_1.

Evaluation of equation (27) shows that for $M_1 > 1$, M_2 is less than 1; while for $M_1 < 1$, M_2 is greater than 1. There is nothing in the foregoing analysis to suggest that both these flow situations are not possible. But knowing p_2/p_1 and T_2/T_1 from (28) and (29), it is a simple matter to calculate the change in entropy of the gas from

$$\frac{\Delta s}{c_v} = \gamma \ln \frac{T_2}{T_1} - (\gamma - 1) \ln \frac{p_2}{p_1}$$

For values of $M_1 > 1$ we find that the values of T_2/T_1 and p_2/p_1 yield positive values of Δs, whereas when $M_1 < 1$ the change of entropy is negative. The latter situation is physically impossible because it contradicts the Second Law of Thermodynamics. Thus the normal shock functions are only tabulated for values of $M_1 > 1$.

From the foregoing we may deduce that a normal shock wave can only arise in a supersonic flow, and that it produces a sudden reduction in velocity to a subsonic value. Thus the normal shock is a compression process and $p_2/p_1 > 1$ for all $M_1 > 1$. Owing to the irreversibility introduced by the shock wave, however, there is a loss of *stagnation* pressure, i.e. $p_{02}/p_{01} < 1$. All these characteristics are immediately obvious from the tables or the graphical representation in Fig. A9. We note also, from the $M_2 = f(M_1)$ curve, that M_2 is asymptotic to a definite value as M_1 increases, and that M_2 is close to this lower limit when M_1 has increased to about 5. By dividing the numerator and denominator of equation (27) by M_1^2 and

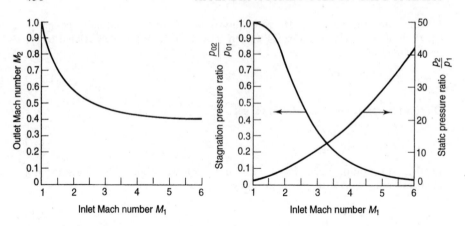

FIG. A.9 Changes in Mach number and pressure across a plane normal shock wave

letting M_1 tend to infinity, the lower limit of M_2 is seen to equal $[(\gamma - 1)/2\gamma]$ which is approximately 0·38 for air.

It is clearly of interest to see how efficient the normal shock is as a compressor, because it can be made use of at the front of intakes of turbojet and ramjet engines when these are operating at supersonic flight speeds. The simple 'pitot' intake is illustrated in Fig. A8(b). The efficiency of the process can be defined as the ratio of the ideal (isentropic) temperature rise to the actual temperature rise for a given pressure ratio p_2/p_1. In which case we have the isentropic efficiency η, where

$$\eta = \frac{T_2' - T_1}{T_2 - T_1} = \frac{[(p_2/p_1)^{(\gamma-1)/\gamma} - 1]}{T_2/T_1 - 1}$$

p_2/p_1 and T_2/T_1 can be found from the normal shock table for a range of M_1, and hence η can be plotted versus M_1 as in Fig. A10. The efficiency falls off rapidly as M_1 is increased. Nevertheless, when it is realized that $p_2/p_1 = 4.5$ for $M_1 = 2$ (from Fig. A9), and that this pressure ratio has been achieved in a negligible length of intake, an efficiency of 78 per cent (from Fig. A10) is not too low to be useful. For this example, M_2 is 0·58 (Fig. A9) and after the shock wave a further pressure rise

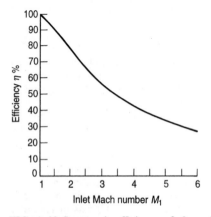

FIG. A.10 Isentropic efficiency of plane shock wave

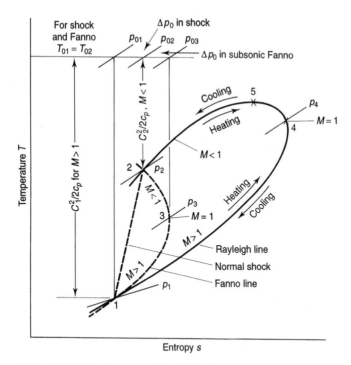

FIG. A.11 Main features of Rayleigh, Fanno and normal shock processes

can be obtained by ordinary subsonic diffusion in a divergent duct. Higher efficiencies can only be obtained by designing supersonic intakes to operate with a system of oblique shocks. This type of shock forms the subject of the next section.

The main features of Rayleigh flow, Fanno flow and flow through normal shocks can be summarized neatly by drawing the processes on a T–s diagram as in Fig. A11. Such a diagram is a useful mnemonic. The Fanno and normal shock processes are shown as dotted lines because they are essentially irreversible processes. The three lines are all drawn for the same value of mass flow per unit area (i.e. ρC) which we have seen is constant for all three types of flow. This is why states 1 and 2 on either side of the normal shock coincide with the points of intersection of the Fanno and Rayleigh lines. There are four features which perhaps are not emphasized sufficiently by notes on the figure. (i) A Fanno process can occur only from state 2 towards state 3 or from 1 towards 3. It cannot pass *through* 3 without a decrease in entropy which would contravene the Second Law of Thermodynamics. (ii) A Rayleigh process can occur between 1 and 4 in either direction or between 2 and 4 in either direction, but it cannot proceed *through* state 4 in either direction. In practice, the picture is modified because friction is present simultaneously with heating or cooling; but also it would be physically difficult to suddenly change from heating to cooling at the point in the duct where the gas attains state 4 and, without doing so, passage through state 4 would clearly contravene the Second Law. (iii) It is interesting to note that when a subsonic flow is heated, the maximum temperature is reached at state 5 (where it can be shown that $M = 1/\sqrt{\gamma}$), although the stagnation temperature must continue to rise with energy input until the gas attains state 4. In other words, between 5 and 4 the density decreases, and hence the velocity increases, at such a rate that the static temperature falls. (iv) When point 3 on the Fanno line or point 4 on the Rayleigh line is reached the flow is choked, and further friction (i.e. additional length of pipe), or heating in the case of Rayleigh flow, causes the state to move on to another Fanno line or another Rayleigh line respectively. M remains equal to unity. These would be lines appropriate to a reduced mass flow per unit area, and would lie to the right of those shown because of the increase in entropy due to additional friction in one case and heating in the other.

A.7 Oblique shock waves

In certain types of supersonic flow, shock waves are formed at an inclination to the direction of flow as in Fig. A12. Strictly speaking the analysis of such phenomena involves a study of two-dimensional flow although, as will be made apparent, some of the equations obtained for plane normal shock waves can be used with a little modification. The analysis of *plane* oblique shock waves shows that for any given incident Mach number,

(*a*) there is a limiting flow deflection angle δ_m which cannot be exceeded;
(*b*) provided $\delta < \delta_m$, at any given value of M_1 there are two physically possible plane shock waves having different values of the shock angle β. The shock wave having the smaller β is the weaker shock, i.e. it produces the smaller pressure ratio p_2/p_1 and, although the flow velocity is reduced by the shock, the flow can still be supersonic downstream of it.

Oblique shock waves are formed when a body is immersed in a supersonic stream. If we consider the simple case of a wedge of semi-angle θ, two types of flow are found to occur as depicted in Fig. A13.

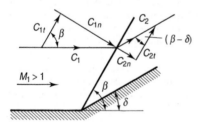

FIG. A.12 Oblique shock

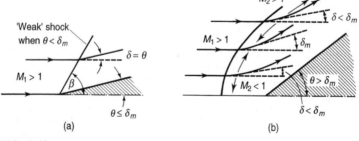

FIG. A.13

(i)   *When* $\theta \leq \delta_m$ a plane oblique shock wave is attached to the apex of the wedge. For $\theta < \delta_m$ it is the weak shock which is formed, while for $\theta = \delta_m$ there is no distinction between the two shocks and only one value of β is possible.

(ii)   *When* $\theta > \delta_m$ a plane oblique shock wave is not possible, and it is replaced by a detached curved shock wave as in Fig. A13(b). (The same type of flow would be set up by a round-nosed body such as a compressor blade of aerofoil shape.) On the axis of the wedge the shock is normal, $\delta = 0$ and $M_2 < 1$. For a certain distance from the axis, $\delta < \delta_m$ and the curved shock can be regarded as being made up of a number of plane oblique shocks of elemental length and of successively increasing δ. These are always the stronger of the two possible shocks, i.e. those having the larger β. Downstream of these shocks $M_2 < 1$, and thus the flow in the region near the body is subsonic. Further out in the mainstream, beyond the point where δ has increased to δ_m, β decreases until it is zero in the free stream (strictly speaking at an infinite distance from the body). In this region the curved shock can be regarded as a number of elemental plane shocks of the weak type (small β), and the flow remains supersonic with $M_2 \to M_1$ as the distance from the body increases.

Reverting to Fig. A12, consider the components of the velocities C_1 and C_2 in the tangential and normal directions relative to the plane shock. On either side of the shock there is no change of state along the shock, and with no force acting in this direction there can be no change in the tangential component of momentum or velocity across the shock. Thus

$$C_2 \cos (\beta - \delta) = C_1 \cos \beta \qquad (31)$$

From the point of view of the normal component, the shock is a plane normal shock which we have already analysed. Thus equation (27) can be used but with $M_2 \sin (\beta - \delta)$ substituted for M_2 and $M_1 \sin \beta$ for M_1. Similarly, equations (28) and (29) for p_2/p_1 and T_2/T_1 will apply if $M_1 \sin \beta$ replaces M_1. To relate δ, β and M_1 we apply the continuity equation in the form

$$\rho_2 C_2 \sin (\beta - \delta) = \rho_1 C_1 \sin \beta$$

and substitute C_1/C_2 from (31) to give

$$\frac{\rho_2}{\rho_1} = \frac{\tan \beta}{\tan (\beta - \delta)}$$

But $\rho_2/\rho_1 = (p_2/p_1)(T_1/T_2) = f(M_1, \beta)$ from the modified (28) and (29), so that we have an equation for δ in terms of M_1 and β which finally reduces to

$$\tan \delta = \cot \beta \, \frac{M_1^2 \sin^2 \beta - 1}{\dfrac{\gamma + 1}{2} M_1^2 - (M_1^2 \sin^2 \beta - 1)} \qquad (32)$$

For given values of $\delta (<\delta_m)$ this equation has been shown to have two real positive roots, yielding the two values of β corresponding to the weak and strong shocks. After differentiating, and some algebra, equation (32) also yields the maximum deflection δ_m in terms of M_1. The results are often plotted as in Fig. A14. They are also given in the oblique shock table of Refs (1) and (2). Such a table gives the two possible values of β_1, p_2/p_1, T_2/T_1 and M_2, for each value of M_1 and a series of values of δ. The curve which divides the weak and strong shock regions in Fig. A14 gives the maximum deflection δ_m which the flow can undergo with a shock wave attached to the apex of the wedge. From the chain dotted curve we see that although the Mach number downstream of a strong shock is always less than unity, it is not necessarily always greater than unity for a weak shock.

It is of interest to compare the efficiency of a compression by an oblique shock with that of the normal shock considered in section A.6. We took $M_1 = 2$, and

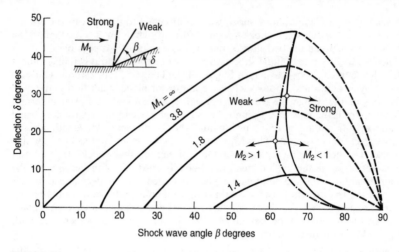

FIG. A.14

found that a pressure ratio (p_2/p_1) of 4·5 was obtained with an efficiency of 78 per cent. From the oblique shock table of Ref. (1), with $M_1 = 2$ and an arbitrarily chosen value of 10°, we obtain

	Weak	Strong
β	39·3°	83·7°
M_2	1·64	0·60
p_2/p_1	1·71	4·45
T_2/T_1	1·17	1·68
η	0·98	0·79

The last line is calculated from

$$\left[(p_2/p_1)^{(\gamma-1)/\gamma} - 1\right] \Big/ \left[\frac{T_2}{T_1} - 1\right]$$

with $\gamma = 1\cdot403$ as used in the tables.

Since a pressure ratio of 1·71 is obtainable from a weak plane oblique shock with an efficiency of 98 per cent, it would appear that a supersonic intake designed to make use of one or more such shocks might be more efficient than the simple type of 'pitot' intake referred to in section A.6; and this has been found to be the case. In the Oswatitsch intake, which makes use of a conical centre body as shown in Fig. A15, the flow is decelerated to a low supersonic velocity by several successive oblique shocks (two are shown) with the final transition to subsonic velocity taking place via a normal shock. The main problems in the design of such supersonic intakes are (a) the establishment of a stable shock pattern insensitive to minor

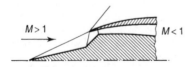

FIG. A.15 Oswatitsch intake

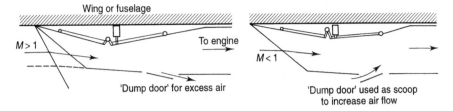

FIG. A.16 Intake with variable geometry

changes in the flow downstream (e.g. in the engine), and (*b*) the maintenance of a good performance under off-design conditions. For aircraft engine intakes which have to operate over a complete speed range from $M_1 = 0$ to $M_1 \gg 1$, variable geometry is essential. This is accomplished by incorporating an adjustable centre body and/or cowl, and bleed slots. Figure A16 illustrates a variable geometry intake of the 'scoop' type, which is of rectangular cross-section: it may be slung under the wing or run along the side of the fuselage.

A.8 Isentropic two-dimensional supersonic expansion and compression

Perhaps the foregoing sections have left the reader with the impression that isentropic expansion and compression of a supersonic stream is impossible in principle. A moment's reflection on the existence of successful supersonic aircraft wings, and efficient nozzles and diffusers for supersonic wind tunnels, should dispel this illusion. Processes which would be isentropic apart from the effect of viscous friction are possible if the duct walls, or immersed body as the case may be, are correctly shaped. Certainly it is possible to avoid the large loss of stagnation pressure due to breakdown of the flow which was illustrated in Fig. A4.

Consider supersonic flow that is initially parallel to a surface but which encounters a small change of direction of the surface; it may be a convex or concave deflection as in Fig. A17(a) and (b). If the change in direction is infinitesimal, the corner is the source of an infinitesimal disturbance which is communicated to the main flow isentropically along a Mach wave. From section A.1 we know that this wave will make an angle $\mu = \sin^{-1}(1/M)$ with the direction of flow. And from the velocity triangles of Fig. A17 we see that the convex corner produces an increase in velocity, dC, which must be accompanied by a pressure drop dp; while the concave corner leads to a decrease in velocity, implying a compression.

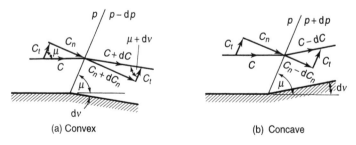

(a) Convex   (b) Concave

FIG. A.17 Expansive and compressive Mach waves

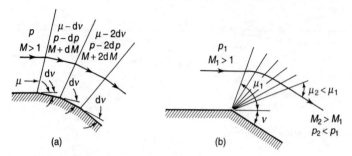

FIG. A.18 Isentropic expansion

Any rounded convex corner giving a finite deflection can be regarded as the source of a series of infinitesimal deflections as illustrated in Fig. A18(a). The Mach waves, or *characteristic lines* as they are often called, diverge and do not interfere with each other, so permitting the flow to accelerate smoothly and isentropically to the downstream pressure. In the limit the same thing can occur at a sharp finite corner: in this case the flow expands smoothly through a fan of Mach waves as shown in Fig. A18(b). This is known as *Prandtl–Meyer flow*, and the evaluation of finite changes of pressure and Mach number associated with a finite deflection is possible with the aid of tabulations of Prandtl–Meyer 'expansion angles'. We shall proceed no further with this topic, but say merely that the analysis of supersonic flow patterns using a step-by-step method moving from one Mach line to the next, is referred to as the *method of characteristics*. When it is realized that in most situations the Mach waves will suffer reflections from neighbouring surfaces or jet boundaries, or intersect and interact with Mach waves from an opposing surface, it will be appreciated that the analysis is complex and beyond the scope of this Appendix.

To conclude, let us consider briefly what happens in a finite concave corner. If it is rounded as in Fig. A19(a), the Mach waves will appear as shown. They must converge because the Mach number is decreasing and hence the Mach angle μ is increasing. For a sharp concave corner there can be no compressive equivalent of an expansion fan, because this would involve successive Mach lines marching upstream of one another. What happens, since no corner is truly sharp, is that the Mach lines converge to a point in the stream near the surface and coalesce to form an oblique shock as in Fig. A19(b). The compression process is then no longer isentropic. In other words, an isentropic diffusion of a supersonic stream can only occur in a passage bounded by gradual concave curves.

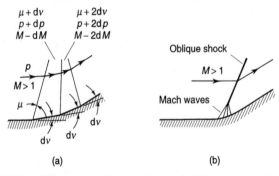

FIG. A.19 Compression: (a) isentropic (b) non-isentropic

Appendix B
Problems

The following problems are numbered by chapters. Unless otherwise stated the following values have been used:

For air:	$c_p = 1 \cdot 005 \, \text{kJ/kg K}, \gamma = 1 \cdot 40$
For combustion gas:	$c_p = 1 \cdot 148 \, \text{kJ/kg K}, \gamma = 1 \cdot 333$
For air and combustion gas:	$R = 0 \cdot 287 \, \text{kJ/kg K}$

Readers who wish to carry out more detailed design point and off-design studies, allowing for such factors as variable fluid properties, bleeds, and power extraction, will find the commercially available program GASTURB useful. This can be obtained from Dr J. Kurzke, Max Feldbauer Weg 5, 85221 Dachau, Germany.

2.1 In an ideal gas turbine cycle with reheat, air at state (p_1, T_1) is compressed to pressure rp_1 and heated to T_3. The air is then expanded in two stages, each turbine having the same pressure ratio, with reheat to T_3 between the stages. Assuming the working fluid to be a perfect gas with constant specific heats, and that the compression and expansion are isentropic, show that the specific work output will be a maximum when r is given by

$$r^{(\gamma-1)/\gamma} = (T_3/T_1)^{2/3}$$

2.2 A simple gas turbine with heat-exchanger has a compressor and turbine having respective isentropic efficiencies η_c and η_t. Show that the combined effect of *small* pressure drops Δp_{hg} (in gas side of heat-exchanger) and Δp (total in combustion chamber and air side of heat-exchanger) is to reduce the specific work output by an amount given by

$$\frac{\gamma - 1}{\gamma} \times \frac{c_p T_3 \eta_t}{r^{(\gamma-1)/\gamma} p_1} \left[\Delta p_{hg} + \frac{\Delta p}{r} \right]$$

where $T_3 = $ turbine inlet temperature, $p_1 = $ compressor inlet pressure and $r = $ compressor pressure ratio.

Assume that c_p and γ are constant throughout the cycle.

2.3 Consider the ideal cycle for a gas turbine with heat-exchanger and separate LP power turbine. When the heat-exchanger is fitted in the normal way, using the exhaust from the LP turbine to heat the air after compression, the ideal cycle efficiency is

$$\eta = 1 - (c/t)$$

where $c = (p_2/p_1)^{(\gamma-1)/\gamma}$ and $t = T_3/T_1$. Another possibility is to pass the gas leaving the HP turbine through the heat-exchanger before it enters the LP turbine, and to let the LP turbine exhaust to atmosphere. Derive an expression for the ideal cycle efficiency of this scheme in terms of c and t, and hence show that for all values of $c > 1$ the efficiency is higher than that of the ordinary scheme.

Finally, by referring to sketches of the cycles on the T–s diagram, say in what respects the normal scheme might be superior in spite of the lower ideal cycle efficiency.

2.4 In a gas turbine plant, air is compressed from state (p_1, T_1) to a pressure rp_1 and heated to temperature T_3. The air is then expanded in two stages with reheat to T_3 between the turbines. The isentropic efficiencies of the compressor and each turbine are η_c and η_t. If xp_1 is the intermediate pressure between the turbines, show that, for given values p_1, T_1, T_3, η_c, η_t and r, the specific work output is a maximum when $x = \sqrt{r}$.

If this division of the expansion between the turbines is maintained, show that

(a) when r is varied, the specific work output is a maximum with r given by

$$r^{3/2} = (\eta_c \eta_t T_3 / T_1)^{\gamma/(\gamma-1)}$$

(b) when a perfect heat-exchanger is added the cycle efficiency is given by

$$\eta = 1 - \frac{T_1 r^{(\gamma-1)/2\gamma}[r^{(\gamma-1)/2\gamma} + 1]}{2\eta_c \eta_t T_3}$$

Assume that the working fluid is a perfect gas with constant specific heats, and that pressure losses in the heater, reheater, and heat-exchanger are negligible.

2.5 A gas turbine plant has a compressor in which air is compressed from atmospheric pressure and delivered to two turbines arranged in parallel, the combustion gases expanding to atmospheric pressure in each turbine. One of the turbines drives the compressor, to which it is mechanically coupled, while the other develops the power output of the plant. Each turbine has its own combustion chamber, the fuel supply to each being capable of control independently of the other.

(a) The power output of the plant is reduced by varying the fuel supply to the combustion chambers in such a way that the inlet temperature to the turbine driving the compressor is kept constant, while the inlet temperature to the power turbine is reduced. If the mass flow through a turbine is proportional to $p/\sqrt{T}$, where p and T refer to the turbine inlet conditions, show that the mass flow through the compressor is proportional to $r^{1/\gamma}$ and the power output of the plant is proportional to

$$r[r^{(\gamma-1)/\gamma} - 1]/[1 - kr^{(\gamma-1)/\gamma}]$$

where r = pressure ratio
$k = T_1/(\eta_c \eta_t T_3)$
T_1 = compressor inlet temperature
T_3 = temperature at inlet to the turbine driving the compressor
η_c = isentropic efficiency of compressor
η_t = isentropic efficiency of turbine driving the compressor.

The isentropic efficiencies of the compressor and turbines may be assumed to remain constant under all conditions of the problem. Pressure losses, mechanical losses and the increase in the mass flow through the combustion chambers due to the mass of fuel burnt, may be neglected. c_p and γ may be assumed to be constant.

(b) Derive corresponding expressions for the alternative method of control in which the fuel supply to the combustion chambers is reduced in such a way that the inlet temperatures to both turbines are always equal.

(c) Calculate the percentage of full power developed under each of the foregoing methods of control when the pressure ratio of the compressor has fallen to 30, on a plant designed to give full power under the following conditions:

pressure ratio of the compressor	4·0
inlet temperature to the compressor	288 K
inlet temperature to both turbines	1100 K

$\eta_c = 0.85$, $\eta_t = 0.88$, $\gamma = 1.4$

[0·521, 0·594]

2.6 A compressor has an isentropic efficiency of 0·85 at a pressure ratio of 4·0. Calculate the corresponding polytropic efficiency, and thence plot the variation of isentropic efficiency over a range of pressure ratio from 2·0 to 10·0.
[0·876; 0·863 at 2·0 and 0·828 at 10·0]

2.7 A peak-load generator is to be powered by a simple gas turbine with free power turbine delivering 20 MW of shaft power. The following data are applicable:

Compressor pressure ratio	11·0
Compressor isentropic efficiency	0·82
Combustion pressure loss	0·4 bar
Combustion efficiency	0·99
Turbine inlet temperature	1150 K
Gas-generator turbine isentropic efficiency	0·87
Power turbine isentropic efficiency	0·89
Mechanical efficiency (each shaft)	0·98
Ambient conditions p_a, T_a	1 bar, 288 K

Calculate the air mass flow required and the *SFC*.
[119·4 kg/s, 0·307 kg/kW h]

2.8 A gas turbine is to consist of a compressor, combustion chamber, turbine and heat-exchanger. It is proposed to examine the advantage of bleeding off a fraction $(\Delta m/m)$ of the air delivered by the compressor and using it to cool the turbine blades. By so doing the maximum permissible cycle temperature may be increased from T to $(T + \Delta T)$. The gain in efficiency due to the increase of temperature will be offset by a loss due to the decrease in effective airflow through the turbine. Show that, on the following assumptions, there is no net gain in efficiency when

$$\frac{\Delta m}{m} = \frac{\Delta T/T}{(1 + \Delta T/T)}$$

and that this result is independent of the compressor and turbine efficiencies. Assumptions:

(1) No pressure loss in combustion chamber or heat-exchanger.
(2) The working fluid is air throughout and the specific heats are constant.
(3) The air bled for cooling purposes does no work in the turbine.
(4) The temperature of the air entering the combustion chamber is equal to that of the turbine exhaust.

A plant of this kind operates with an inlet temperature of 288 K, a pressure ratio of 6·0, a turbine isentropic efficiency of 90 per cent and a compressor isentropic efficiency of 87 per cent. Heat transfer calculations indicate that if 5 per cent of the compressor delivery is bled off for cooling purposes, the maximum temperature of the cycle can be raised from 1000 to 1250 K. Find the percentage increase in (a) efficiency, and (b) specific work output, which is achieved by the combined bleeding and cooling process. Make the same assumptions as before and take $\gamma = 1.4$ throughout.

What results would you expect if these calculations were repeated for the plant with the heat-exchanger omitted?

[25·1 per cent, 47·8 per cent]

2.9 An auxiliary gas turbine for use on a large airliner uses a single-shaft configuration with air bled from the compressor discharge for aircraft services. The unit must provide 1·5 kg/s bleed air and a shaft power of 200 kW. Calculate (a) the total compressor air mass flow and (b) the power available with no bleed flow, assuming the following:

Compressor pressure ratio	3·80
Compressor isentropic efficiency	0·85
Combustion pressure loss	0·12 bar
Turbine inlet temperature	1050 K
Turbine isentropic efficiency	0·88
Mechanical efficiency (compressor rotor)	0·99
Mechanical efficiency (driven load)	0·98
Ambient conditions	1 bar, 288 K

[4·78 kg/s, 633 kW]

2.10 A closed-cycle gas turbine is to be used in conjunction with a gas-cooled nuclear reactor. The working fluid is helium ($c_p = 5.19$ kJ/kg K and $\gamma = 1.66$).

The layout of the plant consists of two-stage compression with intercooling followed by a heat-exchanger; after leaving the cold side of the heat-exchanger the helium passes through the reactor channels and on to the turbine; from the turbine it passes through the hot side of the heat-exchanger and then a pre-cooler before returning to the compressor inlet. The following data are applicable:

Compressor and turbine polytropic efficiencies	0·88
Temperature at LP compressor inlet	310 K
Pressure at LP compressor inlet	14·0 bar
Compressor pressure ratios (LP and HP)	2·0
Temperature at HP compressor inlet	300 K
Mass flow of helium	180 kg/s
Reactor thermal output (heat input to gas turbine)	500 MW
Pressure loss in pre-cooler and intercooler (each)	0·34 bar
Pressure loss in heat-exchanger (each side)	0·27 bar
Pressure loss in reactor channels	1·03 bar
Helium temperature at entry to reactor channels	700 K

Calculate the power output and thermal efficiency, and the heat-exchanger effectiveness implied by the data.

[214·5 MW, 0·429, 0·782]

3.1 A simple turbojet is operating with a compressor pressure ratio of 8·0, a turbine inlet temperature of 1200 K and a mass flow of 15 kg/s, when the aircraft is flying at 260 m/s at an altitude of 7000 m. Assuming the following component efficiencies, and ISA conditions, calculate the propelling nozzle area required, the net thrust developed and the SFC.

Polytropic efficiencies of compressor and turbine	0·87
Isentropic efficiency of intake	0·95
Isentropic efficiency of propelling nozzle	0·95
Mechanical efficiency	0·99
Combustion chamber pressure loss	6 per cent comp. deliv. press.
Combustion efficiency	0·97

[0·0713 m², 7896 N, 0·126 kg/h N]

3.2 The gases in the jet pipe of the engine considered in problem **3.1** are reheated to 2000 K, and the combustion pressure loss incurred is 3 per cent of the pressure at outlet from the turbine. Calculate the percentage increase in nozzle area required if the mass flow is to be unchanged, and also the percentage increase in net thrust. [48·3 per cent, 64·5 per cent]

3.3 A naval aircraft is powered by a turbojet engine, with provision for flap blowing. When landing at 55 m/s, 15 per cent of the compressor delivery air is bled off for flap blowing and it can be assumed to be discharged perpendicularly to the direction of flight. If a propelling nozzle area of 0·13 m² is used, calculate the net thrust during landing given that the engine operating conditions are as follows.

Compressor pressure ratio	9·0
Compressor isentropic efficiency	0·82
Turbine inlet temperature	1275 K
Turbine isentropic efficiency	0·87
Combustion pressure loss	0·45 bar
Nozzle isentropic efficiency	0·95
Mechanical efficiency	0·98
Ambient conditions	1 bar, 288 K

The ram pressure and temperature rise can be regarded as negligible.
[18·77 kN]

3.4 Under take-off conditions when the ambient pressure and temperature are 1·01 bar and 288 K, the stagnation pressure and temperature in the jet pipe of a turbojet engine are 2·4 bar and 1000 K, and the mass flow is 23 kg/s. Assuming that the expansion in the converging propelling nozzle is isentropic, calculate the exit area required and the thrust produced.

For a new version of the engine the thrust is to be increased by the addition of an aft fan which provides a separate cold exhaust stream. The fan has a bypass ratio of 2·0 and a pressure ratio of 1·75, the isentropic efficiencies of the fan and fan–turbine sections being 0·88 and 0·90 respectively. Calculate the take-off thrust assuming that the expansion in the cold nozzle is also isentropic, and that the hot nozzle area is adjusted so that the hot mass flow remains at 23 kg/s.
[0·0763 m², 15·35 kN; 24·9 kN]

3.5 Extending the example on the turbofan in Chapter 3, with the additional information that the combustion efficiency is 0·99, determine the *SFC*. Also, calculate the thrust and *SFC* when a combustion chamber is incorporated in the bypass duct and the 'cold' stream is heated to 1000 K. The combustion efficiency and pressure loss for this process may be assumed to be 0·97 and 0·05 bar respectively.
[0·0429 kg/h N; 55·95 kN, 0·128 kg/h N]

4.1 The following data refer to the eye of a single-sided impeller.

Inner radius	6·5 cm
Outer radius	15·0 cm
Mass flow	8 kg/s
Ambient conditions	1·00 bar, 288 K
Speed	270 rev/s

Assuming no pre-whirl and no losses in the intake duct, calculate the blade inlet angle at root and tip of the eye, and the Mach number at the tip of the eye.
[48·20°, 25·43°, 0·843]

4.2 An aircraft engine is fitted with a single-sided centrifugal compressor. The aircraft flies with a speed of 230 m/s at an altitude where the pressure is 0·23 bar and the temperature 217 K. The intake duct of the impeller eye contains fixed vanes which give the air pre-whirl of 25° at all radii. The inner and outer diameters of the eye are 18 and 33 cm respectively, the diameter of the impeller periphery is 54 cm and the rotational speed 270 rev/s. Estimate the stagnation pressure at the compressor outlet when the mass flow is 3·60 kg/s.

Neglect losses in the inlet duct and fixed vanes, and assume that the isentropic efficiency of the compressor is 0·80. Take the slip factor as 0·9 and the power input factor as 1·04.
[1·75 bar]

4.3 The following results were obtained from a test on a small single-sided centrifugal compressor:

Compressor delivery stagnation pressure	2·97 bar
Compressor delivery stagnation temperature	429 K
Static pressure at impeller tip	1·92 bar
Mass flow	0·60 kg/s
Rotational speed	766 rev/s
Ambient conditions	0·99 bar and 288 K

Calculate the overall isentropic efficiency of the compressor.

The diameter of the impeller is 16·5 cm, the axial depth of the vaneless diffuser is 1·0 cm and the number of impeller vanes (n) is 17. Making use of the Stanitz equation for slip factor, namely $\sigma = 1 - (0·63\pi/n)$, calculate the stagnation pressure at the impeller tip and hence find the fraction of the overall loss which occurs in the impeller.
[0·75; 3·35 bar, 0·60]

4.4 The following design data apply to a double-sided centrifugal compressor:

Outer diameter of impeller	50 cm
Speed	270 rev/s
Mass flow	16·0 kg/s
Inlet temperature	288 K
Inlet pressure	1·01 bar
Isentropic efficiency of impeller only	0·90
Radial gap of vaneless space	4·0 cm
Axial depth of vaneless space	5·0 cm
Slip factor	0·9
Power input factor	1·04

(*a*) Calculate the stagnation pressure and temperature at the outlet of the impeller, assuming no pre-whirl.

(*b*) Show that the radial outlet velocity at the impeller tip is about 96 m/s and hence find the Mach number and air leaving angle at the impeller tip. (In calculating the circumferential area at the tip, the thickness of the impeller disc may be neglected.)

(c)   Assuming isentropic diffusion in the vaneless space, find the correct angle of the leading edges of the diffuser vanes, and also find the Mach number at this radius.

[(a) 4·40 bar, 455 K, (b) 1·01, 14·08°, (c) 12·40°, 0·842]

4.5   A single-sided centrifugal compressor is to deliver 14 kg/s of air when operating at a pressure ratio of 4:1 and a speed of 200 rev/s. The inlet stagnation conditions may be taken as 288 K and 1·0 bar. Assuming a slip factor of 0·9, a power input factor of 1·04 and an overall isentropic efficiency of 0·80, estimate the overall diameter of the impeller.

If the Mach number is not to exceed unity at the impeller tip, and 50 per cent of the losses are assumed to occur in the impeller, find the minimum possible axial depth of the diffuser.

[68·9 cm, 5·26 cm]

5.1   An axial flow compressor stage has blade root, mean and tip velocities of 150, 200 and 250 m/s. The stage is to be designed for a stagnation temperature rise of 20 K and an axial velocity of 150 m/s, both constant from root to tip. The work done factor is 0·93. Assuming 50 per cent reaction at mean radius, calculate the stage air angles at root, mean and tip and the degree of reaction at root and tip for a free vortex design.

[$\alpha_1 = 17\cdot07°$ $(=\beta_2)$, $\beta_1 = 45\cdot73°$ $(=\alpha_2)$ at mean radius; $\alpha_1 = 13\cdot77°$, $\beta_1 = 54\cdot88°$, $\beta_2 = 40\cdot23°$, $\alpha_2 = 39\cdot43°$ at tip; $\alpha_1 = 22\cdot25°$, $\beta_1 = 30\cdot60°$, $\beta_2 = -20\cdot25°$, $\alpha_2 = 53\cdot85°$ at root; $\Lambda = 11\cdot2$ per cent at root and 67·4 per cent at tip]

5.2   Recalculate the stage air angles for the same data as in the previous question for a stage with 50 per cent reaction at all radii, and compare the results with those for the free vortex design.

[$\alpha_1 = 28\cdot60°$ $(=\beta_2)$, $\beta_1 = 48\cdot27°$ $(=\alpha_2)$ at tip; $\alpha_1 = 1\cdot15°$ $(=\beta_2)$, $\beta_1 = 44\cdot42°$ $(=\alpha_2)$ at root]

5.3   The first stage of an axial compressor is designed on free vortex principles, with no inlet guide vanes. The rotational speed is 6000 rev/min and the stagnation temperature rise is 20 K. The hub–tip ratio is 0·60, the work-done factor is 0·93 and the isentropic efficiency of the stage is 0·89. Assuming an inlet velocity of 140 m/s and ambient conditions of 1·01 bar and 288 K, calculate:

(a)   the tip radius and corresponding rotor air angles β_1 and β_2, if the Mach number relative to the tip is limited to 0·95
(b)   the mass flow entering the stage
(c)   the stage stagnation pressure ratio and power required
(d)   the rotor air angles at the root section.

[(a) 0·456 m, 63·95° and 56·40°, (b) 65·5 kg/s, (c) 1·233, 1317 kW, (d) 50·83° and 18·32°]

5.4   An axial flow compressor has an overall pressure ratio of 4·0 and mass flow of 3 kg/s. If the polytropic efficiency is 88 per cent and the stagnation temperature rise per stage must not exceed 25 K, calculate the number of stages required and the pressure ratio of the first and last stages. Assume equal temperature rise in all stages. If the absolute velocity approaching the last rotor is 165 m/s at an angle of 20° from the axial direction, the work-done factor is 0·83, the velocity diagram is symmetrical, and the mean diameter of the last stage rotor is 18 cm, calculate the rotational speed and the length of the last stage rotor blade at inlet to the stage. Ambient conditions are 1·01 bar and 288 K.

[7, 1·273, 1·178; 414 rev/s, 1·325 cm]

5.5 A helicopter gas turbine requires an overall compressor pressure ratio of 10:1. This is to be obtained using a two-spool layout consisting of a four-stage axial compressor followed by a single-stage centrifugal compressor. The polytropic efficiency of the axial compressor is 92 per cent and that of the centrifugal is 83 per cent.

The axial compressor has a stage temperature rise of 30 K, using a 50 per cent reaction design with a stator outlet angle of 20°. If the mean diameter of each stage is 25·0 cm and each stage is identical, calculate the required rotational speed. Assume a work-done factor of 0·86 and a constant axial velocity of 150 m/s.

Assuming an axial velocity at the eye of the impeller, an impeller tip diameter of 33·0 cm, a slip factor of 0·90 and a power input factor of 1·04, calculate the rotational speed required for the centrifugal compressor.

Ambient conditions are 1·01 bar and 288 K.

[Axial compressor 318 rev/s, centrifugal compressor 454 rev/s]

6.1 (Chapter 2 also refers) The reference hydrocarbon fuel for which the combustion temperature rise curves of Fig. 2.17 have been drawn contains 13·92 per cent hydrogen and 86·08 per cent carbon, and the relevant value of the enthalpy of reaction, ΔH_{298K}, is $-43\,100$ kJ/kg of fuel. An actual adiabatic steady flow combustion process employs this fuel, with fuel and air entering at 298 K in the ratio 0·0150 by weight. A chemical analysis of the products shows that 4·0 per cent of the carbon is burnt only to carbon monoxide, and the combustion temperature rise ΔT is found to be 583 K.

Using Fig. 2.17, calculate the combustion efficiency based on (a) ratio of actual to theoretical f for the actual ΔT, (b) ratio of actual to theoretical ΔT for the actual f. Compare these values with the efficiency based on the ratio of actual energy released to that theoretically obtainable, given that ΔH_{298K} for CO is $-10\,110$ kJ/kg.

[0·980, 0·983, 0·981]

6.2 (Chapter 2 also refers) A gas turbine combustion chamber is supplied with liquid fuel at 325 K and air at 450 K. The fuel approximates to $C_{10}H_{12}$, and five times the quantity of air required for stoichiometric combustion is supplied. Calculate the fuel/air ratio, and estimate the fuel products' temperature assuming the combustion to be adiabatic and complete.

In addition to the following data, use appropriate values of c_p from p. 17 of the abridged tables of Ref. (8) in Chapter 2: the combustion temperature rise curves of Fig. 2.17 may be used to obtain an initial approximate value of the products' temperature.

Data:

> $C_{10}H_{12}(liq) + 13O_2 \rightarrow 10CO_2 + 6H_2O(liq)$; $\Delta H_{298K} = -42\,500$ kJ/kg $C_{10}H_{12}$
> For water at 298 K, $h_{fg} = 2442$ kJ/kg
> For the liquid fuel, mean $c_p = 1\cdot945$ kJ/kg K
> Composition of air by volume: 0·79 N_2, 0·21 O_2

[0·0148, 984 K]

6.3 The overall pressure loss factor of a combustion chamber may be assumed to vary with the temperature ratio according to the law

$$\frac{\Delta p_0}{m^2/2\rho_1 A_m^2} = K_1 + K_2 \left[\frac{T_{02}}{T_{01}} - 1 \right]$$

For a particular chamber having an inlet area of 0·0389 m² and a maximum cross-sectional area A_m of 0·0975 m², cold loss tests show that K_1 has the value 19·0. When tested under design conditions the following readings were obtained.

Air mass flow m	9·0 kg/s
Inlet stagnation temperature T_{01}	475 K
Outlet stagnation temperature T_{02}	1023 K
Inlet static pressure p_1	4·47 bar
Stagnation pressure loss Δp_0	0·27 bar

Estimate the pressure loss at a part-load condition for which m is 740 kg/s, T_{01} is 439 K, T_{02} is 900 K and p_1 is 3·52 bar.

Also, for these two operating conditions compare the values of (a) the velocity at inlet to the chamber and (b) the pressure loss as a fraction of the inlet stagnation (i.e. compressor delivery) pressure, and comment on the result.
[0·213 bar; 70·2 m/s, 67·7 m/s; 0·0593, 0·0597]

7.1 A mean-diameter design of a turbine stage having equal inlet and outlet velocities leads to the following data.

Mass flow m	20 kg/s
Inlet temperature T_{01}	1000 K
Inlet pressure p_{01}	4·0 bar
Axial velocity (constant through stage) C_a	260 m/s
Blade speed U	360 m/s
Nozzle efflux angle α_2	65°
Stage exit swirl α_3	10°

Determine the rotor blade gas angles, degree of reaction, temperature drop coefficient $(2c_p\Delta T_{0s}/U^2)$ and power output.

Assuming a nozzle loss coefficient λ_N of 0·05, calculate the nozzle throat area required (ignoring the effect of friction on the critical conditions).
[37·15°, 57·37°, 0·29, 3·35, 4340 kW, 0·0398 m²]

7.2 The following particulars relate to a single-stage turbine of free vortex design.

Inlet temperature T_{01}	1050 K
Inlet pressure p_{01}	3·8 bar
Pressure ratio p_{01}/p_{03}	2·0
Outlet velocity C_3	275 m/s
Blade speed at root radius, U_r	300 m/s
Turbine isentropic efficiency η_t	0·88

The turbine is designed for zero reaction $(\Lambda = 0)$ at the root radius, and the velocities at inlet and outlet $(C_1$ and $C_3)$ are both equal *and* axial. Calculate the nozzle efflux angle α_2 and blade inlet gas angle β_2 at the root radius.

If the tip/root radius ratio of the annulus at exit from the nozzle blades is 1·4, determine the nozzle efflux angle and degree of reaction at the tip radius.

Assuming a nozzle blade loss coefficient λ_N of 0·05 calculate the static pressure at inlet and outlet of the rotor blades at the root radius and thereby show that even at the root there is some expansion in the rotor blade passages under these conditions.
[61·15°, 40·23°, 52·25°, 0·46; 1·72 and 1·64 bar]

7.3 The following data apply to a single-stage turbine designed on free vortex theory.

Mass flow	36 kg/s
Inlet temperature T_{01}	1200 K
Inlet pressure p_{01}	8·0 bar
Temperature drop ΔT_{013}	150 K
Isentropic efficiency η_t	0·90
Mean blade speed U_m	320 m/s
Rotational speed N	250 rev/s
Outlet velocity C_3	400 m/s

The outlet velocity is axial. Calculate the blade height and radius ratio of the annulus from the outlet conditions.

The turbine is designed with a constant annulus area through the stage, i.e. with no flare. Assuming a nozzle loss coefficient λ_N of 0.07, show that continuity is satisfied when the axial velocity at exit from the nozzles is 346 m/s. Thence calculate the inlet Mach number relative to the rotor blade at the root radius.

[0.0591 m, 1.34; 0.81]

7.4 The example of a free vortex turbine design considered in section 7.2 yielded the following results for the gas angles.

	α_2	β_2	α_3	β_3
tip	54.93°	0°	8.52°	58.33°
mean	58.38°	20.49°	10°	54.96°
root	62.15°	39.32°	12.12°	51.13°

The values of the radius ratios in plane 2 were $(r_m/r_r)_2 = 1.164$ and $(r_m/r_t)_2 = 0.877$.

Using the same mean-diameter angles, calculate β_2 at tip and root for a constant nozzle angle design in which α_2 and $C_{w2}r$ are constant over the annulus. Compare the two designs by sketching the velocity diagrams and commenting qualitatively on such aspects as the radial variation of degree of reaction and blade inlet Mach number.

To satisfy radial equilibrium with constant nozzle angle, the constant angular momentum condition should strictly be replaced by constant $C_w r^{\sin^2\alpha}$. Show, by recalculating β_2 at tip and root, that this refinement has only a small effect on the required blade angle.

[β_{2r} 35.33°, β_{2t} 0°; β_{2r} 33.50°, β_{2t} 3.20°]

7.5 In certain designs the maximum mass flow passed by a turbine may be determined by choking in the turbine annulus at outlet instead of in the turbine nozzles. The maximum mass flow will then depend not merely on the inlet conditions as in the case of expansion in a nozzle, but also upon the work output per unit mass flow. Given fixed values of the inlet conditions p_{01} and T_{01}, the temperature equivalent of the specific work output ΔT_{013}, the annulus area A at the turbine outlet and that there is no outlet swirl, show that for an isentropic expansion to a varying outlet static temperature T_3 the maximum flow can be expressed by

$$\frac{m\sqrt{T_{01}}}{p_{01}} = \left\{ \frac{\gamma}{R}\left[\left(\frac{2}{\gamma+1}\right)\left(1 - \frac{\Delta T_{013}}{T_{01}}\right)\right]^{(\gamma+1)/(\gamma-1)}\right\}^{\frac{1}{2}}$$

Comment on the effect this will have on the turbine mass flow versus pressure ratio characteristic.

8.1 The following data refer to a single-shaft gas turbine running at its design speed.

Compressor characteristic			Turbine characteristic		
p_2/p_1	$m\sqrt{T_1}/p_1$	η_c	p_3/p_4	$m\sqrt{T_3}/p_3$	η_t
5.0	32.9	0.80	5.0	14.2	0.845
4.7	33.8	0.79	4.5	14.2	0.850
4.5	34.3	0.77	4.0	14.2	0.842

The combustion pressure loss is 5 per cent of the compressor delivery pressure and the ambient conditions are 1·01 bar and 288 K. Mechanical losses can be neglected. The 'non-dimensional' flows are based on m in kg/s, p in bar and T in K, all pressures and temperatures being stagnation values.

Calculate the power output when operating at a turbine inlet temperature of 1100 K. Comment briefly on the variation in thermal efficiency as the load is reduced at constant speed.

[264 kW]

8.2 The following data refer to a gas turbine with a free power turbine, operating at design speed.

Compressor characteristic			Gas-generator turbine characteristic		
p_2/p_1	$m\sqrt{T_1}/p_1$	η_c	p_3/p_4	$m\sqrt{T_3}/p_3$	η_t
5·2	220	0·82	2·50	90·2	0·85
5·0	236	0·83	2·25	90·2	0·85
4·8	244	0·82	2·00	88·2	0·85

Assuming that the power turbine is choked, the value of $m\sqrt{T_4}/p_4$ being 188, determine the design values of compressor pressure ratio and turbine inlet temperature.

Neglect all pressure losses and assume the mechanical efficiency of the gas-generator rotor to be 0·98 and take the ambient temperature as 288 K. The 'non-dimensional' flows quoted are based on m in kg/s, p in bar and T in K, all pressures and temperatures being stagnation values. The suffixes 1, 2, 3 and 4 refer to the following locations:

        1—compressor inlet
        2—compressor delivery
        3—gas-generator turbine inlet
        4—power turbine inlet.

[5·10, 1170 K]

8.3 When running at a low power condition a gas turbine with free power turbine operates at a compressor pressure ratio of 2·60. The combustion chamber pressure loss is 4 per cent of the compressor delivery pressure and the exhaust pressure loss can be ignored.

The turbine characteristics are given below:

Gas-generator turbine			Power turbine		
p_3/p_4	$m\sqrt{T_3}/p_3$	η_t	p_4/p_a	$m\sqrt{T_4}/p_4$	
1·3	20·0	0·85	1·4	60·0	
1·5	44·0	0·85	1·6	85·0	
1·8	62·0	0·85	1·8	95·0	

The 'non-dimensional' flows are based on m in kg/s, p in bar and T in K, all pressures and temperatures being stagnation values.

Calculate the gas-generator turbine pressure ratio at this condition. Assuming the compressor characteristic to be known, indicate briefly how you would calculate the turbine inlet temperature.

[1·61]

8.4 A gas turbine with a free power turbine gives the following results when tested at ambient conditions of 1·0 bar and 288 K.

$N\sqrt{T_1}$ (% design)	$m\sqrt{T_1}/p_1$	p_2/p_1	η_c
100	454·5	4·60	0·859
95	420·0	4·00	0·863
90	370·0	3·60	0·858

The 'non-dimensional' flows are based on m in kg/s, p in bar and T in K, all pressures and temperatures being stagnation values.

The power turbine remains choked for all of these conditions. The gas-generator turbine inlet temperature at 95 per cent design *mechanical* speed was found to be 1075 K. Assuming single-line turbine flow characteristics, constant turbine efficiency, and constant mechanical efficiency of the gas-generator rotor, calculate

(a) the gas-generator turbine inlet temperature at design mechanical speed with the same ambient conditions;
(b) the compressor power absorbed when running at 95 per cent design mechanical speed with ambient conditions of 0·76 bar and 273 K.

Sketch the operating line on the compressor characteristic, and discuss the effect of ambient temperature on net power output for a fixed gas-generator mechanical speed.
[1215 K, 3318 kW]

8.5 A simple gas turbine is to be used as a source of compressed air. The turbine produces just enough work to drive the compressor which delivers an airflow, m_c, greater than that required for the turbine by an amount m_b. The design point operating conditions are

$$\frac{m_c\sqrt{T_1}}{p_1} = 22\cdot8, \quad \frac{p_2}{p_1} = 4\cdot0$$

and at the design point we also have $T_3/T_1 = 3\cdot3$, $\eta_c = 0\cdot8$ and $\eta_t = 0\cdot85$.

A rough estimate of the equilibrium running line is required on the compressor characteristic for the conditions where T_3/T_1 is maintained constant and the amount of air bled off, m_b, is varied. For this purpose the efficiencies of the compressor and turbine can be assumed to be constant, the mechanical transmission loss and the combustion pressure loss can be neglected, and the turbine 'non-dimensional' mass flow can be assumed independent of speed and related to the pressure ratio r by the expression

$$\frac{m\sqrt{T_3}}{p_3} = k\sqrt{\left(1 - \frac{1}{r^2}\right)}$$

where $m = m_c - m_b$ and k is a constant.

As an example of the calculations required, find the operating point on the compressor characteristic for the condition where $m_b\sqrt{T_1}/p_1$ is reduced to three-quarters of its value at the design point. Take $\gamma = 1\cdot40$ throughout.
[$p_2/p_1 = 2\cdot1$, $m_c\sqrt{T_1}/p_1 = 13\cdot0$]

Appendix C
References

Chapter 1

(1) CARCHEDI, F., GALLIMORE, S., SIEMIERNIK, A. F. and GOODMAN, D. A. Development and early field experience of the Typhoon gas turbine, *American Society of Mechanical Engineers,* paper 91-GT-382, 1991.

(2) MORRIS, R. E. The Pratt and Whitney PW100—evolution of the design concept, *Canadian Aeronautics and Space Journal,* **28**, 1982, 211–21.

(3) BRANDT, D. E. The design and development of an advanced heavy-duty gas turbine, *Transactions of the American Society of Mechanical Engineers,* **110**, 1988, 243–50.

(4) SCALZO, A. J., BANNISTER, R. L., de CORSO, M. and HOWARD, G. S. Evolution of heavy-duty power generation and industrial combustion turbines in the United States, *American Society of Mechanical Engineers,* paper 94-GT-488, 1994.

(5) ROCHA, G. and ETHERIDGE, C. J. Evolution of the Solar Titan 130 industrial gas turbine, *American Society of Mechanical Engineers,* paper 98-GT-590, 1998.

(6) PICKERELL, D. J. Rolls-Royce RB211-535 power plant, *Journal of Aircraft,* **20**, 1983, 15–20.

(7) RUFFLES, P. C. Innovation in aero engines, *Aeronautical Journal, Royal Aeronautical Society,* 1996, 473–83.

Chapter 2

(1) ROGERS, G. F. C. and MAYHEW, Y. R. *Engineering Thermodynamics, Work and Heat Transfer, 5th edition* (Longman, 1994).

(2) McDONALD, C. F. Emergence of recuperated gas turbines for power generation, *American Society of Mechanical Engineers,* paper 99-GT-67, 1999.

(3) BANES, B., McINTYRE, R. W. and SIMS, J. A. *Properties of Air and Combustion Products with Kerosene and Hydrogen Fuels* (AGARD, 1967).

(4) FIELDING, D. and TOPPS, J. E. C. Thermodynamic data for the calculation of gas turbine performance, *Aeronautical Research Council,* R&M No. 3099 (HMSO, 1959).

(5) LITTLE, D. A., BANNISTER, R. L. and WIANT, B. C. Development of advanced gas turbine systems, *American Society of Mechanical Engineers*, Cogen Turbo Power, **93**, 1993, 271–80.
(6) LUGAND, P. and PARIETTI, C. Combined cycle plants with Frame 9F gas turbines, *Transactions of the American Society of Mechanical Engineers*, **113**, 1991, 475–81.
(7) CATINA, J. L., FORTUNE, H. J. and SOROKA, G. E. Repowering Chesterfield 1 and 2 with combined cycle, *Transactions of the American Society of Mechanical Engineers*, **110**, 1988, 214–99.
(8) ROGERS, G. F. C. and MAYHEW, Y. R. *Thermodynamic and Transport Properties of Fluids* (Blackwell, 1995).
(9) McDONALD, C. F. Enabling technologies for nuclear gas turbine power conversion systems, *American Society of Mechanical Engineers*, paper 94-GT-415, 1994.

Chapter 3

(1) HILL, P. G. and PETERSON, C. R. *Mechanics and Thermodynamics of Propulsion, 2nd edition* (Addison-Wesley, 1992).
(2) Relationships between some common intake parameters, *Royal Aeronautical Society*, data sheet 66028, 1966.
(3) SEDDON, J. and GOLDSMITH, E. L. *Intake Aerodynamics* (AIAA Education Series, 1985).
(4) ASHWOOD, P. F. A review of the performance of exhaust systems for gas turbine aero-engines, *Proceedings of the Institution of Mechanical Engineers*, **171**, 1957, 129–58.
(5) YOUNG, P. H. Propulsion controls on the Concorde, *Journal of the Royal Aeronautical Society*, **70**, 1966, 863–81.
(6) ROGERS, G. F. C. and MAYHEW, Y. R. One-dimensional irreversible gas flow in nozzles, *Engineering, London*, **175**, 1953, 355–8.
(7) FROST, T. H. Practical bypass mixing systems for fan jet aero-engines, *Aeronautical Quarterly*, **17**, 1966, 141–60.
(8) LEWIS, G. M. The next European engine for combat aircraft, *Aeronautical Journal, Royal Aeronautical Society*, Feb. 1984.
(9) STEVENSON, J. D. and SARAVANAMUTTOO, H. I. H. Simulating indirect thrust measurement methods for high bypass turbofans, *Transactions of the American Society of Mechanical Engineers*, **117**, 1995, 38–46.
(10) SARAVANAMUTTOO, H. I. H. Modern turboprop engines, *Progress in Aerospace Sciences*, **24**, 1987, 225–48.

Chapter 4

(1) WHITTLE, F. The early history of Whittle jet propulsion gas turbines, *Proceedings of the Institution of Mechanical Engineers*, **152**, 1945, 419–35.
(2) CHESHIRE, L. J. The design and development of centrifugal compressors for aircraft gas turbines, *Proceedings of the Institution of Mechanical Engineers*, **153**, 1945, 426–40.
(3) WRONG, C. B. An introduction to the JT 15D engine, *Transactions of the American Society of Mechanical Engineers*, 69-GT-119, 1969.

(4) STANITZ, J. D. Some theoretical aerodynamic investigations of impellers in radial and mixed-flow centrifugal compressors, *Transactions of the American Society of Mechanical Engineers*, **74**, 1952, 473–97.

(5) KENNY, D. P. A novel low-cost diffuser for high-performance centrifugal compressors, *Transactions of the American Society of Mechanical Engineers*, Series A, **91**, 1969, 37–46.

(6) FERGUSON, T. B. *The Centrifugal Compressor Stage* (Butterworth, 1963).

(7) HANKINS, G. A. and COPE, W. F. Discussion on 'The flow of gases at sonic and supersonic speeds', *Proceedings of the Institution of Mechanical Engineers*, **155**, 1947, 401–16.

(8) ENGINEERING SCIENCES DATA UNIT: Fluid Mechanics—internal flow, 4—Duct expansions and duct contractions, Data Sheets 73024, 74015, 76027.

(9) CAME, P. M. The development, application and experimental evaluation of a design procedure for centrifugal compressors, *Proceedings of the Institution of Mechanical Engineers*, **192**, No. 5, 1978, 49–67.

(10) HERBERT, M. V. A method of performance prediction for centrifugal compressors, *Aeronautical Research Council*, R&M No. 3843 (HMSO, 1980).

(11) CAME, P. M. and ROBINSON, C. J. Centrifugal compressor design, *Proceedings of the Institution of Mechanical Engineers*, **213**, Part C, 1999.

Chapter 5

(1) CONSTANT, H. The early history of the axial type of gas turbine engine, *Proceedings of the Institution of Mechanical Engineers*, **153**, W.E.P. No. 12, 1945.

(2) HOWELL, A. R. Fluid dynamics of axial compressors, and Design of axial compressors, *Proceedings of the Institution of Mechanical Engineers*, **153**, W.E.P. No. 12, 1945.

(3) HOWELL, A. R. The present basis of axial compressor design. Part I—Cascade Theory, *Aeronautical Research Council*, R&M No. 2095 (HMSO, 1942).

(4) JOHNSEN, I. A. and BULLOCK, R. O. *Aerodynamic Design of Axial-flow Compressors*, NASA SP-36, 1965.

(5) HORLOCK, J. H. *Axial Flow Compressors* (Butterworth, 1958).

(6) LIEBLEIN, S. and JOHNSEN, I. A. Resumé of transonic compressor research at NACA Lewis Laboratory, *Transactions of the American Society of Mechanical Engineers, Journal of Engineering for Power*, **83**, 1961, 219–34.

(7) TODD, K. W. Practical aspects of cascade wind tunnel research, *Proceedings of the Institution of Mechanical Engineers*, **157**, W.E.P. No. 36, 1947.

(8) GOSTELOW, J. P. *Cascade Aerodynamics* (Pergamon Press, 1984).

(9) CARTER, A. D. S. Blade profiles for axial flow fans, pumps, compressors, etc., *Proceedings of the Institution of Mechanical Engineers*, **175**, No. 16, 1961, 775–88.

(10) MILLER, G. R., LEWIS, G. W. and HARTMAN, M. J. Shock losses in transonic blade rows, *Transactions of the American Society of Mechanical Engineers, Journal of Engineering for Power*, **83**, 1961, 235–42.

(11) SCHWENK, F. C., LEWIS, G. W. and HARTMAN, M. J. A preliminary analysis of the magnitude of shock losses in transonic compressors, NACA RM E57A30, 1957.

(12) KERREBROCK, J. L. Flow in transonic compressors, *American Institute for Aeronautics and Astronautics Journal*, **19**, 1981, 4–19.

(13) GREITZER, E. M. Review—axial compressor stall phenomena, *Transactions of the American Society of Mechanical Engineers, Journal of Fluids Engineering*, **102**, 1980, 134–51.

(14) STONE, A. Effects of stage characteristics and matching on axial-flow-compressor performance, *Transactions of the American Society of Mechanical Engineers*, **80**, 1958, 1273–93.

(15) CARCHEDI, F. and WOOD, G. R. Design and development of a 12:1 pressure ratio compressor for the Ruston 6 MW gas turbine, *Transactions of the American Society of Mechanical Engineers, Journal of Engineering for Power*, **104**, 1982, 823–31.

(16) WADIA, A. R., WOLF, D. P. and HAASER, F. G. Aerodynamic design and testing of an axial-flow compressor with pressure ratio of 23·3 for the LM2500 + gas turbine, *American Society of Mechanical Engineers*, paper 99-GT-210, 1999.

(17) DAVIS, W. R. and MILLAR, D. A. J. A comparison of the matrix and streamline curvature methods of axial flow turbomachinery analysis, from a user's point of view, *Transactions of the American Society of Mechanical Engineers*, **97**, 1975, 549–60.

(18) DENTON, J. D. An improved time marching method for turbomachinery flow calculation, *American Society of Mechanical Engineers*, Paper 82-GT-239, 1982.

(19) McNALLY, W. D. and SOCKOL, P. M. Review—Computational methods for internal flows with emphasis on turbomachinery, *Transactions of the American Society of Mechanical Engineers, Journal of Fluids Engineering*, **107**, 1985, 6–22.

(20) FREEMAN, C. and STOW, P. The application of computational fluid mechanics to aero gas turbine compressor design and development, *Institution of Mechanical Engineers*, Conference Publications (1984—3) C70/84.

(21) DUNHAM, J. A.R. Howell; father of the British axial compressor, *American Society of Mechanical Engineers*, paper 2000-GT-8, 2000.

Chapter 6

(1) SOTHERAN, A. The Rolls-Royce annular vaporizer combustor, *American Society of Mechanical Engineers*, paper 83-GT-49, 1983.

(2) *Technical Advances in Gas Turbine Design*, Institution of Mechanical Engineers Symposium, 1969.

(3) SPALDING, D. B. *Some Fundamentals of Combustion* (Butterworths Scientific Publications, 1955).

(4) ROGERS, G. F. C. and MAYHEW, Y. R. *Engineering Thermodynamics, Work and Heat Transfer, 4th edition* (Longman, 1994).

(5) LIPFERT, F. W. Correlation of gas turbine emissions data, *American Society of Mechanical Engineers*, paper 72-GT-60, 1972.

(6) LEONARD, G. and STEGMAIER, J. Development of an aeroderivative gas turbine dry low emissions combustion system, *American Society of Mechanical Engineers*, paper 93-GT-288, 1993.

(7) DAVIS, L. B. and WASHAM, R. M. Development of a dry low NO_x combustor, *American Society of Mechanical Engineers*, paper 89-GT-255, 1989.

(8) MAGHON, H., BERENBRINK, P., TERMUEHLEN, H. and GARTNER, G. Progress in NO_x and CO emission reduction of gas turbines, *American Society of Mechanical Engineers*, paper 90-JPGC/GT-4, 1990.

(9) SATTELMEYER, T., FELCHLIN, M. P., HAUMANN, J. and STYNER, D. Second generation low-emission combustors for ABB gas turbines: burner development and tests at atmospheric pressure, *Transactions of the American Society of Mechanical Engineers*, **114**, 1992, 118–24.

(10) ETHERIDGE, C. J. Mars SoLoNO$_x$-lean pre-mix combustion technology in production, *American Society of Mechanical Engineers*, paper 94-GT-255, 1994.

(11) BAHR, D. W. Aircraft engine NO$_x$ emissions—abatement progress and prospects, *International Society for Air Breathing Engines*, paper 91-7022, 1991.

(12) SEGALMAN, I., McKINNEY, R. G., STURGESS, G. J. and HUANG, L. M. Reduction of NO$_x$ by fuel-staging in gas turbine engines, *AGARD Conference Proceedings 485*, 1993.

(13) SUMMERFIELD, A. H., PRITCHARD, D., TUSON, D. W. and OWEN, D. A. Mechanical design and development of the RB211 dry low emissions engine, *American Society of Mechanical Engineers*, paper 93-GT-245, 1993.

(14) CORBETT, N. C. and LINES, N. P. Control requirements for the RB211 low emission combustion system, *American Society of Mechanical Engineers*, paper 93-GT-12, 1993.

(15) SCARINCI, T. and HALPIN, J. L. Industrial Trent combustor–combustion noise characteristics, *American Society of Mechanical Engineers*, paper 99-GT-9, 1999.

(16) BAMMERT, K. Operating experiences and measurements on turbo sets of CCGT—cogeneration plants in Germany, *American Society of Mechanical Engineers*, paper 86-GT-101, 1986.

Chapter 7

(1) HAWTHORNE, W. R. ed. *Aerodynamics of Turbines and Compressors* (Oxford University Press, 1964).

(2) HORLOCK, J. H. *Axial Flow Turbines* (Butterworth, 1966).

(3) AINLEY, D. G. and MATHIESON, G. C. R. An examination of the flow and pressure losses in blade rows of axial flow turbines, *Aeronautical Research Council*, R&M 2891 (HMSO, 1955).

(4) JOHNSTON, I. H. and KNIGHT, L. R. Tests on a single-stage turbine comparing the performance of twisted with untwisted rotor blades, *Aeronautical Research Council*, R&M 2927 (HMSO, 1953).

(5) ANDREWS, S. J. and OGDEN, H. A detailed experimental comparison of (compressor) blades for free vortex flow and equivalent untwisted constant section blades, *Aeronautical Research Council*, R&M 2928 (HMSO, 1953).

(6) ISLAM, A. M. T. and SJOLANDER, S. A. Deviation in axial turbines at subsonic conditions, *American Society of Mechanical Engineers,* paper 99-GT-26, 1999.

(7) SAWYER, J. W. ed. *Gas Turbine Engineering Handbook* (Turbomachinery International Publications, 1985).

(8) SMITH, D. J. L. Turbulent boundary layer theory and its application to blade profile design, *Aeronautical Research Council* C.P. 868 (HMSO, 1966).

(9) AINLEY, D. G. and MATHIESON, G. C. R. A method of performance estimation for axial-flow turbines, *Aeronautical Research Council*, R&M 2974 (HMSO, 1951).

(10) DUNHAM, J. and CAME, P. M. Improvements to the Ainley–Mathieson method of turbine performance prediction, *Transactions of the American Society of Mechanical Engineers, Journal of Engineering for Power*, **92**, 1970, 252–6.

(11) KACKER, S. C. and OKAPUU, U. A mean line prediction method for axial flow turbine efficiency, *Transactions of the American Society of Mechanical Engineers, Journal of Engineering for Power*, **104**, 1982, 111–19.

(12) BENNER, M. W., SJOLANDER, S. A. and MOUSTAPHA, S. H. Influence of leading-edge geometry on profile losses in turbines at off-design conditions: experimental results and an improved correlation, *Transactions of the American Society of Mechanical Engineers, Journal of Turbomachinery*, **119**, 1999, 193–200.

(13) AINLEY, D. G. Internal air-cooling for turbine blades—a general design survey, *Aeronautical Research Council*, R&M 3013 (HMSO, 1957).

(14) HAWTHORNE, W. R. Thermodynamics of cooled turbines, Parts I and II, *Transactions of the American Society of Mechanical Engineers*, **78**, 1956, 1765–81.

(15) BARNES, J. F. and FRAY, D. E. An experimental high-temperature turbine (No. 126), *Aeronautical Research Council*, R&M 3405 (HMSO, 1965).

(16) *Technical Advances in Gas Turbine Design*, Institution of Mechanical Engineers Symposium, 1969.

(17) PRICE, J. R., JIMENEZ, O., PARTHASARATHY, V. J. and MIRIYALA, N. Ceramic stationary gas turbine development program; 6th annual summary, *American Society of Mechanical Engineers*, paper 99-GT-351, 1999.

(18) MOWILL, J. and STROM, S. An advanced radial-component industrial gas turbine, *Transactions of the American Society of Mechanical Engineers, Journal of Engineering for Power*, **105**, 1983, 947–52.

(19) DIXON, S. L. *Fluid Mechanics, Thermodynamics of Turbomachinery* (Pergamon Press, 1975).

(20) *Aerodynamic Performance of Radial Inflow Turbines*. First (1963) and Second (1964) Reports, Motor Industry Research Association. (Alternatively, the work is summarized in: HIETT, G. F. and JOHNSTON, I. H. Experiments concerning the aerodynamic performance of inward radial flow turbines, *Proceedings of the Institution of Mechanical Engineers*, **178**, Part 3I(ii), 1964.)

(21) BENSON, R. S. A review of methods for assessing loss coefficients in radial gas turbines, *International Journal of Mechanical Science*, **12**, 1970, 905–32.

(22) BRIDLE, E. A. and BOULTER, R. A. A simple theory for the prediction of losses in the rotors of inward radial flow turbines, *Proceedings of the Institution of Mechanical Engineers*, **182**, Part 3H, 1968.

(23) BENSON, R. S. Prediction of performance of radial gas turbines in automotive turbochargers, *American Society of Mechanical Engineers*, paper 71-GT-66, 1971.

Chapter 8

(1) EBELING, J. E. Thermal energy storage and inlet-air cooling for combined cycle, *American Society of Mechanical Engineers*, paper 94-GT-310, 1994.

(2) MEHER-HOMJI, C. B. and MEE, T. R. Inlet fogging of gas turbine engines, *American Society of Mechanical Engineers*, papers 2000-GT-307/8, 2000.

Chapter 9

(1) MALLINSON, D. H. and LEWIS, W. G. E. The part-load performance of various gas-turbine engine schemes, *Proceedings of the Institution of Mechanical Engineers*, **159**, 1948, 198–219.

(2) TREWBY, G. F. A. British naval gas turbines, *Transactions of the Institution of Marine Engineers*, **66**, 1954, 125–67.

(3) SWATMAN, I. M. and MALOHN, D. A. An advanced automotive gas turbine concept, *Transactions of the Society of Automotive Engineers*, **69**, 1961, 219–27.

(4) COX, J. C., HUTCHINSON, D. and OSWALD, J. I. The Westinghouse/ Rolls-Royce WR-21 gas turbine variable area power turbine design, *American Society of Mechanical Engineers*, paper 95-GT-54, 1995.

(5) KAYS, W. M. and LONDON, A. L. *Compact Heat Exchangers* (McGraw-Hill, 1964).

(6) RAHNKE, C. J. The variable-geometry power turbine, *Transactions of the Society of Automotive Engineers*, **78** [i], 1969, 213–23.

(7) MAYER, A. and van der LINDEN, S. GT 24/26 advanced cycle system power plant progress for the new millennium, *American Society of Mechanical Engineers,* paper 99-GT-404, 1999.

(8) YOUNG, P. H. Propulsion controls on the Concorde, *Journal of the Royal Aeronautical Society*, **70**, 1966, 863–81.

(9) SARAVANAMUTTOO, H. I. H. and FAWKE, A. J. Simulation of gas turbine dynamic performance, *American Society of Mechanical Engineers*, paper 70-GT-23, 1970.

(10) FAWKE, A. J. and SARAVANAMUTTOO, H. I. H. Experimental investigation of methods for improving the dynamic response of a twin-spool turbojet engine, *Transactions of the American Society of Mechanical Engineers*, **93**, series A, 1971, 418–24.

(11) FAWKE, A. J. and SARAVANAMUTTOO, H. I. H. Digital computer methods for prediction of gas turbine dynamic response, *Transactions of the Society of Automotive Engineers*, **80** [iii], 1971, 1805–13.

(12) SARAVANAMUTTOO, H. I. H. and MacISAAC, B. D. An overview of engine dynamic response and mathematical modelling concepts, *AGARD Conference Proceedings No. 324 'Engine Handling'*, 1982.

(13) MEHER-HOMJI, C. B. Gas turbine axial compressor fouling: a unified treatment of its effects, detection and control, *International Journal of Turbo and Jet Engines*, **9**, 1992, 311–34.

(14) DIAKUNCHAK, I. S. Performance deterioration in industrial gas turbines, *American Society of Mechanical Engineers,* paper 91-GT-228, 1991.

(15) SARAVANAMUTTOO, H. I. H. and LAKSHMIRANASIMHA, A. N. A preliminary assessment of compressor fouling, *American Society of Mechanical Engineers,* paper 85-GT-153, 1985.

(16) SJOLANDER, S. A., ISAACS, D. and KLEIN, W. A. Aerodynamics of turbine blades with trailing edge damage: measurements and computations, *Proceedings, 11th International Symposium on Air Breathing Engines*, **2**, 1993, 1327–34.

(17) AKER, G. F. and SARAVANAMUTTOO, H. I. H. Predicting gas turbine performance degradation due to compressor fouling using computer simulation techniques, *Transactions of the American Society of Mechanical Engineers, Journal of Engineering for Power*, **111**, 1989, 343–50.

(18) SARAVANAMUTTOO, H. I. H. and MacISAAC, B. D. Thermodynamic models for pipeline gas turbine diagnostics, *Transactions of the American Society of Mechanical Engineers*, **105**, Series A, 1983, 875–84.

(19) MUIR, D. E., RUDNITSKI, D. M. and CUE, R. W. CF-18 Engine performance monitoring, *AGARD Conference Proceedings No. 448, 'Engine Condition Monitoring—Technology and Experience'*, 1988.

Appendix A

(1) HOUGHTON, E. L. and BROCK, A. E. *Tables for the Compressible Flow of Dry Air* (Arnold, 1970).
(2) KEENAN, J. H. and KAYE, J. *Gas Tables* (Wiley, 1948).
(3) SHAPIRO, A. H. *The Dynamics and Thermodynamics of Compressible Flow* (Ronald, 1954).
(4) LIEPMANN, H. W. and ROSHKO, A. *Elements of Gas Dynamics* (Wiley, 1957).

Supplementary texts

BATHIE, W. W. *Fundamentals of Gas Turbines, 2nd edition* (Wiley, 1995).
CUMPSTY, N. A. *Compressor Aerodynamics* (Longman, 1989).
CUMPSTY, N. A. *Jet Propulsion* (Cambridge University Press, 1997).
JAPIKSE, D. and BAINES, N. C. *Introduction to Turbomachinery* (Concepts ETI Inc./Oxford University Press, 1994).
LAKSHMIRANYANA, B. *Fluid Dynamics and Heat Transfer of Turbomachinery* (Wiley, 1996).
LEFEBVRE, A. H. *Gas Turbine Combustion* (McGraw-Hill, 1983).
MATTINGLEY, J. D., HEISER, W. H. and DALEY, D. H. *Aircraft Engine Design* (AIAA Education Series, 1987).
McKENZIE, A. B. *Axial Flow Fans and Compressors* (Ashgate, 1997).
SAWYER, J. W. and JAPIKSE, D. (eds) *Gas Turbine Handbook, 4th edition* (Turbomachinery International Publications, 1990).
SMITH, M. J. T. *Aircraft Noise* (Cambridge University Press, 1989).
WALSH, P. P. and FLETCHER, P. *Gas Turbine Performance* (Blackwell Science, 1998).
WILSON, D. G. and KORAKIANITIS, T. *The Design of High-Efficiency Turbomachinery and Gas Turbines, 2nd edition* (Prentice Hall, 1998).

Index